AF678310

PHYSICAL ABUSERS and SEXUAL OFFENDERS

Forensic and Clinical Strategies

SCOTT ALLEN JOHNSON

Vincennes University
Shake Learning Resources Center
Vincennes, In 47591-9986

CRC Taylor & Francis
Taylor & Francis Group
Boca Raton London New York

CRC is an imprint of the Taylor & Francis Group,
an informa business

363.25
J68p
2007

CRC Press
Taylor & Francis Group
6000 Broken Sound Parkway NW, Suite 300
Boca Raton, FL 33487-2742

© 2007 by Taylor & Francis Group, LLC
CRC Press is an imprint of Taylor & Francis Group, an Informa business

No claim to original U.S. Government works
Printed in the United States of America on acid-free paper
10 9 8 7 6 5 4 3 2 1

International Standard Book Number-10: 0-8493-7259-3 (Hardcover)
International Standard Book Number-13: 978-0-8493-7259-9 (Hardcover)

This book contains information obtained from authentic and highly regarded sources. Reprinted material is quoted with permission, and sources are indicated. A wide variety of references are listed. Reasonable efforts have been made to publish reliable data and information, but the author and the publisher cannot assume responsibility for the validity of all materials or for the consequences of their use.

No part of this book may be reprinted, reproduced, transmitted, or utilized in any form by any electronic, mechanical, or other means, now known or hereafter invented, including photocopying, microfilming, and recording, or in any information storage or retrieval system, without written permission from the publishers.

For permission to photocopy or use material electronically from this work, please access www.copyright.com (http://www.copyright.com/) or contact the Copyright Clearance Center, Inc. (CCC) 222 Rosewood Drive, Danvers, MA 01923, 978-750-8400. CCC is a not-for-profit organization that provides licenses and registration for a variety of users. For organizations that have been granted a photocopy license by the CCC, a separate system of payment has been arranged.

Trademark Notice: Product or corporate names may be trademarks or registered trademarks, and are used only for identification and explanation without intent to infringe.

Library of Congress Cataloging-in-Publication Data

Johnson, Scott A. (Scott Allen)
Working with physical abusers and sexual offenders : forensic and clinical strategies / by Scott Allen Johnson.
p. cm.
Includes bibliographical references and index.
ISBN 0-8493-7259-3 (alk. paper)
1. Sex Crimes. 2. Offenses against the person. 3. Sex offenders--Rehabilitation. 4. Sex offenders--Psychology. 5. Forensic psychology. I. Title.

HV 6556.J64 2006
363.25'953--dc22 2006042560

Visit the Taylor & Francis Web site at
http://www.taylorandfrancis.com

and the CRC Press Web site at
http://www.crcpress.com

Dedication

This book is dedicated to all the people who work with physical abusers and sexual offenders. Abuse and sexual offenses are difficult and challenging issues to work with, and I applaud those willing to embark on the journey!

Table of Contents

SECTION II Characteristics of Physical and Sexual Offenders

SECTION III Interviewing Strategies: Detecting Truth from Deception

SECTION IV Treatment Issues

SECTION V Investigation Strategies for Law Enforcement, Probation, and Parole Officers

Preface

I wrote this book to provide a text to educate professionals about what I consider to be important issues when assessing, investigating, and treating physical abusers and sexual offenders. The material in this book is derived from my counseling, clinical, and forensic experience working with physical abusers and sexual offenders and with the victims. I have conducted over 850 forensic assessments of sexual offenders for the State of Minnesota, as well as numerous assessments and treatment for physical abusers and sexual offenders in the community. I recognize that there are several notable authors and wonderful texts available addressing physical abusers and sexual offenders, and I believe that any professional working with these populations should read the available literature. This book provides an opportunity to understand how physical abusers and sexual offenders share many of the same characteristics, engage in much of the same behavior, and require similar treatment interventions.

I have been frustrated that the two fields (abuse and sexual offender) tend to keep to themselves and as a result it is often difficult to understand the vast similarities between the two types of offenders. In addition, the law enforcement and the psychology fields tend to have important knowledge to share concerning abusers and sex offenders yet often do not readily exchange that information. So in this book I am bringing together various aspects of physical abusers and sexual offenders in one text.

The Author

Scott A. Johnson, M.A., L.P., has a Master's Degree in Counseling and Psychological Services from St. Mary's College. He is a licensed psychologist, a Diplomate of the American Board of Psychological Specialties in Forensic Psychology, Academy Certified Diplomate of the American Academy of Certified Consultants and Experts, and a member of the American College of Forensic Examiners. He is also an adjunct instructor at St. Mary's College Graduate Center in Minneapolis, Minnesota.

Johnson is a counseling and forensic psychologist, with extensive experience working with abusers, sexual offenders, victims of violence, as well as providing therapy that addresses depression, anxiety, family, marital, relationship, and general issues. He also provides crisis counseling and critical incident stress debriefings. His expertise is in the areas of domestic and relationship violence and sexual offenses. Johnson is nationally recognized for his work with abuse cases, sexual offenses, and with victims. He provides presentations at national and international conferences, and training on issues involving sexual offenses and domestic abuse. His audiences have included members of law enforcement, mental health professionals, county attorneys and prosecutors, abuse and rape advocates, and educators. Also, he is the author of several book chapters, booklets, and professional materials. He is the author of *When "I Love You" Turns Violent*, 2nd edition, New Horizon Press, 2005. His booklets include *Man-To-Man: When Your Partner Says No* and *Detecting Truth from Deception* (self-published). He has been published internationally for his research on abusers and sexual offenders.

Acknowledgments

I would like to thank the following people:

Pat and George Johnson, my parents, for all of the support they provided; my colleagues for their ongoing support, time taken to review my manuscript, and their caring friendships: Alan Listiak, Ph.D., LICSW; Billy Anderson, M.A., LICSW, L.A.D.C.; Lynn Clarin, B.S., RN; Jennifer Dease, M.S., L.P.C.; and Judy Rand, B.S., RN.

And special thanks to the true experts in the violence field who I have spoken with over the years, who answer their own phone when you call them, who take the time to review their research with you on the phone on the spur of the moment, who gladly share their work, and for their ongoing commitment to understanding sexual offenders and violence: Gene Abel, M.D.; Karl Hanson, Ph.D.; Robert Prentky, Ph.D.; Ray Knight, Ph.D.; Roy Hazelwood, M.S.; and Veronica Valliere, Psy.D.

And, of course, to Fay Honey Knopp who never stopped supporting new authors, who always had a smile to share, and for her wealth of knowledge about offenders. She was perhaps one of the most intelligent and knowledgeable professionals in our field. *In her memory, thanks for your caring work. You are missed!*

A special thanks to everyone, all of my family, friends, and colleagues, who helped me in so many ways.

I pray for an end to relationship and sexual violence. I thank and deeply respect all of the people who donate their time as volunteers, victim advocates, clinicians, law enforcement personnel, supervising agents and court personnel, and other professionals working to address an end to relationship violence.

Introduction

BATTERING AND SEXUAL ASSAULT: THE CONNECTION

It is amazing how many treatment professionals continue to view domestic abuse and sexual violence as two unique and distinct fields. I have consistently said that the two fields need to merge and share their information. Most batterers also engage in sexually violent behavior (e.g., use of sexually derogatory language, expecting sex from their partner) and many sex offenders have engaged in the use of physically, psychologically, and verbally violent behavior as well. In addition, batterers and sex offenders share many of the same personality characteristics and life experiences.

Current literature includes information on battered women who have experienced sexual violence. Studies have suggested (e.g., Campbell, 1989; Campbell and Soeken, 1999; Finkelhor and Yllo, 1985) that a large proportion of battered women admit they have experienced sexual violence within their relationships and as part of the battering. Dutton, Kaltman, Goodman, Weinfurt, and Vankos (2005) indicated that forced sex appears to be a marker for violence severity in battering relationships, as it is associated with increased frequency and severity of violent episodes and increased risk for homicide (Campbell 1989; Campbell et al., 2002).

A survey sent to 8000 women and 8000 men found that approximately 50% of the women raped by an intimate partner and 66% of women physically assaulted by an intimate partner had been victimized multiple times by that partner (Rand and Saltzman, 2003). Female rape victims reported an average of 4.5 rapes by the same partner and female physical assault victims reported an average of 6.9 assaults.

Biderman (1975) suggested that rather than look at each specific act of battering or sexual violence that occurs within a relationship as a separate and distinctive incident, it may be more appropriate to view such relationship violence as conditions that endure. The majority of women in one study reported that the psychological abuse was worse than the battering (Follingstad, Rutledge, Berg, Hause, and Polek, 1990). Dutton et al. (2005) and Coker, Davis, Arias, Desai, Sanderson, and Brandt (2002) found psychological abuse to be a significant and independent predictor of poor mental health outcomes.

Dutton et al. (2005) found that battered women do not experience one type of violence or abusive behavior alone. They identified three violence patterns: (1) involved exposure to moderate levels of violence across violence types; (2) involved high levels of physical violence, psychological abuse, and stalking, but low levels of sexual violence; and (3) involved high levels across all types of violence including sexual violence.

WHY BATTERERS AND SEX OFFENDERS COMMIT SEXUAL VIOLENCE

Batterers and sex offenders, like other types of criminals, commit their chosen crimes as a result of many interacting variables. It is at times very difficult to identify and determine the motive of a specific offender for a specific crime. Sex offenders and batterers use sex as a weapon. Sex offenses are not committed solely for the purpose of sex, or orgasm. Several experts in the batterer and sex offender assessment and treatment professions have identified typologies of batterers and sex offenders and these typologies will be discussed in Section 2. It has become clearer to us that some offenders do not fit neatly into one category or another (although most do).

The following are some of the most common reasons why batterers and sex offenders engage in sex offense behavior. Understand that any given offender may have multiple motives for engaging in his crime, especially sexual crime.

Sex offenders and batterers may commit sexual offenses to deal with their own sense of frustration. They may feel sexually unfulfilled or may become sexually aroused and may believe that people who make them sexually aroused should please them sexually, even if force must be utilized. Offenders may experience difficulties with intimacy, in general, which may be even more true for batterers and sex offenders. They may find it difficult to connect with others socially and/or intimately. They may be in relationships that are very casual or, then again, very emotionally intimate.

Sex offenders and batterers may commit sexual violence out of anger. They suppress anger for some time before becoming violent. Others become enraged at their victim, believing that the victim is leading them on or teasing them, playing a game, or cheating on them. Some offenders become angry and use the rape to punish the victim. Some sex offenders rape to vent anger toward someone other than the victim, and then some vent their anger directly toward the person they believe wronged them or led them on sexually.

Interestingly, having been a victim of childhood sexual, physical, or psychological abuse, or even neglected as a child, does not seem to play a significant role in an offender's choice to offend (Harris, Rice, and Quinsey, 1998). The studies that have been conducted to assess the role of these factors have yielded mixed results. Unfortunately, many of these studies appear to be of poor research design and, therefore, the results are not very useful. I have yet to work with a sex offender who did not experience some form of childhood abuse or neglect. Others have found that only about 15 to 30% of their offenders experienced such trauma. But we know that the vast majority of sexual offenses, child abuse, and cases of child neglect go unreported, and that many thousands of children experience abuse and/or neglect. Yet only a small fraction appears to become offenders themselves. What this seems to indicate is that having been abused or neglected does not increase the likelihood of engaging in the same behavior. Some offenders use their own experiences of childhood abuse as an excuse for abusing others and, for some of these offenders, it may have played a role in their decisions to offend. However, for the majority, the decision to offend is based on many other factors, and the bottom line is that in each offense situation, the offender makes a conscious decision to offend, regardless of past issues or trauma.

Many sex offenders and batterers are under the influence of drugs or alcohol at the time of the offense. However, if alcohol and drug use played a significant role in the offense, then every time the offender was under the influence of chemicals, they would likely be committing new offenses. Alcohol and drug use certainly acts as a disinhibitor; that is, being under the influence allows one to more easily justify engaging in offense or other uncommon behavior. But a decision is still made by the offender, whether sober or intoxicated, to either commit or not commit an offense. This is further discussed in Chapter 19.

Other factors include intimacy issues. I believe that when a young child is not allowed physical and emotional nurturance and intimacy with the parents, then the ability of the child to develop healthy and appropriate intimacy boundaries and a capacity for empathy is severely impaired. Intimacy and nurturance needs tend to be acted out in unhealthy ways through the misdemeanor-type sex offenses, including voyeurism, exhibitionism, telephone scatology, and frottage. These types of sexual offending behaviors were once thought to be innocuous and safe, never leading to much more than a nuisance to the victim or to society. We now know that many violent sexual offenders began their sexual offending careers with such behavior, and that these paraphiliac behaviors help to strengthen violent fantasies and, therefore, tend to lead to and encourage the offender to engage in more serious and violent sexual behavior.

THE IMPORTANCE OF THE OFFENDERS ADMITTING THEIR OFFENSE BEHAVIOR

Some offenders readily admit to their offense/violent behavior. Many, however, deny or significantly minimize their behavior. One can hardly blame them. After all, to admit to an offense to the police would most certainly result in a conviction, likely resulting in being sent to prison. So, it is common that offenders deny and minimize their behavior during the investigation, prosecution, and sentencing stages. Offenders, like every other citizen, have the right not to incriminate themselves.

Once convicted and sentenced, however, the offender who continues to utilize denial and minimization poses more risk for reoffense. Continued denial and minimization suggest more severe pathology and a lack of remorse and empathy. What does this mean? Offenders possess pathology like many nonoffenders. Pathology refers to premorbid problems that an offender experiences. For example, underlying depression may have been a problem long before the offender committed his first sexual offense. (Premorbid refers to preexisting problems, problems that were around prior to the offense behavior.)

Remorse refers to the degree to which an individual experiences unpleasant thoughts and emotions as a result of engaging in a specific behavior. Remorse indicates that an offender appreciates, to some degree, that his offense behavior was wrong. The degree to which an offender has the ability to appreciate the consequences of his behavior and experience appropriate emotions suggests the degree to which he may be treatable. For some offenders, they blame the victim or circumstances and refuse to accept responsibility for committing the offense. These individuals lack remorse or possess minimal degrees of remorse. Because of this, they fail to appreciate the control they have on their behavior and are at more risk for reoffense than those with true remorse.

Empathy refers to the ability to appreciate the consequences that your actions have on another. This requires an ability to put oneself in the other person's shoes. Empathy has three features or characteristics that I believe are important to demonstrate genuine empathy. These aspects are *cognitive recognition*, *emotional connection*, and *behavioral demonstration*.

Cognitive recognition refers to understanding the definition and concept of empathy. Some sex offenders and batterers may understand and be able to verbalize the concept of rape or abuse; that their behavior has caused harm to their victim. Another feature of cognitive recognition involves understanding the concept of being a victim, that is, to appreciate what victims of rape and abuse experience.

Emotional connection refers to experiencing and being able to identify the guilt, shame, and fear of what your behavior has caused your victim. This requires an understanding of your own behavior and a belief that you are not out of control, but rather have made choices to behave physically and/or sexually abusively, and to do so means that you did not care about the welfare or safety of your victim. Or perhaps you truly intended to physically, psychologically, and/or sexually harm them. Emotional connection also requires an understanding of the possible consequences of your behavior. Specifically, the most severe consequences may include the possibility of losing a relationship or marriage, or losing freedom by being incarcerated.

Behavioral demonstration refers to practicing what you preach. This means to do what you say you will do; to behave in ways that live up to and demonstrate your beliefs, morals, and values. This involves believing that your behavior is abusive or sexually violent and to behave in ways that demonstrate this understanding and belief. Sex offenders do not believe that forcing sex is wrong and, therefore, have limited capacity for empathy at the time they offend. Abusive and sexually violent behavior demonstrates quite loudly and clearly that you do believe that abuse or rape is appropriate in certain situations. If you have true empathy, then you do not force sexual contact for any reason.

Obviously, the more traits of remorse and empathy offenders possess, the more they are motivated to address their physically or sexually abusive behavior. In fact, many batterers and sex offenders do complete anger management and sex offender treatment programs never again to reoffend. Others complete the programs and still reoffend, although the majority do not reoffend. This fact is based on rearrest and reconviction data. Those who fail to enter a sex offender treatment program are more likely to reoffend than those who complete treatment, but only about 30 to 35% will reoffend. Again, this information is based on the above-mentioned criteria. However, those who enter a sex offender treatment program and fail to complete it are the most likely to reoffend, at almost twice the rate as those who never enter treatment.

UNDERSTANDING THE ROLES OF PROFESSIONALS INVOLVED IN THE BATTERER AND SEX OFFENDER FIELDS

The public often becomes frustrated or angry at the various agencies and professionals involved in the investigation, arrest, prosecution, and treatment of batterers and sex offenders. It is helpful to understand the major players involved in addressing batterers and sex offenders. The role of law enforcement is to investigate, gather evidence, and apprehend the abuse or sex offense suspect. Their primary goal is to gather evidence to provide the County Attorney's Office with enough standard of proof to convict the offender.

The principal goal of the forensic psychologist is to assess the offender's issues and psychological dynamics; to assess whether the offender is admitting or denying their offense; to determine what type of treatment is recommended (outpatient, residential, prison-based, psychoeducational, short or long term) and whether the offender is amenable to treatment; and to profile the offender in order to offer an accurate prediction concerning the offender's likelihood for reoffense.

The major role of the anger management program staff and sex offender treatment staff is to further assess the offender, to develop an individualized and standardized treatment plan, and to engage the offender in the treatment and therapy process. The therapist assesses the offender throughout the treatment process, paying attention to behavior and attitudes reflective of positive change and those indicating an escalation into future offense behavior. It is never to protect the offender from any consequence for their behavior or risk potential.

The primary role of prevention professionals is to educate youth and adults, not only about what sexual offenses are, but how to deal with a potential batterer and/or sexual offense situation and the prevention of a potential offense. These usually include teachers and community educators who specialize in abuse and sexual offense prevention. Programs that teach "good touch, bad touch" are an example of such a program. It is imperative that children, even as young as preschool, adolescents, and even adults become educated as to what constitutes a sex offense. I worked as such an educator for a year, and found that the majority of my students had at best a minimal and distorted definition of sexual assault, rape, incest, date rape, and child molestation. Many believed that the victim was to blame, even when the victim was a 3- or 8-year-old child. Imagine for a moment that your young child would be to blame for having been sexually molested or violently raped. What could a child possibly do that would force an adult or a person older than the victim to assault them sexually? I can think of no situation. All sex offenses are planned assaults, whether impulsive, spur of the moment, or well thought out ritualistic plans. I have yet to meet the sex offender who had no idea what he was doing. The same is true about the batterer: any type of abuseive behavior is intentional.

It is extremely important to educate everyone, at any age, what abuse and sexual offenses are, how to avoid situations, and, most importantly, that one should always report the offense to his/her parents, teachers, police, minister, or someone who can help should they ever become a victim. Imagine that your child becomes a victim, a statistic, and doesn't know what to do. So out of shame, embarrassment, and eventually self-blame and fear, he or she simply does not tell anyone — not you, or anyone. The child bottles the offense up, keeping it locked inside. Eventually, it gnaws at his or her very sanity, resulting in depression, anger, and increasing fear, as well as a distorted sense of the child's own monster-like being. All of this is possible because he/she did not possess the most basic knowledge of sexual offenses and how to address the issue safely, or he did not know how to deal with a bully or abuser. Please do not believe, even for a moment, that education on this topic is a useless and unneeded process. Everyone deserves the clear-cut facts. Even young children do not become scared as a result of the education. It does not make them fear offenders, but rather helps to encourage a safe sense that they can understand sexual offense if it occurs, and that they can report it and feel good about themselves for doing so.

In lecturing around the country, I find that I have learned a great deal from clinicians and law enforcement personnel. The connection between batterers and sex offenders is significant. It was the Federal Bureau of Investigation (FBI) that appeared to have been the first agency or field of study to merge the two fields in their profiling services. I have learned while profiling and forensically assessing sexual predators, sexual offenders and batterers to appreciate the vast similarities. I have learned from each of the experiences, have listened carefully to their stories, their offenses, their victims. I feel honored to share some of my work findings in this book.

Definitions of Abuse and Sexual Offense

Before we begin our journey to a better understanding of the batterer and sexual offender, it is important to first define what we will be discussing. As is clear in reviewing the literature, defining the topic at hand and specifying the type of abuse or force that occurred is often not done. In fact, many researchers and clinicians label all typologies of batterers into the category of "batterers" or "abusers." Yet not all batterers are the same, and there are many typologies of batterers. Some are far more aggressive than others, some resistant to change, and others perhaps more amenable to treatment interventions. Likewise, many label all sex offenders as "sex offenders." Sex offenders differ as well, some preferring children, some preferring known victims, and others enjoying causing their victim serious physical harm.

Following is a review of the definitions used in this book.

ABUSE refers to any act that is intended to cause harm or results in harm to another person. Included here are verbal put-downs and name calling on one extreme, to pushing, slapping, rape, and murder on the other extreme.

HARM refers to the emotional, psychological, physical, and/or sexual injury that results as a result of abuse. Harm may not always be seen directly, but often is inferred by the presence of certain factors or symptoms. What this means is that you or I may not be able to always see visible signs that abuse has occurred; there may be no scars, bruises, or bleeding. Harm includes fearing for one's safety, for the safety of others, and of imminent injury. Injury includes psychological, physical, and sexual trauma. Fear of embarrassment, of losing one's children, and of being left without any financial resources also constitutes harm.

INTENT refers to what someone wants to do, the plan they have. Intent often causes confusion for people, and it is often believed that abusers do not intend or mean to cause harm. But all abuse is planned and, when abuse occurs, it is because of a conscious choice made by the abuser. The abuser made a decision to give himself permission to abuse. Intent may be difficult to establish or prove, but if the goal was to create fear, injury, or to gain cooperation of someone by the use of intimidation, force, or injury, then abuse was intended. Often abusers claim that the abuse was an accident, and accidents do occur, *but abuse towards one's significant other is always a planned choice, it never occurs by chance*. Intent will be discussed further at a later time.

As for frequency (how often the abuse or sexual offense happens), whether the abusive act occurs one or more times, is still abuse or sexual assault. Too often abusers and sex offenders claim that they did not "*mean to*" or "*intend to*" cause harm or be abusive, but abuse and sexual offense is always a conscious choice and, therefore, intent is implied. Proving this statement is easy. The fact that the majority of abusers and sex offenders tend to assault only a selected few people indicates that a choice is being made, a choice which includes a conscious effort to evaluate situations and adjust how we behave toward others. Toward some, the abuser and sex offender may behave kindly, be well mannered and respectful, while toward others behaving abusively and disrespectfully.

The terms abuse, sexual offense, aggression, and violence may be used interchangeably throughout this book. After all, abuse and sexual offenses are acts of aggression and violence, never an act of love or a mistake. There are four types of abuse — psychological, emotional, physical, and sexual — and each is described below.

PSYCHOLOGICAL abuse (also referred to as verbal and emotional abuse) refers to the attacking of someone's self-esteem and self-worth. Over time, the attacks are made against the whole being of the other person. Examples of emotional abuse include name-calling, put-downs, belittling accomplishments, and swearing. This is the most commonly occurring form of abuse, and usually occurs in all of the other types of abuse. Often people overlook emotional abuse and may accept it as normal behavior. But it is a type of abuse and, in most cases, the abusers will eventually progress to abusing their significant other in other ways, the abuse becoming increasingly more severe over time. Sexual offenders may begin their victim grooming by becoming emotionally abusive and demeaning before engaging in any sexual aggression.

EMOTIONAL abuse marks the beginning of an abusive relationship and should be taken just as seriously as psychological, physical, and sexual abuse. Examples include:

- Threatening to harm someone
- Threatening to or actually destroying someone's belongings
- Threatening to or actually taking or harming the children
- Harassment (any unwanted comments, gestures, or behavior, including monitoring the other's actions (who they talk to, where they go, who they see, etc., termed *stalking*)
- Withholding and/or controlling the other's money, including taking money, making them buy items for someone else, etc.
- Verbal attacks and name calling
- Interrupting eating and/or sleeping patterns
- Forcing someone to do anything degrading, humiliating, or dangerous
- Criticizing the other's thoughts, beliefs, or behaviors
- belittling accomplishments
- Treating the other as if inferior or a servant
- Behaving jealously (e.g., possessiveness, guarding the significant other)
- Hitting or destroying property (e.g., doors, walls, chairs, pictures)

As a result of psychological abuse, the victim may experience fear (of the abuser, being harmed, or of others finding out about the abuse, etc.) as well as feeling trapped, worthless, and exploited. Sexual offenders frequently use psychological abuse to season their victim into compliance as well as encouraging the victim to feel powerless against the offender's advances.

Psychological abuse involves objectification. Name-calling and jealousy are part of objectification. This occurs when the abusers and sex offenders begin to view their significant others as possessions, objects they own rather than as persons. The abuser and sex offender objectifies his significant other/victim to make it easier to abuse them. It is much easier to abuse an object than a person. When an abuser has progressed to psychologically abusing their significant other, they will eventually progress to physical and sexual abuse. It is clear at this point that the abuser and sex offender does not care for, respect, or love his significant other or anyone else, with the exception of maybe himself.

Psychological abuse is an attack upon someone's identity, self-esteem, and self-worth, and both are abuse. On a scale of severity (i.e., not resulting in serious physical injury or death), this ranks at the bottom, but the effects are often more devastating and longer lasting than any other form of abuse. The Continuum of Force (see following page) illustrates how psychological abuse may progress over time. The examples of psychological abuse that I list in the Continuum of Force do not necessarily occur in any order, that is, they may occur in different orders for different offenders, taking into account individual differences. But offenders do not simply wake up one day and begin to utilize serious psychological abuse; they increase the intensity over time.

PHYSICAL abuse involves actually physically attacking someone, which includes any unwanted bodily contact. This includes rough housing, any type of physical aggression, and, of course, any physical touching that results in harm to the victim. Genuine accidental incidents of

CONTINUUM OF FORCE

PSYCHOLOGICAL ABUSE	THREATS	PHYSICAL & SEXUAL ABUSE
➢ Yelling, swearing ➢ Name calling ➢ Put downs ➢ Using personal information to hurt or blackmail ➢ Objectifying ➢ Lying/omission of information ➢ Driving recklessly ➢ Not taking medication as directed ➢ Criticizing, belittling ➢ Harassment or stalking ➢ Sarcasm ➢ Ordering, controlling, commanding ➢ Jealousy ➢ Explosive anger ➢ Interfering with calls ➢ Limiting friendships ➢ Misspending or withholding financial support	➢ To harm self ➢ To harm partner ➢ To harm children ➢ To harm pets ➢ To divorce ➢ To spread rumors ➢ To withhold money ➢ To withhold visitation ➢ To assault others ➢ Displaying weapons ➢ Insinuating harm or use of weapons	➢ Any physical contact that can or does cause harm ➢ Hitting, slapping, or punching ➢ Choking, strangling, grabbing by neck ➢ Hairpulling ➢ Use of weapon or object ➢ Restraining or blocking ➢ Talking into sex ➢ Pressuring into sexual contact ➢ Coercing into sexual contact ➢ Taking advantage of intoxication for sex ➢ Using threatening statements to get sex ➢ Using sex to make up or calm down ➢ Physically forcing sex ➢ Refusing to accept "no's", hesitations, lack of response ➢ Expecting sex ➢ Exerting the "male privilege" to get one's own way or sex ➢ Tricking or being deceitful to get one's own way or sex

Overtime, the intensity of the violence will increase……
from verbal & psychological abuse to threats, to sexual & physical abuse, even murder.

injury from horsing around are not necessarily abusive; accidents do occur. But physical abuse is the intentional causing of physical injury or pain, and, again, intentional involves the decision to engage in the behavior, regardless of whether more harm results than was "intended." Common examples of physical abuse include the following:

- Slapping
- Punching
- Pushing
- Kicking
- Grabbing
- Choking
- Throwing objects at the other person or in their general direction
- Spanking or physically punishing in any way
- Wrestling (against the other person's will)
- Carrying the other person against their will
- Pinching
- Biting
- Spitting
- Scratching
- Throwing the other bodily
- Burning
- Pulling of the hair
- Restraining (e.g., holding the other person down, not allowing them to leave)
- Using any object or weapon against the other (this includes a stick, belt, paddle, whip, rope, knife, gun, or any object that can cause injury or can be used to create fear through threatening to use or actually using the object, including a pillow when used to smoother)

Psychological abuse is inherently involved in physical abuse/physical aggression. It is during physical abuse and aggression that the immediate safety of the victim is in jeopardy. When the abuser and sex offender has progressed to using physical abuse, they have, for the most part, totally objectified their significant other/victim. What this means is that the victim is viewed as a possession, owned by the offender, not as a person. For example, the offender believes that his significant other/victim is "*his*," just like he owns a car. It is always easier to physically attack an object than another human being.

Examine the Continuum of Physical Abuse (see next page). I have found that most offenders begin with lesser forms of physical violence and progressively become more violent over time. Not all offenders will progressively become more violent, but most will in some way increase the violence they use. Again, the examples of physical abuse I list in the Continuum of Physical Abuse do not necessarily occur in any order, that is, they may occur in different orders for different offenders, taking into account individual differences. Batterers and sex offenders do not simply begin to utilize serious physical violence, but rather increase the intensity over time.

SEXUAL abuse involves attacking someone sexually. This includes any forced sexual contact, whether by coercion (psychological force), physical force, threats, or by ignoring the other's rights and requests. Examples include:

- Unwanted sexual comments or gestures
- Touching someone sexually without their consent
- Forcing someone to do any sexual act against his/her will, including having sex with others, video taping, use of objects, etc.
- Ignoring the other's "no's," hesitations, or resistance
- Pushing, challenging, or breaking sexual limits or boundaries

CONTINUUM OF PHYSICAL ABUSE

	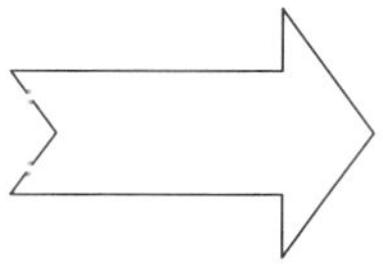	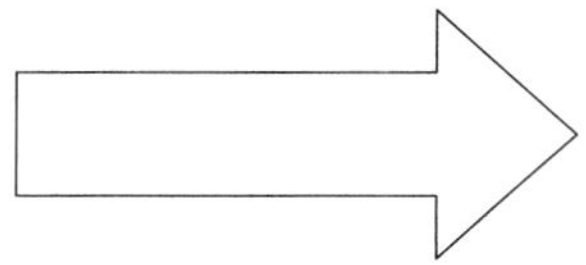	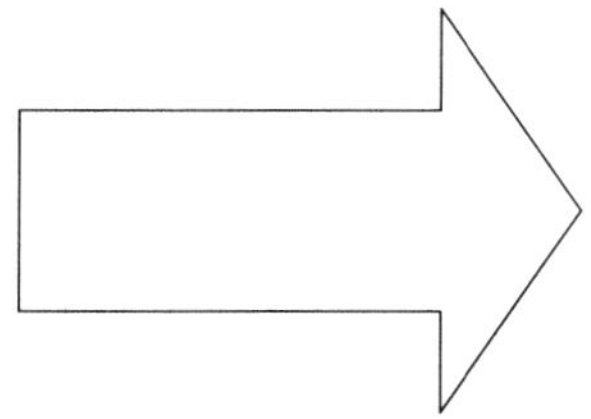
➢ Bumping ➢ Blocking ➢ Pushing ➢ Throwing objects ➢ Breaking objects ➢ Restraining or blocking ➢ Threatening more serious physical violence	➢ Slapping ➢ Punching on the arm or back ➢ Throwing objects directly at someone ➢ Hair pulling ➢ Physical abuse in the home or when alone ➢ Threatening to use a weapon ➢ Pressuring/talking into sex ➢ Refusing to accept "no's" for sex ➢ Taking advantage of intoxication ➢ Using sex to make up/calm down	➢ Hitting or punching on the face ➢ Repeatedly hitting or punching ➢ Using a weapon or object to assault ➢ Occurs when the children are present or in the residence ➢ Expecting sex ➢ Exerting the "male privilege" ➢ Threatening to physically assault or reminding of past physical assault	➢ Repeated punching of the head causing injury ➢ Choking ➢ Use of lethal weapons ➢ Gratuitous violence ➢ Children are physically abused ➢ Children involved in the abuse ➢ Begins to occur in public ➢ Torture ➢ Kidnapping ➢ Murder ➢ Physically forcing sex ➢ Sadistic sexual behaviors

Over time, the intensity of the violence will increase……
from minor physical & sexual abuse to serious, life-threatening, physical & sexual violence.

- Engaging in intimidating behavior
- Begging or using emotions (e.g., love, any affectional pressure) as a means of forcing sexual contact (e.g., "*Prove you love me*")
- Engaging in sexual activity with a person who is impaired (e.g., sleeping, injured, under the influence of drugs and/or alcohol, or unable to say no for any reason — does not speak the same language, hearing/speech impairment.

Sexual abuse also includes all other types of abuse because physical contact is made, and emotional and psychological abuse are involved. Sexual abuse is RAPE, it is aggression, and violence. Sexual abuse is an extremely intricate and complicated topic and, for the purpose at hand, will be dealt with only as it relates to violence within dating relationships.

Examine the Continuum of Sexual Abuse (see next page). I have found that most offenders begin with lesser forms of sexually aggressive and violent behavior and many progressively become more sexually aggressive over time. Again, not all offenders will progressively become more violent, but most will in some way increase the violence they use. Again, the examples of the sexually abusive behavior I list in the Continuum of Sexual Abuse do not necessarily occur in any order; that is, they may occur in different orders for different offenders, taking into account individual differences. Batterers and sex offenders do not simply start off by engaging in more aggressive sexual violence, but rather increase the intensity over time.

I do not in any way believe that any form of abuse or sexual violence is less or more damaging to a victim. Victims are a heterogeneous group, with each victim experiencing differing degrees of harm from the next. Some victims having been exposed to or been briefly fondled may experience Posttraumatic Stress Disorder and require years of therapy. On the other hand, victims of a rape may not require much of any therapeutic intervention and may heal within a short amount of time. Victims have many individual differences and differing coping skills. Any abuse or sexually violent behavior is serious, regardless of the degree of harm or injury suffered by the victim.

VICTIM refers to the person whom the violence/abuse is directed toward.

SIGNIFICANT OTHER and ***PARTNER*** refers to the person with whom the offender is involved with in an intimate or familial relationship.

BATTERER, ABUSER, SEX OFFENDER, and ***PERPETRATOR*** refers to the person committing the abuse or sex offense, and all terms may be used interchangeably. It is important to point out that abusers and sex offenders may be male or female, just as both males and females may be victims. Most often when abuse or sex offenses are discussed, the male is the abuser and the female is the victim. This is due in part to the manner in which statistics on abuse are gathered, as well as the fact that most males will not readily admit to being abused or sexually assaulted/molested. Abuse and sexual offenses also occur in both heterosexual and homosexual relationships. Violence occurring within homosexual relationships may not be as readily reported due to additional stigmas that contribute to keeping the abuse and sexual violence private.

One last distinction will be made. When putting a label on abuse, the type of relationship is taken into account. There are three common labels — domestic abuse, child abuse, and dating or partner abuse. *Domestic violence* refers to abuse occurring within a marriage. Sometimes this term also includes engaged couples or those living together regardless of marital status, especially when used in legal terms. *Child abuse* refers to violence involving an adult toward a child. *Dating and partner violence* refers to abuse occurring within a dating/intimate relationship and can include intimate partners residing together or engaged. This involves any dating relationship, regardless of sexual orientation or age. Many similarities exist between the above labels, in that the three types of abuse may occur within each; the main difference to be made involves the dynamics of the relationship involved.

Sexual offenses include statutory rape, child molestation, incest, rape (stranger or acquaintance), sexual assault, and gang rape. Statutory rape involves any sexual contact that occurs with a person who is not of legal age of consent, generally under the ages of 16 to 18. State statutes differ and

CONTINUUM OF SEXUAL ABUSE

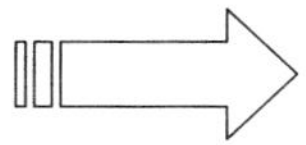	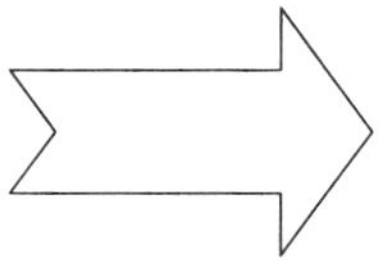	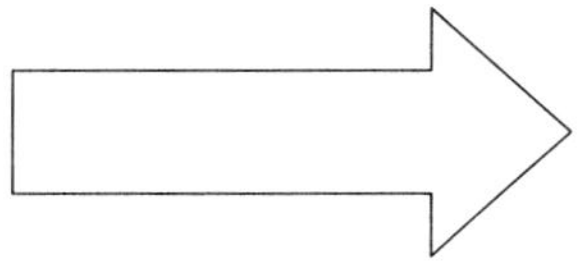	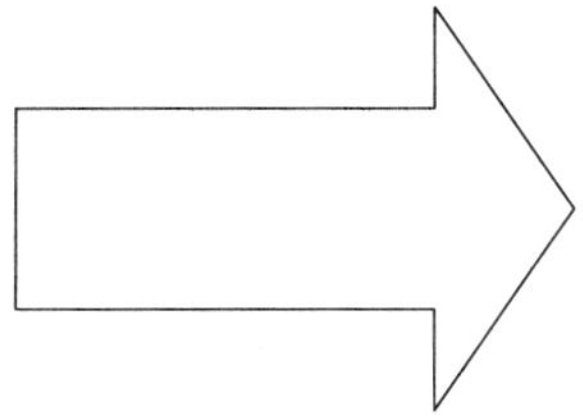
➢ Talking into sex ➢ Taking advantage of intoxication ➢ Guilting into sex ➢ Using pornography to deal with stress or anger ➢ Frequently using sexual Internet sites & pornography ➢ In any way comparing partner to others sexually	➢ Pressuring or coercing into sex ➢ Refusing to accept "no's" for sex ➢ Using sex to make up/calm down ➢ Tricking or using deceit ➢ Unwanted touching or grabbing of sexual areas when in private ➢ Making sexual comments ➢ Openly using pornography ➢ Requests to view pornography together ➢ Begins to withhold sex when angry or to "get back" ➢ Engaging in chat rooms, dating services, or Internet sites for the purpose of meeting other people	➢ Threatening to take sex ➢ Threatening to physically harm ➢ Expecting & demanding sex ➢ Exerting the "male privilege" ➢ Unwanted touching or grabbing of sexual areas when in public ➢ Making direct, personal sexual comments and use of sexually derogatory language ➢ Use of pornography before, during or after sex ➢ Demanding that they act out what was seen in the pornography ➢ Engaging in sexual affairs	➢ Engaging in humiliating or painful sexual acts ➢ Physically forcing sex ➢ Sexual contact occurs in public ➢ Sexually demeaning behavior in private as well as in public ➢ Sexually sadistic behaviors ➢ Torture ➢ Kidnapping ➢ Murder

Over time, the intensity of the violence will increase……
from minor sexual abuse to serious, life-threatening sexual violence.

should be consulted to determine age definitions for your state or country. It does not matter whether the identified victim consented to the sexual contact. By definition, an under-aged person is not able to give consent to engage in sexual contact. Child molestation involves sexual contact by an individual who is significantly older than the child; generally offenders are adults and the victims are under the age of 12. However, even children and adolescents can molest other children of similar ages or younger children. The term *hebephrenia* has been used to describe a sexual interest in adolescents aged 13 to 17.

Incest refers to sexual contact between family members or relatives, regardless of age or relation to the victim. In general, incest victims tend to be children and adolescents, though adults have been the victims of incestuous sexual contact as well. *Sexual assault* generally refers to coerced sexual contact without the use of physical force, and *rape* refers to the use of force or threat of force to gain sexual contact. *Gang rape* refers to forced sexual contact that occurs with several offenders acting in concert. The relationship between the offender and victim (e.g., stranger, acquaintance, intimate partner, or family member) does not change the fact that a sexual offense occurred.

Physical and sexual violence is not a new topic, but rather a new focus for professionals. It would be simplistic to believe that abuse only occurs within marriage or between parents and children, just as it would be simplistic to believe that only perverts or strangers commit sexual offenses. Abuse (battering) and sex offenses are pervasive throughout our society and as a result has presented us with a serious chronic problem which, if left untreated and unaddressed, will lower the quality of our relationships, decrease the safety of our families and communities, and cost enormous amounts of money.

Section I

The Offending Cycle

1 Domestic Abuse and Sexual Assault

INTRODUCTION

I have spoken throughout this country on the topic of psychological force. It is amazing that professionals often can identify psychological force and psychological abuse, but do minimal work with the perpetrator and victim concerning the impact of psychological force and abuse. Psychological force and psychological abuse truly are similar phenomenon and I will use the terms interchangeably throughout the book. Force technically is abuse, and it seems needless to make a distinction between these terms.

In our country, domestic and physical abuse are misdemeanor offenses. We have not learned from history; that is, if we as a society consider physical violence as a minor infraction, then we are in a sense directly and indirectly condoning and supporting it. In Minnesota, it is a felony to fondle the breast or buttocks of a nonconsenting person over the clothing. I believe the victim of fondling can suffer significant trauma from being touched; however, physical violence continues to be a misdemeanor, while fondling is a felony. Lawmakers and lobbyists need to change the domestic violence laws to equal the sexual offense laws. Domestic and physical abuse are the major contributors to the decay of the family unit and, as a result, a decay of the American society as a whole. So, we do develop laws that would make physical and domestic abuse felonies, but we will still be stuck without developing a more thorough understanding of psychological force/abuse. There are several steps involved in adequately addressing and identifying the issues of psychological force used in domestic and sexual assaults.

The first step is to identify the types of force the perpetrators utilize in the commission of their sexual assault. The therapist should carefully read all collateral information to determine the types of force used in the assault. The perpetrator should be asked to identify not only the types of force used, but also to give several examples of each type of force used. The victim can also benefit from this exercise.

I strongly believe that the perpetrator and victim will not heal from their traumas if they fail to understand the type of force involved in the assault. What you will find is that the types of force used in the physical abuse and/or sexual assault are occurring in the perpetrators' everyday lives. If they utilized physical force, they likely have a history of fighting or using threats and intimidation in their relationships and at work. If they utilized manipulation or game playing, they likely continue to engage in manipulation in their relationships and at work. Look and you will find the types of force the perpetrator used and continues to use. They may have different examples of how they use each type of force in their daily lives, but regardless of how the force is used, using force is still using force. The process of identifying the types of force used in the sexual or physical assault will be assessed ongoing as the perpetrator becomes less defensive and more invested in therapy.

Therapists working in this field should be aware that if physical abuse, threats, or intimidation are occurring, then rape is occurring as well (marital, date, and/or acquaintance). Consent involves the ability to safely give or refuse permission to be sexual. How could anyone who is the victim of intimidation, threats, and/or physical abuse safely give consent? If victims refuse the demands or requests of the abusers, the expected outcome would be further abuse. It is easier and safer for them to agree to do whatever their perpetrators demand than to risk being physically abused. I cringe when therapists from domestic abuse programs fail to identify sexual assaults and rapes as occurring in their clients' relationships. How naive it would be to believe that if you were afraid of refusing the sexual requests of your partner because you might be physically abused for doing so, that you actually had a choice. You did not have a choice, at least not one that was worth considering. If you refused, it was likely that your abuser would coerce or physically force sexual contact anyway. To refuse the perpetrator's requests would be to ask to be physically and psychologically abused.

The second step is to correctly label the psychological force as abuse and place the force in the perpetrator's cycle, as different strategies of force would be used at different stages within the cycle. Again, perpetrators should be giving specific examples of each type of force they engage in. Observe perpetrators in group and individual sessions to assess first-hand how they use force, and to document noticeable progress for dealing with problems without force. Confront them when they use force. I find using sarcasm and humor is extremely helpful to point out to the perpetrators when they are using psychological force. The sarcasm helps the perpetrators experience the sense of being put down, as well as the loss of power, which for them has become paired with their process of using psychological force. Using empathy to confront psychological force only serves to reinforce the perpetrators' persona by offering a gentle form of nurturance in the form of empathy. This is not the goal. Use sarcasm and confront the perpetrators immediately when they engage in the use of force. Remember that you are either part of the problem or part of the solution; you either support the perpetrators' healing or the perpetrators' persona. You can undo months of therapy with one moments support of force.

One effective method of aiding perpetrators in understanding how they use psychological force and the impact on others is through role playing. Have group members role play the incident when the perpetrator engaged in the use of psychological force. Examine the perpetrator's needs, that is, what did he really want from others, what emotions was he struggling with? Have the person playing the role of victim discuss the impact.

Let's discuss a typical situation in which the psychological force/abuse continues to affect the victim and the entire family. A perpetrator of domestic abuse is generally required to leave the home. When the victim of incest reported the father to the authorities, she had to leave the home. Do not believe for a moment that the mother and siblings are 100% supportive of the victim. After all, consider the ramifications of her reporting the assault. Her father went to court, was charged with child molestation and/or incest, and spent time in jail. He then had to enter treatment. All of this takes a toll on the family. Money is scarce as a result of the extraneous bills — sex offender treatment for the father, therapy for the victim, family therapy, court costs, and fines. Let us not forget the other effects on the family. What do other family members think, what about the neighbors, classmates, and co-workers? And the list goes on.

The perpetrator has psychologically abused his entire family, as listed above. But it is the victim who often inadvertently becomes the target of blame by the perpetrator, mother, and siblings. Often the father is not allowed to have any contact with his victim, and this prevents him from visiting the home and other siblings. Resentment begins to build. I have been involved with several cases where the victim leaves the home to live elsewhere. It appears on the surface that the victim's intentions have nothing to do with the sexual abuse that occurred, but rather other issues. If the victim lives elsewhere, the perpetrator is then often allowed to return home.

This constitutes an enormous power play. The victim returning home would once again result in the father leaving, an overt punishment and negative consequence for the mother and other siblings. Do not for a minute believe that the victim's decision to live outside of the home was in anyway unaffected by the sexual assault. To prove my point, often the victim would request to leave the home long before the sexual abuse was discovered and, at times, before the actual sexual touching began. Everyone in the family refuses to believe that the sexual abuse played any role in the victim's decision to leave; after all, she voiced her discontent before being touched. What has been grossly overlooked is the impact of the grooming process, which began months before the actual touching began. The victim was reacting to her sense of boundary violations, and recognized that she needed to escape the family unit long before the actual touching began.

When a perpetrator I am working with requests to move home because the victim is living somewhere else, I refuse his request. I require that the significant other/spouse regularly attend therapy and support sessions to become more educated, aware, and to develop more effective strategies to protect herself and the children from possible further incidents of abuse. I also require the victim (if the victim is currently involved in a dating or romantic relationship with the perpetrator) or spouse to meet with me and the perpetrator to discuss the effects of the perpetrator moving home, especially as they relate to the victim not being able to return home as she wishes. One other common problem for incest cases is that when other siblings have contact with the father, but the victim cannot, the victim is further victimized. I cautiously approach this topic with the perpetrators I work with. At times, I do not allow contact between the perpetrator and his children, both male and female children, to avoid the use of this contact as a power play against the victim. No research has demonstrated any long-term or permanent effects on children not having contact with their father for 1 to 3 years. Restricting contact of the father (perpetrator) and his children gives a clear message about the seriousness of his crime. A therapist should carefully examine all of the possible effects of contact between the perpetrator and his children, and the effects of the victim leaving the residence.

I also encourage a comprehensive evaluation to occur for the entire family. Why does the victim feel so unsafe at home that she wants to live elsewhere? It could be that her relationship with her mother and siblings is troubled or abusive. The mother may blame the victim, and the victim needs to leave the residence for her own sanity and health. However, the relationship between the victim and other family members needs to be carefully scrutinized. If incest occurred, the family system supported it occurring, and you can be sure that the perpetrator manipulated and took advantage of any dysfunction or problems within the family.

In addition, there should be no reason for the rush or hurrying of the perpetrator returning home. Most perpetrators I have worked with find numerous reasons why they need to return home. These often include the children missing them, having to pay for a separate residence, and being lonely. The victims often want to rush the perpetrator returning home out of a sense of guilt, fear if they don't support the perpetrator returning home, and financial reasons. But there is no positive impact financially for the perpetrator returning home before it is therapeutically safe. The money that is spent with different residences is far less than the cost of having another incidence of abuse or sexual assault occur, which might result in divorce or prison time for the perpetrator, again impacting the entire family economically as well as emotionally. Rushing an abuser or sex offender into returning home also diminishes the trust the children have for the victim when the victim was the mother. The children continue to fear their father, but the only other adult in the home that could offer protection is rushing the abuser home without regard for the children's emotional and physical safety. I never rush this process, even if the abuser is faced with bankruptcy or other financial difficulties. Money should not drive the reunification process.

2 Control Explored

PSYCHOLOGICAL FORCE, COERCION, AND ABUSE

Psychological abuse (including verbal and emotional) is perhaps the most common type of force and abuse used against victims of battering and sexual offenses. Psychological abuse is also the most damaging type of abuse, and is frequently found to occur with physical and sexual abuse, but also occurs as a free-standing form of abuse (Johnson 1995, 1997, 1998). Interestingly, many male and female adolescents indicated that they condone males using physical or psychological force in certain situations to force sexual contact from a female. These findings were similar to what other researchers have found (e.g., Koss, 1987; Muehlenhard, 1988). Interestingly, even victims included in the Johnson studies condoned the use of force to gain sexual contact in certain situations. This acceptance of the use of force for adolescents certainly sets the stage for how these individuals will act as adults.

In my studies (1995, 1997, 1998), I identified seven types of force used by abusers and sex offenders. They include: emotional blackmail, game-playing/manipulation, pressuring/begging, boundary violations, lying, intimidation, and threats. I provided specific examples of the definitions used in the cited works. In the 1997 and 1998 studies, I utilized a questionnaire with adult male sex offenders incarcerated in prisons in Minnesota. The results were that 53% of the offenders engaged in the use of begging/pressuring and boundary violations, 34% utilized lying, 36% engaged in game playing/manipulation, 35% engaged in intimidation, 29% engaged in emotional blackmail, 17% engaged in the use of physical force, and 15% utilized threats to gain compliance of their victim(s). Collateral documentation verified the majority of the offenders claimed to utilize psychological force to manipulate their victim into sexual contact vs. the use of physical violence. The importance of these findings are paramount. Clinicians, law enforcement personnel, supervising agents, and judges need to take into account that the more common type of force used by sex offenders and even by batterers is unseen, does not leave physical evidence, yet is more damaging than direct physical or sexual abuse.

CONTROL

Control is a form of abuse and below are examples of "control." The list is not inclusive and every person can find ways to control in his or her own sneaky or direct way. The purpose of control is to force another person to behave in a specific way. It is almost always abusive. Whether you are the victim or abuser, it is important to identify what type of control has occurred so that it can be directly addressed. Following are some of the most common forms of control that my clients identify. Take the time to explore your own relationship to determine what types of control are occurring.

Manipulation

Manipulation involves setting up your partner to be the bad guy, to take the blame for *your* behavior, and/or to be forced to respond to you in a way that will appear abusive to others. Manipulation is controlling the other's behavior. Examples include:

- Refusing to return the children as per the visitation order/agreement, and when the other parent attempts to pick up the children, refusing to let the children go with the other

parent, and then in front of the children, saying to the other parent, "If you want me to force them to go with you I will. Is that what you want?" This makes the other parent look mean or unfeeling of the children's wishes.
- Forcing you to question him about what is going on because he refuses to talk or to disclose his problems.
- Bribing the children with gifts, especially extravagant gifts (e.g., cars, trips, special privileges).
- Undermining your efforts, parenting, or other behavior.
- Sulking or assuming a "victim role" to coerce you.

Overt Aggression

Overt aggression involves direct assaultive comments and behaviors. The primary goal is to instill fear in the partner and to gain the upper hand in order to control. Examples include:

- Threatening harm.
- Raising his voice.
- Raising his hand as if to hit.
- Any type of physical abuse or assault.
- Driving recklessly (e.g., speeding, tailgating, unsafe lane changing, driving too fast for the weather conditions, driving while under the influence of drugs or alcohol or when too tired, driving while extremely agitated, or when involved in intense argument).
- Slamming doors or throwing objects.
- Forcing you to engage in humiliating behavior (e.g., forcing you to apologize in public when you did nothing wrong, forcing you to get down on your knees to submit or apologize, forcing you to do things with others you do not want, such as sexual contact or waiting on or serving as a slave or maid).

Subtle Aggression

Subtle aggression involves behavior that may escalate or that has led to overt aggression in the past. Examples include:

- Insinuating that you may escalate the situation.
- Gunning the accelerator of the car.
- Talking in a threatening or condescending voice tone, but not necessarily raising the voice or yelling.
- Purposefully forgetting.
- Coming home late to avoid taking responsibility for household chores or parenting responsibilities.

Harassment

Harassment is any behavior that causes fear, intimidation, or nuisance to another person. Harassment is a direct form of abuse. This includes continuing to engage in some type of behavior or contact that one has been told not to do, or that he is aware is unwanted by the other person. It is important to understand that if a reasonable person would be aware that the contact or behavior is unwanted or inappropriate, then the person cannot justify that he was not aware of this fact. Harassment may include:

- Making repeated telephone calls when told not to do so, or when aware that the calls are unwanted.

- Being where the other person is when you know you should not be there or, again, when you are aware that your contact is unwanted.
- Violating any court order, including Orders for Protection and Restraining Orders.
- Entering the other person's space, office, room, car, etc.
- Making repeated requests of the other person after being told or are aware that they are not interested.

Stalking

Types of stalking include:

- Repeatedly pursuing someone.
- Intruding on the other person's space (e.g., home, work, recreational activities, family).
- An implicit or explicit threat evidenced by the behavioral intrusion.
- A willful, malicious, and repeated following and harassing of another person that threatens their safety.
- Following the individual, tailing them, or having them followed in certain situations.
- Showing up at places where the person is when your presence is unwanted or not allowed.
- Monitoring, taping, or recording another person's telephone conversation.
- Monitoring, taking, reading, or interfering with the other person's communications (e.g., mail, messages).
- Gathering personal information on the person.
- Making certain threats.
- Venting anger or hostility toward the person.
- Exerting power, control, and intimidation.
- Breaking into the person's home; rape; property damage; unwanted telephone calls; surveillance of the home; following the person; sending letters, packages, or emails or other messages through others or in person; spreading rumors; showing up at the person's work; leaving things or rearranging things; physically assaulting the person; injuring or killing pets; kidnapping.

Using the Children

Using the children involves including the children in things they should not be involved in, things that children typically are not allowed or invited to participate in. This occurs frequently when separation or divorce occurs. This may include what I refer to as "adultifying" and "parentifying" the child, which involves giving the child adult power to make decisions that children typically do not make for themselves. Common examples of this includes:

- Allowing the children to set their own curfew.
- Involving the children in discussions involving household finances.
- Involving the children in discussions about the marital concerns.
- Attempting to pit the children against the other parent.
- Not enforcing punishments, refusing to discipline, or failing to support the other parent's disciplinary action.
- Asking the children with whom they would rather live and sabotaging any expression of desire to be with the other parent.
- Failing to follow visitation orders or agreements.
- Involving the children in decisions that only the parents should make (e.g., when the other parent attempts to pick up the children per the visitation order/agreement, refusing to let the children go with the other parent, and then in front of the children, saying to the other parent, "If you want me to force them to go with you, I will. Is that what you want?"

Using the children can also involve what I refer to as "incestuosity;" that is, befriending the child as if the child were an adult and involving intimate or romantic behavior, regardless of whether the behavior reaches direct sexual contact. Examples of this include:

- Engaging in romantic behavior with the children, or more commonly with one of the children of the opposite sex.
- Using intimate physical behavior more frequently than prior to the marital problems becoming significant, or prior to separation or divorce (e.g., overly huggy; sitting closer to the child or children and having their legs or arms touching; kissing more frequently, especially if on the lips; changing clothes, bathing, or using the toilet and allowing the child or children to enter the room, or vice versa).
- Buying or allowing the child or children to wear sexy lingerie, skimpy clothing, etc.
- Giving the child or children special nick names.
- Giving the child or children extra gifts, or more expensive gifts, or more adult gifts, especially in exchange for ignoring the other parent.
- Sleeping in the same room or bed.
- And, of course, any direct sexual contact between the parent and the child or children.
- Telling the children intimate and personal information about themselves or the other parent.
- Telling the child or children lies about their other parent, especially personally embarrassing things (e.g., alleging the other parent is having an affair, telling them that their other parent is not good in bed or interested in sex).

Not Fulfilling The Role of Parent

This involves the failure of one parent to behave as a parent. Neglecting parental responsibilities is serious and results in harm not only to the other parent, but to the children as well who witness their parent failing to demonstrate any sense or serious responsibility or care. Examples include:

- Failing to pay child or spousal support.
- Failing to spend adequate time with the children.
- Telling the children that you cannot spend time with them because they are too busy.
- Promising the children you will do something with them without first consulting the other parent.
- Making excuses to avoid spending time with the children.
- Being uninvolved in parenting activities (e.g., taking the children to the doctor or dentist, going to the children's school or other activities, avoiding helping with homework, failing to attend church with the children, but telling them they must go).
- Not enforcing punishments, refusing to discipline, or failing to support the other parent's disciplinary action.
- Allowing the children to do as they please, to come and go as they want, etc.

Refusing to Talk

This involves avoiding communication with you, and refusing to tell you what they want or need, or when something is wrong. This also includes telling you that something is wrong, but failing to share it with you, leaving you wondering. Silence is a wonderfully effective form of abuse. Without ongoing, direct communication, the relationship is doomed to fail. It is nearly impossible to meet each other's needs when one is not expressing his or hers. Common examples include:

Demanding to Talk or Demanding to Resolve the Problem

Demanding to talk or to solve the problem is abusive and controlling. This occurs when they relentlessly badger you into discussing something that you have already told them that you do not want to discuss. It is fair to say that there are issues that we may attempt to force others to address because of the serious nature of the issues or problems. But if the other person is refusing to discuss the issues, then we do not have the right to continue to badger them. Common examples include:

- Following you around the house.
- Not allowing you to fall asleep.
- Repeatedly calling you at home, work, or on the cell phone.
- Physically abusing you to get you to listen and talk.
- Continuing to discuss the problem after you had reached an apparent solution or compromise.
- Forcing you to stay awake or at home until the argument is resolved.
- Forcing you to discuss the problem with someone who is biased, someone that will side with the abuser.

Blaming

Blaming involves your partner placing the responsibility for the marital and other problems on you. They fail to acknowledge their own role, and even when they do so, they accept a very minor or insignificant piece. This may also include blaming you for their problems and for things that you could not possibly have had any control over. Common examples include:

- Blaming you for their bad day.
- Blaming you for their being upset and subsequently for them getting a speeding ticket because of how upset you made them.
- Blaming everyone and anyone for their misfortunes, failing to accept responsibility for their actions.
- Blaming you for their inabilities to earn more money or for the marital problems.

Making You Appear Crazy

This involves your partner doing and saying things that leads others to believe that you are crazy, mentally unstable, or in need of some type of professional help. This often goes hand-in-hand with manipulation. Common examples include:

- Changing plans at the last minute and telling you or others that you had been informed of the change long ago.
- Canceling plans without telling you and then blaming you for not remembering, when in fact they just canceled the plans without discussion with you.
- Making plans with others or with the children that should involve you, and everyone was told that you would participate, but you were never informed and, therefore, either miss the event or appear uninterested by your apparent forgetting of the event.
- Making promises to the children that require you to drop them off, and then you fail to do so because you were never aware of the plans or because you were informed too late; the abuser will likely make it appear to others and to the children that you purposefully missed the event or intended to sabotage the children getting together with their other parent.

Playing Mind Games

Mind games are another form of manipulation that involves deliberately misreading situations, intentions, or comments. Common examples include:

- Doing the opposite of what you asked.
- Repeatedly making the same comments or requests and making it appear that they have not made those comments or requests before.
- Setting you up to get mad or have to discipline the children and then taking a victim stance when you do so.
- Choosing times to ask something of your partner when the timing is bad or inappropriate.

Denying the Problem or Their Role in the Problem

This is another form of manipulation of those who play mind games involving a denial of personal responsibility for their own behavior and fate. This includes denying that a problem exists and/or making you appear to be blowing things out of proportion. Common examples include:

- Continually denying the existence of a problem despite the presence of evidence or complaints from you or from others.
- Refusing to accept responsibility for one's role in the marital problems.
- Assuming the victim role when you bring problems up or when others attempt to discuss the problems.
- Blaming others (e.g., bosses, police, family) for consequences experienced, such as getting a speeding ticket ("I wasn't speeding; that cop has it in for me"); being disciplined or fired from a job ("I didn't do anything wrong," "I was never late and always did my work; my boss and coworkers just had it out for me.").
- (*My personal favorite from abusers*) Blaming you for calling the police when they have been abusive, refusing to understand that they forced you to do so by *being* abusive.
- Failing to understand why you do not want to be physically or sexually intimate with them; unwilling or unable to believe the effects of abuse on the victim and relationship; minimizing the impact of their abusive behavior or minimizing their abusive behavior.
- Expecting you to simply "be finished" with healing from their abusiveness or from prior trauma.
- Blaming you for having to abuse you (e.g., "if you would just stop nagging;" "I wouldn't have to yell at you if you would just listen and do what I tell you;" "You're supposed to please me; I don't care if you're in the mood, so if you don't want me to force sex next time then next time give me sex when I want it.").

Ignoring

Ignoring is a powerful form of abuse. This involves refusing to listen to you, failing to respond to your needs or requests, and looking right over you when you are in the room. This is a form of discounting the worth of another person. Common examples include:

- Refusing to acknowledge your presence or wishes.
- Appearing to be listening and then discovering they heard nothing of what you said.
- Having to repeat oneself because they are not listening.
- Talking to you or turning up the volume of the radio or television when they are aware or should have been aware that you are on the phone or talking with someone.

Purposeful Forgetting

Almost all forgetting is purposeful. If something is really important, then we go out of our way to remember it. We use notes, calendars, Post-It notes, anything to help us remember something of importance. When your partner regularly forgets what you requested or what was important to you, then the pattern of forgetting is intentional. This may include:

- Forgetting special and important dates (e.g., anniversaries and birthdays).
- Making plans to do something else on the date when you had definite plans.
- Forgetting plans made and then using forgetting as a way to make you appear crazy or that you were the one who failed to communicate the plans.
- Consistently forgetting the same thing (e.g., picking things up at the store, picking the children up at school or an activity, the children's sporting events or church, coming home early because of your obligations).

Prejudice

Prejudice is a form of hatred based almost exclusively on ignorance. This involves hating someone as a result of race, ethnicity, gender, age, or religion. It is not based on any sound facts, only partial information and biased and distorted information. This fits under control and abuse because your partner's prejudice impacts who you socialize with, how you live your life, and your values and beliefs. You may find yourself trying to leave certain situations before your partner begins to make prejudicial remarks or engages in other disruptive, embarrassing, or assaultive behavior. Common examples include:

- Making comments about someone's abilities based on their gender, race, age, religion, or ethnicity (I choose not to give examples here out of respect for all, as there is no way to give examples without offending someone).
- Treating people of other races, genders, ethnicities, age, or religions differently or with violence.
- Refusing to help those of different races, genders, ages, or ethnicities.
- Blaming those of different races, genders, ages, or ethnicities for the marital problems, or for social problems.

Criticizing

Criticizing involves the repeated putting down of your beliefs, behavior or abilities, or of your worth as a human being. Common examples include:

- Criticizing your ability to parent.
- Questioning your knowledge about certain things.
- Doing whatever they can to make you doubt your own abilities and worth.
- Putting you down for your statements or behavior in private or in public.

Discounting

Discounting involves belittling your accomplishments, skill, and opinions. This is a way for your partner to attempt to take a superior stance over you by discrediting what you do and what you say. Common examples include:

- Telling you that you do not know what you are talking about.

- Making comments such as: "Like that really matters," "Another nation heard from, so what;" and "But can you answer this… ."
- Having to demonstrate your weakness in a specific knowledge area by pointing out how limited your knowledge is; yet it is irrelevant to the discussion at hand.

Limiting Your Friendships and Support Networks

This involves controlling how you spend your time and with whom. This may involve sabotaging friendships, attempting to indirectly or directly discourage others from wanting to be around you, or may take the form of blatant control by telling you directly that you cannot spend time with certain people. Common examples include:

- Not giving you messages or mail from others.
- Telling people not to contact you or that they cannot see you.
- Telling lies to others (e.g., telling someone that you do not want to talk with them when, in fact, you do).
- Telling people you are not at home or are unavailable to take a call when you are.
- When out with others, your partner behaves abusively or in such a manner that others choose not to spend time with him.
- Prohibiting you from spending time with others unless your partner is with you; this becomes an increased problem when others refuse to spend time with you if your partner is there.
- Telling people lies about you to discourage continued contact.

Making It Appear To Others That You Are Crazy

Making you appear crazy to others results in others questioning your credibility. Many abusers are kind and outgoing with others and treat others much better than they treat their spouses. When your support network and family begin to side with your spouse, believing whatever they say, you have lost vital support. The abuser then uses this against you. The results are increased isolation and new ammunition that they can use against you, with comments such as, "See, even your family or friends don't want to be around you or do not believe you." Common examples of making you appear crazy include:

- Choosing to begin a fight or argument when around others, but doing so quietly so that others do not hear that your spouse is beginning the argument; but when you respond, it appears to others that you are the problem and that you are the one beginning the argument.
- Changing plans with others at the last minute, but failing to tell you about the changes; when you are late or unable to attend, they play it up to others as though you just didn't care, or that it was not all that important to you, or that you are just acting out anger towards them (your spouse).
- Canceling or changing plans at the last minute to avoid attending activities with you, and then telling others that you failed to tell them about the plans.
- Quickly calming down after an argument or abuse, leaving you agitated and shaken; to others, you appear to be the unstable one. (This is a common occurrence. When the police arrive at the scene of a domestic assault, quite often the abuser has deescalated and appears calm because they have vented their anger. The victim, on the other hand, is not able to just let go and calm down, nor should she. But the result is that the abuser appears to be the credible one and, in some cases, the abuser convinces the police that he is, in fact, the victim.)

Boundary Violations

Boundary violations refer to the breaking or violating of your partner's personal boundaries. This includes both verbal and nonverbal challenges to your boundaries. Examples include:

- Challenging the limits that you have set (repeatedly testing or pushing your stated boundaries, including sexual, physical, and other boundaries).
- Engaging in any behavior that your partner has not given you permission to do or that he/she has already told you not to do.
- Being in your space, office, bedroom, residence, car, etc., without your permission or when you have told them not to be there.
- Utilizing others to pass messages after being told not to do so (e.g., using friends, family, or the children to pass on messages).
- Giving the children power to make adult decisions (e.g., when to go to bed, what to eat, how to dress, curfew).

Violating Agreements or Court Orders

This involves the deliberate and intentional violating of contact restrictions. Make no mistake, any violation of a No Contact Order, an Order For Protection, or any other agreement of restricted contact is intentional and rarely if ever a mistake or misunderstanding. In fact, the court makes sure that the other person understands the limitations. It is not difficult to understand that "no contact" means no contact. Yet abusers are very good at finding exceptions to the rule, and finding loopholes, which only exist in their heads. Any violation of a restraining or no contact order should be taken very seriously and reported to the authorities. Such violations may include:

- Showing up wherever you go, such as at work, a restaurant, friends, etc.
- Having someone contact you for them.
- Mailing, e-mailing, or sending messages to you, whether directly from them or through a third party.
- Passing messages through the children or other family members.

Remember that even if the other person shows up at a restaurant where you are, it is his responsibility to immediately leave, not yours, even if he was at the restaurant first. Court orders limiting or preventing contact are the responsibility of the person being restricted, not the victim's. What this means is that even if the victim invites the abuser to get together, the abuser still goes to jail for violating the court order because there is no exception to the order, unless the court makes exception for a specific contact prior to the contact.

Emotional Blackmail

Emotional blackmail refers to using emotions and personal information to pressure or force your partner into compliance. Common examples include statements such as, "If you love me, prove it," "If you love me, you would do what I am telling you to do," and "I've been dating you for so long and/or have spent so much money on you that you now owe me sex." The goal of emotional blackmail is to equate feelings of care and love with compliance and dominance. For pressuring sexual contact, this involves pairing sex with love, again attempting to trap your partner into being sexual. Sex is never something that is owed, but rather is a shared privilege. Common examples include:

- Using guilt, shame, or other pressuring emotions to obtain compliance.
- Using feelings of abandonment or fear to obtain compliance.

- Threatening to end the relationship or to be unfaithful if your partner refuses to give in.
- Taking advantage of your partner's emotional state or prior emotional problems to coerce compliance.
- Threatening to withhold emotional support if you do not do what is demanded.
- Threatening to have an affair if you don't concede and do as you are told.

Playing Psychologist

Playing psychologist involves assuming a helper role or analyzing what you say and do. Whenever one person assumes the helper role, the power in the relationship changes from fairly equal to a one-up/one-down situation. Of course, crisis can occur and there are times when one person may need psychological support from the other. But these times are few and far between, and occur because both agree to the power exchange in the roles. When two people meet and, at the time, one is in crisis and the other in a helping or supportive role, it is unlikely that a healthy relationship could ever be established. Healthy relationships require that both people have equal power, and that both are emotionally stable. Playing on the other's emotional needs or trauma, the abuser takes advantage and uses the emotional weaknesses or trauma against the other person. Examples include:

- Telling the person that they have problems and that is why the relationship is in trouble.
- Intentionally playing on the other's emotional trauma or problems to begin an argument, to demonstrate to others how emotionally unstable the partner really is, or to justify further abuse.
- To create confusion to cause the other person to doubt her sanity, self-esteem, or self-worth.

Intimidating and Threatening Behavior

Intimidation and threats are clearly controlling and abusive behaviors. They cause the victim to have to protect herself and are often attempts to have the victim back down and give in to the abuser's demands. Common examples include:

- Raising their voice, yelling, and name calling.
- Swearing and insults.
- Raising one's hand or leg as if to hit or kick.
- Slamming of doors, throwing or hitting objects.
- Reminding you of past incidents of abuse.
- Threatening to spread rumors or lies, or telling others personal information about you.
- Making direct or indirect threats of harming you, the children, pets, or others.
- Threatening to take the children.
- Insinuating that harm may occur.
- Driving recklessly with you, the children, pets, or any others in the car.
- Threatening to tell others your personal information (e.g., your medical or mental health history, past legal or financial problems, your mistakes or embarrassing moments; again, this is not an exhaustive list of examples).
- Threatening to have affairs.
- Threatening to divorce or to withhold financial support.
- Threatening to not fulfill parenting expectations and duties.
- Threatening anything that challenges your sense of safety, security, or emotional well-being.

Sarcasm

Sarcasm generally involves the use of slanted, indirect comments that are the result of some degree of anger or discontent. Many use sarcasm to joke, but even then people may sense some degree of insult, regardless of whether intentional or not. When sarcasm becomes a pattern of interaction or language, it becomes abusive. Common examples include:

- Comments such as. "Is that the best you can do," "Here we go again," and any other cliché or line.
- Using jokes to get your message across; the jokes having a "bite" or jab.

Competition

Healthy relationships require a large degree of equality in most areas of the relationship. When one person becomes overly competitive, it becomes impossible to share intimacy and prevents the relationship from growing in a healthy direction. Having to always be right takes a toll on the other person who can never seem to do enough to be competent or be as competent as the other person. Common examples include:

- Always having to win or out-do the other person.
- Having to do everything "right," "properly," or without error.
- Going for a walk, which turns into a fast paced walk/run.
- The other person unable to lose at anything (e.g., unable to lose at a board game, unable to be the last to finish dressing or arriving somewhere).
- Always competing with others, including strangers (e.g., while driving, refuses to allow others to pass them, races to the check-out line in the grocery or department store, has to be the first out of the door).
- Appears to be unable to accept anything other than first place (e.g., fails to recognize achievement if it was not "the best" or perfect; becomes upset for coming in second place, despite a good effort; belittles others for doing less than perfect or for coming in second).

Inequality

Inequality means just that, an unequal value given to the other person in the relationship. This often has to do with resentments or narcissism. The other person must somehow be better than others and is highly competitive. They assume that they are somehow better, more competent, and more important than their partner. Common examples include:

- Attempts to make most decisions for the other person.
- Attempts to belittle the other's decisions.
- Attempts to take charge of activities, even when they are the guests.
- Demonstrates an arrogant and superior attitude.

Sense of Entitlement

This involves an attitude that males are superior to women, and that women should be submissive and subordinate to men. Men are viewed as being more competent and capable than women in most all arenas except maybe childcare and household duties. This sense of entitlement is demonstrated when the man attempts to think for his partner, tells the female how she should feel and how she should complete tasks, and expects to be served by her. Common examples include:

- Making comments, telling jokes, and expressing beliefs, which are sexist or demeaning to women (e.g., commenting that women are moody and irrational, commenting that a women's place is in the home).
- Refusing to allow or discouraging the woman from working outside of the home.
- Viewing himself as the main breadwinner and, therefore, the one with the power to make all important decisions.
- Controlling how money is spent, controlling the bank accounts and other financial matters.
- Dictating household chores and responsibilities.
- Often assuming few if any childcare duties, including refusing to attend to children's and infant's needs at night.
- Flirting with other women, but becoming jealous and angry when his partner talks with other men.
- Treating her like a servant (e.g., demanding and ordering her to do things for him, expecting that she do demeaning things in front of others to show how he is in control).
- Expecting sexual submission (e.g., demanding that she perform unwanted, painful, or humiliating sexual acts; forcing her to allow him to view pornographic materials, attend strip shows, or have affairs; demanding that she view pornographic material with him; or perform sexual activity with others or with others present; forcing her to be photographed during sex or while nude; any form of sexual abuse).

Abandonment

This involves failing to invest in the relationship and refusing to assume the responsibility for the relationship in any way. The abusers are sort of just there, going with the flow and nonassertive. They are emotionally uninvested in the relationship. Common examples include:

- Allowing or demanding that the other person take charge for the relationship or for any major decisions.
- Spending more time with others or with other members of the family, or working too many hours, thus avoiding spending any significant amount of time with the spouse.
- Establishing new routines or activities that require more time out of the home, and possibly requiring a significant financial investment.
- Emotionally detaching from the other person and from the relationship.

Neglecting

Similar to abandonment, neglect involves withdrawing emotional energy from the other person and from the relationship and reinvesting it elsewhere. Neglect is controlling because it forces the other person to take drastic action to get the abuser reinvested or forces the victim to take on responsibilities that are almost unrealistic for one person to assume or fulfill. Common examples include:

- Failing to fulfill family or social obligations.
- Failing to fulfill financial obligations.
- Refusing to find time to spend with the children or other family members.

Lying or Withholding Information

Deceit is an obvious form of control. Let us be very clear here; lying, withholding information, distorting the facts, and telling after the fact or only when caught or confronted are all forms of lying. Deceit and lying are the same thing and both are abusive. Lying occurs when the abuser makes a conscious decision to mislead the other person. It is not an accident or a situation where

"I just forgot" or "I just did not think it was important" are acceptable. If the information was really not that important, why not share it with your partner anyway? If something was important, how could you justifiably forget it?

The mind is a wonderfully amazing organ. It does not forget like some would like to believe. The mind remembers even the smallest, most insignificant detail of an event or situation. Whether we choose to remember, is really the question. Only those with true brain injury, organic brain injury, or those with severe, chronic alcohol or drug problems can justifiably use forgetting as a justified excuse. When abusers claim they are unable to recall because they experienced a "blackout" from their drinking, they are lying. Blackouts involve the inability to recall something that occurred in the very recent past, and involves only a short period of time. However, in my experience, abusers can almost always recall every detail of something that happened when pressured to do so.

Distorting the truth and withholding information are very controlling behaviors because abusers attempt to control another's reaction to the information and attempts to avoid the consequences they would likely experience as a result of the information becoming known. Examples of lying and withholding information include:

- Withholding information until confronted about it.
- Telling only when it becomes apparent that the information will be made known anyway.
- Sharing only the details they believe will be found out, and withholding or even blatantly denying other facts because they believe they will not be found out.
- Telling only after something has occurred.
- Misplacing messages or mail until a deadline has past.
- Purposefully distorting the truth or information.
- Forgetting only information that works against them or only information that would benefit the victim.

Misspending Money

The misspending of money is also controlling because most people are on budgets. Extravagant spending and failure to pay financial obligations place one in possible dire straights. The threat of having a negative credit history/report, accumulating significantly more financial debt than you and your spouse can realistically handle, or not having enough money to pay the necessary bills can be very overwhelming and constitutes abusiveness. When one person spends money on whatever he wants, without regard for the needs of the spouse and the family, he is controlling and abusive. When one person has chosen not to work fulltime or not to work at all, to remain home with the children, money becomes even more a weapon when misused or withheld. Common examples include:

- Withholding or hiding money.
- Opening bank accounts without telling the spouse. (This is not abusive or controlling when the victim of abuse is secretly establishing financial security to utilize leaving in a hurry.)
- Buying whatever they want despite money being tight, or when they do not allow the victim to buy what he/she wants.
- Claiming that the money is theirs, and they do not have to explain to the spouse how or when they spend it.
- Expecting the victim to report all of the income to them, but they refuse to do the same.
- Expecting the victim to use her own income on household expenses, yet refusing to contribute anything significant from their income, therefore relying and using only the spouse's income. (This is not abusive when the one paying the household expenses is earning far more than the other person.)

Jealousy

Jealousy involves protecting objects and belongings that you own from being taken away. People are not objects and, therefore, fearing that someone will steal your spouse from you is unrealistic. If your spouse decides to have an affair or to terminate the relationship, he will do so, not because someone stole him, but because he made the conscious decision to do so. If others are staring at or flirting with your spouse, there is nothing to worry about. If your spouse is flirting back, then it is an issue to be discussed. If he continues to flirt after you have expressed your concerns and dislike for that behavior, then perhaps he is giving you a clear message that he is not happy in the relationship and either marital therapy is necessary or perhaps the marriage is finished. Remember that when others are flirting with your spouse, that is a compliment. Would you really want to be married to someone who was not attractive to others? And remember that those who are validating the attractiveness of your spouse are only getting to glance; your spouse is going home with you, not them. To demonstrate jealousy is abusive and controlling. Jealousy is generally the result of insecurity and frustration.

Sexual Abuse

Sexual abuse involves any behavior or comments that are intended to or do result in the denigration of another person. This includes gestures, comments, insinuations, and direct behavior. Many forms of sexual abuse are against the law and should be prosecuted and treated separately from any other abusive act. Sexual abuse involves the unwanted touching of another's body, forcing someone to listen to sexual comments, or viewing sexual gestures. Included here are harassment, fondling, groping, and coerced and forced sexual contact. Common examples include:

- Forcing any unwanted sexual contact, including kissing, fondling, watching, or any sexual act.
- Staring at or flirting with other women.
- Telling jokes or making comments that involve sexually explicit or sexually derogatory language or inferences.
- Using sexually derogatory names when angry or swearing (e.g., slut, whore, cunt, bitch, fucker). These names become threatening to the victim.
- Insulting a woman's body or comparing her body to other women.
- Comparing her sexual performance to others.
- Using pornographic material of any kind without expressed consent of the woman (e.g., internet porn sites; strip shows, videos, and tapes; magazines, including soft core).
- Forcing her to view pornographic material.
- Insisting that she dress in a certain way (e.g., dressing sexually provocative or revealing or on the other extreme, forcing her to dress in very conservative ways, not allowing her to show her body tone or shape).
- Groping or touching her body in public or in the presence of others.
- Touching her in sexually demeaning ways or ways that she has expressed she dislikes.
- Forcing her to have sexual contact with others or in front of others.
- Forcing or expecting that she engage in sexual contact seen in pornographic material.
- Forcing or expecting her to engage in sexual contact that she doesn't want to do; that is painful, embarrassing, or simply unwanted.
- Forcing or engaging in sexual contact with her when she is not giving or is unable to give consent (e.g., when she is intoxicated or under the influence of drugs or alcohol; when she is sleeping; when she is injured or in a condition where she is not capable of safely saying "no;" if she does not speak the same language to understand what she is expected to do; when saying "no" would result in harm to herself or others).

Explosive Behavior

When a spouse demonstrates explosive behavior, he is telling you loud and clear that he is dangerous and is not going to consider your safety when he becomes angry. Individuals with explosive behavior are in need of specialized psychological intervention, including anger management treatment. When a spouse is demonstrating angry, explosive behavior, one likely experiences fear and a sense of worthlessness. After all, they are choosing to abuse you, and why would you abuse someone you loved? You would not.

Withholding of Feelings

The withholding of feelings is very controlling. How can a relationship grow without both people sharing their feelings? It would be impossible. When one person chooses not to share their emotions, the other is left wondering what they feel, wondering what may be wrong, and questioning the stability of the relationship.

Treating Others Respectfully, But Not You

Not treating a spouse with the respect that is afforded to others occurs commonly with abusers. This is one of their strategies to keep you under their control, doubting your sanity, and successfully making others believe that they are nice, nonabusive people. When they treat everyone but you with respect, you may feel less respected and may have a lowered self-esteem and sense of self-worth. In addition, this may result in isolating you from supportive relationships because others portray you as the problem, perhaps believing that you are overreacting or really the reason why the relationship may be failing. Common examples include:

- Treating others with respect when the abuser is alone with others or when you are together in the presence of others, yet treating you with disdain, contempt, sarcasm, ignoring you, or treating you with some other abusive behavior.
- Purposely engaging in conversation or contact with others when the intention is to portray a calm and respectful demeanor, or establishing contact with your support network before you can contact them in an attempt to give your side of the story first, and the abuser's version is biased and untrue and abusive.

Complaining to Others about You While Portraying Self as the Victim

This involves complaining to others about you when you are not present. The purpose is to present their version of the problems first in an attempt to make your version appear more an excuse or retaliation for them. Others may question why you did not tell them sooner than your spouse did, thereby appearing more guilty or defensive. This becomes controlling because it limits the availability of support you may access. In addition, you now have to explain to others why your spouse would be complaining or making allegations, perhaps making you appear vindictive.

Name-Calling

This involves calling you names that are derogatory, inappropriate, or names that you have indicated are unwanted or abusive. Name-calling demonstrates their willingness to abuse you, to objectify you, and to put you down when angry, rather than attempting to discuss the problem in a respectful manner. This becomes controlling because you are forcibly put down and disrespected. Victims of name-calling tend to doubt their self-worth and self-esteem, and may begin to blame themselves for the problem. The abuser insists that their choice to name call was justified, and likely apologize at some point. Their apology is no less abusive; expecting that you will forgive and forget, and that things will get back to "normal" because of their apology.

Ordering and Pressuring

This involves the abuser being demanding and ordering the spouse to do tasks. Rather than making requests or politely asking, the abuser makes demands. This is controlling and abusive because no one should be ordered to do anything. The abuser orders because he believes that he has the power to do so, and that the spouse lacks the power to refuse. Pressuring involves repeatedly reminding the partner of something, such as one's expectations and wants, without regard for her emotional or physical well being. Pressuring also involves begging and pleading until the abuser coerces the partner into submission.

Apologizing

Apologizing is a healthy and respectful thing to do. However, it becomes abusive and controlling when apologies come too quickly, without an understanding of what was said or done that was offensive. Apologies can be a form of sarcasm, a way to rudely and abusively say, "I'm sorry," but without any meaning other than to appease the other person. Abusers often apologize and expect that you simply accept their apology and then move on. However, the controlling aspect is that to assume that you are able or willing to "forgive and forget" the abusive comments or behavior is unrealistic and disrespectful. First, our minds will not allow us to erase what occurred. Second, we should not forget what occurred even if we could because change must happen. If the abuser chooses to cease their abusive and controlling behavior, then remembering when they were abusive and comparing their behavior to the present is a blessing. The memory helps to appreciate the changes that occurred. However, when the abuse and control are ongoing, the apologies are meaningless. It is often the case where the abuser tells the victims that it is their fault and that they should stop bringing up the past. But how can a victim not bring up the past when the same abusive behavior is repeated.

As you may have found, many of the above categories share common characteristics with other categories. The lines that separate one form of control from another are often difficult to see. It is not important to make such distinctions. I have found that when people have become abusive and controlling, they engage in many of the types of control and they become very good and comfortable with their controlling and abusive behavior. There are likely other forms of control that I have not mentioned. The above list was not meant to be all inclusive.

3 Understanding Stalking

I have a lot of respect for J. R. Meloy's work and strongly recommend that *The Psychology of Stalking* (Meloy, 1998) be read by anyone working with batterers or sex offenders. It is perhaps one of the most comprehensive and easy to understand books, which addresses this complex issue of stalking and obsessional followers.

Stalking is perhaps one of the most misunderstood of all of the forms of intimidating and dangerous behaviors batterers and sex offenders engage in. Stalking is difficult to understand because, like other forms of control and psychological abuse, it is often unseen by the professional. In fact, stalkers are very adept at explaining away their behavior and have many rationalizations for engaging in stalking behavior, often portraying themselves as simply being at the same place as the victim by coincidence or by accident. When victims describe their experience of having been stalked, they are often terrified. Professionals may not believe the victims' claims of being stalked because some of their allegations appear crazy, highly unlikely to have occurred, and again, perhaps, just a misunderstanding. But, make no mistake, batterers and sex offenders who engage in stalking behavior do so in a very planned and methodical manner.

DEFINING STALKING

Stalking involves the crime of acts (not just a single act) of pursuing an individual over time. The stalking behavior is of a threatening nature and potentially dangerous. Definitions of stalking basically include three elements: (1) a pattern (course of conduct) of behavioral intrusion upon another person that is unwanted, (2) an implicit or explicit threat that is evident by the pattern of the behavioral intrusion, and (3) as a result of these behavioral intrusions, the person who is threatened experiences reasonable fear (Meloy, 1998). Meloy and Ghothard (1995, p. 259) defined a pattern of threat or harassment as "more than one overt act of unwanted pursuit of the victim that was perceived by the victim as being harassment."

Meloy (1996) found that the primary motivation for stalking is not sexual, but rather conscious anger or hostility toward the victim. Stalkers will often utilize the defenses of denial, minimization, and projection of blame onto the victim. Stalkers form a narcissistically linked fantasy with their victim. What this means is that the stalker develops a scenario in his own mind that he and the victim are involved in a more intimate relationship than reality proves. Stalkers have justifications and explanations for their behavior and will not admit that they are intending to create fear in their victims. I have talked with law enforcement personnel and even experienced clinicians who are conned by the explanations alleged stalkers provide when they are caught. They tend to be believable because they truly believe their delusional thinking.

CHARACTERISTICS OF STALKERS

Stalkers may be perceived as being psychotic or seriously mentally ill. However, most individuals who stalk are not psychotic at the time they are stalking; most are not delusional, and most will evidence an Axis I diagnosis, including mood disorders, schizophrenia, and substance abuse (Meloy, 1998b). Stalking is typically not something that occurs only within a short time period, but rather a behavior that has occurred over a period of months and years. Stalking often includes e-mails, telephone calls, and physically approaching the victim.

Kienlen (1998) indicated that while stalking behavior manifests itself as seemingly benign gestures (e.g., gifts, letters) meant to be symbols of the stalker's affection, the victim's reaction is fear. The victim reacts with fear because of the stalker's inability or unwillingness to accept the reality that the victim is uninterested in a relationship. Repeated rejection by the victim may lead to escalation of the stalker's behavior to more overt threats or violence (e.g., rape, assault, even murder) toward the victim. Remember that stalkers have developed a fantasy relationship with their victim, which may have actually involved a prior relationship that has ended or a relationship that never existed other than perhaps a casual and innocent meeting. The gifts or gestures of the stalker may appear to the professional to be innocent, yet they are unwanted by the victim. It may be possible at times that an individual who lacks social skills or has mental retardation does not understand that his behavior is inappropriate, and this type of stalker might be more easily dealt with to help end the stalking behavior and harassment.

It may appear at times that victims of stalking are reporting the stalking behavior to get even with the perpetrator or to have just cause to obtain a restraining order. However, few victims make false allegations. In fact, only approximately 2% of the victims in one study were found to have made false allegations of being stalked, while not actually being stalked (Zona, Lane, & Palarea, 1997). Stalkers are adept at violating orders for protection, as well as are many batterers and sex offenders. It is common for victims of stalking to report that their stalker has violated a protection or restraining order (Tjaden & Thoennes, 1997). I have worked with victims of abuse who were being stalked by their batterer. Some reported that the batterer has been in their residence when they were not at home. The reason that they were aware that the stalker/batterer had been in their residence is that things had been rearranged, moved, or taken. A common example was that the pillows on the couch were rearranged in a manner that the batterer used to arrange them, yet the batterer was not to be in the victim's residence. Victims have also reported that someone had tampered with their telephone answering machine, having apparently listened to messages or deleted them. Another victim found that the phone lines within the home had been tapped so that the batterer/stalker could monitor her calls.

Walker and Meloy (1998) indicated that batterers attempt to control the life of their victims. In doing so, the batterer does whatever is necessary to isolate the victim from any source of support. When it reaches the point of monitoring, surveillance, and over possessiveness, and induces fear, stalking occurs. The batterer also accomplishes stalking by doing whatever he can to keep in contact with the victim and have her under his control. This may include giving gifts, making unwanted telephone calls, and other behaviors that keep the victim in contact with the stalker. The most dangerous batterers can be identified by their stalking behavior (Hall, 1988). Stalking is a risk factor for future physical abuse or a lethal incident in combination with other factors.

Walker and Meloy (1998) further found that most marriage and family therapists ignored (40%) or minimized (91%) the danger cues presented by their abused client. It is convenient at times to ignore and minimize the potential impact of stalking and harassment. In fact, the victim's description of being harassed and stalked often appears to be either an exaggeration of intent and circumstances (e.g., the offender just happening to be in the right place at the right time) or simply unbelievable.

Meloy offers the following examples of stalking (1998a, p. 9, Table 1 reprinted with permission):

Sexualization: "I'm going to fuck you and fuck you up."
Displacement: "Your new boyfriend is dead."
Devaluation: "You deserve to die and you will."
Idealization: "If I can't be with you forever and fulfill our perfect destiny, no one will be with you."
Grandiose Elaboration: "Anytime … anywhere."
Projection: "Tell them I'll be waiting for them … ."

Projection Identification: "I'm watching you all the time and, if you keep taunting me, I'll hurt you."
Denial: "I didn't threaten you."
Minimization: "I was just kidding."
Rationalization: "She deserves it; look what she did to me."

Understand that offenders are selfish and do nothing by chance. Everything they do has some degree of planning involved and stalking is a very deliberate crime of violence. Imagine being followed by someone that has harmed or may harm you. It would be unnerving to see them wherever you go, or to find that they have been in your residence or are literally following you everywhere. Others may minimize this as the offender's interest or love for you while ignoring the violence involved in stalking.

It is imperative that clinicians, police officers, and prosecutors and judges do not minimize or ignore any claim of harassment or stalking. When in doubt, err on the side of believing the victim and take the time to have appropriate professionals investigate. Detectives or a private investigator may be required to help prove the stalking behavior is occurring. It is never the job of a clinician to determine whether an offender has allegedly engaged in stalking behavior, that is the job of the courts and law enforcement. It is the clinician's job to believe and support the victim; there is a 98% chance that they are telling the truth. A clinician may have the alleged perpetrator take a polygraph, though polygraph has limitations. I have found that when holding an offender responsible for any allegation the victim makes generally turns out with the offender admitting the alleged behavior. Always err on the side of safety. If the offender is engaging in stalking behavior while in treatment, then the offender is not invested in the program and that would appear to be grounds for termination from the program with a recommendation of incarceration as a natural consequence.

4 Decision Making Process

Following the offender's awareness of the use of force is the decision making process. This is the intricate labyrinth of decisions the offender makes leading up to the initiation of the offense and carrying the offender through the completion of the offense. Again I use the term *offender* to refer to abusers, assaulters, and any type of sexual offender.

OFFENSE DECISION MAKING

1. The offender experiences stress, triggers, and a perceived provocation. The offender recognizes that these feelings are escalating. He has identified an arousal, urge, or desire on the positive end of the continuum or identified a sense of irritation, anger, betrayal, or perceived affront.
2. The offender fails to utilize appropriate intervention strategies. The offender decides not to do what he knows he needs to do, or that he is, for some reason, incapable of doing. These intervention strategies include, but are not limited to, the following examples:
   a. **Refuse to take a deep breath**. This is perhaps the simplest de–escalating behavior anyone can do. Taking a deep breath helps settle the body and mind by filling the lungs with air. Imagine that when someone is performing in sports or is involved in a fight, they breathe through their nose, taking shallow, quick breaths. This starves the body of oxygen and forces the body into the fight-or-flight response. Rational thought processes are ignored and the person prepares for the confrontation.
   b. **Refuse to take a time-out**. During a time of conflict, it is often helpful to take a few steps back and think about the problem or situation at hand. Sometimes this involves waiting an hour or several days before continuing the discussion or before making a decision. When an offender wants to take sex or to harm his victim, there seems to be no time for a time-out. The offender wants what he wants when he wants it and will take what he wants regardless of the possible consequences. Offenders: They want to engage in a specific behavior and do so without hesitation. The offender is unable or unwilling to consider slowing down. To offenders, their victim is a toy, an object to do what they want to without concern for the victim's well being.
   c. **Refuse to reframe the situation**. Reframing involves considering alternative meanings or interpretations to a given situation. The offender feels and believes that it is acceptable to do whatever he wants. An offender feels that he has been insulted, put off, rejected, or, in some way, cheated, or simply pissed off or horny. The offender is not willing or able to consider that his identified victim may not have intended to harm him or to frustrate his needs.
   d. **Blame the victim and the world for any injustice perceived or experienced**. Offenders are narcissistic and care very little for the well-being or rights of others in the world. Having reiterated that, offenders blame everyone other than themselves for whatever goes wrong. If they get a speeding ticket, it is because the officer targeted them and, of course, they were not speeding. If someone cuts them off on the road, then it must have been deliberate, the other driver knowing who they were passing and doing it deliberately to get to the offender. For offenders, nothing happens around them that is not perceived as being directed at them. The rapist, who blames the victim

for "getting me turned on" and then refusing sex and being a tease, is justified in his mind because the victim is playing games and needs to be taught a lesson.

e. **Giving themselves permission to be "out of control."** This is the final step in rationalizing the offender's behavior. Individuals who are psychotic or under the influence of certain drugs may genuinely be "out of control." However, I estimate that very few of the offenders committing violent offenses are truly "out of control." An "out of control" attack is likely to appear very frenzied, chaotic, and disorganized. This type of offender does not care where the offense occurs and is not going to stop at merely hitting once or gently taking sex. The injuries would be grossly out of portion to the resistance of the victim and the act far more violent than even what the offender may have intentioned. Psychosis and drugs, such as PCP, can cause this kind of "out of control" behavior. Of course there is one other exception to this rule. A delusional or psychotic offender who is obsessive–compulsive may be considered "out of control" for this offense behavior when the offense involves exceptional obsessive or compulsive behavior. In Minnesota, sex offenders diagnosed as having sexual psychopathic personalities would be deemed as lacking the ability to control their deviant sexual urges and may be civilly committed. But, for these examples of offenders who may genuinely be "out of control," there is absolutely no cure or intervention that has been empirically supported to eliminate the violent behavior. With that having been said, those claiming to have been "out of control" when they committed their abuse, molestation, rape, or other assaultive behavior, should be informed that they will likely require lifetime incarceration or hospitalization as there is no help for them.

3. Therefore, offenders are in control of their violent behavior if they choose to be. And relinquishing control is not something that offenders are very willing to do. They do, however, talk themselves into believing that they are powerless and out of control. This can become a self-fulfilling prophecy. However, self-fulfilling prophecies are not out of our control. So the offenders convince themselves that they are "out of control" and justify their impending offense behavior.
4. Offenders personalize the situation. As mentioned earlier, offenders personalize everything that is said and done to them. They are unwilling to consider the possibility that no hurtful result was intended by their victim, or that it may have simply been coincidence that it happened to them. With this they can now feel vindicated for their impending assaultive and violent behavior.
5. The offender gives up. This is the Abstinence Violation Affect (AVE) and involves the offender giving himself permission to abuse by justifying that he is far too close not to do it, or that he has already begun and that he is "out of control" and must finish what he started. The offender tells himself statements such as:
   "I cannot handle it anymore."
   "They provoked me and asked for it."
   "They made me angry."
   "I am angry and am unable to control my anger."
   "Well, I have already crossed the line, what difference does it make now?"
   "The hell with it" or "fuck it."
6. The offender justifies the escalating and violent response with supporting self-talk. The offender tells himself that he is not abusing or raping, that he is simply responding with just cause. After all, the victim hurt him, so why not hurt them back and teach them a lesson.
7. The offender engages in violent and abusive behavior and talk. He tells himself and convinces himself that it is okay to be abusive and violent; that the victim asked for and deserved to be beaten, raped, or molested.

8. The offender then rationalizes the offensive behavior with statements such as, "The victim had it coming," and that the offender gets "out of control." The offender then blames the victim for having to be abused or raped, for somehow asking for it and for forcing the offender to commit the violent act.
9. Following the offense, the offender attempts to move on and forget the abusive situation. Some offenders apologize for their violent and abusive behavior and may even ask for forgiveness. Offenders may make promises that it will never happen again. The offender expects and demands immediate forgiveness and expects that the victim will forget the incident and move on. If the victim brings up the issue of the abuse or rape, the offender generally becomes increasingly more frustrated and angry. Most victims that remain in relationships with the offender realize this and stop bringing it up in order to survive the relationship.
10. The offender returns to a state of heightened alert, never reaching "normal" or relaxed again. The offender decides to move on and try to forget that it happened. The offender expects that the victim will not mention it again. For abusers and rapists who remain in a relationship with their victims, the offender does not tolerate the victim mentioning the offense behavior again. Threats are often used to remind the victim that if they talk about the abuse to anyone, more harm will come. Offenders are very adept at threatening by insinuation.
11. The offender is ready for the next battle. The self-fulfilling prophecy is alive and beaming here: The offender expects others will harm him in some way, provoke him, or cause some sort of problems for him. And, as a result, the offender is ready and waiting. Over time the modus operandi changes. The offender wants and needs change, so the abuse and rape behavior changes as well.

5 Three-Stage Abuse Cycle

Lenore Walker (1979) first wrote about a three-stage cycle of abuse. Her abuse cycle involved an escalation phase, an explosion phase, and the honeymoon phase. Though this is a simplistic depiction of the cycle of violence, it is accurate. However, I believe that it is important to examine in more detail what occurs before, during, and after an abusive incident rather than simply identify what stage of abuse the offender is in. I provide a more in-depth cycle later in this book.

Walker's (1979) first cycle stage involves tension building and escalation. As I indicated in another book (Johnson, 1993), this stage involves the build up of tension, frustration, and anger. The batterer focuses on how he perceived being wronged, annoyed, ignored, or threatened. This stage may last for several minutes or months. On average, it likely lasts for approximately a few minutes to hours for most batterers.

During the escalation stage, the victim may attempt to calm the batterer or to provoke the batterer. When attempting to calm the batterer, the victim may become nurturing, compliant, and may even accept responsibility for whatever the batterer is complaining about. When provoking the batterer, the victim may either be attempting to get the impending violent incident over and possibly get it over on his/her own (the victim's) terms. When the victim is aware of the impending violence, provoking the batterer allows the victim to prepare for the violence and perhaps lessen the intensity or degree of violence or injury. However, at no time is the batterer justified in engaging in violent or abusive behavior, whether provoked or not. Remember that provoked implies that someone is pushing our buttons, challenging us, and encouraging a confrontational or possibly violent outcome. But at no time is the batterer truly out of control and he can choose a nonviolent reaction if he wanted to do so.

In relation to the more detailed cycle I present later, Walker's (1979) escalation stage comprises Stages 2 to 6 ("triggers" through "revenge planning"). Walker's second stage is the explosion stage. This is where the actual psychological, physical, and/or sexual violence occurs. I include verbal and psychological abuse here as well because it is abusive and results in harm to the victim. It is at this stage that the batterer intensifies the verbal abuse, yelling and using demeaning names and phrases against the victim. This is the stage during which the physical and sexual violence occurs. The batterer may hit or use a weapon to injure the victim, and may engage in unwanted sexual contact with the victim. This is the shortest stage, lasting significantly less time than the other two stages. This stage is intense and violent. In relation to the more detailed cycle I present later, the explosive stage comprises Stages 7 to 10 ("acting out the revenge plan" through "physical and sexual abuses").

Walker's final stage is the honeymoon stage. This follows the violent episode and involves the batterer making some type of genuine or insincere amends and apologies. Over time, however, the batterer means none of what they say when they apologize. They will say anything to the victim to prevent the victim from leaving them and to prevent the victim from telling anyone what has occurred. The batterer's promises to change are always a lie if he has abused more than once in the past. Think about it: If the batterer was genuinely remorseful for this violent behavior, then he would seek appropriate help to aid in his change and healing. If the abuse has occurred more than once and, for every time it occurred after the first incident, I believe that the likelihood for change significantly decreases. The final stage of the abuse cycle is the batterer's last hope at retaining his power and control over the victim. Also in relation to the more detailed cycle I present later, the honeymoon stage comprises Stages 11 to 14 ("sense of relief" through "physical and rationalizes abuse").

6 The Offense Cycle Explored

INTRODUCTION

A recent TV talk show focused on the issue of child sexual abuse and what should happen with the offenders. An offender who had served 8 years in prison was on the show with his mother who defended him. Part of his defense when he was asked if he planned out this crime or if it just happened was, "It just happened." Many in society would like to believe that certain things just happen; that an individual driving under the influence of drugs or alcohol who gets into an "accident," who maims or kills someone, that this was just something that "happened," A person committing physical abuse on a child or adult also claims that "it just happened;" "I lost control." Somehow, we become fascinated with how powerless we appear to be in our lives, when, in fact, we have total control of our behaviors.

Society can no longer accept that crimes just happen, that violence just simply occurs as a result of "losing control." There is absolutely no evidence that would support that any type of violence, whether psychological, physical, or sexual, just occurs. There is an overwhelming amount of evidence, however, to prove that people are in control of their behaviors. Just as persons abusing alcohol choose to put themselves in a situation where alcohol and/or drugs are available, choose to continue to use or pick up another glass of alcohol, and then choose to get into a car to drive home, so also, an offender of physical and sexual abuse chooses to enter into his escalation stage and progress to behaving violently toward another human being.

The audience on this particular talk show was outraged with this offender, who is an untreated sex offender. The response that his sexual offense "just happened, with no forethought or planning," should outrage us all. This is a dangerous man and a man that the judicial system has just set forth to once again abuse other children. This man lacks any insight into his behaviors and, therefore, lacks any chance of not reoffending. He does not have the skills needed to understand the underlying issues that were part of his plan, let alone understand how he can stop the decision to offend. What I am talking about at this point is examining the cycle of violence.

The cycle of violence has traditionally been with battered women. When Lenore Walker (1979) first wrote about a three-stage cycle of abuse, the cycle involved an escalation phase, which was the build up of significant tension and frustration; an explosion phase, which was the actual psychological physical and/or sexual violence; and the honeymoon phase, which was where apol ogies were given and amends were made. Although this cycle does help to point out a progression of thoughts, behaviors, and emotions, I believe we must go a step deeper and break down these three stages into as many stages as possible, so as to understand how the progression, that is, the decision to act out violently, occurs.

Of the numerous cycles that have been available to help explain abuse and violence, I chose a 14-stage cycle (see next page). It is important to stress that at any point in the cycle, an individual could choose to utilize coping strategies and, therefore, the cycle. No one is ever powerless to change the course of his behavior when he is in his cycle. In fact, the offender must invest significant energy to remain in the cycle, just as he would need energy to get out of his cycle. The cycle I am about to describe may be used with physical batterers, child abusers, sexual offenders, rapists, harassment and stalking, and any other form of violence against another individual. This cycle may be adapted to fit any type of violent behavior.

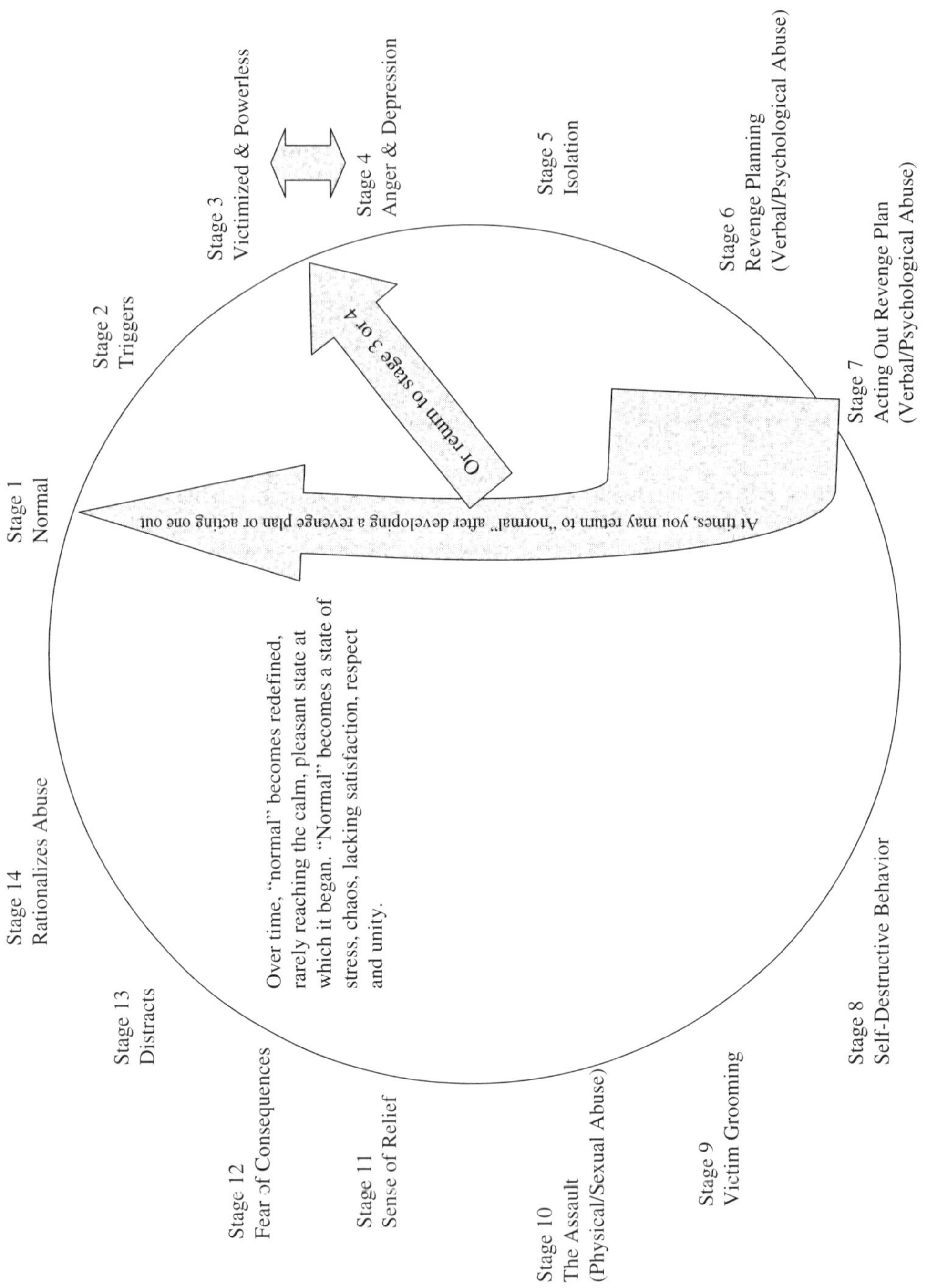
Stage 1
Normal
Stage 2
Triggers
Stage 3
Victimized & Powerless
Stage 4
Anger & Depression
Stage 5
Isolation
Stage 6
Revenge Planning
(Verbal/Psychological Abuse)
Stage 7
Acting Out Revenge Plan
(Verbal/Psychological Abuse)
Stage 8
Self-Destructive Behavior
Stage 9
Victim Grooming
Stage 10
The Assault
(Physical/Sexual Abuse)
Stage 11
Sense of Relief
Stage 12
Fear of Consequences
Stage 13
Distracts
Stage 14
Rationalizes Abuse
Over time, "normal" becomes redefined, rarely reaching the calm, pleasant state at which it began. "Normal" becomes a state of stress, chaos, lacking satisfaction, respect and unity.
At times, you may return to "normal" after developing a revenge plan or acting one out
Or return to stage 3 or 4

The following 14-stage cycle can be compared to the three-stage cycle. Stages 1 through 6 occur in the *escalation* stage. Stages 7 through 10 occur in the *explosion* stage. And Stages 11 through 14 occur in the *honeymoon* stage.

There are stages people go through *before* they act abusively. A complex pattern of emotions, thoughts, beliefs, and decisions occur prior to action occurring. Lenore Walker (1979) originated this Cycle of Violence, a three stage process described in the before, during, and after occurrences of an abusive situation. This simplistic model is easily understood by almost everyone and is a useful tool when dealing with abuse. However, the entire process of choosing to become abusive involves a more intricate scheme, a cycle or pattern that repeats itself. We will examine each of these stages and, by the end of this book, one will have a more thorough understanding of the cycle that abusers go through.

It is important to take a microscopic examination of each stage in the offending (abusive, acting out) cycle. Each stage has specific themes of behaviors, emotions, and thoughts (self-talk). An offender should ask himself: "What do I do, feel, and think?" at each stage to help himself describe and get in touch with his own personal cycle. Each stage becomes a clearly defined and understandable part of his cycle, of his life. Remembering that abusive behavior is always a choice, a person is never out of control — ever!

STAGE 1

Normal

It may be impossible to effectively examine a cycle of acting out without understanding what our everyday and normal life is like. Each individual will have a different description of what constitutes normal for his life. The normal, okay stage is where all cycles begin. We begin with homeostasis; that is, a state where everything seems to be going your way. What is normal for one person may be terribly wrong for another. Each person must define his own normal stage himself; normal may be that you and your significant other are getting along, talking, and pleasing each other. You may be satisfied with the intimacy you both share. You look forward to sharing time with your spouse again. Basically, normal may be feeling satisfied with your life, relationships, work, etc. You may feel respected, valued, and a sense of accomplishment. Although not perfect or stress free, things are manageable. You are at least getting by, if not doing better than that. It is at the normal stage when you appear to utilize healthy, appropriate coping skills to deal with life's everyday stresses and let downs. It may be an experience of a sense of inner peace and contentment, and you see your life goals actualizing.

In order to help you understand what your normal stage looks like, I have an exercise for you. Describe what you are doing and how things are going, your emotions at this stage, and what you tell yourself (self-talk or thoughts). For example, in as simple of terms as possible describe your life; examine and describe the following areas of your life: Relationship with your family, relationship with your significant other, work situation, financial management, medical and health, friendships, etc. The idea here is to paint a picture that best represents how you define your life when it is going well. The more detailed description, the more in-depth of an understanding you and others will have of your normal life.

I will now give several examples of descriptions others have offered to illustrate the normal stage. Socially, you may be content with a circle of friends. You may be on a sports team or member of a club. Work is going fine, the boss is satisfied with your work. When things are going well, you have no need to use or abuse alcohol or drugs. Problems are being met head on and you may have a general sense of being spiritually secure and solid. You are resilient and able to cope effectively with most of what life throws your way.

Behaviors/Descriptions

Relationship with significant other is going well
Feeling of comfortable relationship with the family is going well
Boss is satisfied with your job performance
Bills are getting paid
General health is good

Emotions

A general sense of accomplishment and achievement
Happy
Wanted
Needed
Competent
Trusted
Calm
Loved
Respected
Adequate
Good self esteem

Thoughts/Self-Talk

Thoughts that generally occur here tend to be healthy, positive self-talk because life appears to be going well; little may bother you. Examples may include:

"I'm doing well"
"I am a good parent"
"I am a good lover"
"My significant other wants and loves me"
"I am competent at work"
"People like to be around me"
"I seem to be managing my bills and my money situation rather well"

For some people, their normal stage includes significant levels of stress, fears, and frustration. A general sense of dissatisfaction, a sense of doom, and a low tolerance of frustration may be a daily experience. What I am describing is often how depression masks itself, and many people suffer from depression and have never had a name for their symptoms. If this description fits your life, contact a therapist in your area to discuss your concerns. Depression is easily treated and people can and do learn to live happy lives. My point in describing depression here is that if life is clouded by depression, and that "normal" stage involves being depressed, then trigger events occur much more frequently. By choosing not to address the above issues, you are setting yourself up because even the smallest of stress may send you into your cycle. The best example of this is a person whose fuse is only a finger's length and always flickering with a spark, loaded, "ready to go off" at any time with little warning.

By now, you may have a better understanding of what your "normal" life looks like. Go through the list you have made and refer back to it on a daily basis for a couple of weeks. Add to this list, and again the more depth you give, the better the understanding of your own life. The goal here is not that we have perfect lives, but that we are at a place that we can be content with what we have in life.

Exercise

1. Make a list of all the behaviors you would engage in at this phase and describe "normal."
2. List the emotions that occur at this phase.
3. List the self-talk that occurs here.

STAGE 2

Triggers

This stage involves experiencing stress and provocation. We all experience stress and provocation probably on a daily basis. For most people, coping involves remaining focused on the issue at hand and dealing directly with the problem. This may involve openly discussing the problem or stressor with the person creating it, or perhaps relaxing and letting go of the stressor.

For those who do not deal effectively with disappointment and stress, this stage results in an escalation into obstructive and deviant thinking. The stressor or trigger now results in extreme emotions and a sense of powerlessness over the situation.

Exercise

1. Make a list of the most common triggers and stressors you experience that create problems for you.
2. Make a list of the behaviors that you engage in when you experience triggers and stressors.
3. List the emotions that you experience when your triggers occur and when you are stressed.
4. List the thoughts or self-talk that occurs with the behaviors and emotions of the victimized stage. Ask yourself: "What do I tell myself when I feel victimized?" Examples here may include the following:

    Stressors/Triggers
    - Getting behind at work.
    - Slow drivers who are in front of you.
    - A relationship ending or conflict within your primary relationship.
    - Being rejected emotionally or sexually.

    Behaviors
    - Yelling
    - Sulking
    - Argumentative
    - Making mistakes
    - Avoidance of stressor or person

    Emotions
    - Confused
    - Irritated
    - Frustrated
    - Annoyed
    - Angry
    - Dumped on

Self-Talk
"How could he/she do this to me?"
"Why is this happening to me?"

At this point, the stage is set to move into the next phase unless you intervene with relaxation or utilize reframing or rational self-talk. Without intervention, you move into the next stage.

STAGE 3

Victimized and Powerless

This stage is characterized by the failings of being victimized, betrayed, helpless, powerless, abandoned, tricked, etc. The following are examples of having been left alone while your significant other went out, and the feelings that came up were abandonment, frustratation, being upset, and loneliness. The self-talk here easily understands how we moved into feeling victimized. Again, paying attention to the quotes: "Why me?" "He/she does not love me or care for me," and "How can he/she do this to me?" The self-talk, which occurred at the trigger phase, led you directly into feeling like a victim. Although the self-talk may have been inappropriate and inaccurate, believing in it has led now to feeling or experiencing a sense of victimization. A sense of doom may set in, a sense of powerlessness, and, of course, a sense of being wronged.

1. Make a list of the behaviors that you engage in when you feel victimized and betrayed.
2. List the emotions that you experience when you feel victimized.
3. List the thoughts or self-talk that occurs with the behaviors and emotions of the victimized stage. Ask yourself: "What do I tell myself when I feel victimized?" Examples here may include the following:

Behaviors
Pouting
Sulking
Pacing the floor
Crying
Not listening to other people
Argumentative

Emotions
Used
Abandoned
Betrayed
Powerless
Cheated
Alone
Irritated
Frustrated

Self-Talk
"How could he/she do this to me?"
"What am I supposed to do?"
"Is there a better than this?"
"I deserve to be happy?"

At this point the stage is set to move into the next phase. You may not be able to focus on any other alternatives other than becoming angry. This moves us into our next phase. You may also find that you bounce between Stages 3 and 4 for some time. You may stay stuck at this stage for hours, days, or even months, before progressing to the next stage.

STAGE 4

Anger and Depression

The self-talk from the victimized and powerless phase makes a clear-cut path to this phase — anger and depression. It is here that the resentment, anger, and depression set in and become compelling and over-riding emotions. Anger truly has two components — one directed toward others and one directed toward self. Anger directed toward others leads to some degree of hostility, which may appear as verbal, physical, or sexual comments and gestures or behaviors toward another person. At this stage verbal abuse may be occurring; however, physical or sexual abuse has not yet begun. When anger turns toward oneself, it becomes depression. When this occurs, you question the very worth and esteem of your being. It is common to feel dirty, different from others, and as if other people can see all of your negative qualities.

Exercise

1. Make a list of the behaviors that you engage in when you become angry and/or depressed.
2. List the emotions that occur; the behaviors at this phase. List the thought or self-talk that occurs for the behaviors or emotions. An example may include the following:

Behaviors
- Sulking
- Raising the voice, yelling argumentative with others
- Drug or alcohol use may begin
- Threatening other people

Emotions
- Hurt
- Frustrated
- Adrenaline rush
- Excited
- Sad
- Concerned
- Confused
- Hurt

Self-Talk
- "He/she has no right to do this to me."
- "They push my buttons."
- "They deserve this."
- "I must be no good."
- "Something is wrong with me."
- "They do not love me."
- "I am not loveable."
- "What now?"

Again the self-talk from the victimized, powerless phase leads easily into the next phase of anger and depression. As you complete your own list of behaviors, emotions, and self-talk, be as open and honest as possible. Think of as many situations in which you have become angry or depressed, and use that data to make one general list. If a healthy, coping skill does not occur at this point, the next stage (isolation) follows.

STAGE 5

Isolation

It is at this phase that you would choose to isolate yourself from other people. The anger, which turns to depression, has now left you with a confusion about your own self worth and coping skills. The last thing that may be on your mind is talking to other people or letting others around to see you through this turmoil situation. Isolation may occur in many ways. You may choose to entirely isolate yourself by not being around other people, staying home, driving to an isolated location, not answering the telephone, and not seeking company. You may also be more passive with the isolation by being around other people, but not interacting with them. An example of this would be that you are at a party, but you are the wallflower or picture on the wall, nobody noticing you, and you don't interact with them. You fail to tell anyone what is going on and fail to ask for help. No one may have the slightest idea that you are hurting and need help or support.

Exercise

1. Make a list of all the behaviors you would engage in when isolating and withdrawing from other people.
2. List the emotions that occur with the isolation.
3. List the thought and/or self-talk that occur. (What do I say when I isolate myself from others?)

Examples of behaviors:
- Sitting at home in the dark.
- Not answering the telephone.
- Avoidant of any in-depth discussion with others.
- Being in a crowd, but not participating; sitting in the corner; avoiding contact.
- Driving to an isolated location.
- Breaking off engagements with people.
- Arguing so that people will leave you alone.
- Blaming other people.

Emotions:
- Alone
- Confused
- Hurt
- Sad
- Somehow different from others
- Not adequate
- Low self esteem
- Depression
- No one cares

Self-Talk
- "No one would want me if they really knew me."

"No one would like me if they knew what I was thinking."
"I am no good, but I deserve to be liked; I deserve to have my needs met."
"Let's show them."
"I'll make people meet my needs.:

STAGE 6

Revenge Planning

At this point in your cycle, you have made a decision to seek revenge, not help. It is at this phase where you begin to build "get back" fantasies; that is, to plan how you would like to get even with this person or persons. It may or may not involve illegal behavior, and commonly does not in the initial phases of someone's abuse. However, over time, this phase of building revenge plans may become less used, skipped, or even obsolete. A mental energy at this point is focused on finding a way to get even that would result in the quickest sense of happiness and pleasure for oneself. Another goal here would be to show the person or persons who you thought hurt you that they did, in fact, hurt you and that you have the power to evoke a punishment. Common examples here may include behaviors that could be legal, but that again have the message of somehow psychologically injuring or causing some discomfort to the victim or victims.

Example

Behaviors/Plan: Again, here we are not looking at behaviors that are engaged in, but rather fantasies of how you would like to get even. Examples under behavior may include:
Coming home late several times without explaining or calling.
Having a few beers.
Spending your money on gambling or other recreational items instead of paying bills.
Being late for major or significant appointments.
Not being there when your significant other wants someone to watch the kids, thereby, having to change their plans.
Having an affair.
Destroying the other persons' belongings.
Ending the relationship.
Physically or sexually assaulting the individual.

Emotions
Anger
Power
Victimizing
Justified
Invincible

Self-Talk
"I'll show him/her."
"They deserve this."
"Look what they have made me do."
"This will really show them."
"They'll never hurt me again, once I do this."
"I'll have them begging for mercy."

The purpose of this revenge planning is coming to grips with a real or imagined power or establishing real or imagined power. Although abuse may be contemplated at this point, it is rarely carried out at this phase. When most abusers initially begin their abuse history, revenge planning may often serve as an outlet or event, which restores some sense of equilibrium and normalcy. If this were to relieve enough tension and stress, the individual would likely return to the normal phase. However, the normal phase becomes redefined each time unresolved issues accumulate. Therefore, what was once a calm, respectful, meaningful existence becomes a more stressed, confused, and unfilled one. Over time, however, the get-back fantasies, revenge planning does not satisfy the needs of the offender and they move into the next phase, which is acting or the revenge plan.

Exercise

1. Make a list of your own behaviors/plans that occur here.
2. List the emotions that occur with each behavior or at this stage, and the self-talk that occurs during the revenge planning phase.

STAGE 7

Acting out Revenge Plan

It is at this phase where you would act out or get-back fantasies and/or revenge plan. You may act out exactly what you had planned in the previous phase or you may modify it some to either lessen or increase the amount of aggression used. Again, here the behaviors that occur are not illegal, and would not move beyond verbal abuse.

Exercise

1. Make a list of the behaviors you would engage in at this phase.
2. List the emotions that would occur at this phase.
3. List the thoughts or self-talk that occur here.

Examples:

Behaviors
- Coming home late.
- Drinking, though not getting drunk.
- Getting into verbal arguments.
- Blaming other people.
- Avoiding responsibilities.
- Being late with appointments or duties or obligations.

Emotions
- Powerful
- Victimizing
- In control
- Satisfied

Self-Talk
- "This will show them they will never do that again to me."
- "They will listen next time."
- "They deserve this."

"I am powerful."

For some abusers, acting out the revenge plan and get-backs may again offer a sense of reestablished equilibrium, at which point they would return to the first stage of normalcy. Again, however, normalcy becomes redefined and less fulfilling each time they return to this stage without resolving their issues.

STAGE 8

Self-Destructive Behavior

The self-destructive phase includes behaviors that become more destructive toward oneself than others. Included here would be actually getting into more verbal fights with individuals, arguing with those who could serve consequences upon you.

Examples

Being engaged in conflicts with your boss, significant other, parents, or even the law.
Other behaviors here might be becoming intoxicated or high on drugs.
Speeding.
Driving recklessly.
Not eating properly.
Not taking medications as prescribed.
Not paying attention to your health.
Pushing people away.

It is typically at this phase where individuals will say that they are powerless to stop their cycle, and give themselves permission to move into actually victimizing someone physically, sexually, or psychologically.

Examples

Behaviors
- Drinking heavy amounts of alcohol.
- Driving while intoxicated or impaired.
- Becoming intoxicated and/or using drugs.
- Arguing with authority figures.
- Breaking the law, such as speeding, shoplifting.
- Breaking probation contracts.
- Engaging in sexual compulsive behaviors.
- Engaging in gambling.
- Overeating.
- Not taking medications as prescribed.
- Not paying bills on time.
- Having affairs or use of prostitutes or telephone sex.

Emotions
- Powerless yet powerful
- Confused
- Vengeful
- Sexually aroused
- Physically aroused and agitated

Unable to stop self, yet in control
Invincible
Sly

Self-Talk
"Fuck it."
"It doesn't matter now."
"Nobody cares."
"Who can stop me?"
"I am out of control."
"I cannot stop myself."
"This won't hurt anyone."
"I'm going to get my needs met."

Exercise

1. Make a list of the behaviors you would engage in at this phase. Be as open and honest again as you possibly can.
2. List the emotions that occur at this phase.
3. List the thoughts or self-talk that occur here, as well.

STAGE 9

Victim Grooming

As the self-talk has built up in the previous eight stages, it is easy to see that, at this point, less are inclined to turn back and not follow through on his/her plan. It is at this phase where the victim becomes identified and the offender would engage in behaviors that would bring him closer to his victims. Behaviors here for physical assault may include setting the situation up so that the offender would be alone with his intended victim, and the victim would be less likely to get away from him when he chooses to be physically violent. This may also include scoping out a sexual partner, going to the bars, playground, college, or high school campuses to scope out someone who appears vulnerable. Grooming may also involve building a relationship and trust with the intended victim or victims, buying them gifts, taking them places that they like to go, spending money on them, finally setting up the opportunity so that the final attack may occur. In physically abusive situations, for example, it would be imperative that there would be no one around to intervene. For a sexual assault, or rape, again the same would be true; the victim would be isolated and away from other people so as to lessen anyone intervening in your plan. Over time, however, an abuser is less concerned with others finding out and may begin to commit his crime with others present. A sense of invincibility may set in to the degree that he/she does not even care if the police intervene, believing that he will somehow be exonerated from any charges.

Examples

Behaviors
Cruising hot spots (play grounds, schools, stores, well traveled roads, etc.).
Giving gifts, spending time with the intended victim. Showing affection and concern for the intended victim.
Setting it up so that the victim is dependent on you for money, affection, socializing, etc

Emotions
Powerful

Devious
Sneaky
Ingenious
Invincible
Fear (though very minor)
In control
Powerful
Attractive
Well liked

Self-Talk
"I'll show them."
"This is going to be great."
"They trust me."
"They'll never know what happened."
"They'll want me."
"They are sexually attracted to me."
"They know I am right."
"They owe me."
"They deserve this."

Again, the entire purpose of this phase is to find the victim and set up the circumstances so that you will be successful in your abuse of her.

Exercise

1. Make a list of all the behaviors you would engage in at this phase.
2. List the emotions that occur at this phase.
3. List the self-talk that occurs here.

STAGE 10

The Assault

It is at this phase where the actual psychological physical and/or sexual abuse occurs. Now that the victim has been selected and enough justification, rationalizations have occurred, the time has come for the abuse to occur. Psychological abuse, which can occur here, may include belittling the accomplishments of the victim; calling them names that are degrading; yelling and screaming at them; demeaning them; putting down the person's skills as a parent, lover, friend, etc.; and making threats of harm or threats of damaging their property. Physical abuse would include any type of physical touch, which could or does result in any type of physical harm. This includes: throwing objects (whether directly at the victim or not), pushing, slapping, hitting, pulling hair, biting, burning with cigarettes, using any object or weapon against the person, punching, physically holding the person down/restraining them, etc.

Sexual assault would include any type of unwanted, nonconsensual sexual activity. This may include being nude in front of the other person, taking their clothes off, touching the person's genitals over or into their clothing, or having them touch you in that manner, physically forcing oral, anal, or vagina intercourse with any object or body part, and pressuring these actions. Assault also includes grabbing your partner's genitals; taking sex when they are sleeping; engaging in sex when they have told you "no" or when you know they are not wanting to be sexual; guilting them into sex; and/or forcing them to engage in any sexual activity that is degrading, humiliating, or injurious to them. Psychological force may also be used to gain sexual favors, such as with threats, emotional blackmail, game playing, manipulating, lying, threats, intimidation, etc. (For more on psychological force, see *Man-to-Man When Your Partner Says NO*, Safe Society Press, 1992, and *When "I Love You" Turns Violent*, New Horizon Press, 1993, both by this author). The actual assault may last for several seconds, hours, or even days. It is imperative to understand that this phase did not occur in a split second, but rather the other eight phases occurred well in advance of this incident. Over time, however, an abuser may move from stage one to 10 in a matter of several minutes as he becomes more accustomed to violence as a coping skill.

Examples

Behaviors

Any type of psychological, physical, or sexual abuse described above: threats, intimidation, sexual assault, rape, molestation, physically battering the individual.

Emotions

Powerful
Vengeful
Justified
Physically and sexually aroused
Angry
In control powerful

Self-Talk

"He or she deserves this."
"They like this."
"They want this."
"They asked for this."
"See what they made me do."
"They will never talk back or challenge me again."
"They owe me."

Exercise

1. Make a list of the ways that you have been psychologically abusive, physically abusive, and sexually abusive in the past. List any and all examples, giving specific details because the way you have acted in the past means you are certainly capable of reenacting in the future. Further, the fact that you have been abusive in any way also indicates that you are capable of becoming increasingly more violent down the road. Even if you have not progressed to being legally charged with any offense, understand that any person is capable of becoming violent and warrants an understanding of each individual's cycle.
2. List the emotions that occurred each time at this phase.
3. List the thoughts or self-talk that occurred here.

STAGE 11

Sense of Relief

As the abuse begins to wind down, you are likely to begin experiencing a sense of accomplishment, adequacy, competence, and joy. You followed through with your plan, taught them a lesson, and are now ready to resume your life. You may feel physically and sexually satisfied knowing that you have accomplished your task and met any physical and sexual needs that you have needed to meet. You may feel a sense of re-empowerment, relief, and even a sense of normalcy as you accomplish your task. You may feel vindicated, justified, and happy.

Examples

Behaviors

Making up in some way with the victim.

Telling the victim that you were right and they were wrong. Promising the victim you will not do this again, if they will only meet your needs in the future.

Being engaged in other behaviors to attempt to make up with the victim.

Emotions

Happy

Empowered

Self-Talk

"See what they made me do."

"It was their fault."

"I feel happy that this happened, now they will please me and meet my needs more efficiently." "I am satisfied, etc."

Unfortunately, however, the self-talk that occurs here does not maintain a happy back-to-normal effect. Soon the reality of what has transpired leads to a sense that all is not well, which moves us into our next phase.

Exercise

1. Make a list of all the behaviors you would engage in at this phase.
2. List the emotions that occur at this phase.
3. List the self-talk that occurs here.

STAGE 12

Fear of Consequences

It was at this phase where you would consider the legal, personal, social, and familial consequences of your behavior. You may again fear getting caught by the law, that your significant other would choose to leave you, that they would report it to the authorities, and/or that you would loose face and dignity with all family members as well as from society. You may begin to engage in behaviors that will distract attention from your victim and yourself and focus on other issues. It is also here that you would attempt to make amends, if possible, with the victim so as to lessen the likelihood that she would report it to anyone.

Example

Behaviors

Making amends with the victim.

Giving gifts.

Promising it will never happen again.

Distorting the trust.

Lying about what happened.

Telling the victim she is crazy.
Telling the victim you were out of control and you could not help it and if she could only please your needs, you would not become violent. Blaming the alcohol.
May even threaten additional harm to the victim if she tells others or threatens to tell others.
Threatening that you will go to prison if you tell others.

Emotions
Fear
Guilt
Shame
Doom
Helplessness
Confusion

Self-Talk
"How can I get out of this?"
"I did not mean to do this."
"I must make sure the victim does not tell."
"What will my family think?"
"I do not want to go to prison."
"I will lose my job."
"My friends and family will leave me."
"No one will like me."
"I must do what I can to make sure that no one finds out."
"I'll never do this again."

Despite the self-talk, which occurred here, and the apparent shame and guilt, for most, it is a transitory shame and guilt. The proof of this is that instead of seeking help and talking about the offense behaviors, the offender buries his/her offense behaviors. By doing so, becoming more guilt- and shame-ridden, more isolated from others, and less likely to utilize healthy coping strategies in the future rather than to deal with life stresses. The offender's needs, at this point, are not likely to be met in any healthy way for any length of time, other than for any long length of time. The offender may experience short-term comfort, but it will be, again, short lived.

Exercise

1. Make a list of all the behaviors you would engage in at this phase.
2. List the emotions that occur at this phase.
3. List the self-talk that occurs here.

STAGE 13

Distracts

Unless you would choose to seek help at this point from qualified professionals, it is not likely that you would cease being abusive. Instead, however, you would choose to deal with your offense by distracting and disassociating. By distracting, you would be able to confuse the issues, and relabel the abusive incidents. For example, instead of calling it abuse, you may decide that it was just an argument and to just focus on the other issues going on in your life, not the abuse incidents.

If the victim attempts to talk about the abuse, you would be quick to focus on other issues, thereby distracting her from what had happened.

Disassociating also occurs. This involves distancing yourself from reality, pretending that somehow your sense of reality is different from what it actually is. This may involve trying to convince others and yourself that you would never have engaged in the violent or abusive behaviors. For example, you may make comments like: "I would never do such a thing." On one hand, this is true because the persona that you are acting from right now may not hurt another individual. However, there is a part of your persona that *would* hurt another person. We all have what would be referred to as a "shadow" or "dark side," a more aggressive side of our personalities. This is the persona that can be abusive, but it is part of who you are, it is not a separate entity.

Example

Changing the topic, avoiding talking about the abuse, blaming your victim or victims, blaming external stress, blaming alcohol or drugs, distracting onto other issues to avoid talking about the abuse, blaming a part of you that "you are out of control with or cannot control" or "do not understand these emotions," fear, confusion, power, vindicated, vulnerable.

Self-Talk

"It was a part of me I cannot control."
"I do not understand why I did what I did."
"It is not like me to do that."
"I would never do that (abuse someone)."
"They must be exaggerating what I did."
"I was drunk."
"I do not remember."

Exercise

1. Make a list of all the behaviors you would engage in at this phase.
2. List the emotions that occur at this phase.
3. List the self-talk that occurs here.

STAGE 14

Rationalizes Abuse

We have now come to the point in the abuse cycle where you need to rationalize and justify what you have done. At this phase, you make excuses for what happened. Excuses may include the use of drugs or alcohol, believing you are out of control, other people "make you" behave in certain ways.

Example

Behavior

Argumentative with people.
Offering rationalizations, justifications, and other cognizant distortions rather than admitting and honestly discussing what happened.
Using drugs or alcohol.
Basically blaming forces beyond your control.

Emotions
- Fear
- Confusion
- Victimized
- Out of control

Self-Talk
- "Must not have known what I was doing."
- "I was out of control."
- "He/she made me do it."
- "It was the drugs or alcohol that made me do it."

The offender's rationalizations, excuses, and alibis have distortions that do not appear to be accurate to outsiders. However, to the victim and perpetrator, the cogitative distortions may appear very well grounded and justified. It is often the case that until someone outside the relationship intervenes, that's what it takes to be known.

Exercise

1. Make a list of all the behaviors you would engage in at this phase.
2. List the emotions that occur at this phase.
3. List the self-talk that occurs here.

After Stage 14, return to Stage 1: Normal. However, what was once "normal" has now become skewed and distressed. "Normal" becomes redefined and rarely returns to what "normal" was prior to when any abuse occurred. Without intervention, "normal" becomes fear-based for the victim, who is in a constant state of attempting to survive. For the offender, "normal" becomes a constant state of frustration, disappointment, stress, and angry co-existing with the victim.

CONCLUSION

Having just examined the abuse cycle, I again want to reiterate that this cycle would be true for physical, psychological, and sexual abuse (including sexual assault, rape, child molestation). As you work through this cycle, it is easy to understand that your behaviors are never truly out of your control. As you have seen, your behaviors, emotions, and self-talk contribute significantly to your decision to behave. If you think about the three areas of behaving, emotions, and self-talk as a cycle in and of themselves, it wouldn't matter where you started. If you started with behaviors, you understood that as a result of acting a certain way, you would have emotions and self-talk. If you began with emotions and said, "I right now feel fear or anger," you would then understand that you were experiencing certain self-talk and then react in certain ways to those feelings. And if you began to examine the situation that occurred and started by identifying your self-talk, it is easier to understand how you talked yourself into the abusive or violent behavior.

And once again to compare this cycle to the traditional three-stage cycle: Stages 1 to 6 occur in the *escalation* stage; Stages 7 to 10 occur in the *explosion* stage; and Stages 11 to 14 occur in the *honeymoon* stage. The goods news is that if you completed the exercises in this book, you should have a more in-depth understanding of your own cycle. Even if you have never committed an offense, you now have a better understanding of your cycle. We all go through similar cycles. It does not mean, however, that we all engage in illegal or abusive behaviors, but rather that we would choose to engage healthy, coping techniques and strategies in place of victimizing other people.

I would also like to reiterate that this is not a perfect cycle. As I have worked many years with abusers and victims, I have come to realize that just as there are hundreds of different techniques and different theories to draw from, there are also many, many different cycles different practitioners have devised. I am not suggesting that the one I present in this book is the best or the most accurate. It is, however, a cycle that seems to work with abusers and victims that I have worked with over the years. You may certainly modify this cycle by adding or taking out any phases or stages, as best fits your needs.

What is interesting is that victims, as well, can often times complete this for their abuser. Having been on the reciprocal end, they often see what the perpetrator would not see. When the abuse, both physical and sexual violence within a marital or dating relationship, and both abuser and victim are willing to participate in therapy, I have often had both the abuser and victim complete a separate independent cycle for the most recent abuse that occurred. For the abuser, I asked that he focus on his own abusive behavior. For the victim, I asked that she focus on what she saw the abuser going through, therefore, describing the abuser's cycle. What is amazing is that, with two willing participants, the cycles often come out to be very, very similar and the abuser often gains a significant amount of insight from the victim's perceptions.

One last note, when beginning to have a perpetrator work on his cycle, it is often helpful to have the perpetrator focus on the most recent or most severe abusive incident that has occurred. He can also continue to add to his cycle as he progresses in treatment because he will have become more aware of other incidents and aware of ongoing abuse. I also strongly encourage the therapist to have the perpetrator add to the cycle any and all behaviors, emotions, and self-talk that occur at each phase for all the abusive incidents that he has engaged in. For example, when dealing with the revenge planning/get-back planning or acting out, have the perpetrator list all of the different behaviors that he has engaged in throughout his life, all of the different emotions, all of the different self-talk. What happens is that he begins to develop his own personal encyclopedia of how he behaves, acts, thinks, and feels when he is at this stage. It is also important under self-destructive behaviors, victim grooming, and, especially, imperative under the psychological, physical, and sexual abuse phase to list all of the behaviors, emotions, and thoughts that occurred, and force him to become more aware of his cycle.

Hence, the more the perpetrator is aware of his cycle, the more opportunities he has to recognize that he is in his cycle and, therefore, to utilize healthy, adaptive, coping techniques and strategies to get out of his cycle successfully. His new awareness of his own cycle becomes a new habit; that is, he quickly becomes aware that he is in his cycle long before he gets past Stages 3 to 5.

This chapter is written for the perpetrator. I believe that this is the best way to understand the cycle. Understand that physical abusers and sexual offenders go through the same cycle. At times a perpetrator may skip a stage, but most of the time they experience all of the stages. So most of this chapter is written as if the perpetrator is reading it. This chapter will increase your understanding of all of the thoughts and feelings that occur long before physical and sexual violence even begins.

7 Intent to Abuse: Awareness of the Use of Force

INTRODUCTION

Intent is often referred to as the reason for why people do what they do, the deliberate behavior. Intent is the goal or reward that is expected as a result of the behavior. For example, people work to earn a paycheck. If they are fortunate, they also receive satisfaction from the job itself, and possibly they grow as individuals and achieve higher status in their chosen profession.

The reason I mention the topic of intention is simple, yet so very complex. Or is it? Car accidents are good examples of intent. Car accidents, where no one is really to blame, do occur and usually occur due to poor weather or road conditions. But I believe that someone is usually to blame and at fault for most car accidents. Usually, one of the drivers was speeding, driving recklessly, failing to come to a complete stop, failing to yield, tailgating, or driving under the influence of drugs or alcohol. If my assertions are correct, then the vast majority of car accidents are anything but "accidents." When an individual drives recklessly or under the influence of drugs or alcohol, they are either deceiving themselves that they are actually driving safely and will not cause an accident, or they simply just don't care if they do. Individuals who cause car "accidents" while driving in a risky manner should be held fully accountable for their actions, and their actions imply the intent of engaging in dangerous behavior that has a high likelihood of causing harm to others or of damaging property.

Let's examine another intentional behavior that is often called an "accident." Individuals who drive under the influence of drugs or alcohol have implied their intention very clearly. Let's also examine their decision-making process. First, they make the decision to use drugs or alcohol. They do not simply find themselves using drugs or alcohol, they want to do so and choose to do it. Second, they often use the drug or alcohol at a bar, nightclub, party, or someone else's house. Third, they well know that they have to drive themselves home after the night's festivities are finished. Fourth, they rationalize that they are in at least a realitively decent shape to drive home, that they have the right to drive themselves home, and that they will not cause an accident or get caught. After all, most believe that they drive just fine while intoxicated or high. However, I assert that everyone, all of the time, is aware that they are intoxicated or impaired or high to some degree, and when people feel the impact of the drug or alcohol, they are legally impaired. People under the influence often pay more attention to looking for the police, more proof that they are aware of their impaired condition. Fifth, they were fully cognizant that they would use drugs or alcohol, become impaired, and then drive home. After all, they have done the same thing so many times before. Sixth, they drive home.

When people make decisions, their behavior implies consent. When a sex offender rapes or an abuser abuses, they made a conscious decision to do so. No different than the person who makes a conscious decision to go to work everyday, pay bills, eat meals, or engage in criminal activity. Intent is a conscious process. Therefore, there are really very, very few true "accidents" in this world, yet societies continue to defer the blame from those responsible.

When I hear a criminal say, "I didn't mean to do it," I realize that this is a delusional statement. Nothing we ever do is unintentional. We may not have wanted to hurt someone the way we did, but our actions were intentional. Even the use of sarcasm and humor are deliberate acts to make fun of and humiliate someone. Sarcasm depends on it. If you verbally abuse someone, you know very well that the words are like a knife, that your words will hurt the other person.

Keeping the abuse and sex offense cycle in mind (both the same cycles), and understanding the decision-making process, it becomes more evident that abusers and sexual offenders are never truly "out of control." Abusers invest a significant amount of energy to maintain and increase their escalation. Sex offenders focus their energy on obtaining power and control over their victims. One issue I often have to address in court is the issue of intent. Did someone actually "mean to" or "intend to" abuse, rape, or to injure their victim? This is very easy to answer. Yes, abusers and sex offenders intend to abuse or rape and intend to hurt their victims. Every time they abuse, rape, molest, it is intentional.

AWARENESS OF THE USE OF FORCE

Abusers and sex offenders are always aware that they are using aggression, force, pressure, or coercion against their victims, to make their victims comply with demands or to injure or put their victims down. Certainly abusers and sex offenders may not label their behavior as "abusive," "forced," or in any way "violent" or "coerced," but they are aware that they themselves are in some way escalating and aware to some degree that their victims are not fearful or unhappy or being injured, at least on some level.

Below are facts that will prove my point. After reading this chapter, the statement from offenders that: "I did not know it was wrong," "I did not know what I was doing," and "I was out of control," and "I did not know I was hurting them," will become statements that will no longer be used to allow offenders to avoid taking full responsibility for their offense behaviors.

Let's examine some proof for intent and awareness.

1. ***The abuser and sex offender is aware of his physiological arousal and emotional escalation***. As already discussed, the abuser is always aware that he is escalating. It is nearly impossible to be agitated and escalating and not be aware of it. Even when intoxicated or under the influence of drugs, a person is still aware of his physiological reactions. It may be easier, however, to ignore or minimize his awareness of escalation. When aware of escalation, the abuser now has to make a decision to either continue to escalate, which requires more energy to become increasingly more angry and hostile, or to invest energy in calming down. Either way, the abuser must invest energy and that is a conscious decision, whether intoxicated, high, or sober. Likewise, the sexual offender must first experience the desire to hurt or humiliate his victim. In cases where sexual arousal plays a major role, the sex offender is always aware of his sexual arousal, that he is horny and wanting sex. If he does not care about obtaining consent, then he simply does whatever it takes to force sexual contact. If he wants consent, and consent is not given, he chooses to take the sexual contact by verbal or physical force, unlike nonsexual offenders who would choose to stop.

   Any use of force results in an increased heart rate, hypervigilance, and attempts to control the behavior of the victim. Abusers and sex offenders cannot avoid noticing the effects of escalating or becoming violent. The moment they make the decision to engage in the use of force, they experience physical and cognitive changes. Physically, their body becomes aroused. Signs of physical arousal may include an increased heart rate; faster, heavier breathing; coldness or numbness in the extremities; and hypervigilance on the victim and the current and immediate surroundings. Sexual offenders will also experience sexual arousal, which may include a partial or complete penile erection.

   Cognitive changes include focusing thoughts on the victim and the intended sexual behaviors, fear, anger, and excitement. Sexual arousing thoughts and fantasies occur, which focus on the desired sexual behaviors. Abusers' thoughts focus on perceived affronts, anger, isolation, being ignored, and the thoughts of controlling their partner. Abusers and sex offenders experience thoughts, which justify and condone their assaultive behaviors. They may believe that the victim owes them respect, money, sex, or that the victim should simply bow and cave to any demand made of them. Abusers and sex offenders may believe that their victims wanted to be sexual, despite having used psychological or physical force to gain the sexual contact. Forced sex is common for abusers during the honeymoon stage and when they believe that they can take sex at any time from their property, that is, their partners.

When the physical and cognitive signs of escalation occur, abusers and sex offenders respond to them. Offenders are aware of the effects and use them to further fuel their level of excitement and arousal. As they become more excited and feeling justified in their assaultive behavior, they become increasingly more forceful. The use of force occurs throughout the entire abusive or sexually assaultive incident, and all aggressive incidents will involve verbal and/or psychological abuse, and all physical and sexually abusive incidents will include verbal and/or psychological abuse as well.

The fact that the perpetrator used physical or psychological force to begin with gives the victim a justified fear of the perpetrator. Even if the victim complies with the perpetrator's requests and the perpetrator becomes less forceful, he continues to use psychological force. (Psychological force includes threats and intimidation, as explained earlier in this book.) It would be a losing proposition to argue that the victim would not experience some degree of intimidation and the possibly of threats given that she did not give consent to begin with. Once the physical or sexual assault begins, the victim will continue to experience fear and other effects of psychological force, and may be less likely to resist the perpetrator as the assault continues. This, in part is, one of the reasons why abuse and rape victims blame themselves for not fighting or resisting more rigorously. However, it is unlikely that resisting would have prevented the physical or sexual assault.

2. ***Abusers and sexual offenders isolate their victims from others***. Most abuse and rape occur when others are not present. The perpetrators may even plan to be alone with their victims in order to allow the abuse to occur more easily and in private. Most perpetrators do not abuse or rape in the presence of others or in public places. They abuse and rape in the home, in the car, at an isolated area, or wherever others are not likely to intervene and interrupt the assault. Perpetrators may verbally abuse in public and sexual offenders may con the victim into going with them, but usually wait until they are alone with their partners or victims before assaulting. Over time, abusers and sexual offenders become increasingly more comfortable with their violent behavior and may not care if others witness the abuse, rape, or controlling behavior, but generally this occurs with more chronic abusers and with narcissistic and psychopathic perpetrators. These abusers and rapists are far more dangerous because they have no conscience left and do not fear potential legal and other consequences.

   Abusers and sex offenders also sabotage their victim's support system. They do whatever they can to limit friendships, put a wedge between family relations, and severely limit contact with others outside of the home. Most abuse victims I have worked with have few if any supportive relationships. Even family members are discouraged from supporting the victim because often the abuser either discredits the victim to the victim's own family, portraying the victim as the abuser and even claiming that the victim has psychological and drug/alcohol problems that are the cause of the relationship problems. Abusers are good con artists and are quite capable of convincing the victim's friends, colleagues, and family that the victim is the problem.

   Some abusers will also be verbally abusive towards the victim's family and friends, causing them to stop spending time with the victim. Sex offenders can also interfere with their victim's support network by spreading rumors, claiming the victim has psychological or drug/alcohol problems, and, of course, by claiming that the victim wanted the sexual contact and that the victim is promiscuous.

Perpetrators isolate their victims from anyone who may interfere with their planned assault. This occurs when the victims are invited to a party where they may not know anyone, or when they are not allowed to invite their friends, unless, for example, the friends may provide additional victims. At a party the perpetrator isolates their victim by leaving the party to go for a walk or to another part of the house. Giving the victim a tour of the house is an effective way to isolate the victim without raising suspicion. Another isolative move is to offer to drive the victim to the party or simply to go for a car ride. The only people in the car may be the perpetrator and the potential victim.

Incest perpetrators are usually alone with their victim as well. Often they are babysitting the victim while the victim's mother or care givers are away. The above are the most commonly utilized techniques used to isolate the victim. If the perpetrator's intention is not to force/coerce his victim into unwanted sexual contact, then there would have been no reason to isolate the victim from others. Remember that the vast majority of domestic and dating violence occurs in the residence of either the victim or perpetrator, where they are alone and behind closed doors.

3. ***Controlling their victim requires quick thinking and an ability to react and counter their victims' behavior.*** Maintaining control over someone requires quick thinking, the ability to utilize several types of coercion and psychological/verbal abuse, as well as an ability to adapt to the victim's resistance. This requires a desire to abuse and control. The perpetrator must be ready with alternative methods of gaining control and compliance, and is ready to utilize increasing force and assault to gain compliance.

   Psychological force requires quick thinking, and the ability to use different types of coercion to adapt to the victim's resistance. I cannot overstate this fact. When an abuser or sex offender uses any type of force or engages in the use of aggression, he is always aware of it, as mentioned above. The fact that the perpetrator increases, decreases, or changes the type of force used indicates that he is fully aware of his use of force. Why would you continue using force if your intention was to be respectful and nonviolent? You would not. You would stop at the first indication that your partner or other person was comfortable with your requests. A perpetrator, on the other hand, counters the victim's resistance with more force. Any perpetrator may choose to use verbal, physical force at any time, regardless of whether they have ever used physical or sexual force in the past. When the perpetrator believes or feels that enough justification has occurred, he uses the verbal, physical, or sexual aggression against his victim.

4. ***Attempts to apologize to the victim during or after the abuse***. Why would persons apologize unless they were aware that they did something wrong. Remember, abusers and sex offenders are narcissistic and many possess psychopathic traits. They do what they want, when they want, the way they want, whenever they want, to whomever they want. It is all about taking what they want. So, to apologize implies a sense of having done something wrong, and regardless of whether perpetrators experience a sense of guilt and remorse or not (most loose the ability to experience guilt and remorse and empathy over time), they want to minimize the victim obtaining help or support. In apologizing, perpetrators hope that they can control the victim further by preventing the victim from obtaining support or help. In addition, abusers and sex offenders are masters at conning others, and they want to maintain the image of being caring and supportive persons, and will portray their victim as crazy, if necessary, to maintain their control over the victim. So be careful when perpetrators apologize, it often does not indicate a genuine sense of remorse or empathy or guilt.

   The bottom line is that perpetrators apologize only after their own needs were met, or as an attempt to coerce their victim into staying with them so that they can complete the assault, or to minimize the risk of the victim reporting the assault. If perpetrators evoke guilt or fear in the victim, the victim may be less likely to resist or to report the assault. The victim may even blame herself for the perpetrator's behavior, believing that somehow she should have done what the perpetrator told her to do. The victim may believe that the assault was, in fact, not a sexual assault or not a domestic or dating abuse because of the perpetrator's apparent shame and guilt, suggested by his apology.

Understand that perpetrators are manipulators. If they truly were sorry for what they did, they would have stopped and not completed their assault. In addition, they would seek help and commit to completing an anger management or sex offender treatment program. Many perpetrators promise to complete treatment in an attempt to look good. However, statistically very few perpetrators complete domestic abuse or sex offender treatment unless they are court ordered to do so. Therefore, it is imperative that all abusers and sex offenders are criminally charged for their offense behaviors, and receive stiff, tough consequences. If their behavior constituted a felony charge, then charge them with a felony. Serving some jail or prison time is beneficial by reinforcing to perpetrators that their behavior was serious and wrong and that they alone are responsible for their choice to commit a violent assault or sexual assault.

Without tough consequences, perpetrators minimize their offense, and over time, will likely return to commit more sexual assaults. Beware that most perpetrators experience some degree of fear, shame, and guilt when initially caught and when involved with the court system. But their fear, guilt, and shame are almost always transitory and very few demonstrate genuine remorse. Without a structured anger management, domestic abuse, or sex offender therapy program, perpetrators will not likely make long-lasting changes, and they will not have identified and resolved their underlying issues, which they sought to address with sexually assaultive behavior. The average length of a domestic abuse or anger management treatment program that is effective is 6 to 12 months. The average length of treatment for an effective sex offender treatment program is 18 months. Very few, approximately less than 3% of all criminally charged abusers and sex offenders, will complete an adequate treatment program without court involvement.

5. ***Most abusers and sex offenders choose to injure their victims in areas of the bodies not likely to be noticed by others.*** Interestingly, if the abuser was really "out of control," how could he control where he hit and how he injured his victim? If the perpetrator is truly "out-of-control," then he would be hitting the victim all over the victim's body in a blitz or randomly wild fashion, and causing much more significant and obvious injury. Now, some abusers and sex offenders hit and kick and cause as much injury as possible, and it is these abusers who really do not care about the consequences and who are far more sadistic offenders. But most abusers do not want to get caught or go to jail. They are much more careful how they abuse and injure their partner.

   For example, consider that when the average abuser grabs his victim by the throat, he does not choke his partner much at all. When he grabs the partner by the hair, he usually does not pull out the hair. When he hits his partner in the face, he typically does not repeatedly do so each time. He wants to intimidate and cause fear, he wants to cause some degree of injury, but not enough to require medical attention. Sex offenders can do the same.

   Even though abuse and sexual offenses are very controlled behaviors, more severe injuries can and do result. When abusing or raping, perpetrators may think that they are only using a certain degree of force when, in fact, they are using much more. An example of this is when someone opens a door when angry, and the doorknob puts a hole in the wall. You may argue that the person did not "intend" to make a hole in the wall, and he may not have wanted the doorknob to even hit the wall. But, the fact that he knew he was angry when he opened the door, and he chose not to calm down, suggests that he simply did not *care* if the doorknob caused a hole in the wall. Therefore, the person did not choose to pay attention to how hard he opened the door, which was a conscious decision and the consequences were, therefore, intentional by default. A reasonable person would not expect to open a door with force and not expect that the doorknob just might hit the wall and cause a hole. Most abusers and sex offenders simply do not care how much force they use, except to the degree that they may get caught.

6. ***Abusers and sex offenders are very adept at avoiding getting caught***. They are often quick to flee the scene before anyone sees them there. If they know the police were called, it is common that they may leave the area. They will try to find somewhere to hide or may even have others provide them with an alibi. But intent is implied here. Why would you need to flee the scene unless you knew you did something wrong? You would not need to leave if you were not guilty.
7. ***Abusers and sex offenders often have alibis prepared in advance.*** They are ready to explain to family, friends, and the police just how their victim was "out of control." They are very willing to portray themselves as the victim. In fact, some abusers will even predict and encourage their partners to hit them or to break objects so that they can use that as justification for abusing, and also use their victims' behavior as proof to the police and family that their victims are really the abuser and they are, in fact, the victim. Sex offenders can also frame their victims, making themselves out to be perceived as the victim of miscommunication.

   I have seen many instances where the police have arrived at a residence only to find that the victim is hysterical, crying and angry, and the abuser is calm. But, it is difficult to imagine someone being calm, cool, and collected when abuse or rape has just occurred. In fact, that is simply an indication of a lack of remorse and a lack of empathy on the perpetrator's part. Abusers and sex offenders have to invest a lot of energy in portraying themselves as the victim, and they often practice this "victim" role. Intent is implied here because to have alibis, to provoke their victims into abusive behavior, or to have a rehearsed "victim" role requires a conscious decision and a significant amount of energy.
8. ***Abusers and sex offenders sometimes even post a person to watch for others or to keep others away from the area.*** Abusers and sex offenders may use their children or friends for this purpose. They want others to stay away to allow the abuse or rape to occur. The perpetrator is fully aware that this has an increased psychological impact on his partner. The effect is that the abuser or sex offender can abuse and rape even when help might be nearby, and that the perpetrator can evoke the assistance of his children or others to allow the abuse or rape to occur. Again, this requires a very conscious decision, as well as the risk that those requested to help might betray the abuser or sex offender by calling for help or assisting themselves.

   Whenever the perpetrator has someone acting as a lookout, it would be impossible to ever offer a convincing argument that he was not aware that he was committing a crime. Why in the world would you need a lookout if you were not engaging in illegal behavior? The whole purpose of a lookout is to alert you if someone is approaching, or to prevent anyone from bothering you. Again, the perpetrator's intention speaks for itself when he posts a guard. Furthermore, if the victim is aware of the guard's presence, it serves as a form of intimidation and threat. The guard may actually verbally, physically, or sexually assault the victim after the other offender is finished. The presence of a second perpetrator or guard increases the victim's sense of fear and humiliation. Child molesters often will have the child's parent or other children in the home stand guard, which further shames the victim and lessens the likelihood of the victim relating the experience.

   Some abusers and sex offenders even bring objects and weapons with them in order to intimidate, threaten, and injure their victims. This may include obvious weapons, such as a knife, stick, gun, or belt. But any object that can result in injury is a weapon. This includes threatening or actually throwing a beer can or bottle at the victim, or other objects that can cause injury. Using a weapon or object to threaten or injure a victim requires a conscious decision. The average person would use verbal abuse rather than physical abuse, and even physical abusers and most sex offenders may not all choose to use weapons against their victims. The choice to use a weapon implies a conscious decision, and was not a coincidence or an accident.

9. ***Providing the victim with alcohol or drugs also is a deliberate method of making victims more vulnerable.*** Some abusers and sex offenders want their victims to be high or intoxicated so that they can more easily abuse, rape, and humiliate. They understand that when their victims are impaired, they are not as capable to resist and protect themselves. Encouraging their victims to become intoxicated or high is a choice, not just something that happens. The use of "date rape" drugs or other hallucinogens complicates the situation. The victim may not be able to fully recollect much of what happened, and these types of drugs only remain in the blood stream for a very short time. Therefore, if a victim claims to not recall much about the abuse or rape, always have blood drawn to determine if any illegal drug has been used. I have also worked with abusers and rapists who are drug users themselves, and these perpetrators will force their victims to use drugs with them. The perpetrator then claims that the victim is unreliable and untrustworthy in making any allegations of abuse or rape because of the drug use. However, even when under the influence of drugs or alcohol, and when the effects have worn off, most everyone can recall a vast majority of what occurred, including most of the details of the offense.

Approximately 50 to 75% of all sexual assaults involve the use of alcohol or drugs on the part of the perpetrator or victim (Koss, 1988; Barabaree et al., 1993; Murdoch, Pihl, & Ross, 1990). What we understand about alcohol is that it clouds our thinking and slows our general response time. Alcohol is a depressant, though some individuals experience a sense of increased arousal and agitation when drinking. In order for a male to attain and maintain an erection when using alcohol, for example, he would have to concentrate on being sexual. If the perpetrator believed that his victim was a willing sexual partner and an appropriate sexual partner, then using alcohol or drugs would not appear to be necessary.

A sexual assault that occurs by peers at high school or college can be even more devastating due to the social and political stigma associated with attitudes of blaming the victim. It is even more devastating when the school does not want to make the assault public and/or sides with and protects the perpetrator. School-related rapes and abuse may occur at a party or social gathering. It is a very common phenomenon for the men in fraternities and sports houses to supply alcohol to all who attend. The number of sexual assaults and rapes that occur at these parties is astounding. In fact, many of the fraternity men often brag about their sexual exploits following the party, and plan with others how they intend to have sex with particular women at an upcoming party. Gang rapes are also common at parties. The typical scenario can go like this: Offenders supply the targeted women with alcohol and encourage them to consume large quantities of either drinks or spiked punch. The coercion to become intoxicated is intense. There may be drinking games or expectations and pressure that require participants to consume large amounts of alcohol. Once impaired, the men attack the women. The women are too intoxicated to put up much, if any, resistance and become the victim of a preplanned rape. Again, it is apparent that the abusers and sex offenders were fully aware of their intentions to harm or rape the victim.

It is my professional opinion as a forensic psychologist that the men who participate in gang rapes on campus resemble predatory sexual offenders, perhaps one of the most dangerous types of offenders. A predator does not necessarily have a particular victim identified in advance. They simply decide to commit a sexual assault and groom whomever they find to be a good target, someone who they believe will offer the least resistance. Most sexual assaults that occur on campus involve alcohol or drugs, and the offenders do what they can to get their intended victim drunk or high. Every time they get away with a sexual assault with minimal or no consequences, it further strengthens their belief that they were justified in committing the sexual assault. They do it again and again, and by the time they are caught and appropriately punished, it is not uncommon that they have several victims, most who have never reported the assault and probably never will.

The mere fact that the offenders had to use drugs or alcohol themselves indicates that they were attempting to build up the courage to break the rules and laws. If they were not intending to cross legal, moral, or ethical boundaries, they would not need to use or abuse chemicals, and they would not need to encourage the intended victim to become high or intoxicated. The fact that they provide chemicals to their victims is a clear indication that they were attempting to manipulate their victims by decreasing the likelihood of resistance and decreasing the likelihood that someone will assist the victims. Jodi Foster portrayed a good example of this in the movie *The Accused.*

10. ***Willing to risk the potential consequences.*** Abusers and sexual offenders carefully think before they act. They have considered the potential consequences of their abusive and rape behavior. In fact, many prepare for the consequences. Alibis are a good example. You would not need an alibi unless you were aware of consequences from which you needed to be protected. Abusers and sex offenders are willing to give up their marriage, children, jobs, and even family and friends in order to maintain control over their victims. But to do so requires a decision, and that decision is to either accept whatever consequences may come or to rationalize that there were be no consequences. Regardless, a conscious choice is made.

As you can see, there are no accidental or unintentional acts of abuse. Every behavior we engage in requires some degree of forethought and decision before it occurs. Our bodies do not just act in violent ways. To be truly "out of control" requires either psychosis or extreme use of more serious drugs (e.g., hallucinogens). Psychosis, however, does not simply come and go with episodes of abuse. Psychosis is indicative of serious mental illness and is something that an individual experiences often and with no control. However, abusers or sex offenders who are mentally ill can choose to not take their medication, knowing full well that failure to comply with medications may have serious implications for their behavior. But that is a *choice* some people make. It is within their control to do so.

Many abusers and sex offenders claim that they are unable to recall what was happening when they abused or raped because they were in a "blackout." However, blackouts do not occur at the time, they are an inability to recall events that occurred in the past. For example, an abuser or rapist may have been intoxicated when they abused or raped, but, at the time, they were still well aware of what they were doing. They may experience difficulty recalling all of the previous days events, but they knew full well what they were doing. Consider this example. When people drive drunk, they still know what they are doing, where they are going, and still make a conscious decision about driving. They seem to make it home, having responded to the traffic conditions, turning at the right corners, slowing down when approaching other cars, and usually end up parked at or near their home. How could they possibly do that if they were not aware of their situation and surroundings. Therefore, even when intoxicated, people still know what they are doing.

It has been my experience that when pressured to do so, abusers and sex offenders are able to recall most of what occurred during the abusive/assaultive incident if pressured to do so. Blackouts do not occur at the time of the abuse, but may occur the next day. Again, most are temporary states, not permanent.

8 Impact of Interparental Violence or Experiencing Abuse as a Child

Rosenbaum and Leisring (2003) found a significant relationship between growing up in a violent home and future perpetration. In addition, they found that the presence of a defective or impaired parent–child bond also increased the likelihood of becoming a batterer. Specifically, batterers tended to receive less love and more punishment from mothers and less attention from their fathers. Walker (1979) suggested that the relationship between the batterer and his mother as an "ambivalent love–hate relationship" and indicated that the mother exercised a great deal of control over her son.

Levine (1975) reported that exposure to spouse abuse was likely to impact the children in the home, and that the children would experience problems, including anxiety disorders, truancy, and aggression. Rosenbaum and O'Leary (1981) coined the term "unintended victims" to express their findings that batterers were more likely to come from homes characterized by the occurrence of interparental violence.

Rosenbaum et al. (2003) found that male children exposed to interparental violence demonstrated sympathy and understanding only when children, often losing their sense of empathy as they aged. They also indicated that male children exposed to interparental violence can experience a sense of impotence and helplessness as they were unable to protect their mother from being abused. Rosenbaum et al. (2003) also found that many batterers reported having been the victims of bullying and abuse as children, and that the angriest batterers reported a history of being picked on, humiliated, or excluded. This often included being made fun of, being called names, or being put down by parents, siblings, and peers. This results in a sense of powerlessness and hopelessness and significantly impacts the child's self-esteem. Power and control become a necessary strategy to survival and often underlies numerous psychiatric disorders. Kitzman et al. (2003) demonstrated a strong association between witnessing any type of abuse within the home and problems in the child's life. They demonstrated that witnessing verbal abuse within the home resulted in similar trauma as children who were physically abused.

Children in violent homes often hear, see, and intervene in episodes of marital violence (Fantuzzo et al., 1997; Holden & Ritchie, 1991; Rosenberg, 1987). Children who witness domestic abuse are at higher risk for experiencing multiple forms of abuse (McGee et al., 1997), and especially physical abuse (Appel & Holden, 1998; O'Keefe, 1994). Multiple studies have shown that couples that experience domestic abuse tend to have higher levels of general stress, tend to be younger, and exhibit higher degrees of divorce and single parenting, and have more alcohol-related problems (Fantuzz et al., 1997; Jaffee, Hurley, & Wolfe, 1990; Spaccarelli, Sandler, & Roosa, 1994; Straus, Gelles, & Steinmetz, 1980).

Children who experience the double whammy of both witnessing and being the target of aggression show worse outcomes than children who witness domestic violence, but are not physically abused (Carlson, 1991; Hughes, 1988; McCloskey, Figueredo, & Koss, 1995; O'Keefe, 1994; Sternberg et al., 1993).

Children exposed to interparental violence are at increased risk for problems as a result of their psychological well being (e.g., Cummings, Pepler, & Moore, 1999; Jouriles, Murphey, & O'Leary, 1989; Smith, Berthelsen, & O'Connor, 1997). Problems include externalizing and internalizing behavior problems, such as lowered self-esteem, impaired academic performance, socialization difficulties, poor problem solving skills, and the use of aggression to deal with problems and stress, and general anger issues. Internalization of problems included anxiety and depression. Children that are exposed to marital violence experience adjustment disorders that resemble problems shown by children abused by their parents (Jaffe et al., 1986).

Violence toward women was associated with negative psychological consequences for both wives and children (Street et al., 2003). Marital violence has an additional negative effect on women's beliefs about their marriages and families that are strongly associated with women's level of psychological distress (Street et al., 2003; Beach, Fincham, & Katz, 1998). Street et al. found that the wife's psychological distress was the sole factor linking male-perpetrated marital violence and family functioning to the children's behavior. Therefore, the mother's well being, or lack therefore of, is critical in accounting for the well being or lack thereof of her children.

Approximately 45% of batterers reported that they grew up in homes where they witnessed domestic abuse (Rosenbaum & O'Leary, 1981). Witnessing parental violence has repeatedly been reported to be associated with child depression, anxiety, conduct problems, and aggression (Cummings & Davies, 1994). Child psychopathology is four times more likely in children in martially violent homes (Jouriles, Murphy, & O'Leary, 1989). In addition to witnessing parental violence, Rosenbaum and O'Leary found that approximately 61% of male batterers reported that they were the victims of physical abuse as children. Dutton (1995) found that experiencing childhood physical abuse and having cold and rejecting parents predicted chronic trauma symptoms in adult batterers. Witnessing aggression within the home and being abused as a child have been found to be associated with the development of Oppositional Defiant Disorder and aggressive behavior, and Patterson (1982) and Patterson, Dishon, and Reid (1992) found that this pattern is among the most stable of all childhood behaviors across the lifespan.

In addition, children who have Attention Deficit Hyperactivity Disorder (ADHD) are more likely to receive negative interactions with their parents. Children with ADHD also often have Conduct Disorder and Oppositional Defiant Disorder as well, thereby increasing their chances of having negative interactions with peers, teachers, and parents, and these behavior likely will carry into adulthood (Danforth, Barkely, & Stokes, 1991; Fletcher et al., 1995; Johnston, 1996). Trauma history may also play a role in the development of relationship aggression, as does the existence of Post Traumatic Stress Disorder (PTSD). PTSD has been associated with both generalized aggression and partner-directed aggression in both men and women (Rosenbaum & Leisring, 2003).

Rosenbaum et al. (2003) found that, as children, batterers were into more fights with other children, were more likely to report beating up other children, got into more trouble with the police, and were more likely to skip school. Batterers reported more defective relationships with both parents. They also reported that they were less likely to be told they were loved by a parent, more likely to report being beaten up by their mothers, and less often reported that their fathers came to watch them perform in school activities or sports. To summarize, their parents were less involved in their lives and more punitive and less affectionate. They also found that batters described more dysfunctional and chaotic home environments, they frequently reported that their father was drunk, and were more likely to have witnessed their father assaulting their mother. They were deprived of stability, nurturance, security, and love elements essential to healthy development. As a natural consequence, these children grew up to be adults who lacked self-confidence, had poor self-esteem, were under achievers, and evidenced psychological problems. They also found that batterers with childhood Oppositional Defiant Disorder were more frequently and severely aggressive in their adult intimate relationships than other batterers. These batterers also presented with significantly more generalized anxiety, were prone to suicide, and were more impulsive.

9 Power of Ongoing Victimization

I believe that the power of ongoing victimization is significant and often overlooked. Remembering that most physical and sexual assaults occur within the context of a relationship simply separating a perpetrator and victim may not be easily accomplished. For example, perpetrators and victims may share the same set of friends, attend the same school, work together, and/or may attend the same church. They often have children together. The ongoing victimization could be as simple as someone being around a person that the perpetrator knows, for example, or in a situation similar to that of the rape situation. If the sexual assault occurred at a party and the victim and perpetrator share the same set of friends, then getting together with these friends may bring back the memory of the rape. It is not easy to simply let go of one's support network and begin building new friendships. What can worsen the situation is when the perpetrator and victim know each other's family and interact well with the family.

If the family of the victim does not fully blame the perpetrator, but, rather, places some of the blame with the victim, the victim is revictimized because she is aware of the fact that her family does not support her. If the perpetrator's family does not blame him for his abusive behaviors, this also has an effect, as just mentioned. The same is true when the victim and perpetrator work together or attend the same church or social activities. If the support people in the victim's life are aware of what happened and are supportive, then the victim has a better sense of being able to cope with the physical or sexual assault. If people choose not to believe the victim's allegations, then these people become abusers by condoning or minimizing the abuse.

Victims are often revictimized by their religion. For example, if their place of worship does not condone drinking or dancing and the victim engaged in these behaviors prior to being sexually assaulted, then they may place the blame with her for having acted out in ways that were not condoned by their religion. This is, of course, an ongoing victimization, but of a worse nature. Spiritually, what is happening is that people are portraying God as blaming victims; that when people make mistakes, they deserve to be punished. Ongoing victimization of the perpetrator also occurs. When people do not place responsibility on the perpetrator for his actions, it is now worse than saying simply, "You did the right thing," "You did not break the law," "You did not abuse anyone," "You are an okay person." While we are not debating the integral worth of the perpetrator at this point, it is victimizing because the perpetrator understands that, on some level, what he did was wrong. The fact that perpetrators identify a sense of shame, dirtiness, guilt, and anger with themselves for their behavior, indicates that they are aware that what they did was wrong and abusive. The perpetrator desperately needs people to believe that he is responsible and to hold him responsible for his behaviors. Without the support of being held accountable, without taking 100% responsibility for his behaviors, the perpetrator cannot heal.

One last area of ongoing victimization that is so often overlooked is the relationship between the victim and perpetrator. It is imperative when working with abusers and offenders to examine the relationship between the two involved. Overlooking the dynamics of the relationship simply serves to further victimize both by not attending to what might have been positive qualities in the relationship, as well as making it possible to identify the potential of ongoing victimization between them. Research has proven that abuse occurs in cycles. It is not an ongoing, 24/7 event. The victim and perpetrator chose to engage in a relationship because of the positive qualities that they saw in each other, whether it was on a first date or whether they have been involved in a relationship of several years.

It is imperative that professionals examine and identify the positive, healthy characteristics of the relationship and help the perpetrator and victim understand that they are capable of having a healthy relationship. They chose to get together for many reasons, all of which were not negative or unhealthy. They may not be able to identify all of these positive aspects of their relationship without guidance because of the amount of shame and guilt and sense of distorted or damaged self. Also, it is important to examine what the negative aspects of the relationship were. Where did things go wrong; what cues and warning signs were there that they chose to ignore or minimize? Basically, what I am saying is that the relationship between the victim and perpetrator is a historical event, which holds immense learning potential for both. Another reason why it is important to examine the relationship between the victim and perpetrator is that they may have contact in some other context and it will be important to understand how they got along. Is there a physical safety threat? Is there a threat of more abuse or another sexual assault occurring? Additionally, there is the psychological effect of having interaction following abuse or sexual assault. As we all know, simply apologizing for the abuse or sexual assault and allowing several weeks to pass does not take away the effects, nor does it make it easier for the two, once again, to get back together.

To summarize the effects of abuse and rape, both the victim and the perpetrator experience significant trauma. It is imperative to not only address the initial concerns of the victim and perpetrator, such as the emotional, physical, sexual, or spiritual needs, but also to focus on the psychological force that was used during the assault. Pay attention to how the victim was trapped by emotional confusion because in most abuse or rape situations, psychological, not physical force, was used. Help the victim, not just the perpetrator; identify the cycle of the abuse or rape and other abuse that may have occurred. Be aware that when abuse or sex offenses occur, physical, emotional, and psychological abuse occurs, as well. You cannot sexually assault someone without committing physical, psychological, and emotional abuse, ever.

10 Empathy Explored

EMPATHY: THREE FEATURES OR CHARACTERISTICS

Many people come to accept that empathy simply means to understand how another person is feeling, to understand his or her situation, and to understand how the person has been impacted as a result of your behavior. Although each of these characteristics helps to define empathy, some people have difficulty maintaining empathy. That is to say, they demonstrate some aspect of empathy, but only when convenient or useful to them in specific situations. Abusers have difficulty understanding the concepts of empathy. Abusive behavior as a rule demonstrates the lack of empathy.

There are three characteristics of empathy that I believe are important to demonstrate genuine empathy. These aspects are cognitive recognition, emotional connection, and behavioral demonstration and are discussed below.

Cognitive Recognition

Cognitive recognition refers to the understanding the definition and concept of empathy. For abusers, they may understand and be able to verbalize the concept of abuse, that their abusive behavior has caused harm to their victim. Another feature of cognitive recognition involves understanding the concept of being a victim, that is, to appreciate what victims of abuse experience.

Emotional Connection

Emotional connection refers to experiencing and being able to identify the guilt, shame, and fear that the abuser's behavior has caused his victim. This requires a person's understanding of his own behavior and a belief that he is not out of control, but rather has made choices to behave abusively, and to do so means that he did not care about the welfare or safety of his victim. Or perhaps the person truly intended to physically, psychologically, and/or sexually harm the victim. Emotional connection also requires an understanding of the possible consequences of behavior. Specifically, the possibility of losing a relationship or marriage with a partner; losing a house; damaging the relationship with your children; the loss of your dreams (e.g., of being married, of living a peaceful and satisfying life); and the loss of love, respect, and concern.

Behavioral Demonstration

Behavioral demonstration refers to practicing what you preach. This means that you will do what you say you will do and behave in ways that live up to and demonstrate your beliefs, morals, and values. Believing that your behavior is abusive, for example, yet continuing to verbally abuse your partner, actually demonstrates that you do not believe that abuse is wrong. In fact, your abusive behavior demonstrates loud and clear that you do believe that abuse is appropriate in certain situations. If you have true empathy, then you do not abuse your partner for any reason.

Behavior Indicative of Empathy

1. ***Allowing the victim to vent.*** When abuse has occurred, regardless of the frequency or intensity, the victim has a right and a need to vent. She needs to talk about the abuse

openly and in detail with family, friends, and other support people. The victim also needs to discuss the abuse with her abuser when possible. This allows the victim to express her experience with the person responsible for the abuse. No resolution or marital reconciliation is possible without this healing endeavor.

2. ***Discussing what the victim wants to discuss openly, patiently, and respectively.*** Victims heal at their own rate. Some take weeks, others years. There is no right or wrong amount of time allowed for healing. In fact, most victims will engage in the healing process throughout most of their lives. This is because the victims will never forget the abuse suffered. At different times in their lives they will experience situations in which they are reminded of the past abuse. When this happens, they need to resolve the past abuse in terms of how the current situation is problematic for them. The most common example of this is when the victim runs into one of the abuser's friends or family, someone who condoned the abusive behavior. The victim may then experience some of the same feelings as when she was powerless to stop the abuse.

   Abusers must patiently allow their victims time to deal with the abuse on the victims' terms. The abuse occurred on the abuser's terms, but the healing occurs on the victim's terms. The victim will typically remind the abuser of how his current behavior is similar to past behavior. The abuser often becomes frustrated with the victim, reminding him of the past abuse, but what is important for the abuser to understand and appreciate is that current behavior and situations may well be very similar to the past abusive behavior.
3. ***Engage in appropriate coping skills and problem solving skills.*** When the abuser is empathetic, he chooses to utilize appropriate coping skills to address stress, anger, and frustration. This means keeping calm, taking deep breaths, sticking to the current situation, demonstrating patience for the victim, and actively and openly listening.
4. ***Utilize "time-outs."*** When an abuser is serious about demonstrating empathy, he chooses to utilize time-outs when escalating. Taking time to calm down and prepare for a respectful conversation at times requires a time-out.
5. ***Accept and acknowledge abusive and insensitive behavior and comments.*** No person is perfect. At times, anyone can choose to be insensitive or verbally abusive. But what separates the abuser from someone who engages in occasional insensitive behavior is that the nonabuser is willing to admit his behavior, take full responsibility for his behavior, and choose not to repeat the same behavior. The abuser, however, blames the victim, makes excuses, and rationalizes his behavior, and he chooses to repeat the behavior.

Indicators of a Lack of Empathy

Just as there are several indicators that someone is empathetic, there are indicators that suggest a lack of empathy. Empathy is a complicated skill. There are biological factors that may prevent someone from having empathy. When this is the case, however, a person lacks empathy throughout most of his life. For most people, empathy is a learned skill. We learn from out family, friends, and society. The following are examples of behavior that suggests a lack of empathy. Again, I focus on abusers.

1. ***Apologizes quickly.*** When abusers apologize quickly, they may mean part of what they say. But the truth is in their behavior. Do they quickly forget the abuse? Are they again engaging in verbally and/or physically abusive behavior? If so, then they lack genuine empathy.
2. ***Expects victim to forgive and forget, and/or rushes the healing process ("get it done with" attitude).*** When the abuser expects the victim to "forgive and forget" and to "just get over the abuse," he is significantly minimizing the impact of the abuse and failing to respect the victim's experience. The victim deserves to heal at her own rate, not on

the abuser's schedule. Rushing the victim to heal and to "forgive and forget" is totally lacking any sense of empathy. The reality is that we never forget anything. The brain is like a video recorder. At times, we may experience difficulty recalling information, but it is stored. Abuse should never be forgotten. It is the memory of what happened that can motivate the abuser to invest in appropriate problem solving on a regular basis to decrease the likelihood of abuse occurring again.

3. ***Becoming irritated when confronted on behavior or comments***. Abusers do not like to be challenged or held accountable. But empathy requires an understanding of how another person was impacted, how the other person feels, and respect for the needs of the other person. Victims should confront their abusers, and it is the abusers' responsibility to listen and accept responsibility for their abusive behavior. Becoming irritated when held accountable clearly demonstrates self-centeredness and a lack of empathy.

   Becoming irritable, angry, or insensitive when reminded of past abuse, especially when the current behavior is similar or the same as the current abuse. Being reminded of prior abuse should serve as motivation to not engage in abuse. However, when current behavior resembles past behavior, then it is appropriate to remind the person of the pattern that exists. If the abuser continues to behave abusively, then reminding him of his pattern is very appropriate. As much as being reminded of prior abusive behavior is shameful and embarrassing, it is part of the abuser's history, just like a vacation or any other memories. Empathy requires acceptance of others reminding us of prior wrongs and of prior abusive behavior.
4. ***Giving ultimatums*** (e.g., "Either you take me back or I'm divorcing you"; "If you call the police, I will tell them you are a bad mom and about when you hit me."). Empathy requires an understanding and acceptance of another's feelings and experiences. When ultimatums are given, control is being exerted and it is verbal abuse. Abusive behavior of any kind demonstrates a lack of empathy.
5. ***Failing to follow through with therapy, treatments*** (e.g., chemical dependency treatment, anger management). If empathy is expressed, then compliance with therapeutic interventions occurs. After all, you seek counseling and therapy when you believe that you have a problem and are motivated for change. When an abuser fails to follow through with therapeutic interventions or fails to follow therapeutic recommendations, his behavior is demonstrating loud and clear that he does not believe that he has a problem, and that he does not share the same concerns that others have, including his victim.
6. ***Failing to abstain from alcohol or drug use.*** Alcohol and drug use never causes abuse to occur. However, approximately 70% of abusers are under the influence at the time they abuse. Alcohol and drugs simply provide rationale for the abuser. The abuser believes that if he is intoxicated or high, he is somehow not responsible for his behavior. But consider this. The vast majority of people who use or abuse drugs or alcohol never engage in abusive or violent behavior. Regardless of whether sober or intoxicated, the abuser must choose whether to abuse or not. Remember that alcohol and marijuana are depressants, and they have a calming effect on people. When an abuser uses alcohol or marijuana and chooses to become abusive, he has to work much harder at escalating, to counter the relaxing and sedation effects of the chemicals. To refuse to comply with alcohol or drug use recommendations demonstrates clearly a lack of respect as well as the belief that the chemical use is far more important than the victim or the relationship. Therefore, a lack of empathy and respect exists.
7. ***Justifying, rationalizing, or intellectualizing the abusive behavior or blaming the victim.*** When an offender makes excuses for his choice of being abusive, this clearly is a lack of empathy. Abuse is always a conscious choice and there are no excuses that could ever explain abuse. Many abusers, however, believe that they are the true victims, believing somehow that the victim "made them" behave abusively. Many abusers assume

the victim role, making comments such as: "If only she would leave me alone," "They know how to push my buttons," or "What would anyone else do in that situation?" Abusers often blame their use of alcohol and/or drugs for "causing" them to behave abusively. But, as is discussed elsewhere in this book, alcohol and drugs have no direct impact on an abuser's choice to be abusive; rather, the abuser uses the drugs and/or alcohol to justify his violence. The abuser may also blame the victim for the victim's past abuse experiences from childhood or other relationships, for his own alcohol or drug use, or mental health problems, such as depression. Abusers may even blame their children for creating a stressful home environment or blame their jobs for undue stress. Blaming the victim is an attempt to project responsibility onto the victim. When abusers intellectualize the abuses, they attempt to use intelligence or research to justify their abusive behavior. They may use intellectually superior words to belittle the victim and to appear more intelligent than they really are.

8. ***Indifference to the victim's feelings or concerns.*** This is often evident by the abuser attempting to "move on" and encouraging the victim and others to "forgive and forget." The abuser may also attempt to convince the victim and others that he was abusive in the past, but not today, and he does what he can to distract others from bringing up the past abuse, even when the past abuse occurred earlier that week.
9. ***Telling others or the children that abuse is a 50/50 responsibility, rather than accepting full responsibility for their behavior***. Abusers are quick to give half of the blame to the victim. This helps abusers feel more in control and less responsible for their abusive behavior. They will do anything to minimize culpability.
10. ***Undermining the victim's support network*** (e.g., interfering with the victim's support network, limiting contact with family and friends, not allowing contact with friends without being present). This is a tactic indicative of a total lack of empathy. Telling the victim's support network lies helps to deflect blame and responsibility onto the victim. The victim's support network may take the side of the abuser and begin to belittle the victim. This serves to isolate the victim and makes it less likely that the victim will tell others about the abuse out of fear of being blamed or appearing emotionally unstable.
11. ***Overdoing niceties.*** Abusers look forward to the "honeymoon stage," waiting for an opportunity to smooth over the abuse and to make up. The abuser gives too many compliments, spends more time being nice than usual, and attempts to convince the victim and others that he, the abuser, is a giving and loving partner who at times "loses control." Whatever the abuser does to be nice and giving is simply a way to control the victim. The abuser is not being sincere in this instance.
12. ***Focusing on a victim's need for change rather than on their own need for change.*** Many abusers attempt to focus the attention on the victim as the problem. They utilize the victim's shortcomings and problems, and blame the victim for not changing and for causing stress.
13. ***When separated, wanting to buy or actually buying new furniture, establishing a new residence rather than a short-term residence.*** When abusers separate from victims, they are unwilling to make sacrifices. Rather than focusing on change in order to return home, they complain and make the victims' lives even more miserable. They want to stay at an expensive hotel or to rent a residence rather than find a friend or family member to reside with on a temporary basis. This demonstrates that they are unwilling to do what it takes to heal, and make it clear that they will withhold financial support from the victim and children in order to afford a luxury residence rather than make sacrifices. They punish the victim and the children
14. ***Engaging in romantic or sexual affairs or moving on quickly and entering new romantic or dating relationships.*** When abusers enter new romantic relationships while still married, this is a sign of lacking empathy and a lack of love. When separated or in

the process of divorce, they may quickly enter relationships in order to appear that they are not abusive and to appear attractive and nice. If they really loved their victims, however, they would wait until some time after the divorce until beginning any romantic or sexual relationship. Any relationship they enter will end in disaster, but the worse impact is on the victims and children. Most often the abuser enters new romantic or sexual relationships in order to control others and to gain sympathy from the new romantic partner. The abuser also counts on the victim taking him back and being able to further control and belittle the victim. Having sexual affairs is simply another way to abuse the victim and use the new partner.

15. ***Wanting the relationship to end rather than admit abusive behavior or without engaging in necessary therapeutic services.*** Some abusers threaten to end the relationship with the victim rather than address their abusive behavior. Many abusers threaten to leave the relationship with their victims if the victims tell others about the abuse or expects the abuser to seek professional help.
16. Withdrawing from the relationship/marriage, withdrawing emotional investment or responsibilities (parenting, activities, church, obligations). Abusers are often very self-absorbed and narcissistic. They are very good at diverting their time, energy, and money into selfish things, and neglecting the victim and children in the process. The abuser then utilizes guilt to coerce the victim to back off and leave him alone, either promising to reinvest in the family or selfishly stating that he has a right to do his own things even if what he does is destructive to the family or relationship.
17. ***Seeking therapy from unqualified persons rather than from a therapist specializing in abuse*** (e.g., finding a "Christian" family therapist rather than a therapist specializing in abuse). Many abusers do whatever they can to avoid finding a competent therapist. Abusers are adept at finding the most unqualified therapist, someone who they can control and gain sympathy from. Abusers seek marital and couples counseling in order to have the victim accept at least partial responsibility for the abuse. Couples and marital therapy are not ethical or appropriate until after the abuser has demonstrated significant progress in an abuse program and the risk for abuse occurring has been minimized. Competent therapists will not provide couples or marital therapy until the abuser has, in fact, made significant progress in an abuse program.
18. ***Not cooperating with psychological assessment or testing.*** Abusers tend to be guarded, defensive, and unwilling to be honest on psychological testing. Fortunately, most tests have built in validity scales that identify when an individual has been less than honest. Even the most narcissistic abuser fails to hide pathology. However, when the abuser fails to cooperate with assessments or testing, he is not taking his abusive behavior seriously and is not believing that he has anger issues.

Finding materials (books, articles, research, Internet sites) that in any way condone or minimize abuse, that support the 50/50 blame, or intellectualizing change without behavioral change or attitude change. There will always be someone who blames the victim of abuse for the abuse. For example, the systems approach to therapy often places equal blame for relationship problems with each person. But how a victim can be responsible for being abused has never been adequately explained. Abusers are also very good at finding inexperienced or unlicensed therapists to take their side and may insist that the victim participate with their therapist. However, abuse is a specialty area within the psychology and social work field. Simply being licensed as a psychologist, social worker, medical doctor, or psychiatrist does not in any way imply competence in addressing abuse. Even those with years of experience in general or family therapy, who have dealt with some cases of abuse, may not comprehend the dynamics of abuse. It is imperative that the therapist has specialized training in addressing abuse, not just from the victim's perspective, but the abuser's as well. Abusers are also very adept at finding research and other articles and materials to justify abuse, distorting the truth with fiction.

19. ***Refusing to participate in group counseling, only individual, or only wanting marital rather than individual or group therapy.*** As stated above, abusers want to avoid taking responsibility for their abusive behavior. They want to cooperate in the least possible way. They will go to great lengths to find a therapist to provide individual therapy. But individual therapy with abusers is extremely risky, as even experienced therapists are often conned by the abuser. Group therapy is the only acceptable mode of treatment for abusers unless the abuser is found not appropriate for groups, which is uncommon. Again, if abusers demonstrated empathy, they would do whatever they needed to adequately address the abuse.
20. ***Failing to support the victim's healing process, need for friends and family to know the truth, to have support.*** Often abusers want the victim to keep quiet about the abuse, to tell others that you are experiencing relationship or marital problems, rather than honestly informing others about the abuse. This type of behavior is indicative of a lack of empathy and a lack of understanding of the harm done to the victim. Demanding or insinuating that the abuse be kept secret does not allow healing to occur, and successfully prevents the victim from getting the help they deserve. In addition, the secrecy allows the abuser to continue to abuse and further support abusive thinking.
21. ***Not allowing money for treatment or using money as an excuse for not seeking counseling.*** It is difficult to imagine how anyone who genuinely cares would use money as an excuse for not doing what he should and is able to do to save his relationship. He certainly continues to spend money on frivolous things, such as cigarettes, alcohol, or other misspending. Yet the abuser is not willing to invest money to save the relationship.
22. ***Bitterness towards the victim or towards the victim's slow-paced healing process.*** Abusers are very good at resentment and bitterness, and for making the victim experience as much suffering as possible for making the abuse known.
23. ***Minimizing or denying the abusive behavior.*** Abusers are the best for portraying unrealistic and fantasy-based stories to justify their abusive behavior. Some minimization is common when one begins treatment, and even more common when the abuser is first held accountable for his abusive behavior. However, after the initial disclosure, abusers who significantly minimize their abusive behavior lack remorse and empathy.
24. ***Demanding or pressuring the victim to make up, engage in sex, or engage in pleasurable activities to avoid addressing the abuse.*** Refusing to allow the victim distance and expecting the victim to make up is selfish and abusive. Expecting, pressuring, or talking the victim into any sexual contact is rape — simply rape. Abusers are very adept at coercing the victim into engaging in pleasurable and romantic activities in order to avoid experiencing guilt and responsibility for the abuse.
25. ***Any ongoing abusive behavior.*** This includes engaging in any abusive behavior, even if the current abusive behavior is different than the previous abusive behavior (e.g., yells more now than in the past, but with less physical abuse).
26. ***Engaging in double standards*** (e.g., expecting one thing from the victim, but not holding self to the same expectations). This is selfish and narcissistic. Practicing what you preach is important. Abusers, however, expect others to simply do what they are told. Abusers exert the sense of privilege and expect others to allow them to do whatever they want when they want, the way they want, without exception or having to explain.
27. ***Bothering the victim at work, jeopardizing the victim's job security.*** When abusers harass the victim at work, this is abuse. The abuser is attempting to make the victim aware that he is in control and that he can interfere with the victim's job security. The abuser may believe that if the victim loses her job, she will be forced to remain in the

relationship and accept the abuse because she would have no source of steady income, thereby being financially dependent on the abuser.

28. ***Engaging in one-way communication, not allowing the victim to contact them easily or effectively.*** Another way abusers exert control and dominance is to engage in unequal access to the victim. Preventing the victims from communicating with them allows abusers to maintain the power and control.

Again, the above are only some of the examples of common behavior that demonstrate the lack of empathy on the abuser's part. Any of the above is serious. However, having several of the above examples occurring is indicative of a more serious abuser. A lack of empathy usually indicates that the abuser is unlikely to change. In most cases when the abuser lacks empathy, his relationship ends.

11 Understanding Urges

INTRODUCTION

This chapter is for the offender. In my attempt to rewrite this chapter for professionals, I found that it became a very difficult task. So, this chapter is written, as it would be when used in a treatment program. I also realize that we all have cycles, and that the majority of people have not chosen to include abusive behavior as part of their cycle. Furthermore, everyone experiences urges to do things that they know are not healthy or that they simply would not like to do. Whether the urge is for a cigarette, a drink, abusive behavior, or that third pastry, we all experience urges.

THE OFFENSE CYCLE

Coping with the offense cycle requires the ability to recognize thoughts, feelings, and behaviors that comprise the cycle (discussed in Chapter 6). Many people believe that once they have become familiar with their cycle that they should be in total control of others, or that they should no longer experience urges to act out. However, urges may lessen and may or may not totally go away. You can only control your own behaviors and can reframe urges to be welcome friends as opposed to feared enemies.

The first issue here is the offender establishing a sense of control over your life; a sense of competency in dealing with situations effectively and appropriately. Remember that it is nearly impossible to control another person's behaviors. The offender can choose, however, where he will be and who he will spend time with. For example, he can control the environment by deciding where to go and where not to go.

Offenders can choose to be around certain people and choose to be in only certain situations. For example, they decide who their friends are, when to spend time with them, and when to avoid them. In some situations, offenders may not be able to choose who they are around, for example, when at work or when traveling. But even then, they are in control of their emotions, thoughts, and behaviors. For example, at work they may not always be in control of what situations occur or who is there. However, they have the choice of how to handle situations, how much time to spend with certain individuals, who to avoid, and even how to respond to those who push their buttons. If offenders feel provoked, this is okay. Being provoked simply means that their buttons are being pushed, that they are being challenged, tested, or tempted. But, only the offender can decide how to respond to others. Only the offender can give himself permission to behave violently or to walk away. The offender may also reframe the situation.

Reframing

Reframing is a technique that proves useful in times of stress and when buttons are being pushed. Basically, reframing entails examining other perspectives of the situation in which you are involved. For example, when your boss is giving you a hard time, you might take a couple of deep breathes, then tell yourself: "I do not deserve to be yelled at or talked to in a condescending manner, but this has nothing to do with the person that I am;" "What is it they are telling me? "I know that I am a competent worker and will find something useful in what my boss is saying to me;" "The fact that they are choosing to give me a hard time is out of my control." Then relabel "they're giving you a hard time" to be a benefit to you, that their feedback will help you to rethink and modify your work, or you will defend your work by explaining to your boss why you believe your work is fine.

"Buttons" are an interesting concept. Too many times you hear how someone justifies his violence with the excuse that so-and-so wouldn't shut up or leave him alone. As if that would *cause* him to automatically act in a certain way, like a computer responding helplessly to the programmer's commands. People are not computers. We can decide how to behave at any time, in any situation. Buttons are our friend. They inform us when something is not going well, or when we are hurting and vulnerable. However, buttons are nothing more than a sign, a symptom, just like urges.

An urge is the desire to act in a certain way — an impulse to do something specific. An urge is our friend. It informs us when things are not well. The following example will explain this extemely well. Stuart abuses alcohol and has done so for approximately 15 years. He has successfully completed chemical dependency treatment, and currently attends aftercare. He becomes easily frustrated when, after maintaining sobriety for several months, he continues to have the urge to drink. Not only does he have the urge to drink, but has thoughts of becoming intoxicated. He remembers how good he felt when intoxicated. As a result, he feels a sense of failure. He questions himself: "If I truly achieved sobriety and successfully completed treatment, why does the urge remain? Have I failed?" Not at all. I would expect that Stuart will continue to have urges to consume alcohol for the rest of his life, especially when stressed, frustrated, or hurting.

For Stuart and others who are chemically dependent, alcohol has been a reliable friend. Alcohol helped to ease and numb the pain of reality and sometimes hide the memory of a traumatic event. Alcohol was a reliable friend and not all of the memories of being drunk were necessarily negative. When experiencing similar situations, memories or emotions that triggered the urge to drink in the past occur again; the urges may become stronger than before. This is often due to the added shame, sense of incompetence, and failure that are added as a result of having completed chemical dependency treatment. Again, Stuart and others may believe that they should no longer experience urges to drink or to become drunk. Further, they may now believe that they should not experience difficulty coping with the urges to drink. But alcohol was used to cope and became a primary coping tool. The situations, thoughts, and memories that led Stuart to drink in the past have now become triggers. When these triggers occur today, his memory of how he coped in the past reminds him of how good it felt to be drunk. In the past, Stuart was competent at using alcohol. He now must develop healthy, appropriate coping skills that support his sobriety. That is, he must now become competent at sobriety.

Let's discuss urges in more detail. Urges, in their most basic state, are simply a sign that something is not right; a sign that things are not normal. Urges let us know when it is time to slow down and think about what it is that we really need at any given time. When stressed, people tend to seek a quick remedy, a quick, effortless solution to solve their problem and ease the pain.

Orgiastic State

Fromm (1956, p. 9) theorized that humans seek an orgiastic state in their attempt to achieve love, belongingness, and union. These may take the form of an auto-induced trance and involve a quick, intense, sometimes violent, temporary state in which the mind and body work together to separate from the reality of the world. There are several ways to achieve an orgiastic state, and the three most common are: using alcohol and drugs, engaging in violence, and sexual orgasm.

Drugs and alcohol alter reality and help numb pain and loneliness. As you become high or intoxicated, you may temporarily forget your troubles, loneliness, and frustrations. That is, until you sober up, at which point you are bombarded with even stronger reminders of what it was you were initially running from. However, drugs and alcohol offer a quick way to provide comfort in times of stress.

Violence offers a sense of control and competency. Through violent or abusive behavior you may attain satisfaction. Violence requires a state of arousal, which generally results in an adrenalin rush, which can be euphoric. The violent act may produce a sense of power, competence, and control over your victim. In addition, using violence decreases the chance that your victim will refuse your demands, and significantly decreases the chance that your victim will challenge you in any way while the violence is occurring. The violent behavior feels good, euphoric, much the same as being drunk.

Sexual orgasm is a quick and easy way to feel as though you are cared for and loved. Having sex is sometimes interpreted as a sign of being loved and cared for. The orgasm, which involves intense pleasure and satisfaction, reinforces the erroneous connection between sex and love. Therefore, a promiscuous sexual encounter, a meaningless sexual act only serve to temporarily satisfy the physical urges. You awake the next morning experiencing the same sense of loneliness and emptiness as before the sexual experience. If there was no relationship between you and your sexual partner *prior* to the sexual behavior, then there will likely be no relationship between you afterwards. It was only a sexual experience. The orgasm provided a sense of relief, a sense of competence, and a sense of being cared for, albeit a false and temporary sense.

Sexual orgasm may be attained with a sexual partner or through masturbation. Masturbation involves the self-stimulation of your genitals. Sexual fantasies are often used to attain and maintain sexual arousal. Sexual fantasies may involve fantasying about being sexual with someone you have seen or met, a past sexual partner, or may include pornography. Masturbating to these fantasies may result in a sense of competency of relationships and sexual behavior, as well as love and belongingness. Sexual orgasms may be a healthy experience when it occurs in the context of a respectful, healthy, and consensual relationship. However, when orgasm is attained for the purpose of dealing with issues such as power, competence, rejection, and loneliness (to name a few), orgasm becomes an orgiastic state, which is not healthy.

When pornography is used for the sexual fantasies, a dangerous message is being learned. The people pictured in the pornography are truly objectified. They cannot respond to your requests or to your fantasies, they cannot say "no," and, in most cases, they are in degrading positions. As I will explain later in this book, most of the women pictured in pornography have a facial expression of pain, fear, confusion, or other nonconsensual expressions. This results in nonconsensual facial expressions being reinterpreted as consensual, which is then reinforced by the orgasm. It is no wonder why sex offenders mislabeled their victim's facial expressions as consensual, erotic, and desirous.

All of the above orgiastic states are similar; they are transitory, intense and at times violent. The sole purpose of orgiastic states is to establish a sense of belongingness, love, and union. They require no real or significant relationship between the perpetrator and victim, yet they produce pleasurable feelings. Each and every time an orgiastic state is utilized to cope with your basic sense of alienation from others, the connection between triggers and orgiastic states becomes stronger. The lesson learned is that when lonely, seek a quick orgasm, take advantage of or rape someone, or use drugs or alcohol to get high, to feel good. You are teaching your mind this twisted pattern. Just as you learned the alphabet, you learn to misuse orgasms, violence, and drugs and alcohol. The result is that when you become stressed, you have urges to act out in whatever way worked for you in the past.

Orgiastic states are unhealthy responses you may choose to engage in when experiencing a sense of alienation from others, a sense of being unloved, and/or feeling incompetent. Utilizing orgiastic states serves to bury any unresolved issues. Whenever you reward yourself for the way you choose to respond to your needs, you increase the likelihood of using the same coping strategies in the future. The adrenalin rush and sense of power, control, and competence experienced during orgiastic states are powerful rewards and reinforcers. Over time, if the underlying issues are not appropriately addressed, the urges may become increasingly intense, and over time may become obsessive; the urge running through your mind over and over again like a leaking faucet constantly dripping. Until you identify and resolve the underlying issues, and choose not to utilize orgiastic states, the urge will likely continue to occur. If you choose to pay attention to your urges, then you have chosen to get back the power you projected onto others.

You are never powerless over your urges. Ask yourself the following questions when urges occur: "What is going on? "Why do I have this urge at this time?" "What am I not paying attention too?" At this point you become your own investigator, seeking out the explanation for your urges. Stuart realized that his urge to drink was strongest when he felt alienated from his wife and when he felt alone. He recognized these underlying issues only by paying attention to his urges. Stuart now views his urge to drink as a friend, someone who tells him when things are not going well, or when he has been ignoring an important need.

Another example is Sam. Sam was convicted of sexually abusing his 8-year-old daughter. He has made significant strides in therapy and has successfully identified his sexual abuse cycle. He now understands how he came to the decision to molest his daughter, and has identified many of his high-risk situations as part of his relapse prevention plan. He continues to experience urges to sexually fantasize about his daughter and other young girls. These would be dangerous urges if ignored and allowed to continue or if the urges were reinforced in any way.

Sam, however, recognizes that he is attracted to prepubescent girls, and when he becomes frustrated or experiences a sense of abandonment or alienation from his adult support network, his urges significantly increase. When the urges occur, Sam stays away from girls and immediately contacts a friend or therapist to discuss his current situation. Sam is open to exploring what needs he should pay attention to. For Sam, urges have become a friend and his ability to recognize and openly address his urges decreases the likelihood of him reoffending.

URGES AS OUR FRIENDS

Urges are our friends. They tell us when there is a problem or that things are not going well. Many people have been taught to fear urges and that they are weak for not being able to control them. But, you cannot always prevent the occurrence of an urge. I believe that urges are helpful, therapeutic friends that protect us if we pay attention to their underlying meaning. Urges can represent unresolved issues in our lives, or they can be a signal that we are not paying attention to some important aspect or happening in our life.

SOMETHING IS OFF BALANCE

Urges simply tell us that something is not right. Maybe we are experiencing stress or are ill. Life is most comfortable when we are in a state of homeostasis, which is a stable state of equilibrium. We become comfortable when functioning in certain ways, establishing patterns that become characteristics of our personality. For example, some people are aggressive and have high amounts of energy. They may be at their best when actively engaged in a project or activity. When expected to quietly sit and meditate or relax they may feel uncomfortable and awkward. They may become restless and have the urge to begin another project. If on vacation, they may develop a schedule for every day, carefully planning each activity to avoid any free time. Vacation becomes another project, a work assignment requiring that they rush from one sight to the next without time simply to rest and relax.

For an abuser, feeling out of control may result in urges to take back control. The risk here is that it may seem easier to attempt to control others than to control oneself. A decision must be made to not attempt to control others.

Experiencing too much or not enough love and affection may create an urge to distance oneself from others or to desperately search for anyone to share time with. There are many reasons why people feel smothered by others. Your partner may, in fact, be spending too much time with you, calling several times a day, sending cards and gifts, and requesting or even expecting to spend every free moment together. We all have a need to feel loved and wanted, yet we also need space to remain grounded. Without appropriate space, boundaries become diffused and two people become an extension of each other. When this occurs, individual identities are lost and they are in danger of becoming enmeshed in the relationship, which may begin to spiral to an unfortunate end.

In addition, your personal identity is necessary to enrich the relationship. When smothered, your personal identity is lost, you become one with the relationship. You spend too much time together and often will lose many of your friends. You may also lose your sense of interest in others. To spend time with others means you may be more painfully aware that you are not happy, and that your relationship has become toxic.

Not feeling as though you are receiving enough love and affection from your partner may certainly throw you off balance as well. In this situation, the relationship becomes lopsided, where one person gives and the other takes, but the taker does not give much back in return. The end result is that the givers feel empty; they have given what they had to give and the other did not fill them back up with the basic fuel of affection and love. It is similar to a washing machine that goes off balance because the clothes are piled on one side during the spin cycle.

Not taking time to relax and center yourself with your world most certainly throws you off balance. We all need time alone as well as time to relax. Time alone offers you the opportunity to evaluate your life goals and ambitions. When with others, it is difficult to clearly see and separate your goals from other's goals. Relaxation is a time to respect yourself, to meditate, dream, or lay down to appreciate the weather or accomplishments you have made. If you are always on the run, it is next to impossible to evaluate progress towards goals or to realize your achievements.

Urges may occur when experiencing stress or a lack of competence, or even when feeling overwhelmed, abandoned, or rejected. Being balanced is to experience a sense of grounding. This means something different for each person. For some, being grounded involves having time to relax, read a book, spend time with family and friends, and may also involve a spiritual realm.

At times, urges result when you experience a lack of competence. Competence refers to the belief that you possess the skills necessary to handle and survive certain situations. For example, you may believe that you are a competent parent or that you are competent at repairing cars. When that sense of competence is challenged or threatened, you may begin to doubt you have the skills to adequately complete the task. If you feel incompetent at being a parent or lover, you may choose to resort to violence to regain the sense of power and competence. However, the issue of feeling a lack of competence is not fixed or resolved with violence, it only worsens the problems. We all struggle with feeling incompetent at different times in our lives, but we can address the issue appropriately, if we choose.

When you experience feelings of abandonment or rejection, especially from your partner, you may feel an urge to quickly fix the problem. Generally, counseling or therapy is required to adequately address these issues. When you experience thoughts of abusing or sexually abusing your partner or anyone else, you may feel that this would resolve the issues or teach them a lesson. But again, abuse and sexual violence only push people away and never results in increased intimacy.

Urges simply tell you that things are not going as well as they possibly could. Addressing urges involves several steps. First, it is important to recognize the stressful situation. The next step involves identifying the specific issues at hand. One way to begin to identify the issues at hand is to answer the following questions: "What do I want right now?" "What do I want to do?" and "What do I want from or to give to someone?" These are not necessarily easy questions to answer, especially if you have not yet learned to examine your needs and wants. But at least it's a beginning. The third step is to develop coping strategies or a plan of action. For example, identify the easiest, safest way to meet your needs. Consulting a professional may be of great benefit. The final step is to allow yourself permission to experience the urges and to address the problems at hand. Urges are normal.

Section II

Characteristics of Physical and Sexual Offenders

12 Characteristics of Physical and Sexual Offenders: Similarities

IDENTIFYING AND UNDERSTANDING CHARACTERISTICS

Sexual and physical offenders present with several characteristics. It is important to identify and understand each of the characteristics in order to provide effective intervention. Not all offenders will display each of the characteristics and the fact that several characteristics may be present does not always mean that someone is an offender. However, if someone presents with several of these characteristics, he is more prone to acting out his needs with physical and/or sexual aggression. I will use the term *offender* or *offenders* to include all types of abusers and sexual offenders, and the terms *abuser* and *sex offender* interchangeably. I will also use the term *violence* to include all forms of abuse and sexual offenses.

1. ***Demonstrate poor problem solving and coping skills***. Offenders often have poor social skills; lack healthy, effective problem solving skills; and seek quick solutions to their problems. Effective problem solving requires patience and practice and often requires delayed gratification. Offenders, in general, demonstrate decreased ability to delay gratification and often have a low frustration tolerance.
2. ***May have intense, dependant relationships with their victims***. Abusers develop relationships that are intense. They want their victims to invest 100% of their time and energy into the relationship, and often expect their victims to focus on them at all times. Abusers are selfish and self-centered, and expect their victims to meet their every need and demand. Abusers also become very dependent on their victims. They need their victims to tolerate the abuser and control, and will do whatever they can to make the victims feel a sense of guilt and responsibility for taking care of them. Abusers are very good at living life as a parasite, that is, to live off of their victims.

   In addition, offenders typically associate sexual behavior with intimacy and are unwilling to take time to establish depth in a relationship. They are likely to fall head-over-heals in love with people they just met, and may obsess about their new partners to a pathological degree. They may become fixated with their partners or with someone they just met. It is not uncommon for incest offenders and child molesters and pedophiles to remain in relationships with women they are not sexually or physically attracted to in order to ensure access to the children and to give others the sense that they are "normal" and attracted to age appropriate females.
3. ***May be jealous and possessive of their significant other***. The ability to attain genuine intimacy is often a cause of significant frustration for the physical and sexual offender. Jealousy is, by definition, the energy and behavior exerted in order to protect something from being taken away. Property pertains to objects, such as cars, jewelry, and stereos. People can never be owned; they cannot legally be purchased or sold. This concept is difficult for many men to understand, but especially for physical and sexual offenders. Jealousy is often displayed in the language the offender uses. Words and statements such as: "She is mine," "Don't look at her," "You belong to me," and "I own you." Any slang name and nickname can also be indicative of jealousy. Jealousy becomes dangerous when the offender believes that his partner may end the relationship. This is especially

true if his partner spends time with others who treat his partner with respect. Men who have sexually assaulted their partners and/or who have demonstrated significant jealousy are at significant risk for engaging in future acts of violence (Goldsmith, 1990; Saunders, 1992b; Stuart & Campbell, 1989; Walker, 1989).

Abusers tend to demonstrate jealousy toward anyone who shows any interest in their victim. Abusers come to view their victims as objects that they own, and as objects that they control. Abusers fear that if the victims have friends the abuse may be discovered and the victim may end the relationship. So abusers act to prevent their victims from leaving the relationships. Jealousy may include:

- Questioning the victims on where they went, who they were with, how much money they spent, and who they talked to.
- Following or having someone follow the victim (stalking).
- Monitoring the victim's movements and travel.
- Insisting that their partner discuss any plans with them before decisions are made.
- Checking the victim's odometer.
- Checking the victim's telephone bill.
- Checking the victim's credit card or bank statements.
- Checking on where the victim is or when the victim leaves work.
- Contacting the victim's family, friends, or co-workers to check up on them.
- Limiting friendships (of male and/or female friends).
- Expecting the victim to stay home and be immediately available to meet their every need.
- Damaging or destroying the victim's belongings, especially things that may help the victim establish independence (e.g., car, personal items).

4. ***Loses temper easily, often over reacting.*** Abusers are often referred to as being "hot-headed," "short-tempered," and "out of control." Abusers tend to have very short fuses and are usually always in an escalated state. They tend to over react to the slightest frustration or circumstance, assuming the worst, and blowing things out of proportion. They may become enraged at an event not worthy of such intensity or give meaning to a situation that meant little or nothing. This creates a sense of fear and confusion in the partner because the offender may become violent with little or no visible warning. Offenders are often referred to as being "hot-headed," "short-tempered," and "out of control." Abusers tend to have very short fuses, and tend to most often be in an escalated state. They over react to the slightest frustration or circumstance, assuming the worst and then blowing things out of proportion.
5. ***May be generally impulsive.*** Abusers and sex offenders may appear impulsive in every-day situations. They may not make plans, but rather act spontaneously to situations and events occurring around them. Offenders who are obsessive–compulsive are the most difficult to treat. They may experience consequences for their impulsive behaviors, or may avoid experiencing negative consequences for some time, in some case, many years. The impulsive behavior is in response to offenders' inability to delay gratification. Offenders lack effective coping skills, which would encourage taking time to think about possible ways to respond to their crisis. Another reason impulsive behavior is common is a result of offenders' perceived lack of competence pertaining to their ability of delaying gratification. It appears to them that if they delay gratification they may experience negative consequences. Although this fear is often not support by fact, their perception is distorted to the point of not seeing options. Some offenders have impulse control or attention deficit disorders.

Examples of impulsiveness may include spending large sums of money without any planning or discussion, especially on trivial things; engaging in snap decisions that should have involved discussion; and often appearing to "act without thinking."

6. ***May minimize and deny that they have a problem with violent behavior.*** Most abusers do a great job of minimizing their role in the abuse as well as their role in the relationship problems. They blame their victims, families, friends, and the public in general for their shortcomings and problems. Physical and sexual offenders often are the last to label their behavior as abusive. It is not that they are unable to recognize their behavior as abusive, but rather a selective distortion of the truth. Offenders ignore their egos and pay attention to their perpetrator persona. The ego represents their social being and this is what most people recognize as a nice, easy-going, social person. The perpetrator persona represents impulsive, self-centered, aggressive, and sociopathic traits. We may all have a perpetrator persona, but have learned to behave in appropriate ways and to tame our deviant urges. Batterers are more likely to minimize or deny their violent and abusive behavior (Dutton, 1995; Hare, 1991; Saunders, 1992b; Webster et al., 1985). Abusers who deny or minimize their violent behavior are at increased risk for violent recidivism (Dutton, 1988; Sonkin, 1987). In addition, attitudes that support or condone abuse are associated with increased risk for violent behavior (Goldsmith, 1990; Hanson et al., 1997; Sonkin, 1987).
7. ***Behaves violently toward significant other, pets, and/or objects.*** There is a very strong correlation between the abuse and maltreatment of animals and relationship of violence toward partners, child abuse, and sexual offenses Abusers tend to pick on those weaker than themselves. I believe it is far too common that abusers who abuse their partners also abuse their children, and the pets get abused as well. Abusers and sex offenders have no qualm about injuring anyone or anything that gets in their way once the decision to abuse is made. When angry or experiencing negative or difficult emotions (e.g., feeling unwanted or unheard, left out, discounted), abusers and sex offenders may feel justified behaving violently and blame others for "making them" or "causing them" to behave abusively. They often feel provoked by their victims.

   There are some abusers and sex offenders who choose to abuse or rape victims who are stronger or more competent than they are. These offenders have a need to overcompensate by abusing or raping someone who is actually or perceived to be less likely to be a victim or to be overpowered. To abuse or rape a more competent or stronger victim increases the offender's sense of excitement and self-esteem.

   Any expression of violence is a direct predictor of future violence. When the offender becomes violent, regardless of the severity, he is giving a clear message that he does not have effective coping skills, he lacks a concern for the safety and rights of others, and that he has justified the use of violence. If the offender justifies the use of violence once, he will do it again; hence, the cycle of violence. Violence is indicative of a lack of healthy coping skills, an ineffective and uninvolved support network, and narcissism.

   Offenders with a history of violence are significantly more likely to engage in future violence (Monohan, 1981; Webster, Dickens, & Addario, 1985). Wife assaulters with a history of physical or sexual violence against family members are at increased risk for engaging in violent recidivism (Gondolf, 1988; Hotaling & Sugarman, 1986; Saunders, 1992b; Sonkin et al., 1985; Stuart & Campbell, 1989). Men with a history of assaultive behavior in current or past relationships are at risk for future violence (Fagan et al., 1983; Healy, Smith, & O'Sullivan, 1998; Sonkin, 1987). Recidivism rates for wife batterers ranges from 30 to 70% over a 2-year period (Dutton, 1995); whether the offender was arrested or treated did not matter (Hamberger & Hastings, 1993).

   It is rare for an offender to cease violent behaviors without professional intervention. Many offenders can and do stop offending for a short period of time. Yet, when stressed and unable to meet their needs, they will again resort to the use of violence; after all, it worked before.
8. ***Experiences difficulty identifying and expressing emotions and needs (other than jealousy, anger, and hostility).*** Most abusers and sex offenders are unable to identify

emotions other than the most primitive emotions of anger, jealousy, happiness, sadness, and fear. Abusers and sex offenders do a good job identifying anger, happiness, sadness; the most very common and primitive feelings. Yet, most abusive situations involve feeling of abandonment, rejection, feeling unheard or misunderstood, and frustration. Abusers and sex offenders do not often demonstrate an ability to appropriately identify or express emotions.

Society teaches men to repress their feelings, to not demonstrate weakness or vulnerability. Emotions, such as fear, adequacy, respect, love, caring, and concern, involve risk. For example, these feelings may include the possibility of their partner demanding things from them or even terminating the relationship. Men especially do a great job of avoiding deeper emotions. The problem is that ignoring and burying emotions does not result in the emotions disappearing. The emotions may be repressed and ignored, but will surface again in situations that demand these emotions.

The very fact that a partner could leave the relationship is, in and of itself, a direct threat to the relationship. If the relationship is not based on respect, commitment, and equal power, the offender may easily become angry and jealous, and may experience a sense of threat to the overall security of his primary romantic relationship. Fear of the relationship ending certainly could be a positive emotion to the degree that one would choose to invest more in the relationship, to support growth, and guide the commitment process rather than simply responding as things change. Fear is a wonderful emotion when taken in context of the situation in which it occurs.

Another problem here is that of identifying what offenders' needs are. How could anyone possibly fulfill partners' needs when not knowing what those needs are. And how could they communicate their needs if they are unable to identify them. The problem involving the inability of correctly identifying needs is a characteristic expression of problems involving anger and violence. As needs remain unmet, frustration, fear, and anger intensify. As this occurs, an offender is likely to rationalize the use of violence in response to meeting his needs. The offender may believe that using verbal, physical, or sexual violence to gain compliance from his victim decreases the likelihood of the victim refusing him or leaving the relationship. In addition, many offenders believe that their victim will forgive and forget, and that their victim may assume responsibility for the offender's violent behavior, and therefore, the offender experiences minimal negative consequences. And the victim is likely to remain with the abuser.

9. ***May behave violently toward others***. As mentioned earlier, any expression of aggression and violence indicates that offenders are comfortable utilizing violence to meet their needs. If they direct their violence toward others, it is only a matter of time before they may choose to direct it toward their partners or children. Violence and aggression in response to stress and unfulfilled needs is always toxic. In fact, I am not aware of any healthy use for aggression or violence short of protecting oneself or another from imminent and, possibly, deadly harm.

Abusers and sexual offenders tend to escalate over time the venting of their anger on just about anyone. In the early stages of their abusive history, they may focus on their partners. But over time, others become fair game in their search for new victims. Pets, strangers, other drivers — just about anyone becomes the focus of their anger. I recall a surgeon in Minnesota who forced an elderly woman driver off the road and punched her in the face several times. He was angry because, in his mind, she was driving too slow. He justified assaulting her, despite the fact that she did nothing to deserve having her face punched. Over time, most abusers will abuse others. Many have histories of engaging in violent behavior beginning when in school.

Men who are violent both in and out of the home engage in more frequent and more severe wife assault than other types of wife batterers (Cadsky & Crawford, 1988; Fagan,

Stewart, & Hansen, 1983; Gondolf, 1988; Hanson, Cadsky, Harris, & Lalonde, 1997; Saunders, 1992b; Sonkin, 1987; Stuart & Campbell, 1989).

10. ***May abuse alcohol and/or drugs, and may use this to justify their violence.*** Anytime someone becomes drunk or intoxicated, or uses alcohol or drugs to cope with stress or anything, he then is abusing drugs and/or alcohol. Clear examples of this include a person who uses drugs or alcohol in order to date, dance, or go out. They feel the need to numb themselves before engaging in any behavior that involves intimacy or risk. Most physical and sexual assaults involve chemicals, either the offender or victim or both are under the influence of some chemical. In fact, statistics indicate that approximately 75% of all violence involves someone under the influence of drugs or alcohol.

    People who abuse drugs and/or alcohol often avoid intimacy, especially emotional intimacy. They may engage in sexual behavior, but they avoid any intimate connection beyond the physical acts. When someone abuses chemicals, he is in need of a structured chemical dependency program that will address the misuse and abuse of chemicals. Rarely is the problem of chemical abuse resolved without structured intervention. Anyone can stop using drugs or alcohol for a month or so, but once the pressure and attention is removed, the need for abusing chemicals returns and the abuse of chemicals resumes.

    Alcohol and drugs never cause abuse or rape, but are often used by the abuser to justify the violence. Abusers often use alcohol and drugs to provide a convenient excuse for their abusive behavior. Despite abusers' use of alcohol to justify their violent behavior, I have yet to find the brand of beer that "makes" or "causes" abuse to occur.

    Offenders with a history of family violence are more likely to misuse substances than others with no such history (Dutton & Hart, 1992a,b; Tolman & Bennett, 1990). In addition, recent substance misuse is associated with risk for violent recidivism among wife batterers (Gondolf, 1988; Hanson et al., 1997; Magdol et al., 1997; Saunders, 1992b; Sonkin, 1987; Stuart & Campbell, 1989; Walker, 1989). Saunders (1993) found that alcohol abuse is considered one of the three most prominent risk factors for domestic abuse.

11. ***May have a dual personality (Jekyll and Hyde).*** Many abusers and sex offenders are well liked in the community, have many friends, and are educated beyond high school. Many are professionally employed. Many offenders take people by surprise with their offensive behavior. No one would have thought that they were someone who would engage in assaultive, violent behavior. But what goes on behind closed doors is often shocking. I can recall far too many examples of hearing neighbors of abusers, rapists, and murderers state that: "I just can't believe that they did it." Abusers have a skill of portraying themselves to others in a far more positive light than is evidenced by their immediate family. Most of the time their egos (social persona) are friendly, outgoing, and respectable. Most count on this to justify their abusive behavior and to avoid consequences. By keeping their victims isolated, preventing others from becoming aware of the abusive nature of the relationship, they maintain an almost surreal image to others. Their victims are cautious about telling anyone about the abuse for fear of not being believed, that others would only believe what they observe about the abuser. Again, most abuse occurs in the privacy of the home and most people do not see or hear it. Therefore, others may have a difficult time accepting the victim's allegations.

    When the abuse or sex offense occurs, the offender's perpetrator persona appears and may catch people, especially the victim, off guard. At times the dual personalities appear and become obvious. Usually this occurs when the offender makes inappropriate comments or engages in deviant behavior, revealing his violent side.

12. ***May have been abused themselves or observed others being abused when growing up.*** One of the most powerful predictors of whether someone will engage in future acts of violence is having witnessed or having been the victim of abuse/violence in the home

while growing up (Hotaling & Sugarman, 1986; Caesar, 1988; Saunders, 1992b, 1993; Sonkin, 1987). This suggests that children who witness family members being abused, even verbally abused, are at a significantly higher risk for becoming abusers themselves when they get older.

The vast majority of the physical abusers and sexual offenders I have worked with have been abused or neglected in some way as children. Certainly not all child abuse or neglect victims choose to become abusers themselves. They may choose to deal with their trauma with therapy or may have the skills necessary to survive without formal intervention. However, the majority of childhood abuse and neglect victims learned a powerful lesson: The use of violence and aggression to get what you want works. Even if they were not directly the victim of abuse or neglect, but observed someone else being victimized, the lesson learned may have been just as powerful and effective.

Some victims become perpetrators to take back the power they lost when victimized, while others become violent to express their anger and frustration. There are other reasons why victims choose to become perpetrators, but for the scope of this chapter it is enough to understand that some victims choose to become abusers.

13. ***May have deficits in intimacy and nurturance abilities***. Many offenders grew up in abusive families. Many were physically or sexually abused, and most were at the very least verbally abused. Many offenders also grew up in families that lacked emotional nurturance and lacked healthy intimacy. Many offenders experienced a lack of emotional support, some had parents who failed to spend quality time with them, and others had parents who were abusive toward each other. If children are raised in an environment lacking intimacy and nurturance, they may have serious impairment in their ability to experience, accept, and demonstrate nurturance and intimacy throughout their livers. Although it is always a choice to engage in physically or sexually abusive behavior, children that grow up in these homes experience a significant predisposition for violence.

Some victims become perpetrators to take back the power they lost when victimized, while others become violent to express their anger and frustration. There are other reasons why victims choose to become perpetrators, but for the scope of this chapter it is enough to understand that some victims choose to become abusers.

14. ***Rigid and dichotomous thinking***. When people hold beliefs that are inflexible, their options for coping with problems become limited. In fact, people who ascribe to rigid beliefs are more prone to become biased against others. Examples of this include racism, sexism, ageism, and other forms of bigotry. Offenders who are rigid in their beliefs often are difficult people to get along with because they are less receptive to considering alternative explanations and strategies for addressing problems. People who are rigid in their beliefs tend to be intolerant of the ways of others and less likely to accept compromise in what they want. Abusers often believe that their way is the "right" way, and that it is a black and white issue. They hold their beliefs even to the extreme of developing delusions, which are strongly held erroneous beliefs, but held on to despite proof to the contrary. A common belief sex offenders hold is that they are powerless to resist their urges once becoming sexually aroused, or that their victims enjoyed and wanted the sex.

People who are rigid in their beliefs may assume that their view of the world or any given situation is the absolute truth. They may accept no evidence proving them wrong, and resist modifying their beliefs because to do so would, in part, make them appear weak.

They expect their partners to spend most or all of their free time with them, and to be accountable for reporting where they have been. This involves jealousy and control. Abusers tend to keep track of their partners' whereabouts, spending habits, and friendships. Abusers will even sabotage their victims' jobs in order to keep the victims isolated and dependent on them. Many offenders attempt to control their partners' lives as much as possible. Control is a sign of abuse, and results in emotions such as fear, a sense of losing control over one's own life, and helplessness. Offenders use control to protect their property, that is, their partner. Control and jealousy go hand in hand. When your partner exerts control over your life, this is a clear sign that he believes you are an object that he owns, and that he can do whatever he wants to you because you belong to him.

15. ***May have stereotypical ideas of how men and women should act.*** Ascribing to stereotypical expectations of how men and women should behave can lead to abuse. Unfortunately, one of the stereotypes of women is that they should be submissive to their male partners. If women are viewed as subordinate to men, then abuse and rape are condoned because women become objects. Spending time talking with new friends and new dates before becoming romantically involved is an effective way to ascertain if they hold stereotypical expectations of how men and women should behave.

    Abusers are very keen at expecting their victims to "do as they are told," and not to think for themselves. Abusers want to maintain control at any and all costs, and will often adhere rigidly to the belief of male privilege; that is, that the male is the "head of the household" and the woman is to be subservient.

    A common belief that male sexual offenders may hold is that when they become sexually aroused they must have intercourse. Some actually believe that their genitals may become injured if they do not. Others may believe that women who flirt or dress in sexually attractive attire deserve to be raped. Another common belief is that if a woman gets a man turned on, she owes him sex. These exemplify stereotypical beliefs that could lead to an offender choosing to be physically or sexually aggressive.

    Extremes in behaviors and moods (quick to anger, overdo nice things, overly cruel, etc.). Abusers and sex offenders are the textbook example of emotional instability. They tend to quickly become extremely happy or extremely angry, with few, if any, emotions in the middle. When things are going well, the abuser or sex offender may be overly nice, helping out around the house, taking a more active role in parenting, and being a team player. They may even buy nice gifts and overdo nice behavior in general. When things start to go wrong, offenders tend to blow everything out of proportion. They go from being nice to being intimidating, controlling, demanding, and threatening. Once angry, they tend to hang on to the anger much longer and do not care if they hurt their victims.

    People who vacillate between extremes in mood may be dangerous. This is similar to the dual personality described above. It often requires little provocation for the offender to justify becoming enraged. Some people who demonstrate extremes in their behaviors and moods are sadistic, that is, they become sexually aroused by the pain they cause in their partners. Some offenders may suffer from a genuine mental health disorder, such as a mood disorder (depression or bi polar disorder), paranoia, personality disorder, substance abuse, or an impulse control disorder. However, if a mental health disorder exists, the offender should be under the care of a psychiatrist. Many of the mental health disorders are easily and successfully treated with medication and therapy. Without medication, however, many of the disorders cannot be successfully treated.

    Personality disorders are very common in the offender population (Hare, 1983, 1991). They are even more common among wife batterers (Hamberger & Hastings, 1988; Hare, 1991; Harris et al., 1993; Hart, Dutton, & Newlove, 1993; Sonkin, 1987).

16. ***May have many friends, be well liked by others, and may even have special talents.*** Many offenders are well liked and have many friends. Some even have special talents such as in sports, art, music, etc. Although the majority of their relationships tend to be superficial, they may have many people they spend time with. One common myth about physical and sexual offenders is that they are the undesirable-looking people in society. But, that is not the case. Most offenders fit into society and appear to be normal, average people most of the time.

Anyone can be an abuser or sex offender. Many abusers and sex offenders have respectable jobs, may be well liked at their place of worship, and may even have a pleasant social image and reputation. The problem is this: What goes on behind closed doors is generally not seen by anyone. When the abuser abuses and/or the rapist rapes or molests a child, it usually happens in the home or when away from others. So most of the people who know the offender may never see the self-centered, controlling, abusive behavior. That is why you often hear people come to the abuser's and sex offender's rescue, saying things like, "I just can't imagine him/her doing that." Many offenders do have genuine positive qualities. But abusive behavior tends to make it more difficult to see the positive things.

17. ***May use weapons for protection or to abuse others.*** Abusers tend to become paranoid and controlling. They may carry some form of a weapon with them on their belt, in their pocket, or in their car. They may keep a gun or baseball bat under their bed. It is one thing to protect yourself from real threats, but most abusers have not been attacked, and most do not live or work in situations where their physical safety is at risk. The need to carry weapons, when no real threat is present, is paranoid and a way to intimidate others.

    Certainly anyone who carries or uses weapons for protection may be paranoid and seeking violence. An obvious exception would be someone who is required to carry weapons as part of their job, such as police officers. Be suspicious of anyone who carries a weapon, whether they conceal it or show it. These individuals may suffer from personality or character disorders, and if this is the case, they may become dangerous when they believe they have been provoked. There really is no good reason for most of us to carry a weapon.
18. ***May have poor social skills.*** Many physical and sexual offenders have few skills to relate to their own age group of peers. They may relate better to people who are significantly younger, and this is especially true of child molesters. Many offenders also have poor skills with conversation and boundaries. They may focus the conversation on trivial things or only a single topic. They may challenge boundaries, tell jokes, make comments, or engage in behaviors and touch that are inappropriate.
19. ***Demonstrates anger and frustration with violent language and/or behavior.*** Coping skills involving aggressive or violent language or behavior may be indicative of impending physical and sexual assault, whether the assault occurs that day or weeks later. Individuals who utilize aggression are likely, at some point, to eventually direct their anger and frustration toward people. The fact that they are comfortable enough with their aggression to openly express it is, in and of itself, a potential warning sign. Anger may be displayed verbally, physically, or sexually. Common examples include aggressive driving, irritability at others (e.g., at a checkout line), yelling, name calling, sexually derogatory or prejudicial comments.
20. ***Uses pornography, especially in response to feelings of abandonment, anger, or as a prelude to sexual behavior.*** Research has not established that using pornography causes sexual assault. However, the use of pornography allows for the objectification of the people involved in the pornographic material. When males utilize pornography involving women, they objectify the women and may reinforce the behavior and thoughts with masturbation and orgasm. The use of pornography causes several problems. Pornography portrays people as inanimate objects. The picture never talks back, does not resist, and never places demands or has expectations of the user.

    Pornography, when used during masturbation, reinforces the person pictured, usually a woman, as a sexual object. The danger is that offenders may fantasize about anything they wish and the pictured partner never rejects or denies their command. When involved in sexual contact with a real person, the offender may then continue to believe that his partner should not talk back or refuse requests. If the partner does refuse, the offender may believe that the use of violence or force would be justified. If the pornography portrays minors, violent or deviant sexual acts, or unacceptable sexual acts, they may justify acting out these fantasies with a real person. The fact that they used violent pornography to masturbate to, for example, may fuel rationalizations and justifications for using violence with sexual partners. Simply put, pornography is used to vent sexual fantasies, which at times may be inappropriate or even violent. Pornography is discussed in more detail in Chapter 15.

21. ***May expect sex or special privileges or rewards as a way to punish or make up after an argument.*** When sexual or other pleasant behaviors follow a violent incident, or are usually involved in the making-up process, they may have successfully paired anger, violence, and aggression with sexual behavior. This is a dangerous connection to make. People may then engage in arguments or even violence in order to have the reward of sexual behavior.

I strongly encourage people to reward themselves when they correct a wrong or when they make up following a disagreement. I do not, however, support rewarding anyone for making up following a violent act. Violence should never be rewarded, ever. When a violent act occurs, regardless how insignificant or minimal, professional help should be sought, which is one of the only appropriate rewards following violence. A physical or sexual offender will demonstrate several of the above characteristics. The presence of any of the characteristics warrants concern.

13 Characteristics of Sex Offenders

INTRODUCTION

Hanson and Morton-Bourgon (2004) indicate several models that are associating with sexual offending (Ward & Siegert, 2002; Knight & Sims-Knight, 2003; Malamuth, 2003). These models suggest that some of the predisposing background or factors for sexual offending is an adverse family environment, characterized by various forms of abuse and neglect. Families that are abusive or neglectful often lack nurturance and guidance, and are often chaotic. This often results in problems in social functioning (e.g., mistrust, hostility, and insecure attachments) and, in turn, may lead to social rejection, loneliness, negative peer associations, and delinquent behavior. Hanson and Morton-Bourgon (2004) point out that the type of sexuality that develops in the context of intimacy deficits is likely to be impersonal and selfish. This may lead to beliefs that condone sexual violence. In addition to sexual deviancy and lifestyle instability, three additional characteristics of persistent sexual offenders may include a negative family background, problems with friends and intimate partners, and attitudes tolerant of sexual assault.

The following characteristics of sexual offenders were adapted from Groth (1997) and Knight and Prentky (1990). Abusers may also share similar or the same characteristics; however, these have been found to accurately describe sexual offenders.

INCEST

Most perpetrators of incest have many of the following characteristics. I define incest offenders as perpetrators who sexually molest or rape children or teenagers who are either family members, step-children, or the children of someone they are romantically involved with.

The perpetrator has a relationship with his victim, usually involving a parental or authority role. The perpetrator attempts to build a trusting relationship with the victim's parent. This is part of the grooming process. The parent may trust the perpetrator to baby-sit, take the victim on outings, and even to stay overnight with the victim. The parent usually does not suspect that anything is wrong. Once trust has been established, the perpetrator begins to groom the victim, such as being extra nice, allowing the victim to do things that the parents may not allow or only in moderation, giving gifts, and allowing the victim to talk in more sexually aggressive language and to swear.

The perpetrator then begins to groom the parent and victim by engaging in physical contact with the victim in view of everyone. This may take the form of tickling, wrestling, hugging, cuddling, and kisses on the cheek. The purpose is to make everyone at ease with the physical contact. The perpetrator can then begin to fondle the victim without raising suspicion. For example, the perpetrator may fondle the victim's breast or crotch while wrestling, and if confronted, may claim it was an accident. Unfortunately, accidents do happen when engaged in rough physical play, as we can all attest. However, the perpetrator takes advantage of the trust he has with the parents and continues to push boundaries whenever he can, eventually resulting in the sexual abuse of his victim.

The trust established between the perpetrator and the victim's parents is often what results in the parents becoming defensive and protective of the perpetrator. That is why it is important to educate the victim's parents of how the perpetrator groomed them and their child. The parents commonly experience guilt and shame and embarrassment for allowing the perpetrator to have contact with their child, and the parents may assume responsibility for the sexual abuse occurring.

When the perpetrator becomes romantically involved with the parent of the victim, the parent may be more unlikely to offer protection because he/she does not expect or want to see anything that would indicate that his/her significant other would molest his/her child. When the perpetrator gives extra attention to the victim, the parent is often pleased because he/she wanted the significant other to like and be liked by the child. When examining an incest situation, it is amazing how the perpetrator orchestrates a rather complex situation. The perpetrator may have had unrestricted access to the victim, with the parent's blessing.

In cases where the perpetrator has also abused the victim's parent, the situation becomes more serious. The parent may not be able to protect the child due to fear of further physical abuse. In fact, the perpetrator may have threatened to injure or kill the parent or the victim should the parent do anything the perpetrator does not want. Further, whenever the parent is being psychologically, physically, or sexually abused, it would be impossible not to have abused the children as well. How could the effects truly be hidden from the children, especially when the children are present when the abuse is occurring?

When the victim's parent respects and loves the perpetrator, this creates a more serious danger to the victim. The victim is not as likely to be believed by the parent when the victim discloses that sexual abuse has occurred. Many incest and child molestation victims have commented that they wanted to tell their parents about the sexual abuse, but they did not think that their parents would believe them. Other victims have indicated that when they informed the parent about the sexual abuse, they were punished and their allegations were summarily dismissed by the parent as either a lie or a get-back toward the perpetrator. I have seen many mothers of incest and child molestation victims who have chosen to remain in a relationship with the perpetrator, even when it resulted in the children being removed from the home or loss of custody. Perpetrators and abusers are very apt at manipulating their victim's parents much in the same way they manipulate anyone.

The perpetrator utilizes psychological force/abuse. Psychological abuse that may occur in incest cases involves manipulation and boundary violations. The perpetrator tricks and manipulates the victim into believing that the sexual abuse is normal and a sign of love. Further, the victim is often directly and indirectly substituted for the parent in the sexual relationship. The victim becomes a pawn in the relationship between the perpetrator and parent and has some awareness of this. The victim may believe that if he/she allows the perpetrator to be sexual with him/her, the relationship between the parent and the perpetrator may continue, which is what the victim may want; that is, a mother and a father, as well as to believe that the parent is happy.

The perpetrator may use and abuse drugs or alcohol. The perpetrator uses drugs and/or alcohol to build the courage to sexually abuse the victim. Most perpetrators blame their chemical use as the reason why they committed the abuse. In addition, most claim that they had a blackout and do not remember anything that occurred during the molestation. As I have already discussed, however, they do have the ability to recall 95 to 100% of the details. They simply choose not to. If alcohol truly made them black out and, during the blackout, they sexually abuse a child, then they would have sexually abused many children, that is, whenever they experienced a blackout and had access to children. Once this is explained to them and the fact that they should be committed to a state hospital for the rest of their life if this were true (because they would truly be "out of control" when they used drugs or alcohol), it is amazing how many begin to talk.

Perpetrators tend to have better relationships with those younger than themselves. Most incest perpetrators appear to get along better with children and adolescents. They enjoy games and activities that are more common with their victims, and they may have few friends who appear emotionally and socially mature. Incest perpetrators tend to be immature and lacking in age-appropriate interests and social skills. They tend not to have many, if any, age appropriate friends. They may get along with other adults, but they tend not to be close emotionally with other adults, preferring the intimacy of children or teens.

Incest offenders and child molesters tend to have patterns of engaging in relationships and friendships with individuals who are significantly younger than they are. They appear more "immature" to other adults. They feel more of a sense of competence with children and teens because they have the upper hand. They are older, have access to more freedom and money, and can do whatever, for the most part, they want. When incest offenders and child molesters interact with age-appropriate peers, they often feel inadequate and lack social competence.

Perpetrators tend not to view the sexual abuse as "harmful" to the child or teen. They may describe the sexual contact as consensual and benefiting the victim. Some believe that they need to teach the victim how to be sexual, while others simply believe that the child or teen is there to be sexually used. They tend not to believe that the victim has been harmed nor do they believe that the victim experiences any significant emotional consequences from the sexual contact.

Perpetrators often confuse emotional intimacy with sexual intimacy. Many incest and child molestation offenders have confused boundaries beginning in their childhood. Their limited intimacy skills and experience equate sexual behavior with sexual intimacy, that is, to be sexual is to be loved and cared for. For individuals with healthy sexual boundaries and a healthy sense of sexual intimacy, it is understood that sexual intimacy is generally only experienced within an emotionally stable and nurturing relationship. But perpetrators are unable or unwilling to accept this.

Perpetrators tend to have a psychiatric history. Often perpetrators have histories of experiencing mental health disorders. Some of the more common disorders appear to include depressive disorders, anxiety disorders, attachment deficit disorders, personality disorders (especially cluster B and C), and, of course, paraphilias. However, the presence of a mental disorder in no way ever justifies or causes sexual offense behavior to occur. Perpetrators may be on psychotropic medication, but often they refuse to comply with prescriptions.

It is imperative to obtain psychological, psychiatric, or therapy reports to aid in understanding the perpetrator's mental health history. It is also helpful to talk with the family and support network. The information could help establish whether the perpetrator has supportive relationships or whether his is isolated and a loner.

CHILD MOLESTATION

Most perpetrators of child molestation have many of the following characteristics. I define child molesters as those who engage in any sexual contact with children or teens, and the victims are not related to the perpetrator, and the victims are not the children of the perpetrator's romantic partner. They tend to possess many of the characteristics mentioned above. In addition, child molesters often present with the following characteristics.

Child molesters tend to get along better with and relate better to children and teens.

Many child molesters find it difficult to establish and maintain age appropriate relationships. In fact, they may have only one person they trust and communicate with on a regular basis. They will, however, find it easier to relate to children and adolescents. It is common for child molesters to be involved in activities that involve children and adolescents. Emotionally and intellectually they tend to be slower and may function at a lower intellectual level than expected for their age. The intellectual functioning of most child molesters ranges from average to below average.

Child molesters will associate with children and adolescents in an attempt to win trust not only of their intended victim, but the victim's parents as well. They may offer to babysit, take the child or adolescent to activities, spend the night, or have even been known to move in with the victim's family as a tenant. Most will establish some relationship with their victim prior to molesting them. Only a small percent of molesters actually abduct their victims. Most utilize psychological strategies to gain the cooperation of their victims, but will resort to physical force if need be. When the perpetrator is a stranger to the victim more force is typically used, to the extent of physically harming the victim.

Child molesters will find numerous excuses to engage in contact with children and adolescents in a search for new victims. Unfortunately, many of the grooming behaviors appear to be ordinary, expected behaviors of adults. Paying attention to the children and adolescents, giving them feedback (especially positive comments), buying them gifts, and may even include touching, such as hand-holding, hugs, and rough play (wrestling, tickling, etc). The primary difference is that the perpetrator has an ulterior motive. A nonmolesting person would not expect to gain sexual gratification from the activity. The motives of child molesters can be inferred by the fact that they cannot truly justify why they are engaging in the contact to begin with. They will, however, offer many, many excuses, rationalizations, and justifications in an attempt to condone and gain acceptance for their behavior.

Child molesters often attempt to establish some relationship with the parents of the victim. The purpose of this relationship is to establish an alibi for engaging in contact with the children and to allow easy access to the children. The perpetrator often establishes a trusting relationship with the parents, and establishes a power or authority relationship with the victim (e.g., minister, coach, therapist, teacher, friend). In this capacity, the perpetrator can justify spending time with the victim alone, and uses that relationship of authority to blackmail the victim.

Child molesters may establish a quick, intense relationship with victims and avoid the victims' parents. Some molesters find ways to intrigue and attract children and teens to them. They entice the victims by engaging in similar activities, allowing the victims to do things otherwise prohibited (e.g., drinking, use of drugs, offering them extra money, extra privileges) and attempt to build a wall between the victim and the parents. They attempt to create a sense of secrecy concerning their relationship with the victim, giving the victim a sense of being special and perhaps making the victim feel more appreciated, understood, and respected. The perpetrator avoids as much contact with the victim's parents as possible to avoid detection. The perpetrator can then use the sexual contact as a way to either increase the emotional bond and sense of obligation the victim feels for him, or the perpetrator can use the sexual contact to blackmail the victim into secrecy. Children and teens are unable developmentally to understand the emotions and responsibilities that accompany any sexual relationship. They become more dependent on the perpetrator.

Child molesters who are strangers to targeted children and teens tend to be more violent. If the perpetrator does not need or want to spend time investing in a relationship with the victim, he needs to rely more on the use of drugs, alcohol, or physical violence. The purpose of the violence is simple: to gain immediate victim compliance and obtain whatever sexual contact that is desired. No relationship with the victim is necessary. For some perpetrators, their relationship with the victim is through fantasy. However, the fantasy involving a specific victim generally ends once the sexual contact is obtained. The perpetrator then moves on to another victim.

Child molesters will cruise for victims, hanging out where the victims can be easily found. The perpetrators of child molestation want to find their ideal victim and will hangout wherever there are potential victims. Child molesters are adept at researching to find the largest pool of potential victims. They tend to use bribes and more subtle forms of force initially. This may include offering comments about how good the victim looks or is at something, or actually bribing the victim to engage in sexual contact with them.

One example comes to mind. A client I had worked with is a pedophile. He was in sex offender treatment and knew that he was not to have contact with any female under the age of 18. One day in group therapy, he gives an example of how he deals with situations that may result in contact with female children or adolescents. He stated that he was walking his friend's dog when two girls approached him and asked if they could pet the dog. He allowed them to do so. The girls then left. This was a prime example of how child molesters subtly find and groom potential victims. First, he should not have been walking the dog in the area of the park where there were girls. Second, when they approached him to pet the dog (it is probably more likely that he initiated the contact), he should have said nothing and quickly left the area. Third, several weeks had passed before he disclosed this incident to the group. The bottom line is that the child molester has a goal and that is to find and groom potential victims.

Child molesters will utilize both psychological and physical force against their victim. They are very manipulative and, when all else fails, have no problems utilizing physical force. Some child molesters will give up on a victim who fights and resists, in part because their fantasy of the victim is, in their minds, consensual, and in part because they lack self-confidence and have low self-esteem. For other perpetrators, however, they enjoy a struggle, and they may become more violent with a victim who struggles. Some are sadistic, becoming even more sexually aroused with the victim's resistance.

Child molesters tend to have more victims than incest offenders, and usually their victim pool includes both relatives and strangers. "Stranger" here is defined as having no prior relationship with the victim until they establish a relationship. Once they establish a relationship with the victim, if they continue to have contact with the victim, I still refer to the victim as being a stranger to the perpetrator. This is because the perpetrator had no justifiable or legitimate reason to establish any contact other than to molest the victim. Those perpetrators who target victims who are strangers are far more dangerous because it is nearly impossible to protect potential victims from molesters (anyone can be a victim in the community) and they can find victims almost everywhere. Child molesters are far more likely to abduct their victims because they need time to commit the sexual violence, and because they do not want to get caught. Incest offenders rarely abduct their victims because they have easy access and they do not want to call attention to themselves and risk getting caught. The child molester often believes that he is immune from prosecution and does not believe that he will get caught.

Child molesters are more likely than incest offenders to utilize physical force. Not only are they more likely to utilize physical force, but they are more likely to engage in extreme forms of physical abuse up to and including torture and murder. Again, incest offenders want and need to maintain a relationship with the victim and the victim's parents, and to use physical force or assault would jeopardize future access to the victim as well as resulting in criminal prosecution. Child molesters simply want what they want when they want it, and they take it. They do not care if they physically harm the victim and they have little intention of getting caught. To kill the victim decreases the likelihood of getting caught, and they tend to view their victims as disposable and sex objects to be used and discarded at their will. This is especially true when the child molester has a victim who is a stranger.

Child molesters are more likely than incest offenders to have criminal histories. Child molesters may have the same or similar mental health histories as incest offenders. Child molesters are more likely to have more extensive psychiatric histories than the incest offender. Many are involved in the juvenile justice system or social services at early ages. Reviewing records will shed light on what the problems were at the time the services were rendered. It is my experience that the perpetrator's version of what the problems are tend not to be reliable. But the child molesters, by far, tend to have the more extensive and varied criminal history.

SEXUAL ASSAULT

The victim is often an acquaintance. Most perpetrators of sexual assault know their victim. Some, however, just met their victims, but have established some degree of rapport. I differentiate sexual assault from rape by the type of force used. The use of psychological force to gain compliance I view as sexual assault. The use of physical force, weapons, etc., I view as rape. However, both can be referred to as rape, sexual abuse, sexual assault, etc. Victims of sexual assault and rape are equally traumatized and experience similar and, at times, the same degree of trauma, regardless of the type of force used. The primary motive for perpetrators of sexual assault is to obtain sexual contact with the least amount of force.

Perpetrators of sexual assault utilize psychological force. This can include pressuring, intimidation, threatening, the use of drugs or alcohol to gain compliance, lying, manipulation, and game playing. Many perpetrators of sexual assault will not utilize much, if any, physical force. They may, however, readily engage in power struggles and manipulation and even threats to obtain compliance from their victims. The use of psychological force can often result in the victims experiencing a sense of craziness, as though somehow they should have been able to fend off the attack. But the use of psychological force should not be underestimated in the degree of impact upon a victim. The perpetrators are not intending to stop until they obtain sexual contact, and their insistence and continued pressure often result in the victims fearing that they will be physically assaulted if they fail to comply. So the use of any type of force is a crime.

Perpetrators of sexual assault often use guilt and shame against their victims. They are often adept at creating a sense of obligation and shame in their victims. They count on the victims "giving in" and allowing the sexual contact despite the victims' protests and refusals. Often the perpetrator has some knowledge about a victim's specific trauma or sensitive subject and uses this against her. For others, they play on guilting victims into sexual contact by making it appear that victims are being selfish and not able to please anyone sexually unless they comply. Victims who have a history of prior victimization are especially vulnerable to this tactic. When the perpetrator is older or more mature than the victim, this tactic is even more effective.

Often the victim will talk of being "obligated" to be sexual. The perpetrator behaves as if the victim owes him sex. The perpetrator and victim have some form of relationship, whether casual or intense. Jealousy and anger are often used as justification to commit the assault.

Perpetrators of sexual assault often use alcohol and/or drugs as part of the offense. Perpetrators of sexual assault are often under the influence of drugs and/or alcohol and, in most cases, they make certain that their victim is also under the influence. Perpetrators may blame the effects of the alcohol and/or drugs for their sexually aggressive behavior. It is also easier to manipulate and guilt a victim into sexual contact when the victim is impaired, and the victim is less able to fight or resist. Encouraging the victim to become high or intoxicated may increase the sense of shame and powerlessness experienced by the victim. In fact, many victims blame themselves for being assaulted because they were intoxicated. Once victims become intoxicated or high, they may be less able to resist, and they may blame themselves for getting into the situation and believe that they deserved to be assaulted. However, the perpetrator is always aware of what he is doing, and sexual assault and rape do not occur just because of intoxication or being high.

Most sexual assaults are planned in advance of the assault. Even if only planned minutes before the actual assault, some planning does occur, After all, perpetrators do not want to get caught and they do not want anyone intervening to protect the victims. Perpetrators are aware of their intention to obtain sex from victims before attempting it. (Intent is covered in Chapter 7.) Many perpetrators experience a sense that their victims owe them sex; that, for whatever reason, their victims are not suppose to or allowed to refuse. This is another indication that the sexual assault was planned. Expecting that the victim will be sexual implies intent. The perpetrator of sexual assaults may or may not have preplanned the assault. However, he has made a decision not to accept "no" from the victim.

Many perpetrators of sexual assault utilize anger and intimidation before, during, or after sex to discourage their victims from reporting the assault. This is an important part of the assault — preventing anyone from intervening and to prevent prosecution. The perpetrator may use family and friends to harass the victim, dissuading the victim from reporting the offense or from cooperating with prosecution. Of course, the perpetrator may harass and intimidate the victim as well. Therefore, prosecutors, therapists, and law enforcement need to protect thc victim from further abuse and harassment.

Perpetrators of sexual assault tend to have unresolved anger issues as well as mental disorders. Many sexual assaulters, like rapists, child molesters, and incest offenders, tend to experience mental disorders (many are discussed in the incest section above). They tend to have inadequacies involving intimacy and sexuality, as well as issues related to anger and frustration. They tend not to do well with limitations placed on them from others pertaining to intimacy issues, and may have quick tempers. They have poor understandings of intimacy and relationships and also tend to confuse sex with love. Many have been the victims of childhood abuse, including physical, verbal, and/or sexual abuse.

Perpetrators of sexual assault tend to believe that they were justified in forcing sexual contact and believe that the victim asked for and deserved it. Attitudes of dominance and sex expectancies become more apparent when interviewing perpetrators of sexual assault. They may truly believe that their date must have sex upon demand, and that if the victim refuses, that it is somehow justification for sexual contact. And they then blame the victim. They may also blame the circumstances, such as being alone with the victim, experiencing sexual arousal, and being intoxicated. Perpetrators are very adept at assuming and internalizing a victim stance. They believe and portray themselves as the real victim of the crime. They feel set up by the victim and believe that they are assumed guilty just because an allegation is made. They seem to lose sight of the facts, the most important of which are that they forced, coerced, or manipulated sexual contact.

RAPE

I define rape as the physically forced sexual contact, including the use of any weapon or object, physically taking or engaging in the sexual contact, or forcing the ingestion of any drug or alcohol (including the unknowing spiking of food or drinks) to gain the sexual compliance of the victim. Again, it really does not matter whether one uses the term sexual assault or rape, but I like to differentiate the two based on the type of force used. Perpetrators utilizing physical force or the use of weapons or objects are certainly far more dangerous to their victims, and the most likely to seriously harm, maim, or kill their victims. Psychologically, however, victims of sexual assault and rape share many of the same effects and can be equally impacted.

Most rapists know their victims. It is a common misperception that most rapists rape strangers. In fact, only approximately 10 to 15% of rapes occur at the hands of strangers. Most rapists know their victims. They have some relationship, whether an acquaintanceship or more intimate relationship.

Rapists who are strangers to their victims tend to use far more violence and are more likely to use weapons. Rapists who are strangers to their victims often inflict far more serious harm, mostly because of their motive. Motives can vary based on the typology of the offender (see Chapter 17). One of the main reasons why rapists who do not know their victims inflict more harm, up to and including killing them, is that they do not intend to have further contact with the victim. Once the rape is finished, the vast majorities move on to other victims. They have conquered that victim, humiliated her, and need a new challenge.

Rapists intentionally inflict serious injury and humiliate their victims. That is the point of the rape. Making victims experience the humiliation of having sexual things done to their bodies without any say, and whatever pain or injury is caused is totally in the rapists' control.

Many rapists are sadistic. Sexual sadism involves becoming sexually aroused to their victim's sense of powerlessness, helplessness, pain, suffering, and humiliation. Of course, any type of sex offender or abuser for that matter can be a sadist. However, in my opinion, rapists are more likely to be sadistic than the other types of sexual offenders. Remember that sexual assault, molestation, and rape (all sex offenses) are about exerting power and control over the victim, with intent of obtaining desired sexual contact. Rapists have no concerns about hurting their victim and will do whatever they want to ensure sexual and sadistic pleasure, and whatever is necessary to significantly decrease the chances of being caught. They are far more likely to go to the extreme of murdering their victims.

Rapists tend to blame their victims for the rape. Rapists tend to view their victims as sexual objects, objects that deserve to be used as they please. They blame the victims for turning them on, for not resisting, for not telling them "no." and for a variety of other things. Rapists have an intense sense of entitlement. This means they strongly believe that they can take sex whenever they want, and that they are blameless for the results of the rape. The rapist feels justified utilizing any degree of force necessary to gain sexual arousal and pleasure. Rape is an exceptionally selfish crime.

Rapists tend to use drugs and alcohol to justify their crimes. This is no different than already discussed above. Rapists can use drugs and alcohol to subdue their victims or to help build up their own sense of courage to commit the rape.

Rapists tend to plan the assaults. Rapists, like all of the other types of sexual offenders, plan their offense in advance. They are not going to accept "no" for an answer. The rapist, however, tends to plan out the crime in far more detail.

In general, rapists are dangerous offenders. They make the decision to have sex with their identified victim at any cost. They will utilize physical force and even torture as part of the assault. The motive for some rapists is to exact revenge through the rape. The revenge may be directed at the victim or someone else. It is common that the rapist feels rejected and abandoned by a female and then assaults her, or he may seek another female to assault. Other goals of the rapist are to vent anger and rage, as well as attempt to compensate for perceived inadequacies.

Rapists can also be child molesters and pedophiles. The preferred victim of a rapist can range from infants to the elderly. Many pedophiles are also rapists, and they inflict very serious physical injury to their victims.

Rapists as a group tend to have few if any criminal and/or mental health histories. However, rapists that are younger (under the age of 24) tend to have more mental health histories. It is no surprise that rapists tend to lack concern for others, lack empathy for their victims, and are the second most difficult sex offenders (after child molesters) to treat. They tend to also have unresolved issues of rage and anger, and, no surprise, they often present with abandonment issues. They have learned to pair anger and violence with sexuality at an early age.

SUMMARY

The above characteristics were derived from my experience working with sex offenders as well as from other experts in the field. However, there is no single profile that can discriminate a sex offender from a nonoffender. The above characteristics can be used as a rough guide to help understand characteristics that have been identified with a particular sex offender. While most offenders will demonstrate some of the characteristics mentioned above, some offenders may meet only one or two of the characteristics. The combination of a clinical interview and psychophysiological testing (such as plethysmography, ABEL Assessment, polygraph, and the Hare Psychopathy Checklist) tend to provide the most useful and accurate data. It is necessary to understand that a sexual offender can fit into more than one of the above categories. And many sexual offenders, regardless of what type of offender, share common characteristics.

14 Characteristics of the Perpetrator Persona

INTRODUCTION

A persona is a role or identity that one assumes, a part of one's personality. We all have many personas. Examples of personas include a parent role, a lover, a victim, a perpetrator, etc. It is as though we all have multiple personalities, except that all of the personalities know of the existence of the other, communicate with each other, and are, for the most part, under one's conscious control.

Many people would like to believe that they have no perpetrator persona. Yet, how many of us would admit to having thoughts of what it would be like to take a bag of money from an armored car, just one little bag? Suppose you saw an armored car stopped in front of a bank. You notice the guards leave the car door open and the money is unguarded for a brief moment. Would you honestly say that you would not have a thought of getting a closer look at the bags of money and perhaps even touching one of the bags? Or perhaps you would even have the temptation of taking a bag. But, you resist the temptation and continue on your way without incident. It was your perpetrator persona tempting you.

What about when you took a test in school, and you found yourself gazing around the room? You may have had a thought of taking a brief look at your classmate's test sheet to see if you both marked the question the same. Or, you may have thoughts of taking an extra 15 to 30 minutes of break time without notifying your boss. You get paid for time you do not actually work, and you do not have the time coming. These are common examples of our perpetrator persona offering a temptation.

How boring life would be without the creativity of selfishness and temptation. Knowing that you did not take the bag of money increases the chance that you will appreciate the money you earn each month. Not having borrowed an answer from your classmate results in a feeling of accomplishment when you receive your grade on the test. Even if you failed the test, you did it honestly and fairly. If you got an A on the test, you feel even more confident of your abilities because you did it on your own.

We keep our perpetrator persona in check with our morals, values, and laws. We think of the consequences of acting out and decide that the consequences are not worth the gain. People who have violated societal norms by engaging in abusive behavior had to do a significant amount of distorted thinking before they crossed the boundary of being a law-abiding citizen to being a perpetrator. As their needs are not addressed, they begin to engage in cognitive distortions and eventually justify their acting out behavior. Their perpetrator persona was triumphant. But, it was their *choice*; they were never out of control. They believed that the gains of acting out were more desirable than not acting out.

CHARACTERISTICS OF A PERPETRATOR

The perpetrator persona has several characteristics, which are described below. Each of these traits will be described by the perpetrator persona who has sexually assaulted his partner.

Selfish and Self-Centered

The only person who matters is me. My needs and desires are so important that others should simply want to do whatever I tell them to do. The world centers around me. When I want sex, she should want to please me. How dare they refuse me.

Immediate Gratification

I want what I want when I want it. Others should understand that I do not have the time to wait for them to come around to my way of thinking. When I am horny I want sexual relief NOW. When I am lonely, I expect people to drop whatever they are doing to spend time with me, doing what I want. If I am hurt, I deserve revenge and satisfaction NOW. I will find some way to get what I want, even if it means fulfilling my desires in small ways at first. When I am horny, you should want to please me now, regardless of how you feel. Please me now or I'll take sex from you.

Lacking Empathy

I really do not care how my behavior affects others. I only know how I am affected. When I am meeting my needs, I am not really hurting anyone anyway. They like what I am doing, and want me to do it. Why is everyone making such a big deal about it?

Risk Taker

I will take significant risks at the expense of others. In reality, it may appear that I take risks, but really what I am doing is justified. My forcing sex from you involves little risk. After all, you are my partner and you will never tell on me. Even if you did, who would punish me for having sex with you?

Sense of Entitlement

I deserve whatever I want. To put it simply, people owe me. You owe me. You are my partner and I want sex. I treat you nicely, spend time and money on you. Now I think you owe me sex. If you refuse, I can take sex because you owe me. I deserve it.

Rationalizing

If I force you to be sexual with me, it does not constitute rape. We are dating and you are my partner. If I use a weapon or physically abuse you, then that would be rape. But, pressuring you is okay to do. Sometimes you need to be reminded of what you owe me and what my needs are. It is perfectly acceptable to take sex from you. Beside, you will enjoy it. I know that you really want it. Somehow, whatever I do, however I behave benefits others. I do not hurt anyone. When it seems that I am taking, I am really helping, and it benefits them, too.

Projects Blame onto Others

It is never my fault for forcing you to be sexual. You tease me, lead me on to believe that we are going to be sexual and then refuse to do so. Stop blaming me for your problems, you owe me, I deserve sex, you deserve whatever I do to you. How could you possibly blame me for what I did? I was drinking, and you know how drinking makes me black out, and then I am out of control. You led me on, you know that I must have sex when I get turned on, and you turned me on. When you refuse my demands, you are really telling me to take charge and show you what I want.

CONCLUSION

In summary, the perpetrators' beliefs and values vastly differ from our normal day-to-day beliefs and values. I believe that everyone has a perpetrator persona; however, most choose not to behave abusively or in inappropriate ways. Most people choose to exercise their conscience before acting and choose not to take advantage of others.

15 Abuser Typologies

INTRODUCTION

I find that typologizing batterers and sex offenders rarely occurs as part of a forensic evaluation, treatment assessment, or as part of the treatment program. The problems that result are significant. The primary problem is that this relies on a major erroneous assumption: that all batterers present with the same or similar characteristics; that all batterers engage in verbal, physical, and/or sexual violence for the same reason; that all batterers present with the same degree of risk for engaging in future acts of violence; and that all treatment programs and interventions work equally well with every batterer. Now, the same typecasting is true for sexual offenders. They are all treated as if they present with the same risk issues and etiologies. Realistically speaking, this lack of typologizing is simply unacceptable and unethical.

Imagine going to the surgeon and telling him/her your symptoms. Your surgeon decides that everyone who presents with the same problems or symptoms should be treated the same as every other patient. But this does not happen. Your doctor orders tests, blood work, x-rays. Your doctor gives you a fairly thorough exam. And, your doctor takes into account your personal factors and medical conditions that may be impacting the presenting problem. He/she then decides how to treat the problem. Not everyone is treated exactly the same way for all medical conditions. Some medications that are useful to treat certain disorders may not be appropriate for safe use with patients that have personal risk factors (e.g., high blood pressure, diabetes). So, in short, the medical field takes into account personal factors and differences before engaging in the treatment of many medical disorders.

It is no less important to appropriately categorize batterers and sex offenders. Research is clear that incest offenders tend to recidivate at a significantly lower rate than child molesters, who molest male and female children that are strangers or nonrelatives. So, the reasons why offenders commit their offense and the actual risk for reoffense are dramatically different between these two types of child molesters. And, the batterer who is psychologically violent presents with different needs and risks than the batterer who frequently physically abuses his partner. I cannot understate enough the importance of adequately categorizing batterers and sex offenders. Last, different types of batterers and sex offenders respond to treatment differently, and, for the safety of the community, professionals need to assure that the appropriate measures are taken to lessen the risk for new offenses.

In many ways the sex offender field appears to be years ahead of the domestic abuse field in typologizing offenders. This is disappointing and appalling when considering that domestic and partner abuse will account for significantly more deaths and serious injury than sex offenses ever will. I was disappointed in that much of the published research has inadequate sample sizes for batterers. The low sample size in the literature means that the results of suggested abuser typologies need to be viewed with caution.

POPULAR TYPOLOGIZING RESEARCH

Listed below are some of the more popular research related to typologizing batterers and sexual offenders. Some of the typologies simply identify psychological or familial factors without adequately addressing behaviors. At any rate, this is what I found.

Deschner (1984) mentions research by Harper (1979), Faulk (1974), and Oates (1979). Harper suggested that there were two basic types of batterers. The first are men who had been bullied or spoiled only sons who became intensely jealous partners or spouses. The second were batterers raised in subcultures approving of the use of violence. While these appear to have some truth in characterizing batterers, these appear grossly inadequate as stand-alone theories.

Faulk offers five types of batterers. These include dependent and passive, dependent and suspicious, violent and bullying, successful but domineering, and stable and normally affectionate men who went berserk during a depressive or psychotic episode. The primary problem with Faulk's work is that all of his subjects killed or injured their spouses, and those that killed their spouses would be expected to have different, perhaps more extreme typologies than the batterers who hit but do not kill their partners. Also, in approximately one-third of his sample, the spouses were dead and, therefore, could not provide data about the relationship or history of undocumented abuse. Lastly, his sample size was inadequate for generalization, consisting of only 25 offenders.

Oates is outlined by Deschner as having eight typologies, all pertaining to child abusers, some of whom also batter their partner. Deschner adapted Oates' work and developed an eight typology model. Her typology includes batterers and child abusers, with some of her categories questionably related to domestic abusers who have not abused their children. The majority of her categories related primarily to the batterers who abuse their children, not necessarily partners or spouses.

Walker (1995) indicated that there were three types of batterers that appear to be empirically supported.

1. Those who batter at home; they are motivated by abnormal power and control needs; they may cease their abusive behavior with some psychoeducation about anger management and gender-role attitude readjustment.
2. Those who also have serious psychological problems (including depression, dysfunctional thinking and obsessive-compulsive traits, paranoid disorders, borderline traits, and other serious mental disorders) and need individual and group therapy (Dutton & Starzomski, 1993; Hamberger & Hastings, 1986; O'Leary, 1993; Saunders, 1992a).
3. Those who have committed other crimes as well as assaults within the home, and who could be diagnosed as having antisocial personality disorders for which there is little effective treatment (Dutton & Starzomski, 1993; Fagan, Stewart, & Hansen, 1983).

Johnson and Ferraro (2000) and Johnson (1995, 2000a,b) identified four major patterns of partner violence.

1. *Common Couple Violence.* This involves a specific argument during which one or both partners lash out physically at each other. This does not occur at a high frequency and is not as likely to escalate over time, and involves mutual aggression.
2. *Intimate Terrorism.* This involves a general pattern of control. The violence is motivated by the desire to control one's partner. This generally occurs more frequently, escalates over time, and is less likely to be mutual. This type of partner abuse is likely to result in serious injury and emotional abuse.
3. *Violent Resistance.* This was the least defined by Johnson. This is perpetrated almost entirely by women and is one indicator that the woman is planning on leaving the abuser.
4. *Mutual Violent Control.* This involves a pattern in which both partners are controlling and violent, as if the two are intimate terrorists battling for control. This pattern is rare.

Neidig and Friedman (1984) adapted the work or Gelles and Straus (1979) and distinguished between expressive and instrumental violence. They defined expressive violence largely as a function of heightened emotional arousal and tends to occur in a gradually escalating manner with identifiable precipitants. There is often clear involvement of both partners and mutual and reciprocal violence reported by both partners. In addition, both partners generally express genuine remorse. Instrumental violence is perpetrated by men and is intended for the purpose of control and punishment. There are no clear precipitants and the escalation process tends to be rapid. The abuser shows no remorse.

Lawson (2003) examined the three types of physical abusers that have been empirically identified by Hamburger and Hastings (1986), Holtzworth-Monroe and Anglin (1991), Saunders (1992a), and Dutton (1998).

1. The first type of physical abuser has experienced the most severe childhood physical abuse and is engaging in the most severe battering. They are generally under controlled both inside and outside the family. They exhibit antisocial and narcissistic traits, such as generalized aggression, substance abuse problems, little if any remorse or empathy, extremely rigid views of gender roles, and they often have dismissing attachment styles.
2. The second type of physical abuser has experienced the most severe parental rejection as children. They are also under controlled, but with less instrumental intent (e.g., to control or punish). They are more impulsive in their use of violence and are emotionally volatile, exhibit the highest dependency needs, and follow Walker's (1979) three-stage abuse cycle of tension building, explosion, and honeymoon. They are often viewed as exhibiting preoccupied or ambivalent attachment styles. They most often have diagnoses of borderline personality and schizoid disorder.
3. The third type of physical abuser has typically experienced less childhood trauma than the previous two types. They tend to be over controlled in that they deny their anger and yet experience chronic frustration and resentment. They are passive dependent, compulsive, and emotionally suppressed and present with secure or preoccupied attachment styles. They are often diagnosed with either no personality disorder or avoidant, dependent, or passive–aggressive disorders.

TYPES OF VIOLENT WIFE BATTERERS

Gottman et al. (1995) reported that there were two types of highly violent wife batterers on the basis of heart rate reactivity. Type 1 men ("cobras") were those whose heart rate slowed during conflict and Type 2 men ("pit bulls") were those whose heart rate accelerated during the conflict. Their research found that the physical and physiological processes fit into two groups.

Type 1 Batterers

These men were more antisocial and outwardly calm. Heart rate decreases during conflict. These men are more likely to be verbally aggressive during a marital conflict and are more likely to show antisocial and aggressive personality characteristics. They also tend to be violent outside the home to a wide variety of people. They are more likely than Type 2 men to be assessed as antisocial, drug dependent, and aggressive-sadistic using the Millon Clinical Multiaxial Inventory III (MCMI-III) (Millon, 1983). These men were more likely to have witnessed interparental violence in the home, including physical violence. They were also more likely to have higher elevations than the Type 2 men on the Antisocial scale of the MCMI-III. Type 1 men were significantly more likely to have threatened their wives with a knife or gun than Type 2 men.

Type 1 men may be driven by their wives' reasonable demands for greater cooperation in the marriage, for intimacy, and for respect. They are not threatened by increasing independence moves by their wives and they may even compel their wives to have more independence. In response, the men begin to manipulate their own physiology and become calmer. They then begin to manipulate their victims by becoming threatening and belligerent. This behavior is effective in suppressing their wives' anger and in obtaining compliance.

These men tend to become belligerent within the first 5 minutes of an argument. They tend to be more domineering, show contempt, and defensiveness (with accompanying lowered heart rate). This is a highly threatening and terrifying pattern, and the typical response by wives is fear and submission. They tend to have a separation-divorce rate of zero compared to the Type 2 men who had close to a third of their marriages end within a 2-year follow-up. Overall, they tended to have marriages that appear quite stable. Women may remain with Type 1 batterers because they are genuinely afraid to leave because of the intimidation. A large percentage of women married to Type 1 men are themselves antisocial and may be more habituated or comfortable in a violent relationship than others would be (Waltz et al., 1995).

Type 1 batterers may not have impulse control problems. If anything, they may have the opposite problem — too much control.

Type 2 Batterers

Type 2 men are more emotionally needy and hot-tempered. Their heart rate increased during conflict. They were less likely to engage in violence outside the marriage. They had a lower Antisocial scale score on the MCMI-III than Type 1 men. They are threatened by greater independence on their wives' part and they fear abandonment, so they respond with jealousy, fear, and loss of emotional control as their heart rate rises. Eventually, Type 2 men lose control emotionally and become abusive, not in an attempt to suppress their wives' anger, but in an attempt to engage her and keep her in the interaction. These men are more likely to show dependent personality characteristics.

These men demonstrate a pattern of hostility described as a "slow burn," accompanied by an accelerated heart rate. This has been referred to as a power struggle (Rushe, 1995). These men view the marital interaction as a power struggle. Because of the lack of the fear response on their wives' part, their wives tended to match the men's escalating hostility during the conflict. Type 2 batterers also reported violence between their own parents. Their marital relationships appear to be quite unstable and the marriages are likely to end within 2 years.

The authors commented that criminals with antisocial personality disorder frequently are described as appearing calm, well controlled, and extremely dangerous. They often have lowered heart rates during arguments, a trait similar to those who commit a variety of crimes in addition to assault. The narcissistic and independence of these men may also contribute to their inability to judge the needs of their partners, and their continued contempt and belligerence seem to further lower the women's self-esteem.

Gottman et al. (1995) further found that women in relationships with Type 2 men, the men who overtly display more emotion (especially anger), are more likely to divorce than those married to Type 1 men — the cold, intimidating men who systematically control through bullying and cruel and antisocial behaviors.

It should be noted that despite differences found through research, there was no difference between Type 1 and Type 2 batterers in frequency of violence.

Support for the Work of Gottman and Colleagues

Jacobson et al. (1995) reanalyzed the data from Gottman et al. (1995). They found that Type 1 and Type 2 batterers appear to commit different types of abusive acts. Type 1 husbands commit the most serious, more severely violent types of offense, which include threatening their wives with a knife or gun, and are more likely to have used a knife or gun. Type 1 batterers were also more likely to have kicked, bitten, or hit their wives with a fist. More Type 2 batterers slapped their wives than Type 1 batterers. Type 1 men were found to be more emotionally abusive (e.g., belligerent and contemptuous) than Type 2 batterers.

Jacobson et al. also found that both Type 1 and Type 2 batterers reported much higher levels of parental violence than the general population. Many of the Type 2 batterers (approximately 25%)

reported unilateral father-to-mother violence compared with none of the Type 1 men. The vast majority of the Type 1 men reported bidirectional or unilateral mother-to-father violence, compared with about a third of the Type 2 men. Type 2 men were significantly more likely to have witnessed battering than Type 1 men, even though parental violence was generally less common in Type 2 batterers. In summary, parental violence was very high in both types in comparison to the general population. In support of Gottman et al., they found that the battering among Type 2 men has more to do with issues specifically reflecting attitudes toward women than battering by Type 1 men, and more likely to have had batterers as role models.This may support the findings that the attitudes toward women inherent in battering (and the emotional abuse that accompanies it) were passed on from the fathers to the sons.

Margolin et al. (1995) commented that there appears to be research support for the batterer typologies of Gottman et al. The criminal literature suggests that criminals, compared to noncriminals, may be under aroused, with lower resting heart rates (Raine, et al., 1990).

The work of Hare, Frazelle, and Cox (1978) demonstrated that psychopathic individuals showed increases in heart rate in anticipation of an aversive stimulus. This contradicts the findings of Gottman et al. that Type 1 batterers decrease their heart rate during stressful marital interaction. However, strict comparisons are difficult between the studies. Hare (1982) found that psychopathic individuals only screen out aversive events (with heart rate increases) when the situation forces them to attend to cues predicting punishment, but not when there are alternative coping strategies. These findings appear to support Gottman et al.

Margolin et al. (1995) suggest that Type 2 men who exhibit an acceleration of heart rate may be exhibiting a defensive reaction. They may become overwhelmed or occupied with their own discomfort and attempt to reject or withdraw from the situation. Margolin (1988) indicated that some abusive husbands attempt to withdraw from potentially conflictual situations. The Type 2 men described by Gottman et al. may be aware of how destructive conflict has been in the relationship and have an immediate inclination to avoid direct confrontation with their wives. Type 2 batterers tended to demonstrate less aggression and less anger in the encounters than Type 1 men.

Meehan and Holtzworth-Monroe (2001) did not find support for Gottman et al. (1995) Type 1 and Type 2 typologies. They found that heart rate reactivity was useful for differentiating batterers on anger–hostility dimensions, but not for differentiating psychopath or antisocial batterers. Their criticism of Gottman et al. appears somewhat unfounded to the degree that they argued that nonviolent men may also demonstrate Type 1 and Type 2 heart rate changes. However, as numerous research studies have effectively demonstrated, not all psychopaths engage in violence. And not all people who have the predisposition or other related factors for engaging in physically or sexually violent behavior do, in fact, engage in violent behavior. So, in short, their work appears to support Gottman et al., despite their criticism.

Holtzworth-Munroe and Stuart (1994) used a three-subtype model of batterers. This included the Family-Only (FO), the Borderline/Dysphoric (BD), and the Generally Violent/Antisocial (GVA).

The *Family-Only* batterer engages in the least marital violence. They exhibit the lowest levels of psychological and sexual violence and the least violence outside the home. These men exhibit little or no psychopathology.

The *Borderline/Dysphoric* batterers engage in moderate to severe spouse abuse, which includes psychological and sexual abuse. These men may also engage in extrafamilial violence on occasion and tend to be the most psychologically distressed and the more likely to evidence borderline personality disorder and schizoid and schizotypal characteristics. They were also more likely to be depressed, have dependency traits, and alcohol and drug abuse issues.

The *Generally Violent/Antisocial* batterers also engage in moderate to severe marital violence including psychological and sexual abuse. They can also engage in the highest levels of extrafamilial violence and evidence characteristics of antisocial personality disorder and narcissistic personality traits, including criminal behavior and substance abuse.

Holtzworth-Munroe and Stuart also identified several correlates of male violence as risk factors differentiating the types of batterers. These include historical correlates (e.g., violence in the family of origin, association with delinquent peers) and proximal correlates (e.g., attachment, dependency, jealousy, impulsivity, social skills, hostile attitudes toward women, and attitudes supportive of violence). They found that the FO batterers evidenced the lowest levels of risk factors. The violence for the FO batterers appeared to be the result of a combination of stress (personal or marital) and the low-level risk factors (e.g., childhood exposure to marital violence, poor relationship skills) such that during escalating marital conflicts these men would occasionally engage in physical aggression. After the violence incidents, their low level of psychopathology and related problems and low risk factors would generally lead to remorse that would help prevent their aggression from escalating.

In contrast, the BD batterers were hypothesized to have come from a background involving parental abuse and rejection. As a result, these men would likely have difficulty forming a stable, trusting relationship with an intimate partner. They would be jealous and highly dependent upon their partner. They would also experience fear of losing their wives. They would be impulsive, lack marital communication skills, and have attitudes that were hostile toward women and supportive of violence. This group resembles batterer studies by Dutton (1998) who suggested that early childhood lead to borderline personality traits and anger and insecure attachment, and that when frustrated, may result in violence against the adult attachment figure (e.g., the wife).

The GVA was predicted to resemble other antisocial, aggressive groups. They were expected to have high levels of family of origin violence and to associate with deviant peers. They are impulsive, lack social skills in marital and nonmarital situations. They also have hostile attitudes towards women and view violence as an acceptable means of resolving conflict. Their marital aggression was viewed as part of their general use of aggression and engagement in antisocial behavior.

Holtzworth-Munroe and colleagues (2003) gathered longitudinal data on batterers. Their results indicated that the Borderline/Dysphoric (BD) batterers and the Generally Violent/Antisocial (GVA) group of batterers demonstrated the highest levels of reported husband violence, and the GVA men were the least likely to cease their violent behavior. The Family-Only (FO) batterers engaged in relatively low levels of marital violence and were the most likely to cease their violent behavior. Their study found that overall, the wives of BD and GAV batterers suffered the most injury from husband violence, and the GVA, and a fourth group, the Low-Level Antisocial (LLA) men, which fell between the FO and GVA group of batterers, all engaged in more sexual and/or psychological abuse than the FO men.

Holtzwoth-Munroe et al. also indicated that popular belief was that violence occurring within a relationship would escalate over time (e.g., Pagelow, 1981; Walker, 1979). However, research has shown (e.g., Holtzworth-Munroe et al., 2003; Aldarando, 1996; Feld & Straus, 1989; Jacobson et al., 1996) that not all men become more severely violent over time. However, those researchers found that the men who are the most severely violent initially are the most likely to continue their violence over time.

Holtzworth-Munroe et al. also commented that popular belief was that the FO batterer would also progress to being more severely violent over time (e.g., Gondolf, 1988; Saunders, 1992a). In contrast, Holtzworth-Munroe et al. demonstrated that the FO group of batterers may be a stable one, that is, they may continue to have a low risk of marital violence over time, assuming that other risk factors consistent with the FO type of batterers remain and they do not develop more severe pathology. That is to say, FO batterers who demonstrate GVA risk factors (e.g., criminal behavior, delinquent friends, substance abuse, impulsivity, attitudes) are at high risk to escalate their marital aggression over time.

Holtzworth-Munroe and colleagues also found that the GVA men were the most likely to spend time in jail during the course of their study follow-up period. They also found that the BD men scored highest on measures of jealousy and spouse-specific dependency; they were also the most likely to have been treated for depressive symptoms over the follow-up period. The authors were unable to support that there is conclusive evidence to support that there are distinct differences between the BD and GVA subtypes, at least not enough characteristics to differentiate. The BD and GVA were the most likely to have unstable marital relationships and to experience marital dissatisfaction. The GVA men were also more likely to have ended their relationships at a 2-year follow-up.

Waltz et al. (2000) tested the Holtzworth-Munroe and Stuart (1994) findings. In summary, they found support for Holtzworth-Munroe and Stuart's model, with exception that personality disorder characteristics did not appear to distinguish between the generally violent and pathological groups. The current findings, as well as those from Tweed and Dutton (1998) and Hamberger et al. (1996), did not find clear distinctions between the most severely violent and moderately violent men. Tweed and Dutton did not find any difference between their generally violent/antisocial type (labeled *instrumental*) and the dysphoric/borderline type (labeled *impulsive*) batterers on the MCMI-III. The impulsive batterers reported higher levels of borderline personality characteristics, although the impulsive and instrumental batterers did not differ on antisocial characteristics.

Chase et al. (2001) defined two types of batterers — reactive and proactive. Reactive violence responds to (accurately or inaccurately) perceived threats or frustrations in the context of high affective-physiological arousal and minimal cognitive processing. Proactive violence is defined as planned, methodical, and goal-orientated behavior that is acted out in a context of minimal-to-decreasing emotional and physiological arousal.

They cited Tweed and Dutton (1998) and Gottman et al. (1995) as having mutually supported a two-type typology, the Type 1 (impulsive) and Type 2 (instrumental). Those exhibiting violence of a reactive nature were impulsive and Type 2. Those men exhibiting violence of a proactive nature were instrumental and Type 1. Chase et al. (2001) found that the impulsive men exhibited elevations on personality traits known to be associated with increased reactivity to perceived partner rejection or betrayal (e.g., borderline personality disorder/traits, as well as a fearful attachment style) and high chronic anger.

In contrast, the instrumental men had personality profiles consistent with a readiness to intentionally and purposively use aggression for interpersonal means (e.g., antisociality, narcissism, and aggression; Tweed & Dutton, 1998). They also found that reactive aggressors exhibit more angry outbursts in potentially challenging interpersonal contexts and had personalities that are more prone to reactivity than proactive aggressors (Scarpa & Raine, 1997). For family of origin violence, Dodge (1991) theorized and found that reactive aggressors experienced more physical abuse from parents and witnessed less interparental violence than did proactive aggressors. The primary theory is that parental physical maltreatment promotes those things that underlie reactive violence (e.g., hypervigilance, fear, and rage reactions), whereas exposure to aggressive role models (like witnessing interparental violence) leads to proactive violence. This may be due in part to learning that violent acts lead to positive outcomes.

Chase et al. (2001) found that reactively aggressive batterers would exhibit more anger and less domineering-controlling behavior while interacting with their partners; and they would exhibit more personality features and disorders involving impulsive reactiveness to partner behaviors (e.g., dependent and borderline) vs. a readiness to resolve interpersonal ends through the deliberate use of aggression (e.g., antisocial, aggressive–sadistic and psychopathic). They were found to be significantly higher on anger and significantly lower on dominance than were proactive. They were unable to support the fact that reactive batterers would report having experienced more parental abuse and less witnessing of interparental violence as a child. Reactive batterers were more likely to be angry during a conflict with their spouse and to have a dependant personality disorder.

Magdol et al. (1998) found that the antecedents of abuse include risk factors from all four identified domains. Risk factors for partner violence resemble those for stranger violence (Fagan & Browne, 1994). The results of their study were that childhood and adolescent characteristics in the four domains of family relations, family socioeconomic resources, educational achievements, and problem behaviors can partially predict who is at risk for subsequent partner abuse in early adulthood. Early family relations are associated with partner abuse. Close parent–child attachment at age 15 was consistently associated with low risk for subsequent partner abuse. They further found that young persons who have experienced warmth, trust, and open communication were likely to bring these qualities to their adult personal relationships.

Parents with higher socioeconomic status and growing up with both parents present were related to low risk for subsequent partner abuse. Related to early educational achievements, they found that leaving school early was a consistent predictor of subsequent partner abuse. In addition, they found that early problem behaviors were associated with partner abuse. These included parent reports of conduct problems at age 15, self-reports of aggressive delinquency at age 15, and subsequent substances abuse at age 15. The authors found that young people who have a lengthy history of solving interpersonal problems in a coercive manner are likely to use similar tactics in their primary adult relationships. They also mentioned the correlation between partner abuse and earlier substance abuse is consistent with other research showing that early-onset substance abuse is closely related with aggressive behavior (Zucker, Fitzgerald, & Moses, 1975).

Hazelwood's (1995) sex offender typologies appear to correlate with batterers as well. Following is an adaptation of Hazelwood's (1995) sex offender typologies as they apply to batterers.

Interestingly, I find that all of the abusers I have assessed forensically or clinically fit into the typologies identified for sexual offenders. I have long suggested that physical abusers and sexual offenders share far more characteristics than differences. In fact, many abusers rape or sexually abuse their partners in some way as part of the ongoing abuse or the honeymoon/make-up period following the abusive incident. And the majority of sexual offenders used some form of physical assault against their victims and, at the very least, utilized psychological coercion to gain compliance of their victims.

Consider for example verbal abuse. The vast majority of verbal abuse consists of sexually derogatory words (e.g., fuck, fucker, bitch, whore, cunt). And regardless of the context in which the verbal abuse occurs, everyone is aware of the meanings of the sexually charged words and phrases. So even verbal abusers are sexually abusive to some degree.

I believe that Hazelwood offers a rich typology for batterers. However, it should be clear that Hazelwood has not, to the best of my knowledge, applied these typologies to batterers, but rather it is my endeavor to do so. Hazelwood's typologies were adapted from the work of Groth (1977) and appear to incorporate Prentky and Knight (1989) typologies as well. With this said, I propose the following as a typology of abusers who batter their partners, spouse, or even their children. This is my proposed adapted typology of abusers adapted from Hazelwood (1995).

The Power Reassurance Abuser

- Purpose of attack is to reassure himself of his masculinity by exercising power over women or his partner.
- Lacks confidence in his ability to interact socially and sexually with women.
- Utilizes fantasies and sexual activity to prove himself.
- Does not intentionally degrade or emotionally traumatize the victim.
- Often described as the "gentleman abuser," "polite," or "apologist."
- If he finds a passive victim, may take more time to act out fantasies of power and control.
- May apologize to victim, ask for forgiveness after abusing.
- May recontact victim by phone or in writing.

- If unsuccessful at assaulting his partner, he may quickly attempt to assault someone else, including children, pets, objects, or another person.

The Power Assertive Abuser

- Purpose of attack is to exercise his prerogative as a man to control, dominate, abuse, or rape.
- Has no conscious doubts about his own masculinity, outwardly a "man's man."
- Demonstrates no concern for welfare or comfort of victim.
- Utilizes moderate to excessive force.
- May subject victim to repeated physical and/or sexual assaults.

The Anger Retaliatory Abuser

- Purpose of assault or rape is to get even with women for real or imagined wrongs.
- Pairs anger with retaliation.
- Angry with women and uses abuse, humiliation, and sex to punish or degrade them with intention.
- Utilizes excessive force, appears as in a frenzy.
- The attack is an emotional outburst, little planning, impulsive.
- Selects victims who remind him of someone.
- Utilizes weapons of opportunity (fists, legs), attacks are sporadic.
- Once vented, he is relieved and his tension gone.

The Anger Excitatory Abuser

- Purpose of attack is to obtain emotional or ego gratification via victim's pain and suffering.
- Inflicts pain that will result in desired response of fear and total submission.
- Vicious attacks that result in physical and emotional harm.
- Attacks are well planned and methodical, rehearsed in fantasy.
- Utilizes weapons and objects.
- May torture victim with instruments as well as psychologically reducing her to extreme depths of fear.
- Helplessness of the victim is stimulating.
- Keeps victim for extended periods of time.
- The abuser may practice bondage.
- If the abuser sexually assaults the victim, he may insert objects into the victim's vagina or anus.
- If sexual acts are forced on the victim, the acts will be varied and experimental in nature, intended to cause pain, humiliation, and degradation for the victim.
- May keep records and/or record the abuse or rape (though there is no empirical evidence to support this for abusers).
- The rapist may not directly sexually assault the victim, remaining emotionally detached.

The Opportunistic Abuser

- The primary motivation is violent in nature.
- Typically abuses or rapes as an afterthought during commission of another crime (e.g., the recent hurricane on the Gulf Coast in 2005; many people resorted to breaking and entering, theft, assault, and even rape; there was little chance of being caught and the opportunity was there).
- Utilizes minimal to high degree of force.
- Usually consumes drugs or alcohol before or during crime.

The Gang Abuser

- A group leader as well as a reluctant person.
- Utilizes minimal to brutal degrees of force.
- Again, the hurricane on the Gulf Coast, where multiple persons committed crimes; it is unclear whether they would have committed the crimes if not in the others' company.

The adapted typology above is for addressing abusers. These typologies appear to fit most all abusers. However, some abusers may fit into more than one category either because they are versatile offenders or they are evolving into a more serious type of offender.

The major limitation on my proposed adaptation of Hazelwood's typologies is that there has not been any empirical study to demonstrate that the typologies represent most domestic abusers. From a purely anecdotal view, and from the literature, my proposed typologies appear accurate and appropriate for use to categorize abusers. I would be open to anyone who would gather information on how their abuser populations fit into the proposed typologies.

SUMMARY

It appears that Gottman et al. (1995) and Holtzworth-Munroe and Stuart (1994, 2003) provide the most useful batterer typologies. Hazelwood's (1995) sex offender typology also offers a rich typology for batterers. The others certainly add more detail to the abuser's personality characteristics, but do not stand alone as typologies.

Batterers and sex offenders should be typologized whenever forensically assessed or are involved in a treatment program. Remember that offenders offend for somewhat different reasons and different type of offenders reoffend with different degrees of violence and different risk for reoffense rates. Batterers should not be considered a homogeneous group and neither are sex offenders a homogeneous group.

16 Female Batterers

Female batterers are perhaps less understood than male batterers. This is a result of the fact that female offenders are less likely to be arrested and even if arrested, are less frequently required to attend a treatment program. Female offenders are just as likely to initiate and engage in abusive behavior as male offenders, but far less likely to cause significant injury. It is imperative to understand the characteristics of the female batterer to effectively assess and treat them. Significantly more women are being arrested for domestic abuse mostly due to mandated arrest laws. Many studies have found that women self-report as much perpetration of physical aggression as men, though women do not report sexually abusing, engaging in coercive control, stalking, or causing injuries (e.g., Straus, 1999; White et al., 2000).

Hamberger (2005) found that women initiate violence as often as men. However, they also found that women were disproportionately victimized by partner violence compared to men. A meta-analysis of over 80 studies found that women were slightly more likely than men to use partner violence (Archer, 2000). Straus (1999) found that men and women perpetrate intimate partner abuse about equally. However, men inflict far more injury (Archer, 2000). Few female-to-male assaults come to the attention of the courts, so many surveys and research underestimate the occurrence of female-to-male partner aggression. A number of studies have suggested that women's aggression is primarily self-defensive or reactive in nature (Berk et al., 1983; Saunders, 1986). However, the motivations of women arrested for intimate partner violence in some cases involve the desire to control or cause harm, especially among those women with a prior history of violence (Babcock, Miller, & Siard, 2003; Hamberger et al., 1997). Babcock et al. (2005) found that women were more likely than men to admit to their own perpetration of abuse.

DIFFERENCES BETWEEN FEMALE AND MALE BATTERERS

Let's examine some of the differences between female and male batterers.

Schroffel (2004) found that female batterers often engaged in compulsive and premature disclosure in groups more than did male batterers, and that male batterers were more likely to make minimal disclosures. Female batterers were more likely to portray themselves as perpetrators and were more ambivalent to perceive themselves as victims, compared to male perpetrators, who were more likely to perceive themselves as either victim or perpetrator. The women appeared to have difficulty confronting the reality that they were, in fact, victims. The study also found that the women were more likely to exhibit a devaluation of self vs. the males demonstrating a devaluation of their partner rather than of themselves. Of interest, all of the women reported that they had experienced verbal and physical abuse initiated by their partners. These women were hesitant to leave these violent relationships, and often professed love for their partners who controlled and abused them. They also portrayed themselves as to blame for not meeting their partners' needs and expectations, possibly therefore, blaming themselves for being abused.

Schroffel also found that 25% of the women and 8% of the men had been raped and/or physically abused by a current or former spouse or partner at some time in their lives. In addition, 1.5% of the women, compared to 0.9% of the men, reported that they had been raped and/or physically assaulted by a partner within the past 12 months. One and a half million women and 834,700 men are raped and/or physically assaulted each year by an intimate partner. The women who were raped or assaulted by a current or former intimate partner were more likely to sustain injuries than were the men, in addition to experiencing more emotional and psychological effects (e.g., depression, anxiety, and fear) (Schwartz, 1987; Campbell, 1998; Dobash & Dobash, 2001). Additional research has shown that women are 6 to 10 times more likely than men to sustain serious injuries as a consequence of violent acts by their partner (Straus, 1993; Gelles, 1997).

Dobash and Dobash (2004) indicated that when violence is considered within the context of an intimate relationship, evidence suggests that men's physical and sexual violence against women is often associated with a constellation of abuse that includes a variety of additional intimidating, aggressive, and controlling acts (Pence & Paymar, 1993; Dobash et al., 2000; Gondolf, 2002). Women's violence (lethal and nonlethal) is often associated with self-defense and/or retaliation against a male partner. Dobash and Dobash (2004) concluded that women's violence differs from that perpetrated by men in terms of nature, frequency, intention, intensity, physical injury, and emotional harm. All of the women in the study had been victims of repeated physical violence from their male partners. Just over half of these women had used some form of violence against their abuser. They also found that men who were victims of their women partner's violence reported the abuse as inconsequential and they did not fear for their own safety. Both men and women concur that women use violence infrequently and that the violence from women rarely, if ever, results in serious injuries. This finding may explain why men are less likely to report having been abused by their female partner as well as why the legal system is more lenient on women batterers.

Dobesh et al. (1992) found that compared to men, women are far more likely to kill only after a period of prolonged abuse, when they feared for their lives and exhausted all alternatives. In addition, women rarely present with patterns of wife-killing men, which includes murder–suicides, family massacres, stalking, and murder in response to spouse infidelity. Dobesh et al. (1992) also found that females rarely kill or assault strangers or acquaintances; instead, the female's victim tends to be a male intimate or child, the offense usually takes place in the home, the victim is frequently intoxicated, and self-defense or extreme depression is often a motive. For women to kill, they generally must see their situation as life threatening, as affecting the physical or emotional well being of themselves or their children.

Swan et al. (2005) found that most all of the women batterers in their study experienced violence from their partners. They found that greater frequency of victimization from partners and experiences of childhood abuse increased the likelihood that women would use aggression against their partners. Victimization from partners and childhood abuse also increased the likelihood that women would experience symptoms of posttraumatic stress and depression. Women with more symptoms of posttraumatic stress were also more likely to express anger outwardly towards others. Expressing anger outwardly towards others, in turn, predicted an increased likelihood of using aggression against partners. Women who use violence against their male partners are usually violent in context of violence against them by their partners (Dasgupta, 1999; Hamberger & Guse, 2002; Hamberger & Potente, 1994; Magdol et al., 1997; Swan & Snow, 2002). The more a woman experiences aggression against her by her male partner, the more likely she is to use aggression towards him.

Swan et al. (2005) also found that all women in their sample who had become batterers were victims of trauma either from their current partners or as children. Women who experienced more depressive symptoms were more likely to suppress anger (Kellner et al., 1992; Riley et al., 1989). While neither current victimization or childhood abuse was directly related to anger, these symptoms predicted greater levels of depression, which in turn strongly predicted anger. Women who experienced more symptoms of post-traumatic stress were more likely to express anger outwardly and to experience a lack of control in their expression of anger (Ehlers, Mayou, & Bryant, 1998; Feeny, Zoellner, & Foa, 2000; Mayou, Ehlers, & Bryant, 2002; Riggs et al., 1992). Symptoms of post-traumatic stress and depression did not affect women's aggressive behavior directly, but rather indirectly. Swan et al. (2005) also found that being abused by partners predicted the women's aggressive behavior towards their partners, which is consistent with studies that have found a history of high rates of victimization among aggressive women (Dasgupta, 1999; Hamberger & Guse, 2002; Hamberger & Potente, 1994; Magdol et al., 1997; Swan & Snow, 2002).

Of interest, Schroffel (2004) found that all of the women reported that they had experienced verbal and physical abuse initiated by their partners. These women were hesitant to leave these violent relationships and often professed love for their partners who controlled and abused them. They also portrayed themselves as to blame for not meeting their partners' needs and expectations, possibly, therefore, blaming themselves for being abused.

Hendricks-Mathews (1982) reported that therapy was less successful for women who self-blamed than for those who blamed their partners. Women may lean towards self-blame as a coping strategy to protect their illusion of personal control and believe that they got what they deserved (Miller & Porter, 1983). Schroffel (2004) found that women batterers who were affluent and privileged were more ready to move on from a dependent to a more independent lifestyle. Women of poorer social economic status appeared much more convinced that they could not change their lives or their behavior because of the limitations and problems inherent in surviving in a life of poverty.

Hamberger (1997) found that in his study, women responded approximately 75% of the time that they had initiated a physical assault in their relationship. Also in the study, 77% indicated that their partner had also initiated physical assaults, and 25% of the women indicated that they initiated the physical violence 100% of the time, while 37% indicated that their partners initiated the violence 100% of the time. Moreover, over half of the women indicated that their male partner initiated violence more than half of the time.

Hamberger and Guse (2002) found that 95% of the men reported that their partner had ever struck them first, compared to 91% of women in shelters and 87% of "court-ordered" women (women ordered by the court for treatment as batterers). Men reported that their partners had initiated violence in 50% of the incidents. In summary, they found that women reported that they were more likely to report that the man was the first to strike, and most of the men indicated that their partner initiated the violence 50% of the time.

Phelan et al. (2002) found that men reported significantly higher rates of violence initiation than did women. In the study, 100% of the men reported initiating violence between 50% and 100% of the time. In contrast, 91% of the women reported initiating the violence between 0 and 20% of the time.

Numerous studies have indicated that women were as frequently aggressive toward their partners as men (Straus & Gelles, 1990a; Straus, Gelles, & Steinmetz, 1980; Sugarman & Hotaling, 1998). Straus and Sweet (1992) found that women engaged in just as much verbal and emotional aggression (e.g., name-calling, insulting, sulking, slamming doors, throwing things) toward their partner as men. However, women sustained more serious injuries than men (Cascardi, Langhein-richsen, & Vivian, 1992; Harned, 2001; Stets & Straus, 1990).

BACKGROUNDS OF FEMALE BATTERERS

Dowd, Leisring, and Rosenbaum (2005) found that women batterers in their study tended to be socioeconomically disadvantaged and under-educated, presented with histories of childhood attachment disruptions and victimization, presented more mental health problems and substance abuse. Another major characteristic of women was a disruption of social supports in childhood and adulthood. The majority of the women were raised by a single parent, with a sizable minority having been placed outside of the home as children. In addition, they found that these women had parents with high rates of substance abuse, interparental violence, and childhood physical and sexual victimization. Many of these women grew up in unstable and unsafe social environments, presumably without role models to teach them about appropriate boundary setting, conflict resolution skills, and effective coping strategies. Many had learning, behavioral, and substance abuse problems in school, with an outcome of school dropout. In contrast, the women batterers in Hamberger's (1997) study included women that were more educated, more of the women were married, and a larger percentage employed. They reported lower rates of childhood sexual abuse and witnessing interparental violence, but comparable histories of physical abuse. His sample also reported less rates of alcohol abuse.

Another theme that emerged from Dowd et al. (2005) was that of the women's parenting status. Approximately 75% of the women reported having children, many beginning parenting at an early age without substantial support from a stable partner or extended family. Over a third of these women lost custody of their children either temporarily or permanently due to their own abusiveness or inability to nurture and protect their children. The majority of these mothers were single with numerous partners. This certainly suggests how the vulnerabilities associated with one family may be transmitted from one generation to the next (Stith et al., 2000).

Abel (1999) found that women batterers in treatment looked quite similar to victims of domestic violence, with respect to victimization histories and trauma symptoms. Similar findings occurred with the study by Leisring, Dowd, and Rosenbaum (2000). Female perpetrators reported high frequencies of physical and sexual abuse in childhood and physical abuse by their partners in the past year. In addition, levels of posttraumatic stress symptoms were significantly correlated with levels of the women's own victimization as well as their perpetration of psychological and physical aggression. Dowd (2001) also suggested that victimization, substance abuse, and attachment problems can be highly interrelated factors in the etiology and maintenance of female partner aggression.

Dowd et al. (2005) found that the women batterers in their study tended to be of lower socioeconomic status, many having dropped out of school before completing high school, experienced parenthood at a young age, and had few job skills. The majority of women in their study were unemployed, and those that were employed worked primarily at jobs with low wages and few opportunities for advancement. Straus (1990b) found that current economic status does appear to be a factor in higher rates of aggression, regardless of education level. Police officers may be more willing to arrest women of lower socioeconomic status than women of higher socioeconomic status and women of lower socioeconomic status may also receive less favorable outcomes in court.

The management of anger and aggression is likely to be more difficult for women with unstable moods, trauma symptomology, or the effects of chronic substance abuse. The women in Dowd's study (2005) reported high rates of depression and bipolar disorder, past substance abuse, and trauma experiences, far greater than found in the general population (Kessler et al., 1994; Straus & Gelles, 1990b). Straus (1990b) noted that under low stress conditions, women were only half as likely to be aggressive to their partners as men. Both genders were more aggressive as stress levels increased, but under high-stress conditions, women surpassed men in their aggression by 50 to 150%. Rosenbaum et al. (2001) found that women batterers were more likely to complete treatment when they were mandated and supervised by the court than when attending treatment on a voluntary basis. Women are more likely to drop out of treatment without being mandated and supervised to completion.

PERSONALITY DISORDERS AMONG WOMEN BATTERERS

Borderline Personality Disorder

Borderline Personality Disorder is often associated with a history of trauma (Zanarini et al., 1997) and relationships that are characterized by instability and intensity (American Psychiatric Association, 2000). An understanding of female interpersonal aggression would be enhanced by the clarification of the connections among variables such as victimization, early and later attachment difficulties, low capacity for empathy, and deficits in character structure (Dowd et al., 2005). Thinking errors typical of Axis II disorders would suggest the need for treatment for aggression to be structured around a framework that draws heavily from cognitive-behavioral approaches to self-regulation (e.g., Linehan's Dialectical Behavior Therapy) (Linehan, 1993). Attention to Axis II disorders may also help to differentiate between treatment-resistant women who drop out and those who do benefit from the treatment.

Posttraumatic Stress Disorder

High rates of PTSD were found among battered women and among women batterers (Saunders, 1994; Leisring et al., 2000), which is nearly double of that of male batterers. They found that women batterers presented with higher rates of abuse victimization in their histories. Among arrested women, however, Henning and Feder (in press) found a much lower prevalence of PTSD, and theorized that women batterers may show fewer trauma symptoms than battered women.

Women who use violence against intimate partners typically have symptoms of posttraumatic stress and depression (Swan & Snow, 2003). Factors that predict posttraumatic stress and depression (e.g., childhood abuse) are common among women who use violence (Hamberger & Potente, 1994; Langhinrichsen-Rohling, Neidgi, & Thorn, 1995; Leisring, Dowd, & Rosenbaum, 1999; Magdol et al., 1998; Siegel, 2000). A large number of women who are victims of relationship violence develop posttraumatic stress and depression (Astin, Lawrence, & Foy, 1993; Finklhor & Yllo, 1985; Foa et al., 2000; Laffaye, Kennedy, & Stein, 2003).

Golding (1999) found that results from a meta-analysis indicated that approximately 50% of battered women had depression and 40% developed posttraumatic stress disorder (Dansky, Byrne, & Brady, 1999; Kocot & Goodman, 2003). In addition, high correlations have been found between stuffering anger and depression (Kellner, Hernandez, & Pathak, 1992; Riley, Treiber, & Woods, 1989; Spielberger, 1999). Childhood experiences of abuse are also related to adult anger and depression. One of the most pervasive consequences of sexual abuse is anger (Briere, 1992; Courtois, 1988; Donaldson & Gardner, 1985). Numerous studies have found that childhood experiences of physical, sexual, and psychological abuse are related to adult depression and posttraumatic stress symptoms (Beitchman et al., 1991; Beitchman et al., 1992; Briere & Runtz, 1990; Malinosky-Rummell & Hanson, 1993; Windom, 1999). Childhood experiences have also been found to be related to women's use of aggression. Childhood abuse has been correlated with women's use of both physical and psychological aggression in their intimate relationships (Hamberger & Potente, 1994; Leisring et al., 1999; Magdol et al., 1998). Other studies have found that childhood abuse predicts women's violence with intimate partners (Langhinrichsen-Rohling et al., 1995).

TREATMENT FOR WOMEN BATTERERS

Hamberger (1997) indicated that it is important to address power and control issues related to the women's use of violence. Treatment should also address women's social conditioning to accept blame for problems in the relationship, including their partners' violence and their responsibility to fix such problems and to keep the relationship intact (Dutton, 1992; Goodman & Fallon, 1995). It is also important to differentiate women arrested for battering who may be the victims of abuse from their alleged victims. Women who fight back to defend themselves or their children may require nontreatment interventions (e.g., Hamberger & Potente, 1994).

Hendricks-Mathews (1982) reported that therapy was less successful for women who self-blamed than for those who blamed their partners. Women may lean toward self-blame as a coping strategy to protect their illusion of personal control and believe that they got what they deserved (Miller and Porter, 1983). Women that were court-ordered into treatment completed treatment at a significantly higher rate than those who were not court-ordered (Dowd, Leisring, and Rosenbaum (2005). Noncourt-ordered women batterers tended to have fewer pretreatment arrests and were more likely to drop out after the initial intake.

Rosenbaum et al. (2001) found that women batterers were more likely to complete treatment when they were mandated and supervised by the court than when attending treatment on a voluntary basis. Women are more likely to drop out of treatment without being mandated and supervised to complete.

17 Typology of Sex Offenders

INTRODUCTION

This represents a guide to the perpetrator's patterns of offending behavior. The following offender typologies are offered as a guide to understand the offense and offender's dynamics. Sex offenders may fit into one of two groups — those that commit their offense impulsively and those that commit premeditated offenses. Impulsive sex offenses occur with little or no planning, and usually because the situation presents itself for sexual violence. An impulsive thought about raping may occur at any time. For example, offenders may see someone they find attractive or someone that they believe would be a good victim and they offend. An opportunistic offender may have broken into a home to commit a burglary, find a woman sleeping, and decide at that time to rape. This is an opportunistic rape if the offender did not enter the house with the knowledge that someone was home and without the idea that he would rape if a woman were home. Premeditated offenses are planned, thought through, and often fantasized about long before they occur.

Both the opportunistic and predatory sex offenders are equally dangerous, and both may use violence ranging from minor to lethal in severity. The opportunistic and impulsive sex offender may rape in a frenzy of violence while the predatory engages in fantasized, preplanned, and methodical behavior.

Obviously treatment implications are significantly impacted by the typology of the offender. In summary, the opportunistic and impulsive offender benefits most from learning new coping and problem solving skills, stress management skills, and improved social relationships. He also tends to have personality disorders and other mental diagnoses that benefit from psychopharmacologic intervention. The predatory offender is much more difficult to address because his offense behavior is often obsessive and at times compulsive. These offenders usually require a psychopharmacologic intervention, and require a lot of work to understand their cues and cycles, understanding the meaning of their offense fantasies and behavior, and also require empathy training. However, from my own experience, the predatory offender has the poorest prognosis and he tends to present with more psychopathic and antisocial traits, though in a more refined way. I may have oversimplified the treatment implications, but wanted to briefly introduce how the typologies are related to treatment intervention.

Sex offenders select their victim based on their deviant need. Opportunistic offenders may rape the first person they see, male or female, young or old. On the other hand, they may have preferences that limit their victim selection. Predatory offenders have specific victim characteristics that they seek and they will not often settle for less than their preferred victim characteristics. So if you were to be developing a probation agreement or treatment plan, you would need to understand the offender's typology to fully understand how to best protect the public. Offenders may have chosen a teenage girl from the local mall for their current offense, yet they may select their next victim from church. So, the victim selection involved in the current offense may not offer a true picture of the offenders' potential victim pool. Understanding the typology offers more effective intervention for protecting the public and decreasing sexually violent reoffense.

Following are typologies of Prentky and Knight (1990), Groth (1979), Lanning (1995), and Hazelwood (1995). For further understanding of these typologies, please refer back to the referenced authors. While I have attempted to offer a summative outline of typologies, I believe that reading the following authors' works is imperative to understanding the sexual offender and, therefore, required before assessing, profiling, or treating sexual offenders.

PRENTKY AND KNIGHT TYPOLOGIES*

The Opportunistic Rapist:

- The assault tends to be an impulsive, predatory act.
- The sexual assault represents one among many instances of poor impulse control. Uses only as much force needed to accomplish the assault; no evidence of the use of gratuitous force.
- Seeks immediate sexual gratification.
- Indifferent to the welfare and comfort of victim.
- Generally has an extensive history of undersocialized behavior in many areas of his life.
- Exhibits little anger except in response to victim resistance.
- When he knows his victim, he tends to gratify his immediate needs with little or no concern for how it may impact the victim.

Pervasively Angry Rapist:

- Primary motivation is undifferentiated anger.
- Indiscriminately angry with everyone.
- Aggression is excessive and gratuitous and occurs even if victim does not resist. He often inflicts serious injury on his victims.
- Often inflicts serious physical injury, including death.
- Rage does not appear to be sexualized (e.g., the violence used against the victim is not limited to sexual areas of the victim's body; may include general assault to the face, legs, or other areas of the body not related to the sexual areas).
- No evidence that assaults are driven by pre-existing fantasies.
- Anger not limited to women. He tends to have an extreme problem controlling aggression, which may be evident in many areas of his life.
- From childhood to adulthood, his history is often plagued with difficulties controlling his impulses and anger.

Sexual Sadistic Rapist:

- Enduring, persistent sexual/sadistic fantasies or preoccupations (influence the way in which assaults are executed).
- Poor differentiation between sexual drive and aggression.

  Overt Type:
  - Aggression is manifested in physical damage to victim.
  - Low social competence.
  - Anger, belligerent people who appear similar to pervasively angry rapists except that their assaults involve greater planning.

  Muted Type:
  - Aggression expressed symbolically or through covert fantasy that is not acted out behaviorally.
  - High social competence.

* From Prentky and Knight (1990). Reproduced with permission.

Nonsadistic Sexual Rapist:

- Manifests less interpersonal aggression in sexual and nonsexual contents than other types.
- If victim resists, may flee rather than fight.
- Assaults reflect combination of sexual arousal, distorted "male" cognitions about women and sex, and feelings of inadequacy about his sexuality and masculinity.
- Preoccupation with sexual difficulties; rapes represent attempt to establish sexual relationship he desires but is unable to attain.
- Socially isolated, inadequate.

Vindictive Rapist:

- Women are central and exclusive focus of his anger.
- Assaults are marked by behaviors that are physically harming and appear to degrade/humiliate. Behavior ranges from verbal assault to brutal murder.
- Aggression is not eroticized.
- Lack of preoccupation.
- Lower level of overall impulsivity than pervasively angry or overt sadistic rapist.
- Aggression does not appear to be eroticized.
- Show little or no evidence of undifferentiated anger.
- No preoccupation with sadistic fantasies.

Groth and Groth et al. Typologies*

Rape (Coercion):

1. Approach is one of attack or assault; offender gains access to victim through implied or expressed threat to the physical safety of the victim — verbal threat, intimidation with a weapon, and/or physical force; offender may use a position of authority to intimidate the child.
2. Aggression in the form of power and hostility are major psychological dynamics.
3. Child is object of hostility or domination on part of offender; the child is seen as "weak" and "helpless."
4. More typically a one-time offense with a series of different victims, less likely to be ongoing victimization of the same child unless the perpetrator occupies a role of authority in the life of the victim.
5. Victim is depersonalized by the offender or cast into a negative symbolic role.
6. Offender immediately subjects child to sexual penetration and/or forces child to perform overt sexual acts/rituals.
7. Usually no interest on offender's part in having victim enjoy the sexual acts; self-gratification is primary concern.
8. Sexual abuse of the child.

Anger Rapist:

- Means of expressing and releasing anger and rage.
- Rape is one of retaliatory aggression for perceived wrongs, injuries, or put-downs; child victim may be targeted as a way of "getting even" with an adult whose child is related, or as a way of "teaching a lesson to" (punishing) the child.

* From Groth (1979/1981) and Groth et al. (1978). Reproduced with permission.

- Victim is battered and suffers physical trauma to all areas of body.
- Victim is attacked, physically assaulted.
- May use more physical force than is required to overpower the victim.
- Tends to be impulsive and spontaneous rather than planned; he is episodic.
- Assault is of relatively short duration.
- No use of weapon; if weapon is used, weapon of opportunity and used to hurt, not to threaten the victim.
- May approach victim and appear gentle and friendly and then attack.
- Victim may be caught totally off guard.
- Language is abusive; may call victim derogatory names.
- May make victim perform degrading sex acts.
- May masturbate on victim and ejaculate on her
- May only attain erection by masturbating himself or having her perform fellatio.
- Rapist views sex as dirty.
- Rapist's mood is anger and depressed.
- Victim may be actual person he is angry with or may be symbolic.
- Most attacks are short in duration.
- For child victims, this is more common with incest offenders.
- Usually selects same age or older adult victim, or if a child victim, the child is sexually abused in context of physical assault.
- Prior criminal history includes crimes of aggression (reckless driving, assault, disorderly conduct, breach of peace, etc.).

Goals of a power rapist: to debase and hurt his victim.

Triggers/precipitating events: tends to be upset and distressed, angry, frustrated, and depressed; trigger event results in above feelings; experiences feeling of being wronged, hurt, put down.

Power Rapist:

- Power is primary motive.
- Wants to possess victim.
- Sexuality used to compensate for underlying feelings of inadequacy; wants to feel powerful and denies deep-seated feelings of insecurity and inadequacy to show "who is in control."
- Rape is a sexual conquest.
- Uses physical force to get victim to submit; uses whatever force necessary to gain compliance from victim; victim may be unharmed; physical injury would be inadvertent rather than intentional. May be premeditated.
- Preceded by rape fantasies.
- Offender's mood is one of anxiety.
- Offenses are repetitive and may increase in aggression over time.
- Language is instructional and inquisitive — giving orders, asking personal questions, inquiring as to victim's response, etc.
- Assault may extend over short period of time with victim held captive.
- Weapon frequently used and brought to crime scene for purpose of threat or intimidation rather than for injury.
- Victim selected by vulnerability; usually same age range or younger than offender; child victim easily intimidated by adult authority.
- May experiment with homosexuality (though not necessarily sexually attracted to men).

- Tends to engage in conversation of a sexual nature that is both assertive (giving orders or commands) and inquisitive (questioning her sexual history or having her evaluate his sexual performance).
- Denies the sexual contact was forceful; strongly believes the victim enjoyed it.
- May give gifts or ask to see victim following rape in an attempt to prove it was consensual or to calm the victim, to smooth over the rape with her.
- Prior criminal history includes crimes of exploitation (theft, breaking and entering, robbery, etc.) and/or prior sex offenses (including nuisance offenses and/or rapes).

Goals of the power rapist: to express mastery, strength, control, authority, identity and capability; to validate his masculinity to force victim to like him; to submit willingly to his sexual demands; to capture, conquer, and control victim.

Triggers: being challenged by a female or male, something that undermines his sense of competence and self-esteem.

Ritualistic–Sadistic Rapist:

- Anger and power become paired and aggression becomes sexualized (physical force is eroticized). If power is eroticized, victim is subjected to ritualistic acts (e.g., bondage, spanking, enemas). If anger is eroticized, victim is subjected to torture and sexual abuse.
- Dynamics: eroticized aggression — symbolic control, elimination, or destruction of threat or temptation in order to regain psychological equilibrium and achieve a sense of integration and wholeness.
- Assault is calculated and premeditated.
- Offender's mood is one of intense excitement and dissociation.
- Offenses are compulsive, structured, and ritualistic, often involving kidnapping.
- Language is commanding and degrading; alternating reassuring and threatening.
- Assault may be for an extended duration in which victim is abducted, held hostage, assaulted, and released or disposed of.
- Weapon is generally used to capture victim, together with instruments for restraint and/or torture.
- Victim selected by specific characteristics or symbolic representation; usually complete strangers, trend toward same-sex child victim.
- Pleasure in victim's torment, anguish, distress, helplessness, and suffering.
- Usually involves bondage, torture, and bizarre or ritualistic qualities.
- Tends to focus injury to genitals.
- May use object to penetrate rather than own genitals.
- May include murder and mutilation.
- Victim is often stalked, abducted, abused, and sometimes murdered.
- Fully premeditated.
- Works himself into a frenzy as he assaults her.
- Inflicting pain may provide enough satisfaction so that genital-to-genital contact is not necessary.
- Believes he is omnipotent.
- May have history of fighting, cruelty to animals.
- No known criminal history or history of bizarre ritualistic or violent offense.

Goals of the sadistic rapist: to abuse and torture his victim.

Triggers: obsessions of rape and torture.

Molestation (Persuasion). Two types: Fixated and Regressed.

- Approach is one of seduction or persuasion; offender gains access to victim through deception, enticement, and/or manipulation.
- Passivity and dependency are major psychological dynamics.
- Offender displays a positive emotional investment in child; child is seen as "safe" and "caring."
- Offender typically seeks an ongoing (sexual) relationship with child; involved with child over extended period of time.
- Victim is a prop in offender's fantasy onto which offender's needs are projected.
- Offender's sexual behavior sometimes confined to nongenital acts and/or gradually progresses to increasingly overt and intimate sexual acts.
- Offender typically wants victim to enjoy the sexual activity; experiences sexual activity as acceptance or expression of affection.
- Sexual misuse of the child.

Fixated Child Molester:

- Primary sexual orientation is to children, usually beginning in adolescence.
- No precipitating stressor required to offend.
- Compulsive behavior and persistent interest in children.
- Premeditated, preplanned offense behavior.
- Identifies closely with victim and engages in behaviors consistent with the age of victim. May adopt a pseudo-parental role to the victim.
- Males are primary target (although females are also the victims).
- Little or no sexual contact with age-appropriate partners.
- Emphasis in sexual interaction usually focused on sexually stimulating the child and eliciting a positive erotic response from the child.
- If married, only for the convenience.
- Minimal alcohol or drug use history; alcohol and drugs not commonly involved in offense.
- Characterologically immature, poor peer relationships.
- Offense is a maladaptive resolution of life development (maturation) issues.

Regressed Child Molester:

- Primary sexual orientation involves age-appropriate partners; sexual attraction to children regarded by offender as a temporary lapse of control/judgment due to external, situational influences.
- Pedophilic interests develop in adulthood.
- Precipitating stress usually evident.
- Contacts with victims may be episodic and may increase with stress.
- Initial offense may be impulsive and not premeditated.
- Substitution: Offender replaces conflicting adult relationship with involvement with a child; victim serves as a substitute for adult partner.
- Female victims (opposite sex victims) are primary targets.
- Sexual contact with victim coexists with sexual contact with age-appropriate partner.
- Emphasis in sexual interaction usually focused on offender's arousal, stimulation, and sexual release; child is cast into adult sexual role.
- Offense is often alcohol related.

- More traditional lifestyle, but underdeveloped peer relationships.
- Offense is a maladaptive attempt to cope with specific life stressors.

LAW ENFORCEMENT SEX OFFENDER TYPOLOGIES

Child Molester Typologies*

Regressed Child Molester and the Situational Child Molester. The regressed and situational child molester:

- Lacks a true sexual preference for children.
- Uses children as a sexual surrogate or substitute.
- May range from a once-in-a lifetime act to a long-term pattern.

There are four types of situational child molesters:

1. *Regressed.* This offender has a low self-esteem and poor coping skills. He uses children when an adequate adult sexual partner is not available. This type of offender primarily uses coercion to gain compliance from the victim.
2. *Morally Indiscriminate.* This type of offender is a user and abuser of people in general. He abuses his spouse, children, and others. The primary rationale for using and abusing others is "*why not*?" He takes advantage of vulnerability and opportunity. This offender takes what he wants when he wants and has no problem using as much force as necessary to gain compliance from his victim.
3. *Sexually Indiscriminate.* This type of offender is sexual with whomever and whenever he wants and is sexually promiscuous. Similar to the morally indiscriminate offender, though he limits his impulsive and selfish acting out to sexual behavior. The basic motive is sexual experimentation and may sexually abuse children out of boredom. This type of child molester may engage in group sexual activities, swapping of child and adult sex partners, and engaging in some bizarre ritual. He often has a greater number of victims.
4. *Inadequate.* This type of offender is the social misfit and often includes those child molesters who have mental illness, mental retardation, and eccentric personality disorders. These offenders often seem harmless and often live alone. They view children as nonthreatening objects who allow them to explore sexual fantasies. This type of child molester may torture or sadistically abuse their victims.

Preferential Child Molester. The preferential child molester:

- Has a definite sexual attraction and preference for children.
- Often engage in ritualistic sexual behavior.
- Often have a large number of victims.
- Have age and gender preferences.

There are three types of preferential child molesters:

1. *Seduction.* This type of child molester (identifies with children and seduces them). They court the victims and seduce them over a period of time, grooming them to trade sex for affection. Often these offenders are operating child sex rings. They target children who are emotionally or physically neglected or abused. This offender may use threats and

* From K.V. Lanning (1995). Reproduced with permission.

physical violence to avoid being detected until he decides to dump the victim when they become too old.

2. *Introverted.* This type of child molester lacks interpersonal skills and is most likely to engage in brief sexual encounters involving minimal verbal communication. This offender usually targets very young children or strangers. This molester may marry in order to have children and then begin to molest the children when infants.
3. *Sadistic.* This type of child molester must inflict pain or humiliation for sexual gratification. The violent sexual behavior may be required for the offender to become sexually aroused. This offender is aroused by the child's pain and suffering, not directly by the sexual contact.

R. Hazelwood and A. Burgess* further developed and adapted the typologies of A. N. Groth.

The Power Reassurance Rapist:

- Purpose of attack is to reassure himself of his masculinity by exercising power over women.
- Lacks confidence in his ability to interact socially and sexually with women.
- Utilizes fantasies and sexual activity to prove himself.
- Does not intentionally degrade or emotionally traumatize the victim.
- Often described as the "gentleman rapist," "polite," or "apologist."
- If he finds passive victim, may take more time to act out fantasies.
- May apologize, ask for forgiveness.
- May recontact victim by phone or in writing.
- If unsuccessful, may attempt another rape quickly.
- Likely to keep a record of attacks.

The Power Assertive Rapist:

- Purpose of attack is to exercise his prerogative as a man to rape.
- Has no conscious doubts about his own masculinity; outwardly a "man's man."
- Demonstrates no concern for welfare or comfort of victim.
- Utilizes moderate to excessive force.
- May subject victim to repeated sexual assaults.

The Anger Retaliatory Rapist:

- Purpose of rape is to get even with women for real or imagined wrongs.
- Pairs anger and retaliation.
- Angry with women and uses sex to punish or degrade them with intention.
- Utilizes excessive force, appears as a frenzy.
- The attack is an emotional outburst, little planning, impulsive.
- Selects victim that reminds him of someone.
- Angry with women and uses sex to punish or degrade them with intention.
- Utilizes weapons of opportunity (fists, legs); attacks are sporadic.
- Once vented, he is relieved and his tension gone.

* From R.R. Hazelwood and A.W. Burgess (1995). Reproduced with permission.

The Anger Excitatory Rapist:

- Purpose of attack is to obtain sexual stimulation and/or gratification via victim's pain and suffering.
- Inflicts pain that will result in desired response of fear and total submission.
- Vicious attacks that result in physical and emotional harm.
- Attacks are well planned and methodical; rehearsed in fantasy.
- Utilizes weapons and objects.
- May torture victim with instruments as well as psychologically reducing her to extreme depths of fear.
- Helplessness of the victim is stimulating.
- Keeps victim for extended periods of time.
- The rapist will practice bondage.
- The rapist may insert objects into the victim's vagina or anus.
- The sexual acts forced on the victim will be varied and experimental in nature, intended to cause pain, humiliation, and degradation for the victim.
- Likely to keep records and/or record the rape.
- The rapist may remain emotionally detached during the rape.

The Opportunistic Rapist:

- The primary motivation is sexual in nature.
- Typically rapes as an afterthought during commission of another crime.
- Utilizes minimal degree of force.
- Usually consumes drugs or alcohol before the crime.

The Gang Rapist:

- There is a group leader as well as a reluctant person.
- Utilizes minimal to brutal degrees of force.

Hazelwood and Warren (2000) also indicated two types of violent sexual offender: impulsive and ritualistic types. The impulsive offender is more likely to be psychopathic because of his impulsivity, risk taking, cruelty, minimal use of fantasy, and generally antisocial history. The ritualistic offender appears to be significantly more specific in his intent and criminal behavior, being largely motivated by complex reenactment of specific, sexual fantasies that pair himself and the victim in erotically repetitious and arousing interactions.

SUMMARY

There are many similarities among the above sex offender typologies. I believe the overlap is a result of many years of research and work with sexual offenders and, therefore, are robust typologies that can be depended upon.

18 Female Sex Offenders

There are many women who commit sexual offenses; however, this has been largely ignored. Women are often viewed as less harmful, less violent, and less likely to engage in sexually abusive behavior. Denov (2003) indicated that some authors and clinicians claim that female sex offending is a rare phenomenon (American Psychiatric Association [APA], 1994; O'Hagan, 1989; Robinson, 2001). The *Diagnostic and Statistical Manual (DSM-IV)* (APA, 1994) indicated that paraphilias rarely if ever occur in females. O'Hagan (1989) declared that sexual abuse by females can be considered an "aberration" which has little or no significance. Various sources suggest that female sex offending is a rare phenomenon (e.g., U.S. Department of Justice, 2002). However, this gross under-reporting of sexual offenses committed by females can be attributed to how data are collected, societies that fail to recognize the frequency of female-initiated sexual offenses, and agencies and professionals unwilling to accept that females do, in fact, commit aggressive sexual offenses including child molestation.

In fact, when a female sex offender is caught and charged, there often appears to be sympathy from the public toward her and disbelief that anything she did could have caused any serious harm. Allen (1987) found that professionals tended to absolve female sex offenders by indicating that there was no malicious intent; that despite their sexually aggressive acts, they were portrayed as posing no threat to the victim or community; and often the details of the crime are reconstructed so as to blame the victim for the offense. Denov (2001) summarizes that as a result, professionals may fail to intervene as necessary and the victims may remain at risk for further victimization. In addition, as a result for not being held accountable, the female sex offender gains no insight nor is she held accountable or supported to change her sexually aggressive behavior. Put simply, all too often female sex offenders and female batterers are not taken seriously and not assessed to be the risk that they truly are.

Consider the recent two female teachers who engaged in sexual contact with a male student. Each of them was portrayed as loving and kind, and the two adolescent boys viewed as liking the sexual contact and of suffering no harm. Yet, consider the same scenario with a male teacher engaging in sex with an adolescent female. There is no question that what this man is doing constitutes sexual abuse and the public view is that he should be sent to prison. The adolescent girl is seen to have suffered harm arising from the sexual abuse. Somehow male victims are supposed to enjoy the sexual attention and sexual contact. The confusion about the boundary violation, the coercion used to gain compliance, and the secrecy expected to continue to hide the sexual contact takes a significant toll on the victim's life.

Though women use less physical aggression in the commission of sexual offenses, their use of verbal aggression/abuse at times surpass that used by men. In fact, women tend to surpass men in the use of verbal aggression, rejection of others, and hidden aggressiveness (Frieze et al., 1978; Tavris, 1992). Women often find some way to utilize physical restraint and then sexually manipulate their male or female victim into the sexual contact.

The high incidence of female initiated or co-offender female sexual abuse demonstrates that cases reported to police and child protective agencies are not representative of the true phenomenon (Mendel, 1995). Steffensmeier and Allan (1996) pointed out, like so many others, that the literature is sparse when it comes to examining female offenders. I found this to be the case as well, as most of the studies had small numbers of subjects.

Forced and Pressured Sex by Women

Women self-reported their use of aggressive strategies (physical force, exploitation of a man's incapacitated state, and verbal pressure) to force men to engage in sexual touch, intercourse, or oral sex against their will (Krahe, Waizenhofer, & Moller, 2003). Exploitation of the man's incapacitated state was used most frequently, followed by verbal pressure, and physical force. An additional 5% reported attempted acts of sexual aggression. Studies (e.g., Struckman-Johnson & Struckman-Johnson, 1994; Poppen & Segal, 1988; Shea, 1998; Hogben et al., 1996; Anderson & Aymami, 1993) have found that 2 to 28.5% of the women in their samples admitted to forcing men into sexual contact, most utilizing verbal coercion (e.g., threats, taking advantage of impaired states), only approximately 1 to 7% admitted using physical force to gain sexual contact with the man. Anderson (1998) found higher rates of aggression used by women. In their stury, 43% of the women utilized verbal coercion, and approximately 36% reported having gotten a man intoxicated to make him engage in sexual acts; 27% reported the use of threats; 20% admitted that they utilized force; and 9% reported that they threatened the man with a weapon.

Males reported that they were coerced or forced into sexual contact at higher rates than the women appeared to admit. Muehlenhard and Cook (1988) reported that 24% of the males in their study reported that they had engaged in unwanted sexual contact as a result of threat or use of physical force, and 27% reported unwanted sexual contact as a result of verbal pressure. Pertaining to unwanted intercourse, 6.5% of the men admitted that they were physically forced into intercourse and 13% as a result of verbal pressure. Groth (1979) found that of several hundred male sexual offenders in his study, approximately a third reported a history of childhood sexual abuse; of these, 51% reported a male perpetrator and 42% a female perpetrator. Fritz, Stoll, and Wagner (1981) found that of the men who were sexually abused in this sample, 60% were molested by a female. Others have found higher percentages, with 78% of the men reporting sexual victimization by female perpetrators in his first sample and 72% in the second sample.

Briere and Smiljanich (1993) found that sexually abused men were more likely to report having pressured women into sexual intercourse. In fact, 80% of the male sexual offenders in this study reported having been sexually abused by a female during childhood. Petrovich and Templar (1984) found that 59% of prisoners convicted of rape reported having been molested by a female during childhood. Fromuth and Burkhart (1989) found that 15% of the men in one sample and 13% of the men in a second sample reported childhood sexual contact, and that the majority (72 to 78%) of the perpetrators were females.

Faller (1989) reported that 8% of the perpetration was committed by lone female sexual abusers and 29% by both males and females; therefore, 37% of the cases involved female offenders. Of Mendel's study (1995), 14% of offenses were committed by lone females and 46% of offenses were committed by both male and female perpetrators, for a total of 60% of sexual offenses involving female perpetrators. In summary, 42 to 78% of the self-reported data involved female sexual perpetrators.

Female sex offenders tend to molest female victims far less than male victims when acting alone. Pierce and Pierce (1985) found that of the females that were sexually abused in their study, only 1% reported having been molested by a female. Reinhart (1987) and Faller (1989) found similar results (1 to 2%). Faller (1989) also found that 17% of the females that were sexually abused were abused by both males and females. Fritz, Stoll, and Wagner (1981) found that approximately 10% of the female sexual abuse victims in their study were sexually abused by a female. Allen (1991) found that of the 65 females convicted of sexually molesting children, 72% were sexually abused as children, and out of those, 6% were sexually abused by females. Allen (1991) surveyed males convicted of sexual offenses against children. Of those that were sexually abused as children (36%), 45% reported that the sexual abuser was a female. Mendel (1995) found similar results.

Struckman-Johnson (1988) found that 16% of the males in their study indicated that they had been forced to engage in sexual intercourse at least once in their lifetime. The majority of the male victims reported the use of psychological pressure. O'Sullivan, Byers, and Finkelman (1998) found that 18% of the men in their study reported that they had been the victim of sexual aggression. Of those men, 3.8% reported experiencing unwanted sexual intercourse due to being under the influence of drugs or alcohol by the perpetrator; 1.5% experienced sexual intercourse following threat of physical force; 8.5% reported attempted intercourse due to their own use of alcohol or drugs; and 2.3% reported attempted intercourse due to threat or use of physical force. Larimer et al. (1999) found that 20.7% of their male respondents had been the victims of unwanted sexual contact in the year prior to the survey; 7.9% experienced verbal pressure; 0.6% experienced physical force; and intoxication by 3.6%. However, a limitation of the results in this paragraph is that the men did not specify the gender of their sexual aggressor.

Struckman-Johnson and Struckman-Johnson (1994) found that 20% of men had been pressured or forced to engage in sexual intercourse by a woman and that percentage increased to 27% of the men by their subsequent study (Struckman-Johnson & Struckman-Johnson, 1998). The vast majority of these incidents involved the use of verbal pressure or intoxication. Only 0.5% reported engaging in unwanted sexual intercourse as a result of physical force. Waldner-Haugrud and Magruder (1995) found that 56.9% of the men in their study reported that they had engaged in unwanted sexual contact with a date or lifetime partner as a result of intoxication, 6% as a result of the use of a weapon, and 3.5% for use of physical force.

Krahe, Waizenhofer, and Moller (2003) found that nearly 1 in 10 women admitted to using aggressive strategies to obtain sexual contacts with a man against his will on at least one occasion. They found that exploiting a man's incapacitated state was the most common form of coercion, followed by verbal pressure and physical force. They also found that women's use of sexual aggression was highest toward an expartner, friend, or acquaintance, with nearly none reporting use of sexual aggression toward a stranger. Faller (1995) found that the women in her study usually had multiple victims and most involved what she called intrusive sexual contact (e.g., oral or other penetration or exploitation). Matthews et al. (1989) found similar findings in their sample. Allen's (1991) findings contradicted these findings, though his sample relied only on self-report vs. accessing official sources of information.

Numerous authors (Faller, 1995; Matthews et al., 1989; McCarty, 1986; O'Connor, 1987; Rowan et al., 1990; Travin et al., 1990; Wolfe, 1985) found that the female sex offenders had significant problems in functioning, with high rates of mental illness, mental retardation, substance abuse, and other maltreatment of their children. These rates surpassed those of male clients.

An important finding from Faller's (1995) research was that women sexually abused children both within and outside the family. Matthews et al. (1989) and Wolfe (1985) found similar findings. It was also found that women became involved in sexual abuse with others, usually men, and that the men may play a leadership role when women sexually abuse (Faller, 1995; Mathews et al., 1989; McCarty, 1986; O'Connor, 1987; Rowen et al., 1990; Wolfe, 1985). This means that the women's sexually abusive behavior may, at times, occur in conjunction with or as a result of male dominance. Related to ritual abuse of children in daycares, Finkelhor et al. (1988) found that all cases involved female perpetrators.

Characteristics of female sex offenders who were solo offenders (single mothers, noncustodial mothers, adolescent offenders, and female caretakers) included their unsupervised access to children (Faller, 1995). They appear to use the children to meet their own emotional and sexual needs, at least in part due to a lack of appropriate sexual outlets. In addition, women were more likely to spend time with children and no other adults. Therefore, opportunities to observe and report any abuse is limited. In addition, few of the women tended to receive serious consequences for their sexually abusive behavior and approximately a third had continued access to their victims or potential victims.

Predictors of Sexual Aggression

Childhood abuse, gender role orientation, ambiguous communication of sexual intentions, and the level of sexual activity were found to be predictors of female sexual aggression, as well as peer pressure, toward sexual activity (Krahe et al., 2003).

A victim-to-perpetrator cycle suggests that men with a childhood history of childhood sexual abuse were more likely to become sexually aggressive (Browne, 1994; Senn et al., 2000). Anderson (1996) provided evidence for a similar victim-to-perpetrator cycle for female victims of childhood sexual abuse. He demonstrated that women who were sexually abused as children were more likely to engage in verbal coercion, physical assault, and sexual prevalent among female child molesters (Grayston & De Luca, 1999; Green, 1999; Miccio-Fonseca, 2000). Foo & Margolin (1995) found a similar link for both men and women. Krahe, Waizenhofer, and Moller (2003) also found a link between childhood sexual abuse and subsequent sexual aggression. White and Humphrey (1994) found that sexual victimization in dating relationships was linked to the use of verbal aggression in heterosexual relationships. Boys who experienced sexual, physical, or emotional abuse by primary caretakers demonstrated more sexual aggression later in life (Barnett, Miller-Perrin, & Perrin, 1997).

In addition to finding that a history of childhood sexual abuse predicted subsequent sexual aggression for women, Krahe et al. (2003) also found that the use of ambiguous communication (token resistance, such as saying "no" but meaning "yes" and compliance, such as saying "yes" when they mean "no") by women can increase their use of sexual aggression. They also found that women who reported engaging in sexual aggression had significantly higher numbers of sexual partners. They did not find that engaging in sexual contact at an early age predicted sexual aggression for women. The link between number of sexual partners and sexual aggression corresponds to a similar pattern found for men's sexual aggression (e.g., Abbey, McAuslan, & Ross, 1998; Malamuth et al., 1991). Last, they found that peer pressure toward engaging in aggressive sexual contact increased from both same-sex and opposite-sex peers.

Female offenders are more likely to have been victims as children or adults (Chesney-Lind & Shelden, 1992; Daly, 1994; Gilfus, 1992; Windom, 1989). Female offenders are more likely to have had records of neurological and other biological or psychological abnormalities (Denno, 1994). Daly (1994) reported that female offenders (in comparison to male offenders) had greater childhood and adult exposure of abuse. These females were also more conventional than the males and assumed greater responsibilities for children, were committed to education, and had legitimate sources of income.

Dobesh et al. (1992) also found that females rarely kill or assault strangers or acquaintances; instead, the female's victim tends to be a male intimate or child, the offense usually takes place in the home, the victim is frequently intoxicated, and self-defense or extreme depression is often a motive. For women to kill, they generally must see their situation as life threatening, as affecting the physical or emotional well being of themselves or their children.

Characteristics of Female Sex Offenders

Why women may target younger males to sexually abuse is understandable. Schwartz and Cellini (1995) found that female sexual offenders may be motivated to establish intimacy, but they do so with a distorted and confused definition of intimacy. They may relate better to younger males than to age-appropriate males. Wolf (1985) found that women in her sample tended to commit sexual offenses (1) with co-offenders half of the time, (2) that females used dependency as a reason for their offense behavior and offense dynamics much more frequently than did male sex offenders, (3) females used many of the same cognitive distortions as males, (4) both males and females in the treatment program used coercion rather than physical force, and (5) both males and females were more likely to offend intrafamilialy than outside of the family.

Allen (1991) found that female sex offenders reported (1) less stability and more parental criticism, devaluations, and disappointments, and physical abuse than males; (2) less marital satisfaction with their partner and more satisfaction with their children; (3) more sexual satisfaction and more sexual partners; (4) more physical abuse suffered and more physical abuse perpetrated against their partners; (5) fewer offenses; (6) more of their own victimization; (7) more crimes against strangers; (8) more same-sex victims; (9) more abuse by males; (10) less acknowledgment of guilt and less ability to identify behavior as sexually abusive; and (11) more maintaining of innocence and more anger toward the informant, and less sorrow, guilt, and relief at being caught.

Schwartz and Cellini (1995, pp. 5–12) reported that women perpetrators are more sexually active, more assaultive, and angrier than male perpetrators. They also found that female sex offenders were often raised in families that experienced significant conflict and they were often the victims of physical and sexual abuse. The vast majority of female sex offenders reported having been victims of sexual abuse or physical abuse (McCarty, 1986; Matthews, Matthews, & Speltz, 1989). McCarthy also found that approximately a third had multiple caretakers growing up and had alcoholic families, and, approximately 70% of the female sex offenders who acted alone in perpetrating their crimes had been molested by their brothers. Those who were sexually molesting as a co-offender (in concert with another offender) were abused by adult caretakers. Allen (1991) reported that most female offenders reported that their mother had only one partner, they also reported having negative relations with both parents and a worse relationship with their mothers. Allen also found that over half of the women in his study reported having been physically abused, and that approximately 75% of the women reported having been sexually abused by multiple perpetrators and that the perpetrators were more frequently acquaintances. Females were 15 times more likely to be abused by a male than a female, whereas, in 45% of the cases, men were molested by women.

McCarty (1986) also found that 85% of the female sex offenders in her sample were married while in their teens, with approximately a third being age 15 or younger when they married. Approximately half of the women had a history of promiscuity. Allen (1991) found that approximately 40% of the women in his study reported being married once, and that the women reported average satisfaction with their partners and even more satisfaction with their children. In addition, the vast majority of the women reported moderate to above average sexual satisfaction with their partners. In short, the female offenders were more sexually active than male sex offenders. Allen also reported that most of the women in his study were involved in physically abusive relationships in which mutual physical abuse occurred, and some of the women did not view the abuse as negatively impacting their relationship.

Overall, female offenders tended to have marginal to normal intelligence and functioning, but most maintained regular employment (McCarty, 1986). Most of the women physically and emotionally neglected their own children. Matthews et al. (1989) also found that the women in their sample had steady work histories. They also found that the women tended to be social isolates who engaged in promiscuous sex as a way of getting attention and approval.

Though female sex offenders do not appear to present with serious mental illness, they do present with mental disorders including depression, suicidal ideation, antisocial behavior, and anger problems (Matthews et al., 1989; McCarty, 1986; O'Connor, 1987). Preliminary information on female sex offenders tended to suggest that most have some substance abuse problems, though not specified at this time (Matthews et al., 1989; Allen, 1991). Female sex offenders are often diagnosed as having hysterical or borderline personality disorders (Windom, 1978). Female sex offenders are predominately child molesters vs. rapists of adult women. However, lesbians also sexually assault their partners (Schwartz et al., 1995; Waterman, Dawson, and Bologna, 1989).

Interestingly, Matthews et al. (1989) reported that 16 female sex offenders had a total of 48 victims (20 boys, 27 girls, and 1 adult female). Six of the offenders molested both boys and girls; 11 of the women had engaged in penetration. This information appears similar to that of male child molesters. McCarty (1986) reported that 52% of female perpetrators molested females, 35%

molested males, and 10% molested both. Matthews et al. (1993) confirmed that female offenders more frequently molest females and female rapists also tended to molest female victims (Brown, Hull, & Panesis, 1984).

In reference to attitudes about their crime, Wolf (1985) found that women used dependency on males as an excuse and justification for their offense behavior. Matthews et al. (1989) found that women in their group were empathetic and willing to take responsibility and acknowledge their guilt. Allen (1991) reported that women in his group scored lower than male offenders in issues of responsibility, empathy, guilt, and remorse, and higher in anger. Allen's finding may be impacted on his sample, which included women who appeared to have received more harsh consequences for their sex offense; however, perhaps his sample of women were more extreme and leaning more towards psychopathic individuals.

Typologies of Female Sex Offenders

There are several typologies for female sex offenders. Campbell (1981) suggested that women learn to be deceitful due to societal demanding that they hide their interest in sex and their menstrual periods. He went on to say that women have fewer internal controls and must rely on society to control their behavior, and that they will become more criminal if they gain more equality. This view lacks any empirical support. Windom (1978) identified four types of women sex offenders. Type I is the psychopathic. This type of female offender is hostile, poorly socialized, impulsive, and aggressive. She experiences minimal if any anxiety. Type II is the secondary or neurotic offender. This type of offender resembles the Type I offender, but is more anxious, depressed, and guilt-ridden. Type III is the overcontrolled offender. This female denies problems and has a high degree of control. Type IV is the "normal" criminal. This type of female sex offender is hostile and tense, but not impulsive. This type of offender presents with little psychopathology. Type I and II fit the psychopath as described by Hare (1991) and Vassileva et al. (2005).

Sorrel and Masters (1982) described four types of sexual assaults perpetrated by women: (1) Forced assault, where the sexual assault is marked by the use of physical restraints or believable threats of physical force; (2) the "babysitter," where the abuse is perpetrated on a younger boy by a nonrelative female who uses direct or implied threats; 93) incestuous assaults, where relatives assault children; and (4) dominant women abuse, where a female makes an aggressive sexual approach to an adult male without direct physical force, which intimidates or terrifies the victim.

McCarty (1986) describes three different types of female incest offenders:

1. The Independent Abuser — A single parent usually uses a daughter who is perceived as an extension of herself. This type of female offender has usually been sexually abused as a child and frequently has serious emotional problems.
2. Co-Offender Abuser — This is a highly dependent individual with marginal social functioning. This type of offender offends with a male-perpetrator who dominates her.
3. Accomplice-Colluded Abuser — This type of individual does not actually assault the victim, but either colludes with the abuse or ignores him/her. This individual usually is more functional than the co-offender, but is highly dependent.

Matthews, Matthews, and Spelt (1989, 1993) describe five types of female child molesters:

1. Teacher/Lover — This type of offender views herself as being involved in a romantic relationship with an adolescent male and tends to deny the abusive nature of her behavior.
2. The Predisposed — This female offender tends to have a long history of being sexually abused, usually incestuously. They fear male peers in the same way many male pedophiles fear adult females. They have histories of addictive behaviors and tend to become

involved in destructive relationships. They tend to be angry and abusive, particularly toward their children.

3. The Male-Coerced — This type of offender usually is a dependent individual who is passive, nonassertive, and locked into an abusive relationship with a male perpetrator. These women may later go on to molest on their own. These individuals usually offend against their own children, but they also occasionally offend against adult females.
4. The Experimenter/Exploiter — This type of offender generally is an adolescent or younger girl babysitting for a young boy. These offenders tend to turn to those much younger than themselves because they are socially awkward and uncomfortable with their sexuality.
5. The Psychologically Disturbed — This type of offender is often psychopathic. In Faller's (1987) sample of these types of female sex offenders, all had abused younger female children or relatives.

Impact of Sexual Abuse by Females

Denov (2004) reported that the majority of males and females in her study reported severe sexual abuse by their mothers. The majority reported long-term difficulties with substance abuse, self-injury, suicide, depression, rage, strained relationships with women, self-concept and identity issues, and a discomfort with sex. Take into account that the traditional stereotype of women as being nurturing, caring, and selfless in their care for children and partners, sexual abuse significantly damages the victim's sense of security and nurturance.

19 The Role of Alcohol and Drugs in Abuse

INTRODUCTION

Many abusers claim that they were "out of control" and under the influence of drugs or alcohol when they abused and, therefore, they are not fully responsible for their abusive behavior. Although most abuse occurs when the abuser has been drinking or using drugs, abuse is always a conscious, thought-out decision. Remember that alcohol is a depressant, that is, it generally has a calming effect on the body and mind. From a purely chemical view, alcohol has a relaxing effect. However, alcohol does not relieve tension.

However, for some people, alcohol has a stimulating effect. This can occur when the person experiences a sense of decreased attention to stressors and consequences. Alcohol can easily exacerbate anger, rage, depression, insecurities, and other dysphoric states. The person may feel as though they are becoming increasingly more depressed, angry, or abandoned. Alcohol use makes it easier to abuse by creating an increased sense of agitation and justification and by making it easier to experience a fearless attitude and a lack of concern for any consequence of their behavior. Alcohol use may make it far easier for some people to become violent or reckless. Alcohol use makes it easier for abusers to blame the substance for their abusive behavior. To become angrier, escalated, and violent requires a significant degree of concentration and forethought. Alcohol by itself does nothing to create or cause anger, rage, or violence. In short, alcohol use can serve as a disinhibitor for violent behavior.

Think about it. The abuser verbally abuses the victim and must be quick with the right words and put downs to counter whatever the victim is doing. So the abuser must think and respond to the victim's behavior and responses. When the abuser becomes physically assaultive, he has to force his body and mind to wake-up, become increasingly more aroused and escalated, and then engage in the abusive behavior. This also requires quick thinking, having to adjust the abuse to respond to the victim's behavior. If the victim is running away, the abuser must first catch her to abuse her physically. If the victim picks up the phone, the abuser responds by grabbing the phone and then beating the victim. So abuse requires quick and somewhat flexible thinking on the abuser's part. Therefore, regardless of how intoxicated or high the abuser claims to have been, only he could have concentrated so much on abuse that he abused while under the influence of drugs and/or alcohol.

Alcohol certainly can serve as a disinhibitor. A disinhibitor makes us feel a bit more relaxed, loose, and ready to take far more risks. However, our judgment becoming more clouded does not mean that we are "out of control" or unable to stop ourselves from engaging in abusive or inappropriate behavior. We still must make the final decision to do or not do something, regardless of whether drunk, high, or intoxicated. How we believe we will behave when high or intoxicated is referred to as expectancies; how we "expect" to be effected and as a result how we "expect" to behave. But expectancies are simply the way we believe we will be impacted, not actually how we are impacted. If persons walk on hot coals, they may say that they did not feel the heat. But they did, in fact, feel the heat, they simply expected to be able to numb their feet, and were able to do so. When they experienced the heat or pain, they relabeled the pain and heat sensation as anything but negative, therefore, able to continue to walk on the hot coals. Though they were still walking on hot coals, they still might end up burning their feet. If we believe that when we are high or intoxicated that we become more abusive, then we are more likely to become more abusive. If we believe that when intoxicated we become more relaxed and easy going, then that is how we become. The alcohol did not make us that way, our expectancies did.

Some believe that alcohol causes violent and aggressive behavior. Even some professionals blame alcohol for a person's choice to behave violently. But research has demonstrated that a person's expectations of the effects of alcohol on his own behavior directly impacts his behavioral choice. Very few people, if any become aggressive every time they drink. Anecdotally, I am not aware of any of the offenders I have assessed or treated becoming aggressive each and every time they are intoxicated.

From my own experience, I find that abusers often use alcohol to manipulate their victims into vulnerable situations. Victims are far less likely to effectively defend themselves when they are under the influence of alcohol or drugs. Alcohol appears to provide abusers with a convenient excuse to avoid taking responsibility for their own behavior, and minimizes guilt and empathy for their victim. In addition, abusers and sexual offenders feel more at ease with the alibi that: "I was too intoxicated to know what I was doing." And for some reason, offenders who were intoxicated at the time the crime occurred often receive a lighter sentence.

Interestingly, most sexual assaults and most instances of domestic abuse occur in private. Isolating the victim decreases the likelihood of anyone interfering and decreases the likelihood of the victim asking for or receiving help. Isolating a victim is a conscious choice, but the impact of the isolation is significantly magnified when the victim is under the influence of drugs or alcohol. If the behavior is genuinely *out of control*, then the abuse or rape would occur anywhere at any time, regardless of who was around. If *out of control,* the aggression would not occur solely in private locations. The offender had to delay gratification of his wants to abuse or sexually assault until alone with the victim.

Alcohol use interferes with normal, moral thought processes. Although alcohol is NOT in control of the abuser, crossing moral and legal boundaries becomes easier. Alcohol increases aggressions by making it easier to become escalated; that is, to experience an increase in adrenalin and decreasing attention to disinhibitors and the victim's safety. Alcohol use increases the likelihood of tunnel vision occurring, which allows the abuser to become increasingly focused on gaining victim compliance or for attaining sexual contact from their victim. In short, alcohol can play the role of courage for an abuser, increasing his sense of righteousness and justification.

Abusers blame alcohol use as the reason they committed abuse and sexual abuse. They often make comments such as:

- "I was out-of-control."
- "I was too drunk to know what I was doing."
- "I do not remember what happened."

In addition, males who believe they ingested alcohol responded with greater sexual arousal to scenarios of rape and violence against women. The majority of these men were not sex offenders, but still experienced an increase in sexually aggressive behavior when under the influence of alcohol or when they believed that that they were under the influence of alcohol. It should be noted that these men already had a propensity for engaging in violence or rape behavior. Alcohol merely served to provide the courage and justification for acting out what was already on their minds. Alcohol disinhibits, makes it easier to give permission to use aggression in general, especially with sexual behavior.

For dating violence, physical and sex offenders encourage their potential victims to use alcohol and/or drugs by:

- Spiking their drink without their knowledge.
- Playing drinking games.

- Daring them to drink or get high.
- As initiation rituals.
- To "fit in."
- To impress.

The need to coerce or force the victim into a state of chemical impairment certainly implies forethought on the abuser's part. The goal of the abuser or sex offender is to get the victim into a more vulnerable position (e.g., intoxicated or "high") and then take advantage of that power imbalance to aggress against the victim.

Let us discuss what the literature has to say about the connection between alcohol and violence.

RESEARCH ON THE ROLE OF ALCOHOL PLAY IN VIOLENCE

Below is a compilation of research on how alcohol correlates to violence. Remember that correlating means that there is a relationship between the two factors, that is, a relationship exists between alcohol use and violence. But alcohol use does not cause abuse. Most abusers use alcohol to justify abuse.

It has been found that 25 to 50% of batterers have substance abuse problems (Hamilton & Collins, 1981) and the victim or batterer had consumed alcohol in approximately half of the incidents of domestic abuse (Pernanen, 1991). Bennett (1995) indicated that families where substance abuse occurs and physical abuse of women occurs share several characteristics, including intergenerational transmission of the problem, frequent crisis states, the abuser blaming the victim for his abusive behavior, the abuser forgetting details of the episode, isolation of the victim, impaired emotional development in the family, impulsiveness and low self-esteem among other family members, loss of control used as a coping mechanism, and a short-term payoff in tension reduction (Rogan, 1985-1986).

Perceptual and motor functioning are impacted by alcohol use. This may lead to an over-reaction in situations where aggression or provocation occurs (Pliner & Cappell, 1974).

Cognitively, alcohol makes the drinker less self-reflective and less aware of internal cues, and less concerned about morals and values. The result is that the drinker may decrease appropriate reactions to certain situations. In addition, alcohol reduces the drinker's ability to perceive several situational cues at once. The result is that the person may focus on a single cue, and aggressive cues will likely receive the most attention (Lang, 1993; Taylor & Leonard, 1983). This may lead to misinterpretations of other people's behavior that result in an increased likelihood of aggression (e.g., Collins, 1983, 1993; Lang, 1993).

Alcohol leads to an increase in feelings of personal power among some drinkers, which may increase aggressiveness (McLeelland, Davis, Kalin, & Wanner, 1972). In addition, alcohol may make the drinker less likely to consider consequences, thereby increasing impulsiveness and possibly impulsive aggression (Parker, 1993). This sense of power with reduced or no consideration for the possible consequences makes the aggressive drinker more dangerous.

Pernanen's (1991) data and other survey data (Kantor & Straus, 1987) suggest that alcohol is less frequently a factor in husband-to-wife violence than is some other types of violence. In addition, more severe forms of aggression, such as punching, hitting with a fist, and kicking, were more common in incidents involving alcohol, with alcohol often consumed in anticipation of an altercation (Graham et al., 1996). However, from an anecdotal point of view, it seems that alcohol is related to the majority of spousal assault cases that are adjudicated. I found that 62% (Johnson, S.A., 2001, unpublished) of the batterers I assessed were under the influence of alcohol and/or drugs at the time of the index (current) offense.

Graham et al. (1996) found that the results of studies of treatment samples have supported the link between problem drinking and aggressive behavior, with a higher proportion of substance abusers in treatment reporting violence compared with rates of violence found in the general population (Livingston, 1986); male alcoholics in treatment having a high rate of aggressive behavior to wives (Stith, Crossman & Bischof, 1991); and alcoholic women in treatment being more likely than nonalcoholic women to be victims of aggression from husbands (Miller, Downs, & Gondoli, 1989). In sum, there is considerable evidence that those with a history of heavy or problem drinking report high levels of aggressive behavior.

Gustafson (1993) concluded that alcohol does not affect aggression under nonprovocative conditions. Some of the more naturalistic studies of alcohol-related aggression, where all response options were available (including nonaggressive responses), found increased aggressive behavior with alcohol consumption (Boyatzis, 1974; Takala et al., 1957; Sayette et al., 1993).

Weisman and Taylor (1994) found that intoxicated subjects responded more aggressively than did the sober subjects. Approximately 50% of violent crimes involve an offender and/or a victim who has been drinking (Murdoch, Pihl, & Ross, 1990). Most recent data found that heavy drinking was more common on dates where sexual assault occurred (Muehlenhard and Linton, 1987).

Increases in autonomic stimulation, in a variety of physiological indicates, have been shown to occur immediately before, and during, an aggressive attack (Dengerink, 1971; Donnerstein, 1980; Edguer & Janisse, 1994). Alcohol's stimulant effects have been shown to partially mimic this physiological arousal profile (Levenson et al., 1980; Turkkan, Stizer, & McCaul; 1988). Thereby, alcohol appears to make it easier for some individuals to escalate in certain situations.

AMOUNTS OF ALCOHOL

Even low amounts of alcohol may lead to some individuals becoming more aggressive (Pihl & Peterson, 1993a). Aggressive behavior has been positively associated with higher doses of alcohol (Taylor & Gammon, 1975; Kelly et al., 1988). The research suggested that individuals consuming moderate amounts of alcohol are less aggressive than those who consume low or high amounts.

Most measures showed greater aggression under the alcohol condition. Kreutzer, Schneider and Myatt (1984) found a significant increase in self-reported aggressiveness after alcohol consumption, with aggressiveness highest for the moderate dose rather than the high dose. This was supported by the majority of studies that found that higher doses of alcohol did not necessarily increase the likelihood of an aggressive response.

Finally, Boyatzis (1975) found that men who drank heavily behaved more aggressively than lighter drinkers at experimental parties. Moderate drinkers were more hostile than light-moderate drinkers or nondrinkers on an adjective checklist describing them, but they were not more aggressive in their ratings of a bogus opponent (Graham et al., 1996). Taylor and Gammon (1975) demonstrated that aggressive responding may be related to the quantity of alcohol consumed.

The majority of experimental studies conducted in research laboratories have found a significant increase in aggression with alcohol consumption (Chermack & Taylor, 1995; Gantner & Taylor, 1992; Gustafson, 1992; Laplace, Chermack & Taylor, 1994; Lau, Pihl, & Peterson, 1995; Schmutte & Taylor, 1980; Shuntich & Taylor, 1972; Taylor & Gammon, 1975, & 1976; Taylor, Gammon & Capasso, 1976; Weisman & Taylor, 1993; Zeichner et al., 1982). Taylor's (1993) results indicated that subjects who consumed alcohol set significantly higher shock intensities for their opponents.

Taylor (1993) also found that expressions of pain from a victim have been shown to reduce aggressive responding in numerous studies. Not surprisingly, sober subjects used lower shock settings when opponents expressed their discomfort. However, intoxicated subjects responded more aggressively, by using increased shock intensities, when opponents communicated pain following the receipt of shock. These studies provide yet another dramatic illustration of the aggression-inducing properties of alcohol. At the very least, when under the influence of alcohol, subjects are more willing to utilize increasing/high levels of aggressive behavior compared to individuals who are not/do not believe they are intoxicated.

A major implication of Taylor's (1993) results is that alcohol can initiate aggressive responding among nonaggressive and aggressive people. Subjects with high and moderate aggressive tendencies increased their shock, while intoxicated subjects assessed as nonaggressive did not retaliate against the provocative opponent.

Weisman and Taylor (1994) found that intoxicated subjects responded more aggressively than did the sober subjects. In both experimental conditions, the intoxicated subjects set more intense shocks for their opponents than did the nonintoxicated subjects. It is especially revealing that the intoxicated subjects in the vulnerable performance condition tended to be relatively more aggressive on the first trial.

Zeichner and Pihl (1979) suggested that intoxicated subjects can express more aggressive behavior because they do not process self-relevant information in an efficient manner. This appears to be supported by numerous other studies (some already mentioned) that alcohol use interferes with normal cognitive processes.

Gustafson's (1990) study of college students found that drinking wine did not increase aggression. However, an increase in aggression was found when vodka was used as an intoxicant. Interestingly, more aggression also seems to occur after the consumption of distilled rather than brewed beverages (Takala, Pihkanen & Marranan, 1957; Boyatris, 1974; Pihl, Smith, & Farrell, 1984; Murdoch & Pihl, 1988a, 1988b). Pihl and colleagues found that a distilled beverage and a placebo version of a distilled beverage both resulted in more aggression than a brewed beverage, suggesting the possibility of a powerful expectancy effect regarding beverage type (Lang, 1993).

Again, regarding beverage type, Klein and Pittman (1990) found that drinkers associated all beverage types with adverse consequences; however, among those who did distinguish among beverage types, "fighting and rowdy behavior" was thought to be more associated with beer or hard liquor and less with wine or wine coolers.

EXPECTANCIES

In situations where people consider drinking an excuse for aggressive behavior, it is more likely to occur (Blum, 1981; Pernanen, 1976; Levenson, 1983; Dyck, 1980). Drinkers who expect drinking to lead to aggression are more likely to become aggressive when drinking (Collins, 1981; Fagan, 1990). Drinkers who believe that they can use alcohol to justify aggressive actions and reduce their responsibility for these actions ("deviance disavowal") are more likely to become aggressive when they drink (Abbey, 1991; Collins, 1981; Fagan, 1993). Some people take more risks when they are intoxicated (Pihl & Peterson, 1993b). Drinkers who have a permissive attitude toward intoxicated aggression are more likely than other drinkers to behave aggressively when they drink (Taylor & Leonard, 1983).

Several studies have found that heavy drinking is associated with reports of aggressive behavior *only* among people who expect alcohol to lead to aggression (Derman & George, 1988; Leonard & Senchak, 1993). Aggressive behavior has been strongly associated with subjective perceptions of intoxication (Giancola & Zeichner, 1995).

At least two studies have found that heavy drinking is associated with aggressive behavior only among those who expect that alcohol use will lead to aggression (Dermen & George, 1988; Leonard & Senchak, 1993).

Graham et al. (1998) found that when a person is intoxicated and in the presence of others who are intoxicated, they have an increased risk of engaging in aggressive behavior because of their own intoxication, but may also be exposed to situational provocation from other intoxicated people who also have a higher risk of aggressive behavior. In other words, if alcohol is expected to increase the likelihood of aggressiveness in one drinker, then theoretically, the effect should be multiplicative in a group of drinkers. Brown and colleagues (1980) and Leigh (1987) found that males were more likely to expect alcohol to make them more aggressive; similarly, Crawford (1984b) found that males were more likely to list alcohol as a cause of aggression.

Brown et al. found that individuals who drank less tended to expect positive effects from alcohol, whereas individuals who consumed greater amounts of alcohol were more likely to expect that alcohol use would be associated with increased aggression.

In addition, the role of expectations about the effects of alcohol on aggression is an aspect of drinking that combines both situational and personal characteristics in predicting aggression (Graham et al., 1996). As these studies have shown, the circumstances when drinking that lead to aggression are determined by both the characteristics of the drinker and aspects of the drinking situation.

Graham and Wells (2001) found that respondents reported that opponents were significantly more intoxicated and that alcohol was more likely to have contributed to the opponent's aggression than to the respondent's. This suggests that people perceive themselves to be less affected by alcohol than their opponents are in incidents of aggression. The way that people interpret their own and other's behavior when drinking suggests that they may expect that the alcohol will make the other person aggressive, but not them (Gustafson, 1987; Leigh, 1987; Paglia & Room, 1999; Rohsenow, 1983).

In addition, the expectation that others will respond with aggression when drinking may result in a readiness in the person to respond aggressively to persons who have been drinking. This would be consistent with research suggesting that the escalation of aggression is higher when both parties to the conflict are drinking, compared to when only one person has been drinking (Leonard, 1984).

Graham and Wells (2001) found that respondents in their study indicated: (1) that the main opponent had been drinking, and (2) that the main opponent was more intoxicated than themselves in incidents in which both had been drinking. They also attributed more of a role to the effects of alcohol on the opponent than on themselves in incidents in which both were drinking. Even when the opponent was rated as intoxicated as or less intoxicated than the respondent, there was still a trend for respondents to believe that the opponent was more affected than the respondent by alcohol. Interestingly, in a study of drinking–driving studies, many drinking drivers inaccurately assess their own level of intoxication and believe that they can drink large quantities of alcohol and still drive safely (Foss & Perrine, 1993).

George et al. (2000) found that predrinking expectancy steered post-drinking perceptions along an expectancy-congruent course to shape subsequent behavior. Thus, alcohol's role in stimulating men's sexual responding cannot be construed as occurring through a strictly pharmacological mechanism. Expectations play a significant role in alcohol-related aggression.

In a study that involved forced sex (Norris & Kerr, 1993), subjects who consumed alcohol reported a greater willingness to behave as the same-sex character in the story. Those in the high-dose alcohol condition were less likely to propose nonaggressive solutions and more likely to respond aggressively to further provocation from the hypothetical provoker.

PERSONAL FACTORS

People whose drinking is related to concerns with personal power are more likely to behave aggressively when faced with an adversary, a threatening situation, or a desire to control or overpower another (Collins, 1981; McLelland et al., 1972). Power concerns may be particularly implicated in drinking by young men among whom "hypermasculinity" or "macho" styles of behavior may be associated with aggression (Burns, 1980; Collins, 1983; Fagan, 1993). People who use alcohol in combination with other drugs may be more likely to become aggressive than if they had used either substance by itself (Tinklenberg, 1973). Berkowitz (1993) suggested that the probability of aggressive behavior increases in the presence of cues that are associated with aggression. The encouragement by others in the drinking setting to become aggressive increases the likelihood of violence (Tomsen, 1997).

One study (Graham et al., 1998) identified several personal factors that increase the risk for aggressive behavior. Some of these factors include permissiveness and relationship dissatisfaction

(Leonard & Senchak, 1993), hostility and marital satisfaction (Leonard & Blane, 1992), and the approval of violence and social class (Kantor-Kaufman & Straus, 1987).

Some evidence suggests that drinking immediately preceding the violent crime is much more important than one's drinking history (Graham et al., 1998). Although a relationship has been found between habitual heavy drinking and aggression, Wiley and Weisner's (1995) finding suggest that this relationship may be attributable to the greater likelihood of heavy drinkers being under the influence of alcohol when potentially aggressive situations develop. Their finding supports the assumption of a direct role of alcohol, not simply an association through common personality factors involved in both heavy drinking and aggressive behavior.

Graham et al. (1998) found that intoxicated aggression is the result of the interplay of various influences including societal, personal, situational, and drug effect variables. It is important to recognize that aggressive behavior usually occurs as part of a process, not an isolated act. Often it is not the initial aggressive act that results in the greatest injury, but the escalation of aggressive acts.

Many studies have found general support for the hypothesis that heavy or problem drinkers tend to be more aggressive than other people. Caces, Stinson and Harford (1991) found that nearly all risky behaviors (including a number of measures of aggression) increased with frequency of alcohol consumption, and that husband-to-wife violence increased with quantity and frequency of consumption, with the highest rate among those who consumed the highest quantities per occasion (Kantor & Straus, 1987). Premarital aggression has also been found to be related to the husband's drinking, interacting with various other factors, such as permissiveness and dissatisfaction with the relationship (Leonard & Senchak, 1993).

In studies of children and adolescents, aggressive behavior has been associated with current smoking, drinking, drunkenness, and drug use (Choquet, Menke, & Manfredi, 1991) and with later alcohol use and later alcohol-related aggression among males (White, Brick, & Hansell, 1993; White, Hansell, & Brick, 1993); while for females, early alcohol use was a better predictor than childhood aggression of later alcohol-related aggression (White, Hansell, & Brick, 1993). Hyperactivity diagnosis at age 13 has been found to be related to developing alcohol problems and committing violent crimes (Klinteberg et al., 1993), and childhood aggression has been associated with later aggression and with alcohol abuse (Pulkkinen & Pitkanen, 1993). In a study of university students ages 17 to 22 (Liebsohn, Oetting, & Deffenbacher, 1994), high trait anger was associated with more frequent drinking, drunkenness, and alcohol-related consequences (controlling for consumption).

Among groups identified because of their aggressive behavior, heavy drinking seems to be common. Violent married men reported more frequent drinking and higher endorsement of personal reasons for drinking than nonviolent men (Fagan, Barnett, & Patton, 1988), and a high rate of alcohol problems has been identified among wife batterers (Kantor & Straus, 1989; Stith, Crossman, & Bischof, 1991). Collins and Schlenger (1988) found that violent crimes were more likely than nonviolent crimes to be associated with drinking before the crime.

Bailey and Taylor (1991) found that among moderately to highly aggressive subjects, alcohol increased aggression regardless of provocation, but among low-aggression subjects, a significant effect of alcohol occurred only under low provocation. Graham et al. (1996) found that alcohol consumption increases the probability of aggression only in some situations; namely, situations where there is provocation. Increased aggression is not an inevitable outcome of alcohol use. These studies highlighted the importance of provocation and other situational factors. At the same time, the importance of personal characteristics has been shown by studies of the interrelationships of aggressive behavior and heavy or problem drinking.

In terms of the characteristics of the drinker, Graham et al. (1996) found numerous studies that indicated that people who become aggressive when they drink are more likely to have a history of childhood aggression; to be low on social integration and high on approval of aggression and need for personal power; to expect alcohol to increase aggression; to be part of a subculture that associates drinking with aggression; to be aggressive in nondrinking situations; and to be heavy or problem drinkers (especially consuming high quantities per occasion).

REDUCED ANXIETY

People who drink to reduce anxiety may be more likely to behave aggressively when drinking because the controls over anxiety that would prevent aggressive behavior are reduced (Horton, 1943). Pihl and Peterson (1993c) found that alcohol may increase the capacity for aggression by reducing anxiety and by enhancing motivated psychomotor activity. The combination of impulsivity with alcohol-induced fearlessness and hyperactivity appears prone to produce aggressive acts. Therefore, the more alcohol individuals consume, presumably the less anxious they will be about consequences or harm done to another person.

SEROTONIN

Studies have found that long-term use of alcohol use affects nearly every major neurotransmitter system, including serotonin pathways (which regulate mood and impulsivity) and dopamine, opioid, and cannabinoid receptors (which are important modulators of the rewarding and motivating properties of alcohol) (Hungund & Basavarajappa, 2004; Johnson, 2004; Leshner & Koob, 1999).

Drinking increases serotonin, followed by a decrease, which may increase the effects of dopamine (Graham et al., 1997; Virkkunen & Linnoila, 1993a). An increase of dopamine without sufficient serotonin results in increased stimulation and reduced impulse control (Pihl & Peterson, 1992; Virkkunen & Linnoila, 1992b, 1993; Pihl & Peterson, 1993c).

Pihl and Peterson (1993c) found that deficiencies in the brain chemical serotonin are associated with increased tendency to violence and victimization. Serotonin may play a key role in promoting aggressive behavior. Alcohol is known to facilitate aggression in persons with defects in serotonin. Increased dopamine activity stimulates aggression. Increased serotonin activity suppresses dopamine-induced aggression, whereas decreased serotonin activity increases such aggression.

Pihl and Peterson (1993c) also found that reduced brain serotonin is associated with an increased propensity to exhibit aggressive or impulse behavior, as well as other negative results. Serotonin also modifies the response to threat. When serotonin function is normal, anxiety, the emotional response to threat, inhibits socially inappropriate behavior such as aggression. When serotonin is depleted, anxiety loses its inhibitory effect while retaining its emotional intensity (Spoont, 1992). Thus, a person with low serotonin levels might display aggressive behavior despite intense anxiety induced by the threat of punishment.

Pihl and Peterson (1993c) also found that impaired behavioral control induced by decreased brain serotonin levels may affect a person's response to alcohol. Decreased serotonin may lead to an inability to terminate drinking once it's started. Alcohol transiently enhances serotonin function, perhaps thereby initially moderating violent tendencies. However, serotonin activity may subsequently decrease below baseline levels. This decrease leads to increased impulsivity and reduced threat-induced control of behavior.

IMPACT OF DRUGS

Taylor (1993) found that doses of commonly abused depressants known as benzodiazepines facilitate aggression. Gantner and Taylor (1988) showed that both male and female subjects responded more aggressively after taking diazepam. Lorazepam, another benzodiazepine, also enhanced aggressive responding.

DIAGNOSIS

A general population survey found that aggressive behavior was more common among those meeting a diagnosis of alcoholism, particularly if this diagnosis was combined with antisocial personality disorder or recurrent depression (Bland & Orn, 1986). Diagnosis involving alcohol and drugs appear frequently with aggressive-related offenses. I have found this true in my practice.

TREATMENT MEDICATIONS

Bowes (2005) studied the psychopharmacological effects of alcohol in the brain. He found that patients who received topiramate, which increases activity of GABA receptors, were significantly more likely to remain abstinent from alcohol. Topiramate was also associated with a reduction in alcohol consumption on days of normal heaving drinking and increased rating scale scorers of subjective well being.

Schuckit (2005) noted that the recently approved medication acamprosate (Campral®), which reduces the activity of glutamate receptors, can help to alleviate the impact of alcohol withdrawal. Bowes (2005) indicated that many patients with alcohol dependence are able to control their alcohol use for short periods of time, but are unable to remain abstinent for longer periods. Until recently, only two treatments — disulfiram (Antabuse®) and naltrexone (ReVia®) — were approved by the U.S. Food and Drug Administration (FDA), neither of which markedly improves long-term abstinence. A third agent, acamprosate (Campral), has recently been approved and has shown remarkable effects in reducing the craving for alcohol. Mason (2001) indicated that Campral has been shown to increase the number of patients who remain completely abstinent from alcohol.

Bowes (2005) goes on to state that the drug, naltrexone, has been used for the treatment of alcohol dependence for approximately 10 years. It is an opioid antagonist that is believed to act by reducing the reinforcing effects of alcohol (Kenna et al., 2004). Naltrexone has been evaluated in a large number of prospective clinical trials, most of which have found that it has no effect on abstinence from alcohol or on the time to first relapse with alcohol (Leavitt, 2002). However, naltrexone has been shown to significantly reduce the number of drinks per day and the incidence of heavy drinking. "But it diminishes the likelihood that they relapse to the heavy drinking that is typical of the alcohol-dependent patient" (Leavitt, 2002).

Acamprosate significantly improved the condition in the patients who were completely abstinent from alcohol over the entire treatment duration, which was up to 1 year in some studies (Mason, 2001). "Unlike naltrexone, which did not affect abstinence or time to first drink, acamprosate has its greatest benefits on exactly those outcomes," Mason wrote. In patients who did consume alcohol, acamprosate was associated with a longer latency to the first drink than placebo. There was a higher percentage of abstinence days with acamprosate than with a placebo.

Naltrexone, acamprosate, and the combination were all superior to placebos; combination treatment was significantly most effective than acamprosate alone, but not significantly more effective than naltrexone alone. Combination treatment was also associated with improvements in other outcome measures, including the number of abstinent days and patient reports of alcohol craving.

NOTE FROM DR. MICHAEL L. CESTA, M.D.

The treatment of substance abuse has come quite a distance from the old style of "white knuckling" and attending meetings. Although community supports, including 12 step programs, are a mainstay of substance abuse treatment, we have learned these diseases have a neurobiologic and neurobehavioral component that can be addressed through both psychopharmacologic management and psychotherapy. The view put forth years ago that mediations used to alleviate withdrawal, decrease cravings, and sustain sobriety is "substituting one drug for another" have long passed. In modern psychiatry we recognize the existence of specific brain neurochemistry that can be managed to achieve the desired end result of abstinence. For example, in the diseases of alcohol dependence, three medications have come to the forefront that can make the process easier for a person suffering with cravings and other difficulties with sobriety. ReVia® (naltrexone), Topamax® (topiramate), and Campral® (aAcomprosate) can be used in the maintenance for the symptoms of cravings in alcohol dependence. The use of medications to alleviate the dangers of the withdrawal process have long been the standard of care in medicine. These medications include the benzodiazepams (e.g. Ativan®, Librium®, Valium®) and Tegretol® (carbamazepine) that can be used alone or in combination during the acute withdrawal process. The same holds true for another common disease, opiate dependence. Like alcohol, this disease has a withdrawal process and requires the management of a maintenance stage. It is common to use long-acting opiates, such as methadone, to achieve a graceful withdrawal and to manage maintenance symptoms for an extended period. Recently, the FDA approved buprenorphine for the use of opiate withdrawal and maintenance and this is available from qualified providers in addiction medicine. In addition, there are many areas of research in other addiction diseases, such as cocaine, amphetamines, and others, although these are still in the early stages of development. In summary, patients no longer have to suffer through the stages required to discontinue substance use and maintain abstinence.

Numerous studies referenced conclude that: (1) there is no simple and inevitable relationship between alcohol and aggressive behavior, (2) that situational factors play an important role, and (3) that there appears to be a complex role between expectancies and aggression. However, there is also increasing evidence that the psychopharmacological effects of alcohol play a role in aggressive behavior. The systems most impacted relative to aggression seem to mediate responses to motivationally relevant unconditioned and conditioned stimuli.

Research has shown that intoxicated interpersonal communication and social behavior may be affected by alcohol-induced "myopia" (a short-sighted focus on the immediate situation) (Steele & Josephs, 1990), poorer ability to process incoming information (Pernanen, 1976), and poorer social problem-solving (Sayette, Wilson, & Elias, 1993). Self-awareness appears to be diminished when intoxicated. The role of impaired cognitions due to alcohol has also been implicated in the occurrence of acquaintance rape (Abbey, 1991).

In general, aggression is expected to be linked to drinking by males more than females (Crawford, 1984a,b; Isaacs, 1977), by alcoholics more than social drinkers (Isaacs, 1979), by others rather than oneself (Gustafson, 1987; Leigh, 1987), with the self–other discrepancy being greatest for light drinkers (Rohsenow, 1983).

Pihl and Peterson (1993c) found that in general, "...a person depleted of serotonin is more likely to appear depressed and aggressive, more driven by appetites (more motivated by food, water, sex, and drugs of abuse), and more impulsive (less able to control behavior in the face of threat). Such a person will find it difficult to stop engaging in a behavior once started, unless an alternative source of reward is present. He or she may be more likely to begin consuming alcohol and less likely to stop drinking as long as alcohol in present."

Treatment considerations when alcohol-related violence occurs should address both issues (Graham et al., 1998). Some research suggests that when both problem behaviors are addressed, the risk for relapse with drinking and subsequent violent behavior significantly decrease. The use of anticraving medications appear to significantly improve substance abuse treatment outcomes. Despite many substance abuse providers being overly cautious or stubborn in accepting this data, the use of anticraving medications appears to be a new standard of practice.

Alcohol, Abuse, and Sexual Assault

Here are some of the findings related to sexual offending and alcohol. Approximately 75% of all abuse and sexual assaults involve the use of alcohol on the part of the abuser and/or victim. The percent is probably higher due to victim blaming himself or herself or blacking out. Koss (1988) found that 75% of sexual assaults involved the consumption of alcohol by the victim and/or perpetrator.

Sex offenders use intoxication as a justification or rationalization for committing sexual assault (Barbaree, Marshall, & Hudson, 1993). Abusers intentionally get their victims intoxicated, often without the victims' knowledge (Swisher & Wekesser, 1994). When the abuser was using alcohol, they were more likely to experience and hear "no" as "maybe" (Swisher & Wekesser, 1994).

There is a strong association between drinking and dating and sexual violence (Levy, 1991). Levy goes on to state that "if someone is likely to experience explosive anger or be violent, alcohol and drugs may increase the likelihood, as they reduce inhibitions as well as the ability to make decisions and exercise control over one's behavior." Alcohol has been found to be a consistent marker for men committing physical abuse (Levy, 1991). Makepeace (1987) also found an association between self-reports of having a drinking problem and experiencing dating violence.

Alcohol increases the probability of an aggressive response and has a disinhibiting effect on sexual behavior (Barbaree, et al., 1993). Briddel et al. (1978) found that nonoffending males who believe that they ingested alcohol responded with greater sexual arousal to scenarios of rape and nonsexual violence against women.

Victims are more likely to blame themselves for being physically or sexually assaulted when they were drinking heavily. The percent is probably higher due to victims blaming him/herself or blacking out (Levy, 1991).

Alcohol is a consistent marker for men committing physical abuse (Levy, 1991). Anger, frustration, depression, and rejection are risk emotions and commonly lead to the use of drugs and/or alcohol; abuse or rape may then follow (Barbaree et al., 1993). Barbaree et al. (1993) also found that between 33% and 77% of child molesters report having been intoxicated at the time of the offense. They also found that approximately 30 to 80% of rapists were intoxicated at the time of their offense.

Now lets review a study that was done assessing the role between alcohol, expectations, and sexual offenders.

Relationships between Alcohol Use and Alcohol Expectancies

Veronique Valliere, Psy.D. (1997) studied sex offenders and the impact of alcohol use on how the offenders expected the alcohol to affect them. When people consume alcohol they have expectations of how the alcohol will make them feel and act. Sex offenders use alcohol to justify engaging in sexual contact, even forced sexual contact. What was studied, in summary, was the impact of alcohol on sexual offenders. Specifically, the sex offenders were asked to identify what they expected (expectancies) from the alcohol use. A positive alcohol experience involved expecting to experience positive, good feelings, and behaviors as a result of drinking. A negative alcohol experience was when they expected negative or bad things to happen as a result of drinking. Some of the men were actually given alcohol while others were given a placebo. The placebo looked like alcohol, but was actually alcohol free. With some men drinking actual alcohol and others only believing they were, you can then test how expectancies impact behavior when drunk. I am overly simplifying this research and the results. If you are interested in all the more technical data, you can find the study in the reference section.

The following are findings from the study:

- There were no significant differences between alcohol expectancies of alcoholics and nonalcoholics. What this means is that regardless of whether the man was alcoholic or not, they still had the same expectancies of drinking. For example, if they expected to feel high and more powerful, they did.

- Problem drinkers with high positive alcohol expectancies give less regard to negative consequences and reinforced their own drinking through positive expectations. What this means is that for those men who were problem drinkers or alcoholics, when drinking, they gave themselves permission not to care about the consequences of their behavior. They believed, for example, that if they were drunk, they could do whatever they wanted and did not think about or worry about the consequences.
- The greater the positive expectancies, the less likely they were to be successful at treatment. This means, if the men believed that the drinking was a positive experience for them, and they saw no reasons not to drink, and were not experiencing any negative consequences (in their mind not perceiving negative consequences), then they tended to fail at treatment.
- An increase in positive expectancies leads to an increase in drinking. If the men had increasingly positive expectancies about their drinking, they tended to drink more.
- If the men believed that drinking made them more aggressive (or relaxed), then they became more aggressive (or more relaxed). What this implies is that if the men believed that they would become mean and abusive if drunk, then that is how they behaved when drunk. If they believed that they became more relaxed and easy going, then they became more calm and easy going when drunk.
- Sexual arousal to erotica/pornography is less inhibited by those drinking or believed to be drinking. What this means is that if the men believed that they were drinking or were actually drinking, they became more interested in pornography than when they were not drinking or believed that they were not drinking.
- Males who believed they ingested alcohol demonstrate an increase in sexual arousal to both deviant and nondeviant stimuli, and this was strongest for sadistic and violent rape. Again, when drinking, the men believed they were more sexually aroused to violent and nonviolent pornography and other material.
- Relationship between expectancies and sexual behavior was stronger in those who needed an excuse for sexual behavior. What this implies is that for those men who rape, they may believe more strongly that when they are drunk, that they are actually out of control in regards to their sexual behavior when, in fact, they are not.

The main impact of alcohol appears to be the creation of *Alcohol Myopia*, which implies "near-sightedness." The drinker pays more attention to the strongest impulse they are experiencing and less attention to other issues that are less important to them (e.g., the possible negative consequences for their behavior, the impact on their victim, or self-consciousness). Dr. Valliere offers this example. For those people who are dead set against singing in public, they will not do karaoke no matter how much they drink. But for those who really do want to sing in public, drinking facilitates the "nerve" and helps overcome the sense of embarrassment or self-consciousness. Alcohol does not lower inhibitions, but lowers the control of the negative consequences. After a few drinks, you may not pay attention to the sense of embarrassment if your urge to sing increases. The same is true for rape, abuse, affairs, spending, etc. Your strongest impulse will win after enough alcohol. You become "near-sighted" to that urge, experiencing a sense of "tunnel vision."

For someone who is already angry or upset, after a few beers, they may give themselves permission to abuse or rape. In fact, many abusers and rapists have claimed that they drink so that they can carry out the impulse or urge. Abusers and rapists are also aware of the secondary benefit of using alcohol when they decide to act out their urge or impulse of violence, that is, the decreased accountability that others may place on their actions. Others, including the victim, may blame the alcohol for the abuse or rape rather than the abuser or rapist, which is what the abuser and rapist count on.

Dr. Valliere's research demonstrated:

- That alcohol does not cause rape; I believe this can be applied to abuse as well. If alcohol caused rape or abuse, then everyone who became intoxicated would abuse and rape. The reality is that most people who become intoxicated do not engage in acts of violence.
- Those who are strongly against rape (or abuse) will not abuse or rape just because they are drinking. This is because they had no preexisting impulses, urges, arousal, or interest in violence to begin with.
- How the person is affected by the alcohol is really a self-fulfilling prophecy. If the person expected to become violent when drinking, he usually did. If he expected to become mellow and calm, he usually did.
- For those who already have an arousal to violence, alcohol use increases arousal to violent material, such as pornography and violent movies. That is to say that abusers and rapists had thoughts of abusing and raping *before* they drank, and drinking increases that preexisting arousal.

SUMMARY

Alcohol and drug use increases the probability of sexually aggressive behavior and abusive behavior in people with the preexisting propensity for it. Alcohol and drugs never *cause* violence or sexually aggressive behavior. As Dr. Valliere says, "What brand of beer would make me aroused to a 3 year old?" Well, then, I ask, "What kind of beer makes me want to abuse my partner or to force sex?" Most abusers use and abuse alcohol or drugs, but what the abuser believes that alcohol or drugs will do to them, or how they will behave while under the influence, is primarily determined by the abuser's self-talk, that is, their self-fulfilling prophecy. On the other hand, consider once again "alcohol myopia." The abuser may become more enraged when the victim interferes with his "alcohol expectancy" (e.g., "I just wanted to have a couple of beers to relax and you gotta come in and start bitchin' at me!"). The abuser expected to relax, and the victim failed to comply by engaging in behavior that interfered with the abuser's expectation. When drunk, it is much easier to give oneself permission to be abusive, but only because of the "alcohol myopia."

Problems in Assessing Alcohol and Drug Use in Abuse and Violent Crime

Understanding the role of alcohol or drugs in abuse is at times difficult. Sometimes it is simply easier to believe that the alcohol or drugs made the person behave abusively. But the reality is that all abuse is a conscious, planned choice. Here is support for my beliefs.

1. It is difficult to determine the actual degree of intoxication of the abuser at the time of the offense. For example, how drunk was the abuser when the abuse occurred? Often, the abuser is not arrested for hours or sometimes days after an assault. Therefore, we only have the abuser's word, and sometimes the victim's, on how drunk he actually was. If the abuser says that he was "drunk", what does that mean? Was he feeling "buzzed" or "good," or was he stupidly drunk, wetting his pants, passing out, etc. On one hand, it really does not matter. Whether he was intoxicated or not is a moot point, he still chose to abuse.
2. The victim is not always a good source to judge the degree of the abuser's intoxication. This is because at the time of the abuse, the victim was fearing for her safety and, therefore, not really paying attention to how much the abuser had to drink. Also, the victim was probably not carefully watching how much the abuser was drinking. She had no reason to.

3. Consider this: If the abuser was actually as intoxicated or high as he claims, would he have been able to commit the crime he is accused of? Many abusers claim that they were so drunk or high that they do not remember anything. They were so intoxicated that they did not know what they were doing. But if that is true, then consider Chapter 7: "Intent to Abuse: Awareness of the Use of Force." Could the abuser have actually been so "out of it," yet respond with such precision to his victim's every move? Unlikely. Regardless of how intoxicated he was, the abuser still was aware of what he did when he did it. And he gave himself permission to do it. It was a conscious and planned decision to abuse. He may have told himself that being drunk was an excuse, but even that is a conscious decision to believe it. Consider that a falling-down drunk is not much of a threat.
4. Most abusers were far less high or intoxicated at the time of the abuse than they claimed. And even if they were drunk, they should be able to recall most of the abuse details. Claiming "not to remember" is a conscious cop-out as well. Their memory seems to work well enough to deny their victim's version of the abuse, but not well enough to recall exactly all the significant details. That is not how the mind works. They are purposively and consciously deciding what to "forget," but they could recall the information if they wanted to or were encouraged to.
5. The planning of abuse, the limitations of what they chose to do or not to do to the victim, and how they managed to escape successfully afterward demonstrates planning and a conscious decision. Even if intoxicated, they were still aware enough to abuse and not push it too far.
6. Many abusers went on to engage in "normal" routines after the abuse. How could they do that? How could a person abuse their partner or children and then go on as if everything is just fine and normal? Some even go to sleep after abusing. Well, they can do it because they made a conscious choice to abuse and they feel confident that if they blame alcohol or drugs that they will be "off the hook" and that their victims will forgive and forget.
7. Even when an abuser claims to have a "blackout," they should be able to recall most of their abusive behavior. Some claim that they were in a blackout during the abuse. But that is impossible. How could they abuse so precisely yet not be in control. It would be impossible. Blackouts involve the inability to recall what happened in the past, not something that is happening right now. However, if the abuse or rape behavior is commonplace, the abuser may be less likely to recall specific incidents because they happen so frequently. However, the bottom line is that the abuser has some recollection of what happened. The less they can remember about the abuses or rape, the more likely that it has frequently occurred. Hold them accountable regardless of whether they remember or not.

ACKNOWLEDGMENTS

I would like to offer my sincere thanks to Dr. Veronique N. Valliere, Ph.D., for her insight into this chapter and for her research in this area. I would also like to offer my sincere thanks to Michael L. Cesta, M.D., for his insight into this chapter.

20 Pornography and Abuse

> "If you believe that no one was ever corrupted by a book, you have also to believe that no one was ever improved by a book. You have to believe, in other words, that art is morally trivial and that education is morally irrelevant. Pornography's whole purpose is to treat human beings obscenely, to deprive them of their specifically human dimension."
>
> **Unknown Source**

ROLE THAT PORNOGRAPHY PLAYS IN VIOLENCE

Many people fail to understand the role that pornography plays in violence. We have come to understand how sex offenders use pornography to support their deviant and violent fantasies and understand that pornography should not be used to teach "healthy" or "appropriate" sexuality. Abusers, including those that commit domestic abuse, dating violence, or child abuse, almost all use pornography to feed their deviant and violent fantasies as well.

Pornography seasons men to view sex partners as compliant, powerless objects to do with as they choose. Think about how pornography impacts people in general. People are pleasure seekers as a general rule. We do whatever we can to avoid negative consequences and do whatever we can to maximize the occurrence of pleasurable rewards. Research has shown that pornography does not cause sexual violence, abuse, child molestation, or any form of abuse. However, pornography contributes to negative and unhealthy thinking, and fosters inappropriate and aggressive beliefs about women and sexual partners. Further, the use of pornography supports the deviant sexual and violent fantasies that abusers and sexual offenders possess. At some point, the fantasy is acted out in order to provide the offender with pleasure and satisfaction.

Pornography appears to thrive on featuring social encounters in which women are eager to accommodate any and every imaginable sexual urge of any man in the vicinity (Zillmann & Bryant, 1986). They found that pornography entices actions that result in experimentation, which leads to less than satisfactory results. Beauregard, Lussier, and Proulx (2005) found that there was a link between sexual interests, situational factors, and rapist's modus operandi. The offenders (in their study) who had a significant sexual interest in nonsexual violence showed a higher level of organization in their modus operandi, that alcohol consumption prior to the offense was related to a higher level of coercion, and that a negative emotional state prior to the offense was related to higher levels of victim injury. The offender's personality characteristics impact offense details. The authors found that the use of pornographic material and sexual arousal in the hours preceding the offense related to a reduction of both the level of force used by the offender and a reduction in the degree of injury sustained by the victim. However, their findings do not appear supported by other respected authors, and their population included only federal inmates, a population not generally included in other research studies.

Beauregard and colleagues also found that a negative emotional state prior to the offense is linked with an increase in the level of injury inflicted upon the victim during the assault. This was supported by findings of Marshall et al. (1983) who found that sex offenders under the influence of anger are prone to use violence during a sexual assault. Yates, Barbaree, and Marshall (1984) found that anger increases the level of arousal to coercive stimuli. Several authors (e.g., Proulx et al., 1999; Ouimet et al., 2000) also found that anger prior to the offense was linked to the use of expressive violence and to a high degree of victim injury.

Hazelwood and Warren (1995) indicate that sexual fantasies are an important component of sexual crime. Fantasies serve an important role for the sexual offender, helping to influence the offender's choice of physical, sexual, and verbal interaction with the victim, and the offender's preferred sexual acts and rituals. Several authors have indicated that, for some people, fantasy alone is not satisfactory and that there tends to be a progressive desire to put the fantasy into actual behavior (MacCulloch et al., 1983; Prentky et al., 1989; Ressler, Burgess, & Douglas, 1988; Schlesinger, 2000). Prentky and colleagues indicated that once the restraints for *inhibiting* the acting out of the fantasy have disappeared, the individual is likely to progress to acting out the fantasy and will continue to progressively reenact the fantasy in "trial runs" until the fantasy has been acted out as close to perfect as possible. Several researchers (e.g., Revitch, 1965, 1980; Reinhardt, 1957; MacCulloch et al., 1983) found a similarity between the content of sexual fantasies and the actual crime. In addition, Beauregard, Lussier, and Proulx (2004) found that coercive sexual fantasies were related to a sexual preference for physical violence measured phallometrically, which would suggest that the violent behavior of both batterers and sex offenders are impacted by coercive sexual fantasies. Chene (2000) found that situational factors (e.g., alcohol consumption prior to the offense and anger) were related significantly to the level of coercion in sex crimes.

Russell (1980) documented that women take the brunt of this type of pornography-inspired experimentation. Men, we found, made women comply with their requests to try what had been seen in the pornography. Activities ranged from oral sex to group sex, from being gagged to being beaten, and from urination into the mouth to having intercourse with pets at hand. Requests tended to be backed by brutal force, and many women reported feelings of degradation and humiliation in addition to having been physically hurt.

Zillman and Bryant (1986) found that after massive exposure to pornographic materials, men found pornography less offensive and objectionable. When students were given massive exposure to pornography and had to recommend prison sentences, those viewing massive amounts of pornography tended to request shorter prison terms for rape, likely due to pornography making rape appear to be a trivial offense. The study found that the apparent loss of compassion for women as rape victims, occasioned by massive exposure to pornography, generalizes to a loss of compassion for women's causes. Finally, massive exposure to pornography significantly increased men's sexual callousness toward women.

Zillman's and Bryant's study focused on sexual callousness toward women, and demonstrated that massive exposure to standard pornographic materials devoid of coercion and aggression seemed to promote such callousness, specifically, the trivialization of rape. In addition, massive exposure to pornography is a growing dissatisfaction with sexual reality. Russell found that men readily succumb to the suggestion of hypereuphoria that pornography connects with assorted sexual techniques. The results of this are likely to produce disappointment and dissatisfaction

Cramer and McFarlane (1994) found that 40% of 87 battered women who filed charges reported that their male partner used one or more pornographic materials. Use of the pornographic materials was significantly associated with the women being asked or forced to participate in violent sexual acts, including rape. Malamuth et al. (1980) found that men were most aroused to rape pornography when the victim experienced an orgasm and pain.

Ted Bundy (Dobson, 1989) said, "My experience with pornography that deals on a violent level with sexuality is that once you become addicted to it — and I look at this as a kind of addiction — I would keep looking for more potent, more explicit, more graphic kinds of materials. Until you reach the point where the pornography only goes so far. You reach that jumping-off point where you begin to wonder if maybe actually doing it will give you that which is beyond just reading about it or looking at it."

Pornography includes magazines, movies, videos, and photos depicting people engaging in any type of sexual behavior. This includes people posing naked or partially clothed while exposing at least some of their genital areas. Sexual behavior can include autoerotic, which involves masturbation, a solo sex act that does not involve a partner. Sexual contact also includes acts involving sexual interaction between two or more people. Pornography can be in the form of magazines and photos, or on the Internet. Sex lines and strip shows also constitute pornographic material in my definition. Pedophiles commonly use department store catalogs and newspaper ads depicting children and adolescents wearing underwear or swimsuits. I found that when searching the prison cell of pedophiles that they often had such materials. Yet not one authority figure thought twice about the materials being in the inmate's cell.

Pornography is used to satisfy sexual urges and fantasies. People who say that they buy pornography only to view and read the articles (such as *Playboy*) are simply liars. If they wanted articles, they could have found similar articles and topics in nonpornographic magazines. Pornography serves a sexual purpose each and every time it is used. There is no such thing as a "nonsexual" use of pornography. People view it and usually masturbate to it.

When the user views pornographic material, they observe the entire scene that is depicted. With pornography magazines, many of the facial expressions display anything but pleasure, respect, or consent. In fact, most pornography I have viewed display women and men whose facial expressions clearly are indicative of fear, discomfort, anger, frustration, surprise, or a neutral, void expression. These expressions are contrary to what we would expect of consensual, appropriate sexual partners. Many of the people depicted in pornographic materials are positioned in unrealistic, uncomfortable positions. Often the body is not shown in it's entirety or may be only partially clothed. How many people would engage in sexual contact with clothes or stiletto heels on?

I would imagine that photographers take hundreds of photos of persons depicted in the pornography. It seems to be in the publisher's best interest to show photos displaying bodies in their entirety, with facial expressions depicting consent, comfort, and respect. However, this is not the case. It is as though the publishers and photographers of pornography purposely choose to include in their magazines and movies mostly that which portrays nonconsensual sex and unrealistic sexual interaction. I believe that pornographers purposefully pair the negative, degrading and abusive, because violence and forced sex sells far more than consensual sex would. Consider the images portrayed in the facial expressions of women in magazines that have been labeled as "soft core" pornography. "Soft core" refers to the lack of actual sexual acts occurring between two or more people; basically only one or two people in a state of total or partial nudity posed as if engaging in sexual behavior. So, what they are really selling is sexualized aggression, violence and force with intimacy, and surreal expectations of what to expect from sexual partners. The end result is that they pair anger with sexuality.

The user of pornography becomes desensitized to the humiliation, degradation, injury, and pain of the person depicted in the pornography. Women especially become the objects belonging to men to do with as they please. It was not that long ago when victims of sexual abuse (including incest) were questioned as to whether they enjoyed the sexual abuse, whether they asked for or deserved to be sexually abused, and to have their past sexual histories used to determine if they were promiscuous or virginal, as if it really mattered. Men have long been excused from being held accountable for their physical and sexually abusive behavior. Even today in the armed services, the men are encouraged to find prostitutes when on leave. The men are supported for their physically and sexually abusive behavior by the lack of punishment and the degree of coverup. The military continues to teach men that they must have sex, and that a female (any age) should comply and become subservient to the man who is spending time with her, buying her anything, or in any way showing her special attention.

What would truly happen if the men in the military abstained from sexual contact for several months? Probably nothing. Most men abstain from sexual behavior with others at other times in their lives. Consider marriage, for example. The couple may choose not to be sexually active for sometime for a variety of reasons. And no one dies; no one's genitals fall off during the period of abstinence.

If men developed more respect for women, they would more than likely use far less pornography. This would certainly not please the pornography industry. Recall how the tobacco industry continues to fight as they, in a state of delusional wisdom, attempt to convince people that smoking is a harmless, low-risk activity, despite insurmountable evidence to the contrary. What would the pornography industry do to maintain the right to humiliate and denigrate women? Wouldn't that be a battle to witness?

Other Paraphilic Behavior

Strip shows, viewing strippers, and telephone sex lines involve acting out the sexual fantasies. People who attend strip shows are engaging in sexual acts. If they are involved in a relationship, they are cheating. Think about it. Despite being involved with someone, they choose to attend strip clubs to become sexually aroused, and many will orgasm while there or when they return home. If they engage in sexual contact with their partner after attending the strip show, they will almost always be fantasizing about the stripper they saw, not their partner. In my opinion, attending strip shows, viewing strippers at parties, and use of sex lines is the same as prostitution; you are paying to view the person stripping and possibly engaging in sensuous activity, at times even including a "lap dance." This is a sexual affair.

The Role of Fantasy

Fantasy refers to thinking or imagining something that is not actually occurring at that moment. It may involve recalling and replaying the memory of a prior event, or may involve something that has not yet occurred. For example, recalling your vacation. You recall the ocean, the breeze, the smells and sounds, as well as the comfort you felt. This is likely a healthy fantasy. It brings peace when stressed. Another example is thinking about something that has not yet occurred. Imagining what it will be like to buy a new car. You picture yourself sitting in the driver's seat, cruising down the highway. Or thinking about an upcoming event or date. Imagining what will or might take place (e.g., how you may feel, where you may go). All of these are examples of fantasy. Another example is dreams. Dreams occur when you are asleep, or when you drift off into daydreams during the day.

Fantasizing is a normal human experience. We all do it. It allows a brief escape from reality. There are types of fantasy, basically describing the type of or focus of the content. You may fantasize about loving someone, a vacation, or what you will do at the end of the day. You may think about engaging in revenge, violence, or other destructive behavior as well. When the focus is on sexual behavior, this is referred to as sexual fantasies. Again, fantasies are normal; we control the content (except for dreams), and what matters most is the degree that one utilizes fantasies and whether one reinforces the fantasy.

What you use fantasies for is important. If you focus your fantasy content on sexual behavior with a consenting partner, your fantasies will more likely be positive and healthy. However, if the focus of the fantasy involves violence or engaging in illegal behavior, such as rape or sex with children, then the fantasies become unhealthy or deviant. How often and how much time you devote to fantasy is also important. If you use fantasy to escape reality too often or use it to tune out your partner, for example, then your fantasies are more likely to cause problems in your relationships. If your fantasies become obsessive (the same content occurring several times per week or even per day), and you are unable to control them, you have a more serious problem.

Sexual fantasies involve thoughts, plans, and images that are sexually arousing to you. Sexual fantasies are a form of sexual self-stimulation. If you masturbate during or following the sexual fantasy, then you are reinforcing the fantasy with a pleasurable reward (sexual stimulation and orgasm). You are more likely to repeat the same sexual fantasies when reinforced with masturbation and/or orgasm. If the content or theme is healthy and involving consensual, respectful sexual activity, then you are more likely to act sexual in healthy ways. If your themes are deviant in nature, then you are supporting engaging in deviant sexual behavior. Fantasies have a strong influence on our behavior. All behavior is planned, that is, we think about what we are going to do and then decide to act on it. Therefore, if you are engaging in deviant sexual fantasies often, and reinforcing them with sexual stimulation (masturbation), then you are encouraging yourself to act out.

Pornography rarely portrays men and women as human beings. In fact, I have seen many soft- and hard-core pornographic magazines in which the women appear to have bruises, to have been crying, or expressing fear, hurt or discomfort. And how long does it usually take for a male and female to have an orgasm. According to the pornography, it takes 10 to 30 minutes. How unrealistic.

When men masturbate to pornography, they reinforce that the expressions on the faces of the women, the sexual positions, and the stories portrayed as desirable for women in general. These men become more prone to act out what they see and read in pornography materials with their sexual partners. Pornography use may even become part of the couple's sexual behavior, using the pornography as a precursor to being sexual with each other. The problems that exist here are serious. As the men increase their use of pornography and increase the amount of time masturbating to the pornography, they develop unrealistic expectations of what their sexual partners should be willing to engage in, and without discussion.

I have heard far too many perpetrators of sexual abuse state that their victims displayed facial expressions indicative of pleasure, desire, and consent. After some time in sex offender treatment, they tell a different story. They become increasingly more open and honest about their victim's reaction to the abuse. They admit that the expression on their victim's face was anything but pleasurable and consensual. In fact, most could relate the expression on their victim's face to the facial expression of the women in the pornography materials they used. Many of the perpetrators also discussed how they increased the amount of alcohol and/or drug use prior to, during, or following the sexual abuse to help forget the true expression and reaction of their victim. What insight to finally have, but, unfortunately, too late.

The relationship between sexual fantasy and sexually deviant behavior has been well established (Marshall et al., 1983; Quinsey & Earls, 1990). Remember that pornography provides an unlimited source of sexually arousing material and helps to maintain and strengthen deviant sexual interests. The sexual interests support the deviant sexual behavior (e.g., rape, child molestation, any forced sexual contact). In fact, several researchers (e.g., Malamuth, 1986, 1989; Malamuth, Check, & Briere, 1986) found that deviant sexual arousal patterns do, in fact, differentiate sexually coercive men from nonsexually coercive men. The use of pornography helps shape the offender's sexually deviant and violent fantasies, which then are more likely to be acted out. The presence of deviant sexual fantasies does appear to increase the likelihood of subsequent deviant sexual behavior (Abel & Blanchard, 1974). Marshall et al. (1983) indicated that the success of increasing nondeviant sexual arousal and behavior by applying techniques solely aimed at modifying arousal to deviant sexual fantasies, supports the hypothesis that deviant sexual fantasies not only lead to and maintain deviant sexual behavior, but also impede normal sexual adaptation. Prentky and Knight (1991) indicated that sexual fantasy is complexly related to sexual aggression; the presence and nature of such fantasies may help differentiate rapist subtypes (Brittain, 1970; Burgess et al., 1986; Knight & Prentky, 1990; MacCulloch et al., 1983; Marshall et al., 1983; Prentky et al., 1989; Quinsey et al., 1984). Prentky and Knight further indicated that the degree of deviant sexual arousal may be related to both the frequency of offending and to the amount of violence in offenses (Abel et al., 1977). This is likely due to the fact that the more a batterer or sexual offender views pornography, the more diverse and likely violent the pornography becomes, and, therefore, the subsequent sexual fantasy becomes increasingly more deviant and aggressive. Subsequent violent behavior is then a form of reenacting the deviant sexual fantasies. Prentky and Knight go on to state that detailed studies (Burgess et al., 1988; Hazelwood, Reboussin, & Warrin, 1989) found that the onset of repetitive sexually aggressive behavior coincided with the first appearance of rape fantasies and that these repetitive offenders were also characterized by a high incidence of paraphilias. It is hypothesized that the offender continues to "stage" or practice his fantasy, and the unsuccessful match between the reality of his attempts and the richness of his fantasies contributes to reoffending (MacCulloch et al., 1983).

Pornography and Abuse

I will be referring to women, children, and teens as being portrayed in pornography. However, please understand that men and boys are also portrayed in pornographic material. For the purpose of this book, however, I will be focusing on pornography depicting women, children and teens. It is important to understand the relationship between pornography and abuse. Corne et al. (1992) found that early exposure to pornography was related to subsequent "rape fantasies" and attitudes supportive of sexual violence against women. Pornography does impact your beliefs and attitudes about sexual behavior and women.

The following was adapted from Russell (1999, 1994a,b, 1993) with verbal permission.

1. *Masturbation to pornography is an extremely strong reinforcer for deviant sexual behavior.* Masturbating to pornographic material reinforces the sexual fantasy, and significantly increases the likelihood that the person will use pornography in the future. Over time, the person will be more likely to increase the frequency of viewing pornographic material.
2. *Pornography encourages promiscuity, abuse, bestiality, and paraphilic behavior.* When viewing pornography, you begin to believe the messages portrayed in the material. Pornography encourages unhealthy sexual practices, including sex with children and teens, and encourages and portrays promiscuity and infidelity as desirable and appropriate. Pornography also encourages sexual attraction to deviant and unhealthy sexual interests and practices, which may include sexual attraction to children, the use of force during sex, and developing fetishes (e.g., sexual attraction to objects, such as underwear, stocking, window peeping, sex with animals).
3. Sex lines and strip shows increase sexualization, objectification, and submissive expectations, especially toward women and children. The only purpose of strip shows and sex lines is to provide the viewer with sexual arousal and pleasure. When viewing sex shows or using sex telephone lines, you focus almost exclusively on the sexual aspects of the person on the phone or stripping. You do not even see them as a "whole" person, but rather as breasts, vagina, and as something to please you sexually. Think about it. People rarely refer to the stripper as a nice woman, but they are quick to call her sexually derogatory names (e.g., bitch, cunt, whore). These names are indicative of believing that the person stripping is, in fact, less than human, an object to be used and abused for your sexual pleasure. In addition, strippers and people working on the sex line are submissive, doing most anything you tell them to do. They do not often refuse or argue, it is a one-sided interaction. You are paying for sexual interaction.
4. *Pornography pairs sexually arousing stimuli with rape.* Pornography portrays women and children as being coerced, tricked, and, at times, forced to engage in sexual behavior. Rapes are often portrayed in pornography as leading to pleasurable sex, with the victim being unharmed in the end. But that is an exceptionally distorted view of the way a rape victim experiences the sexual contact. Victims do not end up enjoying the rape later, and most will not have an orgasm. Rape is forced sex. Forced sex includes being talked into or pressured into sexual contact; being threatened with harm if they refuse; physically forcing sexual contact of any kind; taking advantage of someone's degree of intoxication or drug use; taking advantage of someone's mental health issues (e.g., depression, dependency); and in any way engaging in sexual contact without gaining freely given consent *before* the sexual contact begins. Rape also includes engaging in sexual contact with someone who is under the age 18 when you are an adult (and more than 2 years older); if you are or were in a position of power or authority over the person (e.g., boyfriend/girlfriend of the person's parent, therapist, clergy, doctor, attorney, anyone ever in a position of power over them, again, past or present); or when they are blackmailed into the contact. Pornography often portrays subtle coercion and violence leading to

sexual contact, such as between teachers and their students; doctors with their patients; or children and teens and their parents/step-parents/parent's partners.

5. *Become sexually aroused and fantasize about deviant sexual behavior.* Either while viewing or shortly thereafter, the person will masturbate to the pornography. The only purpose of pornography is to provide the deviant fantasy material. The person then remembers what they viewed in the pornography and will either masturbate at some point to the sexual fantasies from the pornography or will engage in sexual contact with his partner while fantasizing to the images from the pornography.
6. *Sexualizes dominance and submission.* Pornography encourages people to sexualize and expect dominance and submission from their sexual partners. Most pornographic material depicts or portrays women and children and teens as being subdued by the male, and the women and children/teens as being submissive.
7. *Creates a desire for increasingly stronger pornographic material.* Over time, regardless of whether it is several weeks or years, the person requires increasingly more violent and deviant pornography to satisfy his needs. The proof for this is that the person buys more and more pornography, usually increasing in deviant content. Movies become more and more hardcore and include more forced sexual scenarios or scenes portraying sex between adults and minors, or sex between someone in a position of power or authority and the women as a client, patient, or in a far less powerful role.
8. *Sexually objectifies females and children.* Pornography portrays women and children as sex objects; objects to be used and abused for sexual gratification. The women and children/teens are referred to as anything but human, being referred to as slaves, "bitches," "cunts," etc.
9. *Increases strength of rape and abuse myths.* Pornography repeatedly uses language and shows scenes that portray sexualized and objectifying language as appropriate and common. The rape and abuse myths most commonly supported include that women and children are supposed to please their partners without regard to sacrifice, pain, or safety. Most of the faces of the women, teens, and children portrayed in pornography are displaying anything other than consensual, happy expressions. You can view any soft-core pornographic magazine to prove my point. Cover the body of the person portrayed, looking only at the face. Identify the expression displayed. Ask yourself, "What would I think about the person if she were in front of me with that facial expression?" Most often the women and children and teens portrayed appear to have been crying, have reddened cheeks, often bruised cheeks, are wearing what appears to be an extreme amount of makeup (appearing to be covering swelling, bruises, etc.). If the expressions of pain, suffering, fear, etc. are paired with sexual reinforcement, then the rape and abuse myths are being supported. These include that women's, children's, and teens' fears and tears are really about lust and wanton sexual desires, and that they deserve and want to be abused and raped.
10. *Increases comfort in the use of violence.* The more frequently persons view pornography, the more at ease they become. They may develop a cavalier attitude about it, and may at some point even leave their pornographic material out for others to see, having no sense of shame about it.
11. *Trivializes rape and deviant sexual behavior.* After a period of regularly viewing pornographic material, the person becomes increasingly more comfortable demanding sexual contact, and may increase the use of sexual profanity and sexualized language. This results in the person becoming desensitized to rape and violence against women, children, and teens. In fact, the narrative stories and statements often shown with the photos are frequently of themes of rape, incest, being too drunk to fight, and other talk of unwanted submission. As the reader jokes about the statements or stories, he is sexually aroused to the photos or scenes, and then strengthens the connection between violence and sex.

The stories help to trivialize rape. I have found many abusers and rapists who will blatantly state that their partners are not allowed to change their minds about sex, nor are they allowed to refuse the man's sexual demands. It is interesting to find that these same men regularly viewed pornography, much more so than anyone was aware.

12. *Users become numb to victim's feelings, desires, and physical safety and well being.* When persons become comfortable with their pornography usage, they begin to act out what they see in the pornographic material. They may force sexual contact when they know their partner is not interested in sex, they may expect their partner to perform degrading sexual contact and demonstrate a disregard for their partner's health and well being. The sexual contact becomes a one-way act, with the focus only on the male's pleasure. Over time, they usually cease taking much, if any, time trying to sexually please their partner. Sex occurs when the male wants it, and only lasts until the male is pleased.
13. *Diminishes fear of social and legal sanctions.* As persons rely more and more on pornography for their sexual pleasure, they begin to demonstrate a fearless attitude about their sexually selfish and sexually violent behavior, letting their partner know that they really do not care of what others would think and believe that they are immune from any sense of criminal prosecution.
14. *Increases comfort that others will accept deviant behavior depicted in the pornography or resulting from the use of pornography.* Over time the person becomes much more comfortable with the pornography use, and becomes much more comfortable practicing what is seen in the pornography. They disregard concerns from their partner or from their victim, and expect compliance. After all, the people portrayed in the pornography rarely refuse, talk back, or make their expectations known.
15. *Places women and children in high-risk situations.* The more the abusers use pornography, the greater the increased risk for acting out with sexual and nonsexual violence toward the women, children, and teens with whom they are involved. They begin to expect that the women, children, and teens in their lives meet their expectations and demands, sexual and nonsexual, because that is what pornography portrays: selfless sacrifice and blind, total submission.
16. *Creating a pornography industry that needs victims.* If you have viewed several issues of any pornographic material and pay attention to the context of the sexual acts and the facial expressions of the women, children, and teens portrayed, you will see what I am referring to. What might have began as a consensual photo shoot or sex movie turns into a living hell when the actresses are not allowed to refuse expectations, and are exposed to violence to force compliance. I am not suggesting that all people portrayed in pornographic material are forced into cooperating, but certainly too many are. Look at their faces. You cannot easily fake fear, hopelessness, and helplessness.
17. *Even "soft-core" pornography such as* Playboy *contains child pornography.* If you view the cartoons and ads in the pornographic material, children are referenced numerous times, included in sexual acts and sexual jokes, all serving to desensitize the viewer to violence and child pornography. Not to mention the centerfold. The centerfold person has numerous photos of herself in the pornography issue, and a foldout as well. Part of the foldout includes photosphotos of the woman as a young child, school-aged child, and as an adolescent. Now picture this. You are sexually aroused (as that is why you are viewing the pornographic material), and now you are viewing photosphotos of this woman as a young child, often wearing nothing more than a diaper. Then you see her as a young schoolgirl in a skirt or swimsuit. Your sexual arousal now includes an arousal for children and teens, as evidenced by the fact you took time to view this centerfold's photosphotos. And then you wonder why it seems that the incidence of incest and child molestation is on the rise.

18. *Pornography offers ever-increasing sexually deviant material.* You can find pornography to fit every desire, from what might appear consensual to rape, sex with animals, incest, and rape leading to murder.
19. *The Internet provides free sexual material, including actual rapes and child pornography.* Anyone can easily access any type of sexual material on the Internet without paying for it. Just surfing around, you can find child pornography, actual rape and incest photos and movies, and all within the privacy of your home or office. Most abusers regularly use the Internet for pornography and much of this occurs without anyone being aware. Sex offenders also utilize the Internet for pornography as well, and, surprisingly, sex offenders do not appear to use any more violent pornography than abusers. Abusers and sex offenders view and regularly use pornography.
20. *Women portrayed as teens and children.* Pornography often portrays adult women as children and teens. This is done with the way the women dress, wear their hair, and the way they act. Even in everyday ads for clothes and perfume, women are depicted as teens to sell merchandise. Think about how young the models are suppose to look, how thin they have to be to succeed as a model.
21. *Children and adolescents portrayed as adults.* Pornography also places children and teens in adult roles, expecting that they will engage in the same sexual contact as adults. Again, look at the ads and cartoons in pornographic magazines, as well as the roles children and adolescents are given in pornographic movies.
22. *Women are often portrayed in unrealistic and uncomfortable positions.* Pornography is notorious for making it appear that women enjoy unrealistic sexual positions. Common examples include portraying women as enjoying sex while wearing stiletto heals, and while still clothed. I am not implying that doing so is wrong or perverted, but I am not aware that most women prefer that. The positions women are often in are unnatural for the body and appear uncomfortable. The men, then, expect that what pornography portrays is reality and that most women would enjoy doing the same as what is being portrayed in the pornography.
23. *Women portrayed performing unrealistic or sadistic sexual acts.* Pornography often has women engaging in unrealistic and sadistic sexual acts. For example, women are often portrayed as enjoying being penetrated anally, vaginally, and orally at the same time and enjoying it. Again, if some women prefer and enjoy this, I am not saying it is wrong or deviant. But I am not aware that most women would say that multiple penetrations at the same time would be desirable or pleasurable. This also implies that women want two or three sex partners at a time. Pornography also portrays acts, such has rape, as desirable and wanted. Other examples of sadistic sexual behavior include women being fisted (someone placing their entire hand and, at times, arm into the women's vagina or anus), sex with animals, or sex with objects that appear too large or unsanitary.
24. *Multiple penetrations portrayed as desirable, enjoyable, and "normal."* So much pornography portrays women as wanting and enjoying multiple penetrations simultaneously. Again, I am not aware that most women would enjoy this. In fact, if you watch the facial expressions of the women in the photos or movies who are engaging in multiple penetrations, they often appear to be experiencing pain, fear, and humiliation.
25. *Women and children portrayed as subservient sexual objects.* Pornography portrays women, children, and teens as being subservient and obedient. Pornography often depicts women, children, and teens being punished for refusing to do what the man tells them to do. The message that men hear loud and clear is that women, children, and teens are simply sexual objects to use as they please. Interesting how many incest dads began molesting their children after viewing pornography for years, eventually preferring pornography depicting or involving actual children or teens. Abusers have the same history,

viewing pornography and then expecting their partner and children to blindly obey them and to do whatever they are told to do sexually.

26. *Women and children portrayed as wanting to be sexually attacked.* Pornography often depicts women, children, and teens as wanting to be raped, abused, and taken advantage of. Again, listening to the story or reading the text highlights this point. Many of the stories involve the women and children and teens as needing to be taught a lesson or that they are scared, but come around to enjoying being raped or abused.
27. *Portraying violence during sex as appealing, wanted, and acceptable.* Pornography all too often portrays violence during sex as being acceptable and wanted, or that the women, children, or teens deserve to be abused as they are raped. Think about it. Most abusers call their partner sexually vulgar names before, during, and following the abuse, and then they expect sex to make up.
28. *Sexual partners portrayed as compliant, powerless, and undemanding.* Pornography is all about the fantasy, that is, what you imagine happens while looking at the pornographic material. The women, children, and teens are expected to be compliant and helpless. They are not portrayed as being able to make demands upon the man; in fact, they are often abused as a result of making even the simplest request. When the man masturbates to this kind of material, which is exceptionally common with pornography, then he becomes increasingly less sensitive to his partner's requests and limits, not expecting nor tolerating any limits or any degree of refusal from his partner in regards to sexual behavior. He expects that she will do what he says, when he tells her to do it exactly the way he expects, without exception.
29. *Sexual behavior portrayed as casual.* Pornography takes away the human connection that is typically present during consensual sexual contact. Sex is portrayed as something that requires little, if any, intimacy or communication. What happens is that the abuser pays attention to his partner only when he is attempting to obtain sexual contact or during sexual contact, but ignores his partner afterwards. He does not want to cuddle after sex because pornography does not portray that aspect of sexual contact. Over time, sex is the only way he will likely be able to demonstrate any sense of intimacy, and only to a very limited degree.
30. *Women and children as sex objects and body parts.* Pornography focuses on the genitals, not on the whole person. Close-ups of the genitalia are common, and often the head of the women, child, or teen is not shown. This serves to dehumanize the women, children, and teens portrayed. Sex becomes about the genitals, not about the person. Men who demand that their partners undergo surgery to change the body (e.g., breast enlargement or reductions; facial surgery) and men who criticize their partner's genitals have long since succumbed to this impact of pornography.
31. *Becomes increasingly obsessive.* Over time, the men will use pornography more frequently, especially when stressed, experiencing perceived negative emotions or when they want intimacy. The man may utilize pornography in place of his partner for sex or may have to use pornography as a prelude to sex with his partner. He becomes cold and indifferent during sex. The pornography is cold and impersonal, and now so is he.
32. *Begins to practice what is viewed in pornography with sexual partners.* Over time, he begins to experiment with his partner, practicing what he has seen in the pornography. This may include new sexual positions, types of penetrations, having threesomes or affairs, and engaging in the use of pornography together. He may require his partner to wear clothing as seen in the pornography, or to act out something seen in a movie or magazine, such as bondage, sadistic acts, or humiliating acts. Sometimes child molesters use pornography as an excuse to molest, having seen it in the pornography materials and feeling justified in trying it out for themselves.

33. *Experiences difficulty relating to others, especially within porn target population (e.g., children, women).* As time passes, the more that a person uses pornography, the increasingly more difficult it becomes to interact with people. Pornography becomes an affair in and of itself, one based on selfish fantasy. The person, usually a man, finds it difficult to relate to and respectfully interact with women during the day when at night he is viewing pornography, which serves to lessen his respect for women. He usually only trusts women in sexual situations, where he believes that he is in charge. To relate to women at work or socially is difficult because he comes to believe that women are only good for one thing ... to sexually please a man.
34. *Pornography becomes the preferred mode of sexual outlet.* Over time, they prefer sexual contact on the Internet and to the pornography rather than with another person. Pornography requires the use of fantasy, that is, to imagine what it is you want to do with the person or people depicted in the pornography, and then imagining that you are doing exactly what you want. Fantasies are perfect, as is the pornography. Neither ever refuses your demands, neither talks back, and pornography never has a body odor or disappoints. People, on the other hand, are far less than perfect. They expect an exchange of sexual and intimate behavior, and may not smell all that nice all the time. People talk back, want to be pleased as well, and may refuse requests. Again, pornography never talks back or refuses. For the sex offender or abuser, pornography becomes the preferred mode of sexual activity. They engage in sexual contact with a person when they want to dominate, get even, abuse, or hurt someone, and the sexual contact is most often lacking any sense of intimacy or connection. It may appear to the sexual partner that the man was far away, cold, detached, and unaware of or not caring about reciprocating any sexual pleasure.
35. *Confuses sexual arousal and behavior with intimacy.* Pornography teaches that intimacy is one thing only, sexual contact. Pornography rarely depicts a respectful relationship between sexual partners. Abusers pair sex and orgasm with intimacy, but come to believe that "love" and "intimacy" can only be shown through sex and orgasm. The abuser confuses sexual arousal with orgasm and love.
36. *Increase sense of shame, embarrassment, perversion, and isolation.* As persons view pornography more frequently, they initially experience a sense of increased shame and perversion. They isolate these needs and concerns and use more and more pornography to fill their emotional emptiness. Initially they do not believe that anyone would understand their use and pleasure in pornography, so they tell no one. Over time, however, they do not care who knows, as they have numbed themselves of any sense of shame. The shame probably still exists, but has been sufficiently buried and denied. The more isolated the person becomes, the more they use pornography, and over time, the pornography becomes increasingly more meaningful, and may even take the place of friends for the abuser.
37. *Replaces personal relationships with pornography, sex lines, prostitutes, and promiscuity.* As time progresses, the abuser eventually replaces most of his relationships with pornography, empty sexual encounters, and even masturbation. He avoids contact with his partner, or his relationship may have ended as a result of his sexual and nonsexual violent behavior. He will protect and hide his pornography collection, Internet sites, and photos, feeling much more attachment to these than to any person.
38. *Results in empty, unfulfilling sex (still alone).* Pornography involves things (movies, Internet, photos) depicting people. But after the orgasm, the person is still very much alone. Unable or unwilling to communicate his needs, he turns to empty sexual encounters with pornography, strip clubs, sexual affairs or prostitutes. No matter how much he uses pornography or acts out sexually, he is still alone. If involved with a partner or married,

he may seem severely cold and distant, unattached to the family, and uninterested in his partner sexually or intimately. He is alone even when around others.

39. *Orgasm provides quick tension relief (orgiastic state).* Pornography provides material to masturbate to. The problem is that orgasm is a powerful reward and reinforcer for the deviant sexual behavior. For couples to adequately address intimacy or abuse problems, it is often recommended that they refrain from sexual contact for a period of time so that the orgasm does not inadvertently reinforce the abusive behavior or abusive thinking.
40. *Develops sense of sexual entitlement, expectation, and male privilege.* Pornography is based on the premise of male privilege and encourages the man to develop a sense of entitlement for sex and control over women, children, and teens. Abusers expect that they are in control, that they make all or most of the major decisions, and that their partners are obligated to please their every want and need without exception and without resistance.

Pornography reinforces unrealistic expectations of sex, which include:

- Screaming, multiple orgasms. Pornography portrays that men and women have loud, long, unrealistic orgasms as a rule. This sets unrealistic expectations.
- Children enjoying sex, with minimal harm, if any. Pornography all too often portrays children and teens engaging in sex with adults, depicts children and teens being raped and abused, all without harming them. This serves to reinforce rape, abuse, incest, and child molestation as acceptable and the victims as unharmed and actually benefiting from the abuse or rape.
- Hours of uninterrupted sex (no refractory period). Pornography depicts men and women engaging in long periods of sex without their genitals becoming fatigued, losing the erection or moisture, and therefore able to continue sex without rest. This is unrealistic and teaches people, especially men, that sex is about achievement rather than a shared intimate experience.
- Unsafe sexual practices. Again, pornography portrays unsafe and questionably pleasurable sexual acts as commonplace and wanted. This may include being fisted, multiple penetrations, sex with strangers, and rape/molestation as acceptable and desirable, as well as being victimless (the belief that the victims are, in fact, enjoying it).
- Deep penetration as enjoyable and tolerable. Pornography often portrays deep penetration as being desirable and pleasurable. This does not realistically portray the limitations of anal and vaginal penetrations, and may lead a person to belief that a vagina and anus can handle any depth of penetration without causing physical harm. Pornography also portrays fellatio that involves inserting the penis into the mouth of the women and down her throat as being desirous, enjoyable, and safe. The reality is that the throat can be injured, and the woman is likely to vomit, not something that is usually associated with enjoyable sexual activity.

Additional Research

Osanka and Johann (1989) examined the relationship between pornography and abuse, battering, sexual assault, and exploitation of persons. The number of subjects totaled 2380 victims and offenders. She found the following:

- Abusers often used pornography portraying women, children, and men involved in different types of sexual contact. The sexual contact included a wide variety of paraphiliac behavior and violence.

- In 68% of the cases, the abuser physically or sexually assaulted the victim or someone else after viewing pornographic material.
- In 58% of the cases, the abusers pointed out pornographic photos or articles to their victims.
- In 47% of the cases, the abuser attempted to force the victim into some sexual act that was in the pornographic material.
- In 23% of the cases, the abusers were influenced to act violently from the pornography.
- In 14% of the cases, the abuser took pornographic photos of the victims or someone else.

Pornography is often used to season women into prostitution (Swisher & Wekesser, 1994). Pimps and "johns" would have the women act out what was depicted in the pornography. Battered women may also be subjected to act out what their husbands or boyfriends find in pornography.

Interestingly, serial rapist killers also utilized pornographic materials. In fact, they collected pornography, attempted to act out the sexual acts depicted in the pornography, and even took photos of their victims. The photos were used to perfect the abusers/rapists fantasy, which, of course, was influenced by the type of pornography used. The abuser/rapist will purchase and collect the type of pornography that provides the most sexual satisfaction.

Hazelwood (1998) indicated the FBI found that violent sexual and violent nonsexual offenders collect pornography. Specifically, 83% of serial sexual sadists had pornography collections, 61% of serial killers, 90% or better for pedophiles, and 33% of serial rapists (28% of the serial rapists also had collections of detective magazines). He indicated that the purpose of pornography was to reinforce the offender's pathology and to provide new ideas. The offenders have favorite photos and will cut them out.

However, the problem with pornography is that at some point offenders need more stimulation than the inhuman pornography can provide. They reach a point where they need to act out what they have found in the pornography and seek more deviant pornography to appease their deviant sexual appetite. Offenders often practice their deviant sexual and violent fantasies on girlfriends and wives. This is because they are aware that their partners are less likely to call the police and because they believe that they can control the victim. How many battered women return again and again to their batterer? Perhaps most do at some point. For sexual abusers and rapists, the same holds true. Victims experience differing degrees of shame, humiliation, disbelief and, of course, psychological, spiritual, or physical harm. The victims experience helplessness and powerlessness, and fear further abuse and harm, perhaps even death, from their abuser if they leave, tell others, or call the police. The abuser or rapist, therefore, often has the perfect situation in his mind, which is a victim to practice anything he wants on, and to perfect the behavior until he is sexually and psychologically satisfied. However, the satisfaction is short-lived, and the need for increased violence and sexual experimentation increases.

What People Notice First about Pornographic Photos or Scenes

When people look at pornographic materials, they first notice and focus on the genitals of the person pictured. They then become sexually aroused. If they are not sexually aroused, why then would they continue to view the pornography? What I hypothesize is that as the men look at faces of women in pornography, their sexual arousal results in the misinterpretation of the women's facial expression as happy, sensual, and consensual, when, in fact, the facial expressions are unhappy, fearful, or it appears that the women are crying or in pain.

Observe the Photo or Scenes

Are there obvious signs of violence? (e.g., swollen eyes, cheeks, or lips; bruising on the face, arms, wrists, legs, or abdomen; heavy make-up suggesting cover up of injury; signs of crying (e.g., red eyes, teary eyes, mascara running down the cheek, or actually crying); anything that suggests violence has occurred.

Observe the positions or situations depicted:

- Is the sex occurring in an unsafe or unlikely place (e.g., public place; public or private place with onlookers; classroom; church; bosses' office)?
- Is the person depicted being physically or forcefully forced into any sexual contact?
- Does it appear that someone is holding the woman's head while she is performing fellatio; if so, does it appear that she is being forced (e.g., does the person holding her head appear to have a tight grip of her head or hair; is she gagging or appearing uncomfortable)?
- Is someone pulling or holding her by her hair or arm?
- If two or more men are being sexual with a women at the same time, does she appear to be enjoying the sexual contact?
- Ask if you would do this to your girlfriend or spouse.
- Is the man or men penetrating her orally and vaginally and/or anally simultaneously; if so, does she appear to be enjoying it?

Sound of pornographic video/movie:

- Often the sound is dubbed in, added later, especially for the women. Why?
- Read the lips of the women. In many videos, you can lip-read the women saying "no," "stop," "ouch," or yelling or crying. Yet the dubbed sound has her saying: "please do it," "come on," or moaning and groaning in enjoyment.
- The woman may be attempting to stop the sexual activity, yet be unable due to the position of the men or the number of men. This is especially true in scenes where the man is penetrating the women from behind her, on top of her. The woman is often on her knees or bent over furniture. I have seen many women attempt to push the man away, yet they are unable to do so. The woman looks back at the man with a negative expression and attempts to stop him, unsuccessfully (often attempting to use her hands to push the man off or when being penetrated from behind).

I have highlighted some of the things that you need to observe in pornography. Remember that the people who use pornography use it for a reason. And the message of power imbalance and rape and objectification are perceived regardless of whether the viewer believes in violence or not.

EXERCISE

If you wish to see what I am talking about in this chapter, look at any pornographic magazine, movie, or pornographic Internet site. This is easier, however, with a photo from a pornographic magazine. Take the photo or video frame and look at the person depicted. I have done this exercise with sexual offenders. It has been a great tool to help in teaching how they objectify women.

1. Show several photos and ask who finds the photos sexually arousing, and/or if anyone would masturbate to the photos. (If doing this exercise with yourself, find thc photo that you find sexually arousing.)
2. Then, cover the bodies of the women, show only their faces.
3. Have the men (or yourself) identify the facial expressions. I keep these simple (e.g., happy, pleasant, horny, interested, wanting sex; or pained, appear to have been crying, sad, fearful, shocked). After the facial expression has been identified, ask that if their significant other, daughter, mother, or friend had a similar expression on their face, how would they react. What would they say to her?

4. Then uncover the body and ask the men to discuss how when they initially saw the photo of the women naked they were aroused, yet when they saw only the women's face, they were concerned rather than aroused.
5. Label any expression that is negative, neutral, or difficult to assess as inappropriate and not consensual.

Follow-up discussion questions may include the following:

1. How would you describe the appearance of the person? Do they appear calm, consensual, or fearful, distant, and hurting?
2. Is there any evidence suggesting abuse? Check for marks indicative of abuse (bruising on the cheeks, signs of tears or that the person has been crying, over doing of make-up).
3. How cooperative does the person appear? Do they appear to be cooperating or somewhat apprehensive? Consensual means wanting the contact from start to finish, not just wanting some of the sexual contact.
4. Does the person appear conscious or intoxicated?
5. What effect is displayed? Identify how you think the person is feeling. Ask yourself, "If this were my friend, lover, child, mother, etc., what would I ask them?" Would you be concerned by the facial expression and, if so, then it is likely not consensual.
6. If a movie or video, does it have dubbed sound (sound added other than from the actors/actresses)? If so, why would they do that if everything was consensual?
7. Ask what percent of women who are in pornography:

    - Attain fame.
    - Make a life-time career in the porn industry.
    - Commit suicide or are murdered.
    - Are married.
    - Enjoy anal penetration.
    - Enjoy the positions.
    - Enjoy or condone of the chosen sexual partner.
    - Enjoy being penetrated by two or more men at the same time.
    - Enjoy being sexual with other women.
    - Were coerced, drugged/threatened, and/or forced to perform in the pornography or to perform specific sexual acts.
    - How does their involvement in pornography affect intimacy and other relationships, especially with men. Are they abused and used more frequently? Are their children more likely to be physically and sexually abused? Do the women abuse drugs/alcohol at a higher incidence?

I hope my exercise helps to illustrate the potential negative impact of pornography.

SUMMARY

I again want to be clear about something. Some people do enjoy more adventurous sexual behavior. Some people may enjoy fisting, multiple sexual partners at once, and even sadistic or masochistic behavior. If the person is comfortable and willing, without fear of being forced or coerced, it may be a consensual decision involving appropriate sexual contact for them. But when the behavior is forced or expected from someone, it is rape.

Pornography serves to strengthen violent attitudes, helps in the development of deviant and unrealistic sexual fantasies, and fosters deviant sexual behavior. Most abusers use pornography as their reality of what is acceptable sexual expectations and put these expectations into practice. Most physical and sexual offenders utilize pornographic material to strengthen abusive, controlling, demeaning, and objectifying attitudes they hold about others. They then act these deviant expectations out on unwilling victims.

The research on the use of pornography by women and by women offenders is far less documented, though one would expect similar findings with women who use pornography. I strongly recommend reading the following books that offer a lot of information on pornography: *Pornography: Women, Violence, and Civil Liberties* edited by Catherine Itzin (1993), and also *Violence Against Women*, edited by Karin L. Swisher and Carol Wekesser (1994).

21 Motivations for Partner Violence

WHY ABUSE?

Saunders (1986) found that of the women reporting having used minor violence, 31% stated that it was always in self-defense, 23% always fighting back, and 11% first strike. Among women reporting using severe violence, 39% reported they did so for self-defense, 32% fighting back, and 3% first strike.

Men reported using significantly more violence than women to show their partners who is the boss, to tease their partners, and to get their attention, and to report that they used violence to get their own way (Barnett et al., 1997). Cascardi and Vivian (1995) found that most wives who used severe violence did so in self-defense, while in contrast, none of the men who used severe violence indicated that they did so in self-defense. In addition, severely aggressive men escalated more quickly than severely abusive wives. In another study, Vivian (1998) reported that the primary motive for violence reported by wives was self-defense and feeling expression; husbands, in contrast, reported using violence for purposes of control or coercion and to dominate their wives. Hamberger (1997) found the primary reason women gave for using violence was self-defense and the second was retaliation.

Interestingly, Hamberger et al. (1994) identified differences between male and female batterers. Male batterers tended to batter in order to punish for unwanted behaviors, professed ignorance of the motive for use of violence. Factors unique to female batterers included retaliation for previous violence, retaliation for verbal and emotional abuse, self-defense, and escape from aggression. Shared factors included control and dominance, anger expression/tension release, and coercive communication/to get the other's attention. However, significantly more males battered for the motivation of domination, control, and punishment. A few males and none of the females indicated that alcohol abuse was a reason for their violence.

In summary, men tend to disproportionately be the first to initiate violence in the history of the relationship, escalate severity and frequency of conflict, and commit more violent prior assaults than women batterers (Hamberger, 2005). Men were found to be far more controlling than women. Men were more likely to use violence to dominate and control their female partners. Women are more likely to use violence in self-defense or to retaliate for prior abuse. In addition, Hamberger (2005) found that women were by far more likely to be injured, to be more severely injured, and to seek health care for partner-related violence injuries.

ROLE OF ANGER IN ABUSE

Swan et al. (2005) found that anger is an important predictor of aggression. Aronson, Wilson, and Akert (2004) found that anger is often a predictor of aggression. Violent men who are violent toward their wives have greater levels of anger compared to nonviolent men (Boyle & Vivian, 1996; Maiuro et al., 1988; Moreno, Fuhriman, & Selby, 1993). One study found that women who had an angry self concept were more likely to use physical aggression against their partners compared to women who were less angry (Giordano et al., 1999).

Women involved in a court-ordered domestic violence program listed "expressing anger" as a common motive for their violence against their partners (Hamberger et al., 1997). In a study of dating violence, over half of the females and approximately 40% of the males reported their motivation for violence against a dating partner was to express anger (Follingstad et al., 1991). Some studies have suggested that women are more likely than men to suppress their anger, turning it toward themselves, which may explain why women are more likely than men to suffer from depression (Sperberg & Stabb, 1998; Thomas, 1991).

The Role of Fear

One of the hallmarks of abuse is the ability of the batterer to instill fear into the victim (Jacobson et al., 1994). Fear functions to motivate the victim to engage in behavior that will reduce the violent and potentially injurious behavior (e.g., Adams, 1989). Therefore, fear plays an integral role in keeping the victim in a violent relationship. Women were found to be significantly more fearful during arguments (Jacobsen et al., 1994). Male batterers, however, were not fearful of their wives. In addition, men were significantly more likely than women to report frightening their partner as an emotional outcome of their aggression (Barnett, Lee, & Thelen, 1997).

Hamberger and Guse (2002) found that both court-ordered women and women in shelter programs reported significantly higher levels of fear than did men. In addition, men reported experiencing significantly higher levels of amusement when their partners initiated violence than did the women in both groups. Langhinrichsen-Rohling, Neidig, and Thorn (1995) also found that women experienced higher levels of fear than did husbands. When the male partner initiated violence, the women reported higher levels of fear than men. Women were also more likely to be intimidated by their partner's size whereas none of the men were.

Factors Maintaining Women as Victims

Women's economic dependency and a history of minor physical violence were predictors of continued assaults on female partners (Johnson, 2003). High levels of marital conflict and low socio-economic status were more important predictors of continued violence than other factors, such as witnessing or experiencing violence within the family of origin (Johnson, 2003). Help-seeking tactics were more likely to be effective in relationships where the violence was less severe and the history less prolonged, and where power imbalances between the couple were minimal (Bowker, 1983).

Isolation also is a factor for continued violence because it reduces the extent to which both batterers and victims are able to utilize and sustain attachments to others. As the degree of support available to wives increases, the likelihood of husbands using violence against them decreases (Baumgartner, 1993). The isolated wife is most vulnerable to continued assaults. Johnson (2003) also found that abusers isolate their victims as important precursors to a continuation of violence, which is also supported by other studies (Baumgartner, 1993; Counts, Brown, & Campbell, 1992; Levinson, 1989).

22 Domestic and Sexual Abuse, Child Abuse, and Pet Abuse

All too often, professionals and even the so-called "experts" are unaware of the correlation between physical and sexual abuse, child abuse, and pet abuse. It is as though these are all separate types of abuses that occur within a vacuum. But to understand the physical and sexual offenders, you must understand that they abuse and control in many areas of their lives. Every time an abuser is verbally abusive towards his partner and the children are in the house that constitutes child abuse. The same is true for the family pet. Notice, for example, how the cat or dog will cower or hide when even verbal abuse begins. Children respond much in the same way depending on their ages. So even verbal abuse significantly impacts children and pets, even when the children or pets are not the target of the verbal abuse.

Now imagine being a child or pet in the house when physical or sexual abuse is occurring. The double impact of witnessing (whether hearing or seeing it) the abuse and the helplessness of not being able to stop it or being unable to intervene and protect the parent or sibling being abused must be traumatizing. Any abuse occurring in the home significantly impacts children and pets. They are never, ever protected from the effects of the abuse and they are never, ever able to just "deal" with it on their own, even if the children are older teens or young adults. Abuse of any kind significantly impacts everyone involved, from family and friends who did not witness the abuse to the community that is negatively impacted by the abuse (e.g., neighbors, law enforcement, churches, schools) because the abuse threatens the safety and integrity of the community.

I get extremely tired of hearing people make ignorant statements to the effect that their children witnessed the abuse, but are not impacted because the children "understand" and are "more mature for their age." Yet maturity has nothing to do with being impacted by violence. Imagine the hurricane that devastated the Gulf Coast in 2005. Professionals, men and women trained to deal with crisis and disaster, broke down in tears. Some quit their jobs and walked away, others committed suicide. My point here is that even well-trained professionals are negatively impacted by violence and disaster. So, why would children of any age or pets, for that matter, somehow be better equipped to deal with violence within the home? That is ridiculous. Children and pets that are present when violence occurs within the home are significantly and negatively impacted by the abuse, each and every time it occurs.

Professionals need to evaluate for collateral impact of abuse each and every time they assess or treat for domestic abuse or sexual offenses. Children and pets are always impacted, but rarely assessed or treated for their victimization.

Research has supported that interpersonal violence is linked with animal cruelty (Faver & Strand, 2003; Lockwood & Ascione, 1998; Ascione & Arkow, 1999). The importance of the human–animal bond to human health and well being has been documented by Faver and Strand (2003). The impact of threatening to harm or actually harming the family pet and pet companion, therefore, is a significantly effective way to abuse and intimidate a family member. Faver and Strand indicate that the coexistence of child abuse and woman battering has gained more attention. There has been documented evidence that animal abuse frequently occurs in families where children are physically and/or sexually abused (Flynn, 2000a; Ascione, 1993; DeViney, Dickert, & Lockwood, 1983). Faver and Strand examined evidence that batterers use threats and actual harm to family pets as a means of controlling and coercing women. Evidence demonstrates that one barrier often preventing a woman from leaving a violent relationship is fear for the safety of family pets. Close to 50% of the women residing in a shelter for battered women reported that their partners had threatened or harmed their pets (Flynn, 2000b; Ascione, 1998, 2000).

Animal abuse is a marker for family violence. Various studies of family violence found that violence against companion animals occurred where there is also physical and sexual violence against: (1) female intimate partners, both in heterosexual (Browne, 1987; Walker, 1979) and same-sex couples (Renzetti, 1992); (2) children (DeViney et al., 1983); and (3) siblings (Wiehe, 1990). Additional studies revealed that sexually abused victims in day care settings had abusers who frequently threatened to harm or actually harmed pets as a way to intimidate, control, and ensure the silence of their victims (Faller, 1990, Finkelhor, Williams, & Burns, 1988). Levinson (1989) found that in societies where animals are treated cruelly, women were more likely to experience severe forms of abuse, torture, or murder. Children who are particularly cruel to animals are more likely to engage in aggressive or violence behavior towards others (Flynn, 2000b; Felthous & Kellert, 1987; Rigdon & Tapia, 1977; Tapia, 1971). Animal abuse may also be a marker of child abuse or other forms of violence occurring within the family (Arkow, 1996; Boat, 1995).

Tingle et al. (1987) indicated that studies of criminals involved in sexually aggressive crimes have also presented with a history of abusing animals. Tingle et al. (1986) found that nearly 50% of rapists and 25% of pedophiles had shown cruelty to animals as children. In a retrospective study of serial murderers, 36% reported killing and torturing animals in childhood and 46% did so while in adolescence.

Ascione (1993) and DeViney and colleagues (1983) found that, after reviewing studies, approximately 60% of the families in which there was child abuse also had pet abuse occur. The rate of pet abuse reached almost 90% in cases where the children were physically abused (vs. sexual abuse or neglect). Numerous other studies reported similar findings. Several studies have specifically examined the occurrence of animal abuse in households where women were battered. Ascione (1998) found in one survey that approximately 70% of the respondents reported that their batterer had threatened to harm their pet, and approximately 60% reported that their batterer did, in fact, harm the pet. Flynn (2000b) found that of the 40% of the battered women who owned pets, nearly 50% reported that their batterers had threatened or harmed their pets. Considering only the women with children in their study, approximately one-third of the battered mothers whose pets had been abused reported that their children had been abused as well. Quinlisk (1999) found that approximately 70% of their respondents of women involved in domestic abuse programs reported that they owned pets and that the pets also were abused. Approximately 90% indicated that the pets were abused in their presence and approximately 80% indicated that their children had also witnessed the pet abuse.

In all of the studies, most of the women reported that the pets were a significant source of emotional support. In most any family, the pet provides a significant source of emotional support. Therefore, the impact of threatening to or actually harming a family pet is an exceptionally powerful violent act.

In summary, it makes absolutely no sense to treat merely the verbal, physical, and/or sexual abuse that occurs against the adult parent/partner while ignoring the verbal, physical, or sexual abuse that is likely occurring against the child or pet. Therapists and law enforcement personnel need to question specifically about any possible victim in the environment, not just the identified victim.

Section III

Interviewing Strategies: Detecting Truth from Deception

23 Questioning Batterers and Sex Offenders

When questioning an abuser or sexual offender, it is imperative to be as thorough as possible. The offender's perpetrator persona is difficult to access, yet when a connection is made, information can be easily gathered. It is no surprise that most offenders plead "not guilty" when first questioned about their offense. Most hang on to the process of denial as a way to cope with stress and anxiety. In fact, it is not uncommon to find that the offender has a history of resorting to denial, minimization, and rationalization when confronted with inappropriate behavior.

When questioning abusers or sex offenders about the details of their offense, listen carefully to what they say, how they say it, and what they omit. Pay attention to the explanations they offer for the victim's allegations, and to the rationalization they offer for their innocence. At times, offenders may even admit to the offense, but claim the sexual contact was consensual or that they were merely defending themselves. I am amazed at the stories offenders tell. They are often unaware of how their distortions and lies actually give up the truth. What I am about to share with you is what I have found to be effective when questioning sex offenders about the details of their offense. The process may take only minutes or perhaps hours of working with them to get to the truth.

Gestalt therapy focuses on the client's feelings, awareness at the moment, body messages, and blocks to awareness. In short, the focus of Gestalt therapy is on how nonverbal cues betray feelings and information the client may be unaware of, denying, or repressing. The goal of the therapist is to help the client pay attention to his feelings and body cues and to resolve the underlying issues. Of course, this is a thumbnail sketch of Gestalt therapy, but will suffice for the purpose at hand. People who interview abusers and sex offenders would benefit significantly from the precepts of Gestalt theory. The offender will often give clues as to the truth; therefore, paying careful attention to the perpetrator as he is interviewed can result in detecting truth from deception. The perpetrator persona has far too much pride not to be seen. In fact, most of the time when a perpetrator offers his explanation for his offense situation, he may truly be unaware of the truth, yet is fully aware of the truth.

So, I offer a contradiction to begin our discussion on assessing truth from deceit. A perpetrator may genuinely believe that he is innocent of the sexual assault he is accused of committing. He may truly believe that he was and is not capable of harming someone with violence. Regardless of how much he represses, distorts, rationalizes, or justifies his offense behavior, he is on some level fully aware of what he did.

Consider for example, when you edited a paper or article. You may have read and reread what you wrote and believed that you did a thorough job proofing your writing. You confidently submit your work to the publisher or professor, only to find that you had made several errors in spelling, grammar, or that you may have even omitted important information. The phenomenon occurring here is one of seeing and hearing what you expect. When you proofread your own work, you already had a finished version in your head. Therefore, when you were actually reading your work, you saw what you thought would be there. It has happened to everyone. When offenders deny or distort their offense behaviors, one part of them (perhaps the moral or social persona) believes that they could have only acted in ways that supported their beliefs, such as being nice, kind, gentle, and nonviolent persons. Another part (the perpetrator persona) is fully aware of the aggression they are capable of doing, and aware of what they actually did do.

Below is a discussion of the types of cues abusers and sex offenders provide, identify the types of deceit and distortion that are most common with abusers and sexual offenders, and discuss ways to approach the resistive offender. I will use the term *offender* to identify the person who committed or is suspected of committing physical or sexual assault. The term *perpetrator persona* refers to the part of the personality that contains characteristics of selfishness, narcissism, and psychopathy; basically, it is the part in all of us that would act in any given way to attain a goal that pleases us without any sense of concern of how others may be affected.

Understand that to detect "truth from deception" requires careful listening, careful observation, an understanding of the dynamics of the type of offense that occurred, an understanding of general psychological principles, and, of course, some degree of luck. The polygraph is designed to assess for anxiety. Anxiety is measured by the changes in physiological processes, including heart rate, breathing, sweat, and voice stress. This is a simplified explanation of polygraphy, but enough for our purpose at hand. But paying attention to the cues or signs of anxiety and other emotional states, and by careful observation of the offender, it is possible with adequate degrees of accuracy to determine whether an offender "appears" to be telling the truth or not. What is more difficult to determine is what they are telling the truth (or lying) about.

And yet a caveat. Even if it appears that someone is telling the truth, there is no full-proof approach to determining the degree of truthfulness unless there is physical evidence, video, or other hard proof. The same is true for detecting deceit. Even if the offender is demonstrating signs of anxiety and, therefore, signs of deceit, that does not mean that he is lying. He may not do well being accused of violence, have a premorbid anxiety disorder, and other reasons that adequately explain or justify an anxious response.

What we are looking for primarily is inconsistencies in the offender's response. For example, inconsistencies may occur in the following ways:

- The offender's response may appear inappropriate for the given situation. A good example of this is an offender who appears very relaxed and comfortable during the interview or assessment. It would be expected that anyone being interviewed or assessed for committing a violent crime would experience some degree of anxiety.
- The offender's emotional reaction does not seem to fit the situation. One example is when the offender's emotional reaction is minimized or exaggerated for the question or situation. An offender who is unable to cry, but is giving indication that he might, or lacking an emotional response when discussing emotionally charged questions. Fake tears come to mind as an example. One offender was attempting to exhibit guilt and remorse, and when I was not buying into the act, he attempted to cry. He strained and looked at me and then to the floor, and squeezed out a couple of tears. The tears abruptly ended when I acknowledged his apparent concern. However, genuine tears do not require strain to flow, and genuine tears do not abruptly end once the question changes.
- The offender's concern for the seriousness of the situation or the welfare of the victim may appear inconsistent. For example, an offender can be more concerned with the financial costs involved in the offense than with the trauma his wife experienced when he assaulted her, not to mention a lack of concern for the children for having to witness the assault. Another example is when an offender arrives for the court-ordered assessment and expects that the assessment will be completed within a 1-hour visit, and that no intervention will be recommended. They are angry or taken back when they learn that the assessment involving testing may take 2 to 3 weeks.
- The timing of the offender's verbal, behavioral, and emotional response to questions may be inconsistent with what would normally be expected. A good example here is a delayed emotional reaction that is so late and out of place that it does not make any sense. Taking longer than normal for that offender to respond to a question when he responded much more quickly to other questions.

Now again a caveat of caution. Innocent people can exhibit any of these symptoms as well as guilty people. So take everything I am saying in this chapter with a cautious reminder that it is nearly impossible to detect trust or deception all of the time. And even when you are accurate, you may only be accurate for certain aspects of a crime, not everything alleged.

24 Detecting Truth from Deception

INTRODUCTION

Detecting truth from deception is in no way a perfect science. I have heard some arrogant profilers, psychologists, and law enforcement officers publicly state that they can detect truth from deception in 90 to 100% of their cases. That is quite an impressive statistic considering the people who are the genuine experts very clearly stated that detecting deceit is a very complicated process that, at best, may have a success rate of approximately 30 to 80%. But arrogance will always foster some individuals making money with false reputations. Imagine that — deceitful people in the field of detecting truth from deception!

Detecting truth from deception is difficult and what I am presenting is a brief discussion about the process. The more you can accept your limitations, the more effective you will be at this. Consider the problems inherent in interviewing an offender.

- Evidence and witnesses may identify the wrong person. This occurs every day. Well-meaning, honest people providing facts that are skewed and inaccurate. Remember that when violence occurs, people are in a state of heightened arousal; they are either escalated and angry or fearing for their immediate psychological, physical, or sexual safety. The victim may also fear for the safety of children. So, whatever evidence is presented, short of DNA, photos, or video tapes that preserve the information more subjectively, all information needs to be taken with a grain of salt. Every statement may well have some aspects of truth as well as aspects of falsehoods, even when the individual is not intending to lie.
- Victim reactions are unpredictable. Again, the victim of a violent act experienced fear for her own and for the children's safety. The victim's short-term memory will not be as effective in the midst of the violent situation. In addition, the victim may vacillate from being fearful to self-confidence, from wanting to press charges and have the offender removed from the home and obtain an order for protection to the other extreme of engaging the offender in contact and dismissing the order for protection. It is even common for victims to recant their statement and to protect and defend the offender.
- Perpetrator's reactions may be misleading and confusing. The offender's behavior and statements may not be consistent with what occurred. He may be inappropriately calm and cooperative or hostile. He may provide facts (and often does) to implicate culpability on the victim's role, portraying himself as the true victim.

INHERENT PITFALLS

There are traps and pitfalls to pay attention to when interviewing or interrogating an offender. Some of the more common ones are discussed below. These need to be paid attention to and taken into account when deciding how to structure an interview, what to ask and when, and how to interpret the interview data.

1. *Sex offenders and batterers sometimes lie.* I know this may be unbelievable, but offenders have been known to lie. They are defensive, deceitful, and more than likely have antisocial and narcissistic personality traits. In addition, they have more to gain when not admitting to the crime at hand.
2. *Victims sometimes exaggerate.* This is generally not a result of the victim wanting revenge or to overly portray the offender in a negative manner. When victims are questioned soon after the offense, shock and general grief-related issues cloud their statements. In short, they are less likely to lie at that time, but more likely to have some degree of distorted perception because of the emotional, physical, and/or sexual offense behavior that occurred.
3. *Personal reactions of the interviewer.* No matter how long you have conducted forensic interviews, your personal reactions will impact your judgment to some degree. In most situations, you easily put aside biases and prejudgments and pay attention to the facts and interview data at hand. But, there will be cases when your judgment is clouded because of stress or even personal reaction to the details of the case or reaction to the specific offender. At times, we simply do not like or respect the person sitting across from us. It is important to pay attention to our personal reactions and accept our need to be vulnerable and to have personal feelings. Take a time-out from the interview to sort out your feelings and reactions and consult with a colleague. It is totally professional to do so, and true experts do not avoid accepting human limitations.
4. *System failure.* This is probably earth-shattering news, but you know, sometimes the criminal justice system simply fails. I know it is hard to believe. Sometimes despite all of the evidence and even with confessions, offenders seem to skate through the system with minimal, if any, consequence. It is important to remember that your role has limitations and expectations. For example, it is the job of the police officer to offer immediate protection and to interview victims as soon as possible to determine the facts of what actually happened. The officers will interview the abuser to obtain his version of the offense. The officers will determine if probable cause exists to make an arrest. It is not their job to take sides or to determine all of the facts, but to simply determine what course of action is needed at that time. The officers complete an Incident Report containing all of the details of the abuse/assault. This will include statements obtained from the victim as well as from the abuser and any witnesses, including any children that were present during the abuse. The officer's report is then forwarded to the County Attorney or District Attorney's office.
5. However, even with all of the duties the officers provide, they are not making a decision on domestic matters, custody, division of marital property, etc., even though the public perception may be that the officers actually make final disposition decisions about convictions and other issues that they are not empowered to do. So, the officers may get the brunt of the public blame when an arrest is not made in a timely manner even though they are diligently doing their job. And this takes a toll on self-image and self-esteem.
6. Now, of course, the County Attorney's Office is charged with deciding if criminal charges are warranted and, if so, to initiate prosecution. Usually the attorneys will make sure that the victim has an Order for Protection or Restraining Order to keep the abuser away from the victim, the victim's house, place of employment, etc. The County Attorney's/District Attorney's Office will investigate the abusive incident and determine the best course of action. If there are alternatives to prosecution, these options are considered. If not, criminal charges are pursued. And through all of their duties, the County Attorney's Office has to base its decision on evidence, who appears to be the more credible witness, and what would be in the best interest of the victim, offender, and community.
7. My point here is that at any point the system can fail. A police officer may fail to ask the right questions or may do a diligent job yet still no arrest is made. The officer may

make an arrest, but the Country Attorney does not prosecute or offers the all-too-common plea agreement. On the other hand, even with good evidence, sometimes the bad guy wins.

8. *Need for SELF CARE.* Professionals need to be aware to take care of personal needs. It is imperative that professionals regularly consult with peers and supervisors. This allows for venting, discussion of alternatives in handling a specific case, and for emotional and collegial support. None of us works in a vacuum, and we all need the support of others do our jobs effectively and to prevent burnout.
9. Professionals are blamed for the offender's behavior. This public point of view permeates the violence field. The public blames the police for not catching the correct offender soon enough, and/or blames them for using "excessive force" when, in fact, the officer's force was in self-defense or equal to the force being used against them. Prosecutors are blamed when plea bargains are offered or when offenders receive what appears to be shorter than expected sentences. This, despite the fact that, in most cases, the laws, not plea agreements, dictate the type of punishment allowed. But the public does not understand this. And the treatment professional, who provides treatment to a batterer or sex offender, is criticized for "siding with the enemy" or for demonstrating too much care and concern for the offender's rights. Professionals are always blamed for something, most of the time for things beyond their control. This is one more significant reason why self care and collegial support is necessary.

Defenses

Offenders utilize denial for a variety of reasons. Few people readily admit that they have committed a violent crime. In fact, why would any sane person admit that they engaged in a violent act that could result in jail or prison time, and social and familial consequences? They would not. Some of the main reasons why offenders use denial are discussed below.

1. *Self-image and ego strength.* It is difficult to accept that they have, in fact, engaged in a specific behavior. To admit would be to destroy or seriously distort their self-image. The offender simply cannot accept the reality that he did engage in the behavior, and may not believe that he ever would or could do such a thing.
2. *Emotional reaction.* The offenders are aware that they committed the behavior, but the resulting feelings are too painful and unforgiving to admit to anyone. For example, feelings of shame and embarrassment may seem overwhelming at the time and it is more calming and safer to deny the offense.
3. *Consequence avoidance.* Offenders are aware that they committed the behavior, but will not admit it because of the possible consequences. Consequences include:
   - Legality — they do not want to be convicted of a crime or of that specific crime; nor do they want to be incarcerated or to pay fines.
   - Familial — the offender is concerned about how their family will respond when they hear about the crime.
   - Socially — the offender is concerned about how the neighbors, co-workers, church members, and friends will react. The offender fears the damage to their reputation.
   - Financially — the offender is concerned about the possible financial cost if he is found guilty.
   - Employment — the offender is afraid that he may lose his job or promotional opportunities.
   - Relationship — the offender is concerned about how his partner/spouse may react to his guilt. He may fear that his spouse or partner will leave him.

4. *Antisocial personality traits psychopathy.* The offenders lack consciences and are not concerned about the ramifications that their behavior has on others. Denying the behavior is naturally comfortable to them. Offenders lacks any sense of remorse or empathy.

It should be noted that denial is a natural response we have when we have done something wrong. It is a conscious choice we make, and it involves lying about or distorting the truth. Unconsciously, denial may take the form of distortion rather than blatant lying. However, the offender should always be held fully responsible for any distortion of the truth. Understanding the causes of denial and the purposes it may serve help to guide out efforts to overcome it.

25 Cues to Deception

INTRODUCTION

I recall hearing a female on television claiming that she could tell if someone was lying 100% of the time, without exception. I must admit that I enjoyed hearing her make a fool of herself, knowing that no forensic professional has anywhere near 100% accuracy in detecting truth from deception based on a brief interaction, or even a lengthy interaction. Some professionals certainly have more skill and are able to detect truth from deception far more often than others.

Effective interviewers and interrogators pay attention to everything going on in an interview. Follow the guidelines for interviewing discussed in Chapter 37. Understanding the physiological and psychological symptoms of deceit is, of course, rule 1. However, as with most rules involving interpreting the behavior of others, be cautious: not all signs that have been identified by the experts indicating deceit are true for every offender. Not all offenders will give the same signs as expected. Most will, however, give consistent signs of deceit. The best and most effective method of proving truth from deception is not just a confession, but finding collateral sources of proof supporting the offender's disclosure. I have had a small number of batterers and predatory sex offenders confess to crimes and other deviant and violent behavior that later turned out to be false. Though this may not occur all that frequently, offenders may admit to offenses they did not commit due to a factious disorder to gain attention or at times because they may not fully understand what is being asked. If the offender has committed numerous offenses, he likely does not recall each and every offense. He may, therefore, admit to something that he did not do or that he is not certain he engaged in.

CATCHING THE LIAR

DePaulo et al. (2003) found that liars are less forthcoming than truth tellers and that they tell less compelling lies. They also make a more negative impression and are more tense. Their stories include fewer ordinary imperfections and unusual content. Cues to deception were stronger when lies were about transgressions vs. minor events. The authors also indicated that liars can fail to notice some of the ways in which the targets of their lies are reacting. The more the liar engages in deceit, the more energy the ongoing lies require, and the greater the preoccupation with the lies and thoughts about the lies occur, which serve to weaken the liars' ability to effectively engage in deceit.

In addition, DePaulo and colleagues also indicated that the more the liars attempt to control expressive behavior, the more likely they are to fail to regulate expressive behaviors (e.g., tone of voice) and other behaviors that may be more unconsciously controlled. They also indicated that another way the liars' behavior serves to self-betray is to direct efforts at expressive control at the wrong level.

In sum, they predicted that to the extent that the liars deliberately try to control their expressive behaviors, thoughts and feelings, their performances would be compromised. They would seem less forthcoming, less convincing, less pleasant, and more tense. They also found that prepackaged lies could be shorter and less detailed than truthful answers. This is because truthful accounts are based on events that actually occurred and that the person experienced. Lies fabricated from scratch are likely to be shorter and more internally inconsistent than truths and to be preceded by longer latencies. These could include increase in pauses and speech disturbances, more pupil dilation, decreased blinking, and decreased eye contact. People who are preoccupied with the formulation of a complex lie may appear to be less involved and expressive as well as less forthcoming. Liars may also fail because they know less about deceit than their targets do.

DePaulo et al. (2003) also indicated that there is no documented facial expression that is specific to guilt, but rather more general cues to negativity and distress (Keltner & Buswell, 1996; Keltner & Harker, 1998). Shame does seem to have a characteristic demeanor that includes gaze aversion, a closed posture, and a tendency to withdraw (Keltner & Harker, 1998).

They also found that liars made more negative statements than truth tellers and appeared more tense. Lies concerning transgressions were significantly more detectable than lies that were not about transgressions. Pupil dilation and pitch were noticeably different for liars as well. Again, lie telling is more cognitively challenging than telling the truth and, therefore, requires more energy and attention. Liars tended to have more pauses in their speech and had silent pauses. Their responses were also briefer.

Motivated liars had even higher pitches and were more preoccupied about their lies. In short, the liars, in their attempt to anticipate the kinds of communication that targets would find credible, fall prey to their own misconceptions about the nature of truth telling. People who spontaneously correct themselves and who admitted that they could not remember everything about the story they are telling, were more likely to be telling the truth than lying.

LEAKAGE CUES AND DECEPTION CUES

Ekman and Freisen (1969) identified two categories of cues to deception. The first is leakage cues. This pertains to cues that reveal what the liars are attempting to hide; typically, how they really feel. For example, when deceivers attempt to conceal a facial expression, the facial expression may still occur on a macro level and it is usually briefer than it would normally be. The second type of cue to deceit is deception cues that indicated that deception is occurring, though it does not tell the nature of the information being concealed.

THINKING CUES AND FEELING CUES

Ekman (1985, 1992) described two major categories of cues — thinking cues and feeling cues. Ekman indicated that liars who prepare their deceptions inadequately or who cannot keep their stories straight produce inconsistencies that betray their deceits. Those who overproduce create stories that seem rehearsed. If liars need to think carefully about their lies as they tell them, they may speak more slowly than truth tellers. This is an example of thinking cues. Ekman also indicated that the thoughts and feelings may also burden liars more than truth tellers. These include thoughts about whether the performance is succeeding, feelings about this (e.g., anxiety), and feelings about the fabricated performance or about discreditable past acts that the liar is trying to hide.

Feeling cues may include fear cues and this consists of speaking in a higher pitch, faster and louder speech, pauses, speech errors, and indirect speech. The greater the liars' apprehension, the more evident these fear cues may be. Lairs experiencing guilt about their lies, especially when telling lies to people who trust them, should show more behavioral cues. These may include sadness, such as lower pitch, softer and slower speech, and downward gazing. At other times, liars may have the experience referred to as "duping delight," which includes an excitement about the challenge of lying and the pride in succeeding at the lie. This may appear as a higher pitch, faster and louder speech, and use of more illustrators. The guilt reactions and "duping delight" are still being researched. DePaulo et al. (2003) offers an example in support of these cues. Imagine people who are only pretending to enjoy a movie. They demonstrate fewer enjoyment smiles and more faked smiles than people who are really enjoying the movie.

DETECTING CUES OF DECEPTION

Zuckerman et al. (1981) identified four factors that were useful in detecting cues to deception: generalized arousal, the specific effects experienced during deception, cognitive aspects of deception, and attempts to control behavior so as to maintain the deception.

They proposed that liars can experience greater undifferentiated arousal than those who are telling the truth. The arousal may be evident by liars' greater pupil dilation, increased blinking, more frequent speech disturbances, and higher pitch. The affectational aspects relate primarily to the response of guilt. Liars may fidget more than truth tellers, and they may also sound more unpleasant. Guilt and anxiety could become apparent in the liars' distancing of themselves from their deceptive communication. They predicted that liars would communicate in more evasive and indirect ways than truth tellers and that they would maintain less eye contact with their interactional partners.

The cognitive aspects of deception pertain to the fact that lying is a more complex task than truth telling. The liar is forced to create facts that are internally consistent and that may be consistent with what others already know. The greater the cognitive challenges involved in the act of lying, the more likely the liar would demonstrate longer response latencies, more speech hesitation, greater pupil dilation, and fewer illustrators (e.g., hand movements that accompany speech).

Liars' attempts to control their behaviors to maintain deception can paradoxically result in cues that betray it. For example, the behavior of the liar may appear less spontaneous than truth tellers. In addition, the liars' inability to control all aspect of their behavior equally effectively may result in verbal and nonverbal discrepancies.

26 Understanding the Psychopathic Personality

PSYCHOPATHIC PERSONALITIES

Psychopathic individuals present challenging interview subjects. Psychopaths are significantly different from everyone else. What I am presenting here is a summary of information relevant to the psychopath.

HARE'S PSYCHOPATHY CHECKLIST

Dr. Robert Hare (1991) developed the Hare Psychopathy Checklist-Revised (PCL-R). It is the standard by which true psychopathy is scored. Professionals must undergo specific training to administer and interpret the test, which consists of 20 items that are strongly supported empirically. The PCL-R correlates very high with violent and sexual recidivism. The tool assesses specific factors (Factor 1: Affective/Interpersonal Features and Factor 2: Socially Deviant Behavior, and there are several additional factors as well). The most significant drawback of the tool is that it does require specialized training to use accurately. However, there is no other accepted standard to assess for psychopathy. And when working with violence offenders, it is imperative to assess for this affliction.

GENERAL ISSUES IN DEALING WITH PSYCHOPATHIC OFFENDERS

Psychopathic individuals present with several factors that make them challenging and difficult to assess and work with, such as:

1. Psychopaths present with a smooth approach and are often cold and emotionless.
2. Psychopaths generally present with and readily offer explanations for their offensive behavior that may appear very plausible.
3. Psychopaths are either: totally abrasive, resistant, provocative, angry, or cooperative, calm, cooperative, agreeing.
4. Psychopaths are clever and resistant. They are exceptionally convincing and are usually several steps ahead of the interviewer. Whatever books and training the professional has read or completed, the psychopath may well have read and completed as well.
5. If in treatment, psychopaths either fail several times due to resistance or completely cooperate to complete the program as the star pupil and therapist's pet.
6. Psychopaths learn to superficially comply, yet internally maintain and strengthen their own distorted beliefs. This means that treatment interventions do not work; treatment offers psychopaths improved knowledge to commit their offense behavior better than ever, with fewer mistakes. They become smoother and more effectively clandestine.
7. Psychopaths perfect deviant behavior through practice and experience; their MO (modus operandi) is ever changing and will become more perfect with each crime and violent incident.
8. Psychopaths may have convinced victims that it is they (victims) that are at fault; they may have also effectively convinced the victim's support network and even law

enforcement and treatment professionals that they are the true victims. They are very adept at making the victim appear crazy, unstable, and the aggressor.

9. Often have many allies and supporters as a result of their ability to con.
10. Psychopaths often have above average IQs, are middle class or higher, and appear to fit right in within the community, church, work, anywhere they are.
11. May maintain a professional job or often menial, unstable employment history; one extreme or the other depending on IQ and social sophistication.
12. May have stable marital history, or unstable, turbulent relationship history.
13. Often have many children with many different women, but fail to support them. It is as though psychopaths are rabbits with a very high need to procreate or simply not caring to prevent pregnancy, seeing the responsibility being the woman's all of the time.
14. Often have extensive, promiscuous sexual relationship history, also unable to remain sexually faithful. They can con women and men easily and may have multiple ongoing relationships at any given time.
15. Often have unstable residence history, residing with others much of the time. This prevents them from assuming any responsibility and allows for moving around without strings.
16. Often have parasitic lifestyles, living off others. They take advantage of everyone, yet their victims may experience a sense of obligation to the offender and may strongly defend the offender when criticized. In short, the psychopath lives the life of a king, at times totally on the financial support of others.

When Dealing with Suspected Psychopathic Offenders

Following are specifics on dealing with suspected psychopathic offenders:

1. Know your facts in advance. Do not begin an interview or interrogation without having obtained and read all available information and prepare your questions in advance. A common mistake is to enter the interview with a "wait and see" attitude. Throughout the interview, you will certainly have to adjust the direction of the questioning somewhat to fit the new information; the psychopath is extremely adept at taking charge of the interview in a subtle manner and, before you know it, the offender is directing the interview.
2. Do not fully reveal what information you have or do not have. Just like in poker, you do not reveal your aces. Let the offender wonder as to what you know. Shuffle papers around and look at various pieces of information during the interview to give the impression that you have more than you do. But the offender does not need to know all of what you know unless a confession was already obtained or enough information was obtained to formulate an adequate treatment or incarceration or supervision plan.
3. Let them tell their story and confront them on discrepancies, sharing only some of what you know. Let the offenders ask what they want and to tell you their story, but do not relinquish control of the interview to their storytelling. Remember that psychopaths can tell a long and intricate story to distract you.
4. Maintain control of the interview or the offender will take control (you go for a ride). This cannot be overstated. You direct what questions and what information is discussed, not the offender. This means that the interviewer decides what information is discussed and in what order. Granted, there are times, though few, when the offenders need to tell their story the way they must due to mental health issues. But most can take direction from you. Follow your interview outline because that is why outlines exist. You ask the same questions in the same order for a reason — because the end result is a thorough record of the facts.

5. Show no special favors or attention to the offender. They will take an inch and run the proverbial mile. They do not get to bring food into the interview or treatment area if no other offenders are allowed to do so. They do not get to reschedule or miss an appointment without the standard consequences of paying for the missed appointment and the court being notified of the missed appointment if they are court ordered.
6. Observe cues — cold stares, attempts to intimidate, callous factors, lack of expected emotional reaction, providing effect upon command (but appears fake). All of the physical and verbal cues need to be attended to and noted. A competent interviewer can take notes during the interview without disrupting the flow of the interview process. Document everything you notice, even if seemingly insignificant at the time. Later, you can review the notes to decipher which cues and behavior are significant.
7. Note when an offender takes pride in his offense behavior (smiles, eyes widening, increase in physical or sexual arousal, providing facts to you to increase his own reliving of the offense and proud of it). Offenders will give cues and it is a matter of catching the cues. Psychopaths are less likely to give obvious cues, so the interviewer must pay special attention to subtle shifts in body, facial expressions, wording, and verbal or physical excitement or anxiety. Psychopaths and narcissistic violent offenders will often take pride in their offense behavior and, when caught, may relish in their successful offense. This reaction is especially noticeable when discussing how the victim was injured.
8. Never, ever underestimate the psychopathic offender. They are often more educated than the professionals around them. They study the available literature on crime and violence and pain and escape and are often far more educated about violent crime than one would expect. The FBI does not publish manuals for investigation or interview strategies or make such information easily accessible for a reason: they are aware that the offenders will obtain and read the material. Also, the psychopath will try to get under the skin of the interviewer and will attempt to shock the interviewer. Expect the unexpected. They may even become aggressive or violent with virtually no warning.

In summary, psychopaths are difficult offenders to assess and treat. There is no known effective treatment for psychopathy, no domestic abuse or sex offender treatment that demonstrates any acceptable effectiveness. In fact, all of the research suggests that psychopaths simply get better at committing their crimes with treatment intervention. Interesting dilemma — treatment must be offered but no treatment exists.

27 Physical Cues

LEARNING WHAT TO LOOK FOR AND HOW TO LOOK FOR IT

When questioning the perpetrator of physical or sexual assault, it is important to carefully observe their every movement and gesture. This can be a difficult task when the examiner or interviewer is focused on the task at hand (e.g., obtaining a confession or gathering information to make treatment recommendations). However, after you learn *what* to look for, and *how* to look for it, it does become habit. The more you pay attention to the physical cues, the more effective your interrogations and interviews become.

There are several good books on the market for learning how to effectively interview and conduct interrogations. This chapter in no way is meant to be the final word. This is based on my clinical and forensic experience.

EYE CONTACT

Perhaps the easiest of place to observe is in the eyes. Do the perpetrators make and maintain eye contact with you during the interview? If not, observe where they look. Common eye movements to pay attention to include the following:

- Shifting eye contact
- Intensely staring at interviewer
- Staring or focusing on objects, the wall or floor
- Avoiding all eye contact
- Fluttering eyes or rapid blinking
- Closes eyes
- Pupils dilated

If the offender looks down toward the ground or table, they are often experiencing sadness or guilt. They may also be experiencing shame and possibly depression. This may suggest that the offender is, in fact, guilty. Why would guilt, shame, and depression occur unless they engaged in the suspect behavior? Again, this is not proof, but a strong indicator. If they look up at the ceiling or past you/over your head, they are often more antisocial and likely annoyed or disgusted. They may be attempting to appear innocent, yet cannot adequately explain the behavior they engaged in, and they know it. By looking away from you, they may be attempting to find the answer that will somehow justify their offense behavior. Avoiding eye contact decreases the chance of the interviewer perceiving that the offender is lying. When they are confronted about their behavior, they typically become agitated and annoyed at the fact that they are not being believed.

When eye contact has been made and the offender terminates it, something is up. Breaking eye contact accomplishes two things. First, it allows the offender to search for explanations without the added stress from the interviewer. To maintain eye contact with the interviewer offers little opportunity to conceal deceit. The offenders may display anger or fear through their eye contact when what they actually wanted to demonstrate was concern and innocence. In essence, the eye contact may give the truth away. Second, it may appear as if the offender is genuinely not able to retrieve information concerning his offense behavior. I am amazed at how ignorant offenders are of memory problems. They may look around the room searching for a plausible explanation for their offense behavior, one that hopefully will exonerate them. It is common for offenders to attempt to blame alcohol or drugs for causing them to "black out" during the offense. However, a blackout refers to difficulty recalling a specific time period, usually several hours duration. Not remembering offense behavior, however, does not absolve perpetrators from being held 100% accountable and responsible for their offense behaviors. In addition, the blackout occurs after the offense, not during it. They knew what they were doing when the offense occurred.

When the offender shifts their eye focus, something is up. This behavior is indicative of anxiety. Fluttering and closing the eyes may indicate a disassociative state. When this occurs, question the perpetrator whether they use aliases or whether they have ever experienced hallucinations or a loss of memory. There is some evidence to suggest that certain offenders disassociate when they offend. Multiple personality disorder and Dr. Jekyll and Mr. Hyde are two of the most common terms used to describe such experiences. However, most offenders do not meet the criteria to be formally diagnosed with multiple personality disorder. The perpetrator persona is the chosen "part" of themselves that offenders resort to when they offend. Most describe the perpetrator persona as "the part of me that did it," and they may state that "I didn't do it, it was someone else." It is a good idea to question who this "someone else" may be.

Observe the intensity of the eye contact. Do the offenders intensely stare and/or glare at you or are their stares seemingly gentle? What is your gut reaction to eye contact. If you experience tenseness or irritability, then the eye contact is likely intense and is an attempt to intimidate you into believing their stories. Intensely staring or squinting is often an attempt to intimidate the interviewer or an attempt to take control of the interview. It is imperative that the interviewer remains in control of the interview at all times. Confront the offenders when they attempt to take control of the interviews.

Facial Expression

What is the expression on the offender's face? Smiling in response to questions about their offense is indicative of anxiety and of guilt. I believe that the smile was the perpetrator persona acknowledging being caught. It may have surprised the perpetrator when the interviewer focused on a specific fact that he did not expect the interviewer to be aware of. Observe the degree of tension on the offender's face. Common examples include:

- Gritting of teeth
- Tensing of jaw muscles
- Wetting of their lips with tongue
- Frequent swallowing
- Asymmetry of the facial reaction
- Tightening of the lips, especially the corners
- Small facial or smile tremors

All of the above examples are suggestive of anxiety, anger, and irritability. It is also important to pay attention to attempts to control anger, which may be evident by gritting of teeth and tensing jaw muscles. Ekman (1992) outlines different types of smiles very well. He indicated that genuine facial reactions tend to be of relatively short duration, less than 5 to 10 seconds. So, certainly someone attempting to prolong a facial expression may well be faking. Again, with the caveat — everyone exhibits emotions somewhat differently, and the duration of a particular offender's facial expression or emotion may well be sincere. Ekman also indicated that genuine smiles usually involve the eyes squinting and that this is difficult to fake.

Think about the cues of escalation. The offender's body and mind go through fairly predictable arousal response to real or perceived provocation. Attempts to hide or actively display irritation and escalation certainly involve the mouth (e.g., gritting of teeth, tensing of the jaw muscles) and these signs can usually be easily seen. The frequent wetting of one's lips could certainly be due to legitimate causes (e.g., side effects of certain psychotropic medications, Tourettes Disorder), but, generally speaking, I find this more common when an offender is anxious and defensive. Frequent swallowing can often easily be observed as well. The more the offender becomes anxious or angry, the more these signs are likely to occur.

Arm Position

The position of offenders' arms reveals a great deal of information. Observe not only the position of their arms, but when and how they move them. Common examples to pay attention to include:

- Crossing of arms across chest
- Leaning on arm, using it as a head rest
- Leaning on the table
- Frequent arm movements or changing of positions

When people cross their arms, it is usually due to irritation, anger, or a more subconscious need to protect the self from some perceived harm. When offenders cross their arms during questioning, pay attention to the question being asked. It is likely that they are anxious about that particular question and are fearful or concerned that you may be picking up on their response, which may be incriminating.

Using their arms to lean on during the interview suggests comfortableness with the interview process. This should be suspect given the purpose of the interview. Why would someone be calm and free of anxiety when being questioned about a physical or sexual assault? It has been my experience that offenders who are inappropriately calm and lacking any significant anxiety either naively believe that their version of the offense will be believed or simply do not believe that what they did was wrong or harmful to their victim. The latter is indicative of the psychopath, who cares little for the well being of others.

Hand Positioning

How the offenders position their hands during the interview often reveals significant anger and anxiety. Common hands gestures or movements include:

- Rubbing the fingers together
- Tensing the hand, not necessarily into a fist
- Making a fist
- Concealing their hands, such as under the table
- Placing hands on the table and tensing or making a fist
- Using middle finger to scratch an itch or resting middle finger on face while other fingers are retracted
- Actively pointing fingers/hand at interviewer
- Tapping, hitting, or slamming hand or fingers on the table

Rubbing hands or fingers together may indicate anxiety, guilt, or shame. It is often helpful to ask offenders what they are feeling when these observations are made. The other examples listed above are indicative of anger, rage, anxiety, and most of the time offenders engage in them with the intent to have the interviewer back off. The offender may or may not be aware of the degree they are escalated. Therefore, it is important to confront these behaviors.

Agitation and Anger

Offenders that become angry, defiant, rude, or agitated are potentially more dangerous during the interview or interrogation process. The caution here is that innocent people can become noticeably irate, upset, and uncooperative. The difference is that innocent people tend to maintain their disgruntled effect much longer than guilty offenders. This is discussed elsewhere in this section. I find it important to identify and examine what specific question or offense detail was the offender responding too? This may give some insight into offenders' vulnerability in that whatever provoked their anger or irritability signals cues that the offenders may not be able to hide their feelings, and anger-based feelings can often be used to cover feelings of vulnerability and guilt.

An interesting note here is that the interviewer is in a position of power and authority over the offender, and the interviewer's report will impact offenders' lives. One might expect that offenders would behave in a respectful, cooperative manner. However, when they demonstrate any degree of hostility or anger during the interview, they are truly dangerous offenders who likely have a history of engaging in verbally and physically assaultive behaviors. The interviewer may experience some degree of intimidation given the offender's hostility. Imagine how others would be affected by the same behavior, others including the victim, other children, and adults who may lack any power to protect themselves from the offender's rage. Question the offender on the fact that if he escalates so quickly during the interview, how quickly does he escalate in the community, at home, or with children. In addition, question him on how others would react or be affected by such behavior. Again, the offender often provides much more information than he intended to.

SUMMARY

In summary, it is important to pay close attention to the physical cues perpetrators offer. For example, when questioned, do perpetrators close their eyes or do they make eye contact? When they cry, do the tears appear real and free flowing, or do they appear forced. Does the person appear to be attempting to cry, but few or no tears appear? These are signs that suggest that offenders are withholding information, and the tears and sorrow they are attempting to express are fabricated. When questioned about a crime, do they demonstrate anxiety, such as giggling, smiling, or grimacing? If so, they are likely giving you the message that they know more than they are willing to share, but that their perpetrator persona is not able to accept credit for their crime at that time.

One case example will help illustrate the above. Consider Susan Smith, the young mother who drowned her two children. When she was questioned on television, she made a statement about how her children were abducted. She alleged that a man had kidnapped them. However, as she spoke, she closed her eyes. When she spoke about hoping that wherever her children are that they were all right, she smirked, and then smiled. She knew exactly where her children were, and she knew that they were already dead. She felt the desire to continue to withhold the truth, yet her body and emotions betrayed her. She simply was unable to deceive others about her heinous crime.

The above information is not an absolute guide to the detection of deceit. Exceptions and extenuating circumstances do occur, and the interviewer must rely on sound clinical and ethical principles throughout an interview. Furthermore, training and experience in the areas of ethics, the law, psychology, and interview techniques are essential to establish a basis of expertise for interviewing sex offenders. Interviewers must understand the etiology, symptoms, and treatment of physical and sex offenders before they interview abusers or sex offenders. Without the above knowledge, the interviewer stands to miss essential information, and may arrive at erroneous conclusions. Remember that the primary goal of interviewing abusers and sex offenders is to encourage the offender to reveal the truth so as to allow for proper prosecution and/or treatment.

28 Words and Phrases

COMMONLY USED STATEMENTS

The words and phrases abusers and sex offenders use often give away the truth. I will discuss some of the most commonly used statements below. These are statements I frequently hear when conducting a forensic assessment. Not every person who makes any of the following statements is being deceitful, misleading, or failing to cooperate. With that having been said, let's examine some of the following statements. Offenders answering questions with any of the following statements are likely being deceitful. Remember that many of the questions asked require a "yes" or "no" response, not an explanation or vague answer. When asking an open-ending question, the following answers are strongly indicative of deceit.

"Not That I'm Aware of"

This is perhaps my favorite statement. What exactly is the offender saying here? Who, if not the offender and victim, would be *aware* of what occurred prior to, during, and after the offense? This statement implies a deliberate attempt to disguise the truth with ambiguity. The question may have been, "Did you hit her?" How can the offender not know what happened.

"I've Told You All I Can"

Well, if the offender has told you all he knows, then why respond with such an inviting answer? Who is in the position to allow the offender to disclose the truth, as his statement implies, he was told by himself or others to say nothing more at this time. Most often the offender has consulted an attorney and was advised not to say anything beyond the rehearsed answers. The offender's self-image is also on the line. To admit to anything further threatens his very being and sense of worth. For example, it may be palatable to admit some of the offense behavior, while admitting to other details may mean that the offender has to label himself as a "pervert" or "sicko" or "monster."

"If He/She Says That's What Happened, Then That's What Happened, But I Don't Remember Right Now," or "Whatever They Said Happened, Happened"

This response is common when asking the offender to tell you what occurred. This is a cowardly attempt to admit to the offense behaviors without actually taking responsibility for engaging in the assaultive or sexually abusive behaviors. If the offender assumes a helpless and defenseless role, he is more likely going to evoke sympathy from others rather than anger. In addition, this statement allows for the offender to cast doubts on the victim's statements by blaming the victim for the offense. If the offender does not actually admit to the behaviors, but passively assumes blame, then the victim's credibility may be challenged. In fact, people close to the offender may offer excuses for why the victim may have "exaggerated" his/her statements.

Offenders who were under the influence of drugs or alcohol use this excuse often, especially incest offenders. It is imperative that offenders be held responsible for their actions and be forced to admit to the offense behavior. Allowing offenders to get by with this statement prevents them from taking full responsibility for their assaultive behaviors and again tends to encourage others to question the validity of the victim's statements. If drugs or alcohol caused offenders to be assaultive, then every time they used chemicals they would engage in assaultive behaviors. In addition, they would not have been concerned with hiding their offense behaviors, and would have committed the physical or sexual assault in the presence of others. After all, if they were truly "out of control," they would be powerless to postpone gratification of their violent urges. This response is admitting guilt through default. They do not fully believe that they are culpable and attempt to avoid any confrontation of their own behaviors.

"It Might/Could Have Happened That Way"

This response is common when asking how an offense occurred. Then again, the offense may have been the result of an alien attack? The offender made decisions throughout the offense to force the victim into sexual contact or to assault the victim. The question that was asked was likely a yes/no question. To answer with an explanation is indicative of guilt and, therefore, the answer was "yes, I did it."

"I May Have" or "Maybe I Did It"

This response is common when questioning directly about specific offense details or when confronting the offender. Again, the question of whether they committed the offense or engaged in a specific behavior is a "yes" or "no" answer. The offender is admitting guilt by the fact that he is not denying the offense. How many people would reply with that answer if they were innocent, I can not imagine anyone doing so.

"I Don't Remember"

So, how is it that the offender is unable to recall the offense or offense details? Where was the offender? Is it possible to not recall being physically or sexually abusive? I do not think so. Given the degree of adrenalin released when we engage in behavior that is forbidden, it would be next to impossible not to remember the event. Further, the gratification resulting from the control, humiliation, and orgasm further reinforces the behavior. The offense behavior was positive for the offender and, when behavior is reinforced, it tends to be repeated. When the victim of a physical assault obeys or acquiesces, the offender again wins. We tend to repeat behavior that gives us pleasure, not pain. Even when the offender wants to be punished with some sort of pain or consequence, he enjoys it, justifying the victim wanted it after all. Therefore, when the offender engages in physically or sexually abusive behavior, he does remember it. That is one reason why he made the above statement rather than simply answering, "I did not do it." Also, offenders remember other less significant events in their lives. Prove that they do not remember such a significant event! You cannot, as there is no proof that they would forget a significant event. There is an abundance of research on memory recall, which supports my point. Some of the offense details may be confused with prior violent offenses; that is understandable. So, offenders may indicate that they do not recall all of the details of the current offense as it blends into so many other violence incidences.

"I Thought It Was My Significant Other" or "I Thought They Were Older (16/18)"

This a common claim made by those who have sexually abused a female child. Often the situation involved sleeping in the same bed with the victim, and most often in the offender's bed. Child molesters and those sexually abusing adolescent females often using this excuse are quite frequently under the influence of drugs or alcohol and blame their behavior on the fact that they were drunk or high. At times, their claim of innocence appears to be believable; I mean, accidents do happen, don't they? Well, let's take another look.

When the details of the offense are reviewed, it becomes clearer that the offender made a conscience choice to abuse the child or adolescent. In fact, it would be nearly impossible not to be aware that it was a child, not an adult, laying next to him in bed. If the offense involved a brief touch, perhaps it was an accident. But, if the touch occurred for several minutes and/or involved fondling of the victim's genitals or more detailed sexual behavior, or was repeated, then there is no room for doubt that the offender intended to engage in sexual contact and that he knew it was not his partner.

Imagine what the differences are between the body of an adult, child, or adolescent. There are significant differences in development. For example:

- Body odor/perfume/cologne (most children and adolescents do not wear perfume, and those that do usually wear a lower quality compared to the parent's; children and adolescents will also smell differently than their parent)
- Body size (imagine the physical differences between the child or teen and the parent)
- Type of clothing
- Size of mouth and teeth
- Genital hair
- Size of breasts
- Size of vagina or penis
- Size of vaginal and anal canals

Behavior of victim — if the victim engaged in any sexual contact with the perpetrator or how the victim reacted to the offender's behavior:

- Their general response to the abuse or sexual molestation
- How they kiss
- How they touch
- How they talk, what they say
- How they perform oral sex
- How they perform vaginal or anal intercourse

Ask the offender to explain how the victim was different than his significant other. If he continues to be resistive to the questions, ask to see a picture of the victim and of the significant other. Compare their bodies, faces, height, weight, limb sizes, etc. The differences should be easily identified. It was never an accident. Many offenders who use this excuse believe that they were in love with their victim, and that their victim loved them. The loving relationship they describe is an adult-to-adult relationship, not a child– adult relationship. At times, I diagram this on a white board and illustrate the above differences. It is an effective way to get the offender to admit to the sexual abuse. Ask the offender, "How could you not be aware of the fact it was not your significant other?" Have them describe in detail, step-by-step the offense behaviors.

"I Didn't Know It Was Hurting Him/Her"

The offender may have indeed convinced himself that the victim was not negatively affected or harmed in any way. However, when he is asked to describe the facial reaction of the victim and to describe the victim's behavioral response, the effects become all too clear. For example, common victim facial expressions and behaviors include:

- A numbing look
- Eyes closed

- Looking scared, empty, lost, vacant
- Not responding with equal behaviors (e.g., not returning the fondling or kissing)

Ask the offender, "How could you not be aware of the fact it was not your significant other?" Have him describe in detail, step-by-step the offense behaviors. Then have him describe the victim's reactions to each behavior. Most often the victim made little if any eye contact, and he/she may not have responded back with equal sexual behaviors or comments. Be aware that sexual offenders usually relabel the negative reactions by the victim into more positive and consensual reactions. What was a fearful and shy reaction is often relabelled to be lustful and wanting of the sexual contact.

"He/She Wanted It"

This statement is an admission of guilt, but it is also a statement indicating that the offender is projecting blame onto his victim. The bottom line is that regardless of whether the victim initiated the sexual contact, it was the responsibility of the adult to act appropriately. Further, what did the victim really want? How much grooming, insistence, persuasion, coercion, or force did the offender utilize to gain compliance from the victim? How would a child know how to be sexual, for example, if the offender was not influencing him/her in some way? If the offender provided the victim with alcohol or drugs before requesting sexual contact, what was the purpose of doing so? If the victim was a willing and appropriate sexual partner, why help him/her become impaired?

"I Can't Tell You"

This is often in response to asking the offender for more details of the offense. This statement is indicative that offenders have awareness of what they did, but they are not ready to admit their offense behavior. Most offenders who make this statement have discussed their case with an attorney and have been advised not to say anything concerning their offense. However, the underlying message that is given is that offenders are not going to say too much about the offense at that time. Ask them, "What is it you cannot tell me, and why?" "What would happen if you told me, or what would it mean about you as a person if you discuss this crime or if you admit that you are guilty?" It is also a good idea here to question the offender about the use of aliases or possible disassociative experiences. If they have used aliases or have disassociated, then ask them to describe and explain the details of how and when that occurred.

"I Made a Mistake"

On one hand, this offender admits he engaged in inappropriate behavior, yet on the other hand, he is attempting to minimize the seriousness of his behavior. It is as though viewing the offense as "*a mistake*" somehow lessens the degree to which he should be held accountable, and suggests that a less severe consequence would be in order. Remember that when the offender abused or molested his victim, he made a conscious choice to do so, and he was aware that the victim would be adversely affected in some way. Making *mistakes* does not relieve the offender from being held 100% responsible for his behavior.

"It's All a Misunderstanding"

This statement indicates that the offender was with the victim, otherwise how could a *misunderstanding* or *mistake* occur? Most offenders who use this statement experience difficulty facing reality. They want to believe (as most offenders do) that they are not capable of engaging in abusive behavior, and that they would never choose to abuse others. Further, they do not want to accept the fact that they did commit the abuse or molestation. Some offenders convince themselves that the victim enjoyed the sexual contact, despite the fact that they used force to get the sexual contact in the first place.

Often after claiming the offense was a misunderstanding, the offenders claim that their victims misunderstood their behavior, and that if the victims had only indicated to them that they were being hurt or were afraid, the offenders would have ceased the behavior. Again, the offender wants to protect his image, and he may hold strongly to the belief that he really, truly is a "nice, gentle, giving" person. Most of the time he probably is the kind of person he imagines, but at times he is also the monster he so dreads and hates.

One episode of the television program *Law and Order* provided a good example for the above categories. A teenage male shot and killed a boy. As it turned out, he killed the wrong boy. When the police questioned him, his responded, "So what, it was an accident. I mean, I didn't mean to kill him; it was the wrong boy, so it was an accident." The attitude of the teenager was that if he killed the wrong boy, then he should not be punished. Many sex offenders and abusers utilize a similar thinking style, one in which they believe that they should not be punished for sexually, physically, or psychologically harming their victim because either their victim deserved to be abused, or because the offenders did not "intend" to harm their victim.

"I Would Never Do Something Like That" or "How Could I/Anyone Do Something Like That?"

These are common statements made by offenders who are experiencing shame, guilt, and disbelief about their offense behavior. When questioning these offenders, it is often helpful to ask them to describe how they felt and what names they would call a sex offender or abuser they saw on the news prior to having committed their offense. They cannot believe that they are now one the perverts or monsters that are common to the news daily. Some compassion helps to overcome the pain of the reality that they are now sex offenders or abusers. Further, focus on how their admission will work in their favor with the possibility of a less severe sentence as well as how their admission will help their loved ones, especially their victim.

"People Who Do That Are ... Perverts, Monsters, Sinners, Baby Rapers, Etc."

This type of statement is indicative that perpetrators are struggling with their offense behavior as well as with their identity. They realize what they did and are in disbelief. What I said in the previous paragraph also applies here. Note that in the last several types of statements, offenders are focused on their reputation or image, not on fear of the consequences or how the victim was impacted. They are separated or cut off from the effects of their offense (possibly disassociated). These offenders tend to make good treatment candidates. However, it is important to be aware of and assess offenders' levels of depression and suicidal ideation.

"I'm Not a Child Molester/Monster/Pervert"

This statement is similar to the above two; however, the offender is unknowingly admitting his offense. It is as if he is saying, "Yes, I did it, but I am not a bad person." Again, he is more focused on his image and what the offense means to his image than anything else. The offender may be numb to the impending consequences.

"Ask My Significant Other/Family/Friends, They Will Tell You I Did Not Do It," or "I Am Not Capable of Doing This"

Offenders who make these types of statements are interesting to interview; however, they are also extremely difficult to reason with. It is as though they truly believe that their support people (family, friends, co-workers) will be able to erase or hide the truth. Even if the abuse occurred while others were present, the others were not present the entire time the perpetrator had contact with the victim.

For example, if a child were molested while the other parent was home, the other parent would likely have left the offender alone with the victim at some point. Question whether the other person ever left the room, even if for a moment, say to use the bathroom, get something from another room, fix a snack, or to answer the telephone. Maybe the other person was sleeping when the abuse or molestation occurred. Offenders have an instinct for using their significant other as an alibi or excuse for abusing or molesting others. In fact, I have worked with many abusers and sex offenders who misinterpreted their significant other's lack of concern about the grooming behaviors as a sign that the abuse was normal and condoned. The other person, however, may not have picked up on the clues or the grooming behavior the offender was engaged in.

Another explanation for offenders using this type of excuse is that they truly believe that if their significant other believes in their story, then the authorities will have no option but to believe their story as well. In fact, if you question the significant other and the offender about their relationship, it is not uncommon to find that the offender makes most, if not all, of the decisions for the family, even if it appears on the surface that the significant other is the head of the relationship. Offenders are experts in deceiving those around them into believing that it is they, the offender, who is really the victim, the confused, and innocent one.

With this type of statement, it is common for the offender's support network to stick up and protect the offender at any cost. This is particularly true when the victim was a child or adolescent. The support people may claim that:

- It was the victim's fault, the victim led the offender on or is exaggerating the offense
- The victim was abused by others, and now he/she is trying to get the offender in trouble because the victim is angry
- The victim asked to be abused, he/she was seductive and knew exactly what he/she was doing
- The victim never told the offender to stop, wasn't injured, or never told anyone

All of these examples are statements used by the support network of offenders I have worked with. It is interesting, yet frustrating, when the offender's support network defend, beyond any reasonable explanation, and protect the perpetrator and blame the victim. The support network becomes delusional, offering irrational explanations to explain and justify the offender's actions. The bottom line is that the offender was the adult and had the legal and moral obligation to not take advantage of or abuse another person. The adult had the power to stop the behavior. Yet the offender and his support network place the blame on the victim. In a very direct sense, the support system is responsible for the abuse and was criminally negligent of endangering the welfare of a child or minor.

"I Handled My Wife Wrong"

The above comment was made recently by a Minnesota Vikings football player who was arrested for physically abusing his wife. What this statement implied is that he believed that it was his job to "*handle*" his wife, which in turn implied that she was "*his*" object. We can handle situations and objects, but not people. It is likely that upon further questioning it would become clear that he objectified most, if not all, of the women in his life. "Handling involves an object, not a person. He was abusive. Offenders often make similar statements when questioned about physically or sexually abusing someone. The belief is that by claiming that they simply mishandled the situation and, therefore, the whole offense is being blown way out of proportion.

Use of Derogatory Names, Such as Bitch, Slut, Cunt, Ho, Fucker, as Well as Racial or Ethnic Slurs

Use of words or names that degrade people are indicative of objectification and prejudice. What exactly is a bitch, cunt, etc.? They are simply words that are meant to humiliate, put down, slam, injure, or insult. In addition, regardless of the context in which the terms are used, most involve sexual innuendos and, therefore, are indicative of sexually abusive thoughts and likely behavior as well. They are not a description of a person. When offenders refer to their victims or others in their life with derogatory names, they have given a clear picture of what their value system is like.

Offenders who use derogatory names are often quick to escalate into anger and rage, and to blame the other person for (in their mind) "causing" them to feel angry. What follows is the belief that they must teach the victim a lesson. They believe that their victims deserved and asked to be abused. A good example here is date rape. The couple are engaged in mutual fondling and, at some point, the female says "stop." The offender is frustrated and blamed the victim for "turning me on" and often resorts to manipulation, name calling, and attempting to make the victim feel guilty. And when these techniques fail, the offender used threats or physical force. The offender believed that he was justified to force/coerce sex and he believed that his victim had no right to change her mind or to frustrate him. Even when sexual contact does not occur, abusers frequently use sexually derogatory names against their victims.

In most every violent situation, the offender engaged in verbal and/or psychological abuse *prior* to engaging in more forceful measures. Therefore, when offenders engage in verbal or psychological abuse, they are often very close to engaging in physical force to get what they want.

29 Claims of Memory Problems

I am amazed at how ignorant abusers and sex offenders are of memory problems. Approximately 70% of all physical and sexual assaults involve a perpetrator who was under the influence of drugs or alcohol. Consider for a moment a state of blackout produced by alcohol or drug use. To experience difficulty in memory recall is common for chronic drug or alcohol users. Some of the other attributes of chronic abusers of chemicals include inability to maintain employment or relationships, breaking financial obligations, and poor health and hygiene.

I will argue, however, that at least 90% of the offenders I have worked with were not chronic drug and alcohol abusers. Most of the offenders maintained employment even if through temporary jobs. Most were involved in ongoing dating or marital relationships and had moderate-to-good health and hygiene. Their ability to recall the details of their offense behaviors was intact. It is amazing and perhaps a true miracle that approximately 98% of the perpetrators I have worked with could recall at least 95% of their offense behaviors, quite enough to prove them guilty and to provide issues to address in treatment. The bottom line is this: offenders can recall their offense behaviors regardless of whether they were under the influence of drugs or alcohol at the time of the offense.

It is understandable why abusers or sex offenders would want to block the memory of their offense behaviors. Who would want to have repeated memories of how they hurt their victim, often someone they cared for? At the very least, how could they have acted in a violent and aggressive manner when their self-image was one of compassion, caring, and gentleness? But the mind is a wonderful organ; it retains memories of every experience of our lives. Of course, memory recall can be affected by organic problems as well as by a conscious desire to repress or distort the reality of a memory. Volumes of evidence exist that memory recall can be directly affected by conscience choices.

CONVENIENT FORGETTING

Convenient forgetting of offense details is common. Offenders claim to recall all, some, or none of their offenses, and it is up to the investigator/assessor to guide the offender to full recall. This may be accomplished in several ways. First, have the offender explain the offense situation in his own words. Listen carefully to his explanation. Is he claiming to have interacted with his victim, or is he denying any contact whatsoever with his victim? What physical evidence that the offense occurred is documented? Was the victim physically injured, were weapons or objects involved, did anyone witness the offense, or observe the emotional or physical condition of the perpetrator or victim prior to, during, or after the offense? All of these offer areas to begin challenging the offender's distorted claims of innocence.

ROLE OF ALCOHOL ON MEMORIES

At the time the offense occurred, the offender was in full control of his behaviors. Drugs and alcohol do not cause us to behave in violent, methodical, or complex behaviors. In fact, drugs and alcohol reduce our motor control and decrease sexual arousal. The most common drugs involved in physical and sexual assaults are marijuana and cocaine. Marijuana is a depressant, the same as alcohol. So drug and alcohol use does not cause an offender to forget his offense. However, substance use may impair the formation of new memories.

Gudjonsson (1992) indicated that studies have been conducted on the adverse effects of alcohol on memory. The three main areas identified included (1) short-term retention; (2) the encoding and formation of new memories, which include transfer of short-term memory retention into long-term storage; and (3) retrieval. Studies indicated that moderate amounts of alcohol impair the process of memory formation. However, short-term memory retention and retrieval appear unaffected. Loftus (1980) found that marijuana had similar effects, and that marijuana may also impact short-term retention. Yuille and Tollestrup (1990) found that even in small amounts, alcohol use appears to interfere with the acquisition and encoding of new information, rather than with retrieval. In conclusion, small to moderate amounts of alcohol may interfere with how new information about an event (e.g., an offense) is stored.

CLAIMS OF AMNESIA

Some offenders claim that they experienced amnesia and, therefore, are unable to recall the details of their offense. The American Psychological Association (APA) defines amnesia as being characterized by a disturbance in memory that is either due to the direct physiological effects of a general medical condition or due to the persisting effects of a substance (e.g., use of drugs or alcohol). Individuals afflicted with an amnestic disorder are impaired in their ability to learn new information or are unable to recall previously learned information. Functional amnesia occurs without any evidence of brain pathology and can include situation-specific amnesia where the memory for a specific episode is lost (e.g., for the offense) (Kopelman, 1987). Taylor and Kopelman (1984) suggest that there are three main ways amnesia for offenses occur: (1) as a result of the pharmacological effects of alcohol, (2) as a part of a psychotic mental state (schizophrenia or depression), or (3) as a psychological defense mechanism (e.g., repression or suppression).

Kopelman (1987) goes on to describe three possible mechanisms in psychogenic (functional) amnesia:

1. Faulty memory acquisition (the severe emotional stress or trauma impairs the ability to register and encode the situation).
2. The trauma associated with the situation or offense is unacceptable to the conscious mind and is subsequently repressed.
3. Following the work of Bower (1981), the psychogenic amnesia is viewed in the context of a state-dependent memory. Here, the amnesia is viewed as a retrieval deficit.

Rogers (1997) indicated that some of the criteria for considering feigned amnesia include the following:

1. Impaired attention or immediate memory that is much worse than impairment of overall learning and memory.
2. Standardized scores on tests of recognition memory that are lower than standardized scores on other tests.
3. Reports of dense retrograde amnesia in the absence of other neurological abnormality.
4. Reports of dense retrograde amnesia together with intact new learning and memory.
5. Gross inconsistency between test performance and everyday functioning.
6. Gross inconsistencies across tests or testing occasions.
7. Evasive, uncooperative, or other unusual test-taking behavior.

Meyer and Deitsch (1995) indicated that several patterns have been found of interview and test data to suggest malingering. To counter malingering, it is important to rigorously question any symptom that is being reported, and to do so with open-ended questions. This will help to counter the lie teller's basic knowledge about symptoms of medical and psychological disorders. A malingerer will tend to report symptoms that are generally rather rare, as well as a high number of symptoms that are usually unrealistic given the individual's presentation and situation.

TESTING

Most psychological testing and instruments have built-in scales to assess for minimization, denial, or exaggeration. On those that do not, simply compare known data from the individual's test response. For example, knowing that the individual is court-ordered for an anger management or sex offender assessment, but they report no anxiety, no fear, no anger, and no significant marital or relationship problems. It would be highly unlikely that some degree of elevation occurs given the nature of the assessment. Also, if they are denying experiencing any difficulties resulting from their drug or alcohol use, but alcohol or drugs were involved in their presenting problem, they are being far less than honest. In addition, comparing similar questions from one test protocol to another helps to see if the individual answered similar questions consistently. On one test, for example, an individual may have indicated anger problems or occasional drug or alcohol problems, whereas on another test he flat out denied any concerns in those areas.

SUMMARY

In summary, it is important to understand the genuine symptoms and etiology of memory problems. Offenders may claim to have a wide variety of amnesia and memory difficulties that appear conveniently to help justify their offense behavior. When in doubt, offenders should be referred to a neurologist and for testing if they insist that they possess symptoms that do not appear genuine.

30 Sociopathy, Malingering, and Defensiveness

FEIGNING AND MALINGERING PSYCHOLOGICAL DISORDERS

Clark (1997) writes about sociopathy, malingering, and defensiveness. It is important to understand how criminals feign and malinger symptoms of psychological disorders as well as simply lie about their behavior and thoughts. Meloy (1988) asserts that malingering is seen as obtaining a clear advantage for the person. Travin and Protter (1984) viewed malingering as a "fully conscious sociopathic adaptation." Cleckley (1976) and Doren (1987) have emphasized the sociopath's propensity for untruthfullness and insincerity; they viewed such deception as one of the defining characteristics of the disorder of malingering. Yochelson and Samenow (1976) also view lying as a fundamental pattern for sociopathic individuals and as an essential precondition of their criminality.

Clark states that it is important to be cautious when interpreting the Minnesota Multiphasic Personality Inventory, 2nd edition (MMPI-2) measures of faking because the individual may well be genuinely disordered and not faking. Meloy and Gacano (1995) asserted that psychopathic malingerers (1) often perform worse than true patients with organic impairment on neuropsychological assessments and (2) manifest impairment on psychological measures that are not commensurate with behavioral observations. They support that sociopaths are inferior malingerers and more easily detectable than are nonsociopaths. Kropp (1994) and Rogers (1995) found that psychopaths as a group were not significantly better than nonpsychopaths at simulating mental illness. Rogers (1990a,b) found that "at least within a forensic sample, the presence of Antisocial Personality Disorder does not appear to signal a greater likelihood of malingering.

Rogers and Mitchell (1991) found that offenders arrested for serious crimes can be motivated to report, over-report, or fabricate drug use as a mitigating factor in sentencing. This comes as no surprise to an experienced forensic psychologist or law enforcement agent. The belief that being under the influence of drugs or alcohol somehow absolves an offender from responsibility for his offense behavior is basically public sentiment. Unfortunately, many "professionals" ascribe to this erroneous belief as well.

Sexual Offenders

Kennedy and Grubin (1992)* developed four specific denial patterns specifically for sexual offenders, similar to Langevin's (1988) "degrees of admission" (Patterns 2 and 3 collapsed):

1. Men who admit their offenses but denied causing any harm to the victims, and instead claim to have helped their victims (18% of the sample). They tend to take a political stance toward the law, mostly male child offenders.
2. Men who externalize responsibility for their sexual offense behavior (20% of the sample). They may have admitted their behavior but blame either the victims or others. They complain about being treated unfairly by the legal system. Usually young female victims.

* Reprinted with permission.

3. Men who denied the extent to which their sexually deviant behaviors were ego-syntonic (27% of the sample). This included mostly incest offenders who admit the offense and harm caused. However, they blame a temporary aberration of behavior or altered mental state.
4. Men who totally deny their offense (35% of the sample). Mostly adult female victims, not motivated for treatment.

Sewell and Salekin (1997)* explain the four explanatory models of dissimulation (malingering) for sexual offenders:

1. *Pathogenic model.* Rogers and Dickey (1991) describe the model in which denial and minimization among sex offenders is understood as reflecting a primitive ego. Ward, Hudson, and Marshall (1995) concluded that for a sex offender to admit the full scope of his behavior to others would be to admitting it to himself, which may be psychologically unbearable. Ward et al. (1995) argued that denial processes may be both precursors to offending as well as after-the-fact justifications for allowing (1) offending to continue, (2) self-esteem to be maintained, and (3) negative self-evaluative cognitions to be disputed.
2. *Criminological model.* Rogers and Dickey (1991) found that sex offenders with marked psychopathic traits and sadistic behaviors are more challenging to evaluate and treat (Chaplin, Rice, & Harris, 1995; Ogloff, Wong, & Greenwood, 1990). The use of penile plethysmography (PPG) and the Psychopathy Checklist-Revised (PCL-R) are highly recommended to address the problem.
3. *Adaptational model.* Rogers and Dickey (1991) in which the male offender perceives himself to be in an adversarial situation, believes that something might be gained by being defensive, and conclude that dissimilation is the most effective option. This model conceives denial/minimization as an effort to minimize the anticipated social and legal sanctions.
4. *Socioevaluative model.* Dissimulation is an attempt to constrict the extent to which the private world of the individual can be examined and made public, regardless of the reason of the evaluation. Little thought is given to the possible outcome/consequences, and requires only the perceived adversarial situation.

* Reprinted with permission.

31 Physical Abusers' and Sex Offenders' Responses When Questioned about Their Crimes

THREE PATTERNS OF OFFENDERS

Physical and sexual offenders generally react to confrontation about their offense behavior in similar ways. There are basically three patterns of offenders — those that will admit their offense behavior, those that will partially admit (minimizers), and those that deny their offense.

ADMITTERS

Some offenders confess to the crime with which they are confronted. They may admit the crime for several reasons. For some, they admit because they experience guilt and remorse. It is often very difficult to determine whether the guilt and remorse are genuine (this will be discussed later). To admit the offense frankly allows for more immediate relief from the guilt and shame. Some offenders are simply intellectually slow and do not possess necessary skills to deceive or minimize their offense behavior. For others, admitting to the offense is necessary because they have a conscience and need to be honest to survive psychologically. In either case, whatever they say may be used against them for prosecution. But the good news is that a conviction is obtained faster and the victim may receive services to allow healing much sooner. In addition, a dangerous sex offender is removed from the street and future victims are spared the tragic event of being sexually and/or physically assaulted.

Those who confess typically demonstrate one of two profiles — those that are remorseful and those that are remorseless. Those that are remorseful experience guilt, concern, and possibly depression. They are sorry for what they did, admit to the offense, and are willing to openly discuss the offense. Those that are remorseless demonstrate a shallow appearance of remorse, but are basically unaffected by the offense and do not appear to appreciate or accept the damage caused to their victim.

PARTIAL ADMITTERS (MINIMIZERS)

This category of offender refers to those who minimize their offense behavior. They admit certain behaviors in hopes that they will not be held accountable for all of their offense behavior. For example, an offender may admit that they did not stop being sexual when the victim told them to stop, but they deny using physical force, despite physical evidence that may prove that physical force occurred. They may admit that they were sexual with the victim, but claim it was consensual. These types of offenders tend to make statements such as, "I do not recall everything," or "If the victim says I did that, I must have, but I don't think I did," In reality, offenders can recall what they did and these types of statements suggest that they do know what occurred, but are emotionally not prepared to admit what happened, usually out of fear of the consequences or because of shame and guilt.

Offenders who are adamant about not admitting fully to the offense behavior are more difficult individuals with whom to work. They are aware of what they did, but are not going to fully admit their offense behavior. This is usually a result of the offenders lacking remorse, lacking guilt, or minimally experiencing any of these emotions. Basically, these types of partial admitters tend to be self-centered and are often described as being "cold" or "smooth" in the way they relate to others. These individuals tend to be shallow and are either unable or unwilling to acknowledge anything that does not fit with their misguided sense of reality. They may actually believe that the victim was compliant despite having utilized physical force that resulted in victim injury. They are unable to effectively explain how the victim sustained any documented injuries and unable to effectively explain the discrepancies between their version of the offense and that of the victim and of witnesses.

Deniers

This category of offender basically denies the offense. They may deny that they were even with the victim or deny that any physical or sexual contact occurred. Offenders that deny their offense behavior may do so for several reasons.

The first may be that they cannot believe they actually committed the offense, but recognize that they did, in fact, commit the offense. They may be bewildered by the fact that they were capable of engaging in violent physical or sexual behavior toward a loved one or another person. Of all of the types of deniers, this is perhaps the most workable type because they will likely experience guilt, shame, and remorse at some point. They may make statements to the effect that, "I can't believe that I would do something like that, but if the victim says I did, I must have."

The second most common type of denier is one who lacks any sense of conscience. They are more commonly referred to as psychopaths. They do not care about the well being of the victim, nor do they care about what others think about them. They are incapable of appreciating the consequences of their behavior and act out in whatever way pleases them. To the psychopath, the police and legal system are annoying and are not perceived of as being fair. The psychopath is usually unable to understand why they are not believed and or why their version of the offense is not being accepted as factual.

The third most common type of denier is one who will eventually admit to some, if not all, of the offense behavior at a later time. This offender denies the offense outright from the start. He will not admit the offense until either undeniable evidence is presented to him or until enough time has passed that allows for him to emotionally accept the reality of what he did.

Related Factors

Shame

The impact of shame plays a role in denial for offenders who have a conscience and who have a minimal offense history. I assert that the more frequent and lengthy the abusive or sexually abusive history, the less shame is directly involved. Shame is about how an individual is impacted by a situation, and how the behavior affects one's self-image. An example would be that because "I did something bad, I must be a bad person." Shame is about allowing one's behavior to dictate and define self-worth and self-image.

Guilt

Guilt concerns feeling bad about something someone has done, about one's behavior. An example would be that, "I did something wrong, and the behavior was wrong, but I am still an okay person." In order to experience guilt, one must believe that they did something wrong. This is a good sign in an offender because the lack of guilt suggests more severe pathology, usually psychopathy. But guilt must be dealt with with caution because it is imperative to determine why the offender is experiencing guilt. For example, if the offender experiences guilt because he believes that he actually did something wrong, that is appropriate. But many offenders claim to experience guilt and sorrow for the offense behavior, but truly feel no guilt at all. What they experience is either fear of the potential consequences or arrogance in believing that, by demonstrating and claiming guilt, they may be able to decrease the degree of consequence for them. Again, it is at times impossible to determine whether the offender is demonstrating genuine guilt or simply portraying what his impression of guilt may look like.

Remorse

Remorse is about recognizing how one's behavior impacts another person and feeling sorry for having hurt them and for engaging in the offense behavior. As already mentioned, it is difficult to determine true remorse from feigned remorse. True remorse, for example, may involve an offender feeling that what he did to his victim was wrong and experiencing some degree of guilt and responsibility.

Remorse, like any other emotion, can be exaggerated, faked, or simulated. Most experienced and well-trained professionals can detect insincere remorse, but even professionals can be fooled. Remorse is often confused with guilt. The primary difference between the two is that remorse requires at least some sense that what one did was wrong and assuming responsibility for their behavior.

Empathy

Empathy pertains to being able to perceive and understand how another person feels and is affected. This involves understanding how a victim was impacted by the offense behavior. The longer the offender's offense history, the less remorse he likely experiences.

This may involve, at the very least, the ability to recognize how the victim may have been impacted in at least to some degree (emotional, psychological, physical, sexual, or spiritual). Understanding the impact of the abuse or the full extent of the harm or injury caused may be difficult for the offender to understand or fully appreciate. But remorse involves the ability and willingness to be aware of at least some degree of harm caused to the victim, if even on a superficial level. For example, anyone can look in a book and read about and regurgitate or list the most common effects sexual assault victims experience. But the majority of offenders who lack the ability to experience remorse tend to demonstrate a lack of understanding of the actual effects of sexual assault and are unable to adequately explain the effects in detail. They sound as if they are reading from a book or a list without being able to explain much about the effects. We have all heard at one time or another someone talking about something and it was clear they knew little about the topic. That is what remorseless offenders sound like the majority of the time when they attempt to fake remorse.

Another example comes to mind. When interviewing sex offenders who are in treatment, it becomes clear who the vast majority of offenders are who lack the ability or willingness to demonstrate remorse. They regurgitate what has been taught to them, but they are unable to explain with any degree of depth or any semblance of true understanding. This becomes an important factor in determining whether an offender has decreased his risk for reoffense.

However, with any type of admitter, it is difficult to determine whether offenders are experiencing genuine guilt, concern, and remorse or whether their emotions are the result of concern for their situation. For example, an offender may demonstrate what appears to be genuine guilt and remorse, but upon further assessment, it is discovered that the extent of those emotions were superficial and concerned the unpleasant situation surrounding the prosecution of the offense, not pertaining to what the victim experienced.

Genuine guilt, shame, and remorse involve constant degrees of the awareness and appreciation of the rights and needs of others. To genuinely experience guilt, for example, one would have to believe that they have actually done something wrong, and that they feel bad about hurting the victim. Temporary or tertiary guilt looks something like this. An offender states that he is sorry for what he has done, for what he put the victim through. But he is unable to explain what was wrong with his offense behavior, other than it hurt the victim. He appears disconnected with what would be considered "normal" emotional reactions to guilt. On the surface, the offender appears sorry for what he has done, but upon further assessment and questioning, he appears superficial. The guilt will subside as the pressure and focus decreases, perhaps after he has obtained a plea agreement.

To genuinely experience shame requires a belief that how an individual behaves impacts his identity. For example, if an offender hurts another person, then believes that he himself is a bad person and that what he did was wrong, he is more likely to experience shame and guilt and subsequently remorse. But first, the offender has to believe that what he did was wrong. Temporary or tertiary shame looks something like this. An offender states, "What kind of person would do something like this?" or "I am a bad person or pervert for doing that." However, he is unwilling or incapable of stopping the abusive or harmful behavior. Even for experienced and well-trained professionals, it is impossible to determine with certainty whether an offender is genuine or faking, and even if they appear genuine, whether it is a result of fearing the consequences of their offense behavior or whether it is an appropriate

Another issue is that the degree of remorse does not appear to impact successful treatment or reoffense rates. Pithers et al. (1988) found that low victim empathy was a common predictor for sexual recidivism.

SUMMARY

It is at times difficult to ascertain whether an offender is demonstrating resistance and denial of offense behavior due to narcissism, psychopathy, and a general lack of remorse and empathy or whether the offender possesses traits that will allow successful rehabilitation. Even when an offender admits his offense behavior, that is still no guarantee that he possesses any significant degree of remorse or empathy. Violent offenders may cooperate simply because it makes their involvement with the legal and therapeutic intervention less stressful and less in duration.

32 Pleading Their Innocence

INTRODUCTION

Many perpetrators claim that they are innocent of their offense despite pleading guilty. Some maintain that the only reason why they pled guilty was out of fear of receiving more time in jail or prison, or that their attorney advised them to do so because it was likely that they would lose the case if they went to trail. However, how many people would plead guilty to a felony that would forever change their lives if they were innocent of the crime? Probably not too many.

People tend to engage in behavior that results in positive, desirable rewards. On the other hand, people tend not to engage in behavior that results in negative consequences, or that prevents them from obtaining what we want. So, why would an innocent person plead guilty? They would not. Some offenders are offered what is called in Minnesota the Alford Plea, which is a convenient way to say that they are certain to lose if the case goes to trial, but they do not want to plead guilty, so they make a half-hearted claim of guilt without having to plead guilty *per se* to the offense. Some judges have the offender still plead guilty on the record of the offense details, while others do not.

I find it interesting that batterers and sex offenders tend to make same and similar comments about their innocence, and I have since written their statements down. The most common excuse given by batterers and exoffenders when they plead guilty or take an Alford plea is discussed below.

"My lawyer told me to plead guilty" or "My attorney advised me that if I went to trial I would have lost." The very fact that they did not fight the charges was really due to the fact that they knew they were guilty. Okay, perhaps I am being cynical; there certainly are some batterers and sex offenders that may well be innocent of the charges before them, yet are unable to disprove the allegations. I have worked with offenders in such cases when after some time, they were found not guilty of the offense, though the damage has already been done in that they were required to complete unnecessary therapeutic intervention, in addition to having been traumatized by the entire ordeal.

So with that having been said, let's refocus on the guilty offenders who make continued claims of innocence. When they claim that they would have lost in court if they went to trial, the primary reason they would lose in court is because they committed the offense. If a lawyer advised an innocent person to plead guilty, then the lawyer would be unethical. The bottom line is that enough proof was present to prove that the perpetrator was guilty. The question to ask perpetrators is: why would they plead guilty if they truly believed they were innocent?

The more perpetrators maintain the stance that they are innocent of the offense, and the greater the number of people they have to convince, the more energy they will exert to achieve a reality that they are, in fact, innocent. When perpetrators have convinced their significant others or families of their innocence, for example, it becomes an even stronger battle to continue their denial. To confess the truth would be to risk the loss of their support persons and loss of reputation. In addition, if the support system of the perpetrator understood that they were in fact guilty, then they would have to deal with the issue of being lied to. Deceit can devastate relationships and the perpetrator may fear the loss of support more than the legal consequences.

WHY DENY THE OFFENSE WHEN, IN FACT, THEY ARE GUILTY?

There are several key reasons why perpetrators deny their offense behavior. Some of the more common reasons are discussed below.

They convince themselves that they are innocent. Many perpetrators deny their offense when initially questioned for a variety of reasons. Some may fear the legal, social, and familial consequences, some believe that they may be able to beat the system if they deny the allegations, while others have not yet made a conscious decision to be honest at the time. At some point, however, perpetrators must make a decision to either maintain their innocence or admit some or all of their guilt. The longer they maintain their denial and the more time that passes, the less the chance of them admitting the offense. Also, the more people that they convince of their innocence, the more difficult it will be to admit their offense because to do so would involve informing their support network that they lied. Some perpetrators spend enormous sums of money in an attempt to prove their innocence and believe that they are unable to turn back and tell the truth. I have worked with clients who spent over $40,000 defending themselves, and in the end, they were found guilty by jury, at which point they may begin to admit their offense behavior. The best way to work through this type of denial is to help the perpetrator understand the emotional and intimacy cost of those who they have lied to.

They fear losing the support of family or their significant others. The fear of losing their family or support network can be extremely frightening and overwhelming. This is especially true when perpetrators have invested significant energy into convincing others of their innocence. The threats of losing relationships are real and, at times, relationships end once the perpetrator admits the offense behavior. Involving the perpetrator's support network in the interview process provides more information and often information that would not have been otherwise obtained. However, I do not recommend involving others in the initial interview.

They have invested far too many emotional and financial resources in their denial. When perpetrators invest energy into deceit, they have a lot to lose if they were to admit the truth. They may have spent significant amounts of money on their defense, money that may have come from family or a second mortgage. Deceit also involves redefining one's self-image. On some level perpetrators are fully aware of the heinous crime they committed. They may experience difficulty accepting that they did, in fact, commit the offense. Their life, family, and community will never again be the same.

Attempt to beat the system. Some perpetrators believe that their offense behavior was legal and justified. With this belief as the battle flag, perpetrators assume an aggressive victim stance and do everything in their power to blame the victims. They truly believe that the interviewer and the court will believe their story regardless of the existence of evidence proving their guilt.

Why plead guilty when claiming innocence? The answer is simple. Perpetrators know they committed the offense and they are not ready to admit it to themselves or others, or they truly believe that they will be able to con the authorities. The very nature of the human state is to be self-protective and most people will do anything to avoid pain while, at the same time, do whatever they can to increase the chance for pleasure. Throughout life, the perpetrator has likely fought for what he thought was right, fought to protect himself, and fought to take back anything taken from him. It is difficult to believe that perpetrators would actually accept a plea bargain simply because a lawyer told them to do so. The reality is that they knew the odds were in favor of them being convicted of the offense because they did, in fact, commit the crime.

33 What to Question with Offenders

HELPFUL QUESTIONS TO ASK IN CERTAIN SITUATIONS

When the perpetrator apologizes or wants to apologize to the victim or the family or anyone, for that matter:

"Why would you apologize if you did not do anything?"
"What are you feeling guilty about?"

When the perpetrator removes the victim's clothes or dresses the victim after the assault:

"Why did you take their clothes off?"
"Why did you put their clothes back on?" "Is it normal for you to do this? Do you usually undress and dress your sex partners?"

When the perpetrator gives excuses of how they were never alone with or never near the victim:

"Why cover up being with the victim?"
"What would be so terrible or what would it mean if you were with the victim?"
"Why would you develop an elaborate alibi/excuse?"
"Why would the victim develop an elaborate alibi or story?"
"Why would the victim make everything up?"
"How could you never have been left alone with the victim; didn't anyone leave to go to the bathroom; others were present at all times, every minute?"

When the perpetrator will not protect himself, assumes a victim stance, makes statements, such as:

"Whatever"
"Why not defend yourself?"
"How did you become the victim of this offense?"
"It's your life and assuming a helpless, acquiescing role gives power to me to make decisions that will affect your life? Why not be honest and explain what really happened?"

When the perpetrator blames the victim for the offense or uses the victim's past to either justify the sexual contact or uses it as a reason why the victim would make up such a story:

"What relevance does the victim's past have to do with whether you committed the crime?"
"Did you know they had been victims in the past? If so, when did you find out?"
"Why would you put yourself in a risky position with the victim when you knew about her past?"
"How do you think your victim has been affected by this offense?"

When the perpetrator denies that the physical or sexual contact was consensual or that sexual contact never occurred:

"Why or how would the victim recall such vivid details of the assault and be believed when questioned by strangers if it did not happen?"
"You were the last person to be with the victim, how can you explain their injuries?"
"Who do you think did this to the victim?"
"How do you think she received the injuries?"
"What do you gain from explaining away the allegations or offense details when the physical evidence and victim's statement prove otherwise?"

When the perpetrator claims that others can prove his innocence:

"What evidence do they have to prove you didn't commit the crime?"
"How would they know what you did with the victim when you were alone or in another room?"
"Don't you find it strange that your support people protect you without having heard all of the facts?"

When the perpetrator denies that he was or is intimidating or abusive:

"How do you account for or explain the victim's injuries if you did not assault her?"
"Why are you escalating so easily?"
"I wonder how you intimidate others?"
"Here you are becoming intimidating with me, yelling, making a fist. I wonder how others would be affected by you right now if they were here?"
"How do you think this behavior intimidates others who have less power than you do?"
"How would your victim have been affected by your intimidating behavior?"
"You anger so quickly, and you are intimidating right now. I have power over you right now and am an authority figure, and my report will affect your life. Why would someone demonstrate intimidating behavior in this situation unless he truly had a problem with his anger?"
"Why does the physical evidence place you there, with the victim?"
"If/when we do a polygraph, what do you think it will say? What details would it indicate you are lying about?"

When the perpetrator focuses on how he himself is being affected rather than the victim:

"Why aren't you concerned about the victim?"
"How do you think the victim has been and continues to be affected?"
"Do you think he/she deserved it?"
"How did you become the victim here?"

When the perpetrator pleads guilty and then denies committing the offense:

"Why not fight the charges?"
"Why would your attorney advise you to plead guilty? Because they knew you were guilty?"
"What did you gain by pleading guilty to a felony that would forever change your life if you were innocent?"
"Your attorney told you to plead guilty because there is evidence beyond reasonable doubt that you were guilty. Your attorney was simply telling you straight forward."

"You are basically a selfish person. I mean, you would fight to protect your rights in any situation. Why did you choose not to fight at court, unless, of course, you knew you were guilty?"
"You knew you were guilty and that you would probably lose in court?"

When the perpetrator takes an Alford Plea:

"What would you gain from pleading guilty to a felony criminal sexual conduct charge if you were innocent?"
"Why take an Alford Plea? Why can't you simply answer yes or no to the charges, plead guilty or not guilty?"
"An Alford Plea is a cop-out; however, it is also a plea of guilt. If you are pleading "not guilty," any other plea is one of guilt. Explain this."
"What message does your Alford Plea give to your family and friends? Perhaps that you are guilty and cannot accept responsibility for your behavior?"

When the perpetrator provides little information, or only gives details that the interviewer already knows, or the perpetrator offers no information unless asked and keeps answers minimal (it feels like the interviewer is fishing for answers):

"This feels like a fishing expedition; why don't you just tell me what happened?"
"Explain your answers in more detail; help me understand your version of the offense."

When the perpetrator's story contains inconsistencies:

"How do you explain the differences between your version and the victim's version of the offense?"
"What or how would the victim gain from lying?"

When the perpetrator asks the interviewer to answer the question, throw it back in his lap using statements such as:

"I don't know, you tell me."
"Why are you refusing to answer this question?"
"Given your situation, why would you not cooperate?"
"Why would you want someone else to answer for you?"

When the perpetrator says, "I did not do it:"

"Do what?"

When the perpetrator states that he is not a "monster," "pervert," or "anyone who would do that:"

"What would it mean about you if you did do it?"
"Describe a pervert, monster, or a type of person that would commit such a crime."
"Sometimes people, good people, make mistakes and hurt someone. Do you think you might have hurt the victim?"
"If you were watching the news, long before this offense, and a perpetrator of the same type of crime you committed was being discussed, what would you have called that person? What would you think about that person? Can you in any way relate to that person?"

When the perpetrator avoids answering "yes" and "no" questions with "yes" or "no" responses, and elaborates on each answer or insists on explaining:

"You appear to be searching for answers. Simply answer the questions with a "*yes*" or "*no*."
"You are attempting to justify your offense behavior. Just answer the questions."
"You are attempting to confuse the truth with long explanations that are little more than attempts to justify your offense behavior. Stick to the question being asked and do not add details. I will ask for more detail or clarification if needed."
"Just answer "*yes*" or "*no*." Do not explain, do not think about the answers, simply answer the question."

When the perpetrator blames alcohol or drug use or a "blackout" for his inability to recall offense details:

"How many times have you used drugs or alcohol? How many times when you used drugs or alcohol did you commit sexual assault (usually they say never before)? Then, ask what was different that night."
"People can recall 90 to 100% of what they did when under the influence of chemicals, they do not simply forget. What did you do with your victim?"
"Alcohol and most drugs people use are depressants. The effects actually decrease our ability to be sexual, to maintain an erection, etc. How did you manage to be sexual even when under the influence of these chemicals?"

The above are suggestions on how to address situations, which commonly occur when interviewing offenders. Keep in mind that the above examples may not work with every offender or fit every interviewer's style. The bottom line is that offenders reveal many significant details and messages with their style of defense. Pay attention to how the offender defends his answers and pleas of innocence. Did he become delusional or angry, or is he willing to work with your feedback and more honestly reveal the truth? Confront the perpetrator on any information that does not make sense.

When perpetrators are asked a question, pay attention to how they react. Are they surprised, angry, or confused? How long do they take to answer? Yes or no questions demand immediate answers. When they hesitate, confront them. Tell them, "You do not need to think about the answer, it is simple, yes or no." It is also helpful to keep the questions coming at a fast pace so as not to allow the perpetrator to think about the answers or to anticipate the next question. When he attempts to think too long about an answer, confront him and, if need be, offer some possible answers. Sometimes this technique forces the perpetrator to reveal information he was withholding.

When interviewing perpetrators remember:

1. Identify and explore the stressors and triggers that occurred at the time of the offense. Did they recently end a relationship or lose a job? Have perpetrators identify what precipitated the offense. Some perpetrators will admit to the offense when they believe that the interviewer understands what they were going through in their lives.
2. Explore possible motives. This may include revenge or rage, or an attempt to obtain intimacy or to meet other needs. Again, the perpetrator will provide much information about why it happened.
3. Explore patterns identified. Does the perpetrator's criminal or relationship history indicate specific patterns? Is the perpetrator's life transient or stable, and how often does he engage in criminal behavior, whether or not he was adjudicated? Do not minimize the significance of criminal behaviors that occurred years apart, as some sex offenders may act out every 3 to 7 years, with no apparent criminal behavior in between.

4. Examine criminal history as well as any complaints on file that did not result in criminal charges or were dropped for some other reason. It is strongly recommended that copies of any legal information be obtained, including Pre-Sentence Investigation reports (PSI) and criminal complaints, including juvenile records. Clues indicating other sources of information come from documents being used, such as PSI, and from the perpetrator's statements. Whenever possible, I obtain a copy of said information rather than having someone summarize the details. The most accurate information will come from a written copy of whatever document you are examining. Follow up with the specific therapist, parole or probation officer, or police officer with any questions that may arise from the information. When charges were dismissed, talk with the county attorney or police department to find out why.

   One example of why and how this is beneficial is a case where I was assessing a sex offender. His criminal history included assault (unspecified) and burglary (unspecified). When I reviewed the PSI and criminal complaint (several years old), some interesting and important information was obtained. It turned out that the assault charge involved attacking a female, and this helped to prove a pattern of violence against females. Further, the burglary involved breaking into the home of a female, who was asleep at the time of the break-in. The report indicated that she awoke to find him standing over her bed. Despite not knowing whether he was intending to sexually assault her, the reports provided important information concerning the perpetrator's criminal inclinations. The bottom line is that without obtaining all of the seemingly irrelevant information, you cannot be sure that you will not overlook information that can become very relevant.

   Another example is a case where a sex offender's PSI indicated that he had a prior sex offense conviction in another state. When I obtained a copy of the PSI from that state, another sex offense conviction was listed for yet another state. Keep in mind that not all states report to the FBI, and a National Crime Information Center (NCIC) check does not guarantee that all offenses will be identified. Whenever the perpetrator used alias names or resided in other states, it is strongly recommended that you should contact these states directly to obtain any information concerning assaults or sex offenses.

   Another example is appropriate here. A sex offender had several convictions for sexual assault. He commented that he was accused of another sexual assault in a different county, but that he was innocent and charges were dropped. I spoke with both the county attorney and the investigating police officer. Both assured me that they believed that the perpetrator was guilty, but the victim was unable or unwilling to testify. When I requestioned the perpetrator about the new information obtained, he admitted that he did engage in sexual activity with the victim, but that it was consensual. Although he had not been convicted for that offense, the information proved very helpful in more accurately identifying this perpetrator's profile. In addition, the information gathered can be used in treatment.

5. When first questioning the perpetrator immediately after the offense occurred, listen carefully to the details they provide. It is very common that their statements and choice of words will expose the truth without conscious awareness. The perpetrator may not have had time to prepare his defense and will either admit the offense behavior or will offer explanations that attempt to either justify his behavior or plead for his innocence. It is very common for perpetrators to tell the investigating officer statements such as: "He/she deserved or wanted it," "I don't know what happened," or other statements indicating that he is guilty of the offense at hand. Later, the perpetrator may claim that he never made any such statements and that the officer made the statements up. The reality is, however, that the perpetrator's perpetrator persona gave the truth away and he took pride in the offense behavior. When later questioned, the perpetrator was not ready or able to admit the truth to himself or to others. That is why initial police reports should contain statements and, whenever possible, direct quotations from the perpetrator.

6. Compare information and answers from all data possible. For example, compare statements the perpetrator made when initially questioned by the police to statements the perpetrator made when questioned for the pre-sentence investigation, and then with statements made during the clinical interview. Any discrepancies should be explored. Also pay attention to what the perpetrator omits. Often times, information that is omitted is crucial to prove a pattern. For example, the most common information omitted is past offenses, especially offenses that involved physical or sexual assault.

34 Attitude of Perpetrator

When interviewing batterers and sex offenders, it is important to pay attention to the attitudes they reveal concerning their offense and current situation. The following are common attitudes demonstrated by batterers and sex offenders. When they demonstrate the specific attitudes listed below, assess whether they are concerned about their victim, their family, or themselves. It is important to know whether perpetrators are focused only on their own situation or whether they demonstrate empathy and concern for their victim. Of course, it is often difficult to ascertain the veracity of offenders' claims of or appearance of empathy and genuine concern for their family and/or victims, but assessing the following areas has been more than helpful for me in my forensic practice.

EXPLORE OFFENDERS' CONCERNS

Keep in mind the circumstances that resulted in the interview or assessment. Offenders may demonstrate concern about their own predicament. These offenders tend to be more selfish, narcissistic, and self-centered. Batterers and sex offenders with previous criminal histories may fit more into this category of offender. Examples of concern for self would be statements such as: "I just want all of this to go away so I can get back to normal;" "The victim is ruining my life with their allegations, I did not do anything." Although it would be normal to have some degree of concern for oneself when alleged to have committed or convicted of battering or a sex offense, but an offender that expresses concern for others as well may be more workable in treatment and may invest in the process of change.

Offenders may express concern for the victim and the victim's family. This type of remorseful offenders may present with depression, anxiety, and may accept responsibility for their offense behavior. These offenders are more likely to want to help the victim in their healing process. Offenders may also express concern for their own family and how they are impacted by the offense. Similar to the offender who expresses concern for the victim, this offender may be more open to honesty and cooperation. Genuine remorse would likely require some degree of concern for the victim and the offender's own family. However, empathy has not been shown to decrease a sexual offender's risk for violent reoffense (Hanson & Morton-Bourgon, 2004). However, offenders who demonstrate remorse tend to be more cooperative in treatment.

ASSESS THE BATTERERS' AND SEX OFFENDERS' ANGER

Most people would experience some degree of anger when arrested or convicted of a violent offense, and this would be considered a normal reaction. But it is important to ascertain at what or at whom the offender is angry. Offenders who are angry with themselves for their offense behavior are likely to experience depression, anxiety, and remorse. These offenders may be more cooperative with treatment interventions and more likely to incorporate change.

The more difficult and perhaps more dangerous offenders are those who are angry with the victim, the victim's family, or the criminal justice system. It is an interesting contradiction for many offenders who are angry at the police officer for giving them a speeding ticket. They somehow forget or are unwilling to accept that they were, in fact, speeding, and that the police have the task of enforcing driving laws. In fact, the police are being paid with our tax money to perform their jobs. We all applaud when we witness the police stopping another driver for reckless driving, alcohol-related driving infractions, or for speeding. But somehow many offenders forget these facts and blame the police officers for "targeting" them when, in fact, "others were going faster than they were." The offender forgets that regardless of what other drivers were doing, they were, in fact, speeding and choosing to break the law. The police were doing the job they are paid to do. Offenders who blame others for their own behavior have difficulty changing their problematic behavior because they do not believe that they have a problem and, therefore, they have nothing to change. In short, I find that offenders who blame others for their criminal behavior are more likely to reoffend with same or similar behavior.

HOW COOPERATIVE ARE THE OFFENDERS

Some offenders are blatantly upfront with their motivation for cooperation, stating, "My attorney told me to just take the deal." They have no internal motivation for accepting responsibility nor do they believe that they did anything wrong. Their cooperation is superficial and limited, limited to the point of infringing on their life, work, or activities. These offenders tend to be angrier and less insightful. Some offenders cooperate for the reason that they want to move on with their lives, and they often make statements such as wanting people to "forgive and forget so that I can get on with my life." I find that these offenders may become increasingly more difficult to work with and may drop out of treatment prematurely, and often complaining about their clinicians. Of course, my favorite type of offender are the ones who accept responsibility for their criminal behavior and cooperate because they believe they should and need too for their own moral reasons.

REASONS FOR UNCOOPERATIVE BEHAVIOR

Resistance and refusal to cooperate can easily be misinterpreted. Experienced clinicians and police officers understand this and are cautious in responding to uncooperative behavior. Batterers and sex offenders may deny their offender behavior or aspects of their offense behavior because of a need to protect them psychologically, that is, to protect their self-esteem and sense of worth. These offenders appear more shamed and guilt ridden and some compassion is warranted. This is especially true when first being interrogated for their offense, when they may not have come to terms that they did, in fact, commit the offense. These offenders typically would be first-time offenders, not recidivists.

Some offenders, however, are uncooperative because they are unwilling or unable to recognize the seriousness of their offense behavior. These offenders do not believe that they deserve any legal or any consequence for their behavior, expect others to give them a break, and regardless of degree of intelligence, their reasoning appears immature and childlike. They are almost delusional in their refusal to take the consequences seriously and appear unaffected with any threat of consequence for failing to cooperate. Once a consequence is given for their failure to cooperate, however, I find that they generally become angrier and terminate prematurely from treatment. These offenders are also more likely to complain about their clinician for holding them accountable and for their delusional belief that their clinician is holding them to a different standard and being treated unfairly. I find that offenders that are uncooperative due to their unwillingness or inability to understand the seriousness of their offense behavior are not good treatment candidates and are less likely to learn from their experience.

SUMMARY

To summarize, remember that if the general attitude of the perpetrator is one of blaming others for their offense and current situation, and their cooperation appears superficial and motivated primarily on accepting as little blame as possible, then they are less likely to benefit from a community-based batterer or sex offender treatment program. Perpetrators who demonstrate little concern for their victim and who blame others and/or circumstances for their offense, care little for others and tend to lack empathy and remorse. These tend to be the psychopathic perpetrators who pose the most risk to the community. On the other hand, perpetrators who demonstrate concern for the well-being of their victim and who accept at least some responsibility for their offense behavior tend to benefit from treatment and are more likely to respond favorably to community-based services.

Of course, there are professionals who believe that perpetrators who present with moderate to significant degrees of denial and minimization should be given the opportunity to complete an abuse program or sex offender treatment in the least restrictive environment possible. The result is that these perpetrators are referred to community-based batterer or sex offender treatment programs. Many are placed in denier groups. My own experience working with these individuals indicates that community-based programs are ineffective with this population. At best, only 10 to 15% of these perpetrators complete or come close to completing treatment. That means that approximately 85 to 90% of these deniers fail to complete treatment. Most are terminated from treatment within 1 to 12 months.

The overall success rate indicates that for those perpetrators who present with moderate to significant degrees of denial and/or minimization should be incarcerated. I would argue that it is in the perpetrators' best interest to be placed in a controlled environment, one that would decrease the opportunity for them to continue to deny and lessen the opportunity to reoffend. In the interest of community safety, incarceration of perpetrators who deny or significantly minimize their offense behavior decreases the risk of another citizen being victimized at a time when the perpetrator is at the greatest risk of reoffending. In summary, a general rule of thumb would be to work with perpetrators on a community basis who are accepting responsibility for their offense behavior and incarcerate those who do not.

35 Approaches for Interviewing Batterers and Sex Offenders

INTERVIEWING OFFENDERS

When interviewing a batterer or sexual offender, it is important to keep in mind the purpose of the interview. Several reasons why offenders are interviewed will be discussed. Offenders are interviewed as part of an investigation of a crime or other allegations of misconduct, and this generally involves an attempt to obtain a confession from the offense, or at the very least, help to understand the allegation or offense more clearly. Another reason why offenders are interviewed is as part of a forensic assessment for the Court. This may include an assessment to determine the recommendations for the most appropriate type of consequences, intervention, or treatment. Without a thorough assessment, it is difficult if not impossible to make effective recommendations that take into account the offender's entire life rather than simply the concerns resulting for the offense at hand. Sometimes the forensic assessment is for the purpose of determining whether the sex offender should be committed to a state hospital or prison for life as a sexual predator or sexually dangerous person. In this case, the interview is used to facilitate more information about the offender's entire life; assess the degree of remorse, empathy, and responsibility for the offense behavior; and to determine whether the offender may be treatable in a less severe environment. Offenders may also be interviewed as part of an intake process before entering an anger management, domestic abuse, or sex offender treatment program.

Regardless of why the offender is being interviewed, it is imperative to have the goal in mind. The examiner conducting an intake assessment is far less concerned about obtaining a confession and the police interrogator is less concerned about treatment availability or options. The purpose of the interview sets the tone for how questions are asked, what techniques the interviewer utilizes, and guides the focus of the interview topics. It is important to be clear from the onset the purpose of the interview. An examiner conducting an intake interview needs to be careful not to be assessing for a criminal offense without first informing the offender and obtaining informed consent. No different than the police officer needing to inform the offender of their legal rights before engaging in questioning that could result in a confession.

It is also important that the offender be made aware that they are not being interviewed for usual intervention. Any court-ordered treatment is not ordinary and requires far more structure. The offender has fewer rights in that their supervising agent must be kept informed of progress, failure, or completion of intervention programs. In addition, some states have laws that require offenders to sign releases of information allowing the treating professional to inform the victim and the supervising agent of when the offender misses appointments as well as any change in the offender's status in the program (e.g., completion of the program, missed appointments, suspensions and terminations). The offender must be made aware of these limitations and that safety to the victim and the public in general will outweigh the offender's right to privacy.

There is one last issue to address. I believe that it is nearly impossible for a treating professional to offer risk assessment or forensic services for a client they are working with or have worked with before. The treating professional will always have a bias in favor of the offender, wanting to believe that the treatment intervention had some positive impact. However, a forensic assessment demands neutrality on the examiner or interrogator's part. To have completed a forensic assessment and then becoming the treating professional is far less a concern as long as one process is clean and clearly ended, and the other begins. But treating professionals who offer strong opinions on behalf of the offender are crossing serious lines of conduct. Someone may be "doing well" in treatment and appear to be "responding favorably" to the intervention and yet remain at significantly high risk for violent reoffense. Remember that offenders with psychopathic traits may appear to do very well with program material and assignments, but simply become better criminals in the long run.

Emotional vs. Unemotional Offenders

Gudjonsson (1992) describes offenders as being either emotional or nonemotional. Emotional offenders are those who experience remorse for what they have done. These offenders have a troubled conscience about the offense they have committed. He recommends that the interrogator utilizes a sympathetic and compassionate approach with an emotional offender. He describes a nonemotional offender as being not normally troubled by conscience for his wrongdoing. This type of offender will resist becoming emotionally involved with the interrogation. The most effective approach with a nonemotional offender is a factual analysis approach (e.g., uses common sense explanations and arguments rather than focusing on the suspect's emotions). In this approach, the interviewer confronts the offender with reasons why it is futile to deny the offense and the interviewer informs the offender about all the "evidence" the police have that implicates him in the crime. It is acceptable to exaggerate the strength of the evidence you have against the offender.

Inbau, Reid, and Buckley (1962) point out that the two approaches (sympathetic and factual) should be used during the interrogation, but the emphasis of their use in a particular case will depend on the personality of the suspect. It is always important to match the specific interview techniques with the offender's personality. In addition, interviewers must utilize techniques that fit their own strengths and that they are comfortable using to engage the offender. Most offenders are very adept at perceiving the interviewer's comfort and confidence or discomfort and lack of confidence. How the interviewer approaches and prepares for the interview is important. The interviewer must be competent and have an understanding of the offender's offense as well as an understanding of the characteristics of offenders in general who commit similar offenses.

Approaches for Interviewing Offenders

I have classified the types of interview approaches into three: the Columbo approach, the Sipowitz and sarcastic approach, and the Explanatory approach.

Columbo

Okay, so I like Peter Falk in his role as Columbo. This approach appears to make the offender's story the truth. The interviewer takes the side of the offender and utilizes the appearance of confusion when discrepancies occur. The interviewer gently questions the offender with details, which are inconsistent and allows for the offender to guide the truth process. The overall tone of the interview is one of utmost respect. The offender is believed and gently confronted when necessary.

The interviewer plays dumb, and allows the offender to explain away the offense, only to confront the offender with the inconsistent details, which in the end tend to prove the offender wrong. The strength in this approach is that the offender is given the impression that the interviewer believes his story. Even attempts to explain away inconsistencies appear to be believed. However, at some point the offender has an insatiable need to take credit for what he did and, when caught in his lies, the truth cannot easily be denied. This approach works best with offenders who are slow thinkers, and for those who have a need to explain away any possibility that they are guilty.

This approach works best with offenders who admit most, if not all, of the details of their offense. It is not very effective with moderate or significant levels of denial or minimization. This may be most effective with offenders who have engaged in their first offense, have minimal if any psychopathology, and do not have a history of prior abusive or violent behavior (no prior history of any violent or abusive behavior that was not reported). Remember, however, that an interviewer is not the friend of the offender nor is the interview or interrogation being conducted as part of a voluntary and usual circumstance.

Highly Confrontive (Sipowitz)

A good example of this style is the role Dennis Franz plays as Detective Sipowitz on *NYPD Blue*. This approach utilizes confrontation, sarcasm, and direct questioning. This approach is more closely related to an interrogation. The interviewer assumes an antagonistic, nonbelieving stance right from the beginning. The offender is questioned directly on each fact, and the interviewer expects "yes" or "no" answers rather than long explanations. Questions are asked quickly with little time for the offender to think. This approach works best with offenders who deny or significantly minimize the details of their offense. It works especially well with offenders who make an Alford plea, which is really another way for them to admit that they are guilty without actually admitting anything.

This approach can effectively guide an aggressive offender into making incriminating statements. However, this approach can also easily escalate an offender who is more impulsive and psychopathic.

Explanatory

The explanatory model is currently the most popular with law enforcement. This model offers the offender a way to admit his involvement in an offense while maintaining some degree of self-esteem and dignity. Through this model, the offender is offered a possible scenario for how the offense may have occurred.

Gudjonsson (1992) refers to the explanatory approach as "*theme development*." In this approach, the interrogator must have an understanding and sympathetic attitude in order to gain the suspect's trust. The interrogator suggests various "themes" to the suspect, which are aimed to either minimize the moral implications of the alleged crime or give the suspect the opportunity of accepting "moral excuses" for the commission of the crime. For the emotional type offender, the interviewer might tell the offender that anyone in a similar situation might have acted in the same way. Basically, the interviewer is suggesting a morally acceptable reason why the offense occurred. Another way this might occur is to make it easier for the offender to blame the victim or others for the offense, at least to the degree that the offender admits his involvement in the offense. Offer praise to the offender to help manipulate his confession. Help the offender believe that it is in his best interest to cooperate and learn from his mistakes, that is, from his offense, which he needs to discuss and admit. For the nonemotional offender, Gudjonsson recommends to try to catch the offender in some incidental lie. No matter how small the lie, he argues that the offender is then at a psychological disadvantage trying to keep track of the lie/lies told.

One example for trying to sympathize with the offender goes something like this: An offender punched his girlfriend in the face and then raped her. The questioning might be:

> *"I can understand how a man can be tired when he gets home from work, and when she (the victim) began to demand things from you, you probably felt overwhelmed. I can understand that, any normal man would feel that way. And when you hit her, you felt like you could not stop yourself, somehow out of control. Is this the way it happened?"*

If the offender feels that some of his behavior is normalized, then the sense of guilt and shame ease to some degree and the offender is more likely to offer some degree of confession. The rape part of the example offense might be addressed with something similar to the following:

> *"I can also understand that after you hit her, you felt bad and wanted to make up. Is it possible that you had sex with her to make up?*

Now these sample scripts are simplistic and more than likely the interviewer would have added much more detail and appeared very sympathetic and understanding of the offender and of the offender's situation.

SUMMARY OF APPROACHES

Most interviewers utilize a combination of the above methods and they may use all three approaches in a given interview. Flexibility is the most important tool an interviewer can possess.

In general, it is imperative that interviewers utilize interview techniques that are consistent with their areas of expertise and comfort. Not everyone can utilize the highly confrontive approach. The interviewer must first be comfortable utilizing such a direct and confrontational approach. This approach can also be easily misused in such a way by the interviewer as to be abusive toward the offender. It is imperative that we never lose sight of the goal of a batterer or sex offender interview — to establish the truth while minimizing resistance. Even small admissions need to be recognized and supported by the interviewer. The concept of successive approximations is important and vital to the interview process. Every time the offender admits to some detail or lessens their level of denial or minimization, they should be rewarded with therapeutic and well-timed compliments.

A word of caution is appropriate here. An offender who makes some degree of confession should never be assumed to be telling the whole truth. The offender may offer some minor confession about some insignificant offense detail to simply have the interviewer back off. Some offenders will admit to only those facts that the interviewer is already aware of. The goal is to have the offender admit to offense details that were not made known or that only the offender and victim could have known. So if the offender admits that he was with the rape victim and that he had sex with her, it does not mean that if he denies using force or a weapon during the offense that he is telling the truth. He may be telling you what you already know and refuses to provide any other confession. An admission to a particular offense has never been established as meaning that the offender is being honest about other offenses, many of which may not have been reported.

INTERROGATION TECHNIQUES

Gudjonsson (1992)* describes recommended nine techniques for interviewing offenders. Prior to proceeding through the nine steps, the interrogator should be thoroughly familiar with all the available facts about the case and the suspect. The interrogator must be well prepared. I like Gudjonsson's approach because it makes the most sense and I have successfully used and adapted his strategies.

1. **Direct positive confrontation**. This involves telling the offender that he is considered to have committed the crime. Then the offender is confronted with the accusations a second time. Passive reaction to the accusation is considered to be evidence of deception. Inbau, Reid, and Buckley (1986) indicated "...the easiest confessions are obtained from suspects who fail to make a denial after the positive confrontation."
2. **Theme development**. The interrogator must have an understanding and sympathetic attitude in order to gain the suspect's trust. The interrogator suggests various "themes" to the suspect, which are aimed to either minimize the moral implications of the alleged crime or give the suspect the opportunity of accepting "moral excuses" for his commission of the crime. This kind of theme development is most effective with emotional offenders.

* Reprinted with permission.

3. **Handling denials**. Needs to be stopped by the interrogator. This means that the interrogator does not allow the suspect to persist with the denials. Inbau, Reid, and Buckley recommend the use of the "friendly–unfriendly" technique (when the various attempts at sympathy and understanding have failed). The "friendly–unfriendly technique, also known as the "mutt and Jeff" technique (Irving and Hilgendorf, 1980), can be applied in various ways. This approach may make the suspect more responsive to the sympathetic approach. This is particularly effective with the quiet and unresponsive suspect.
4. **Overcoming objections**. Objections that the suspect may give as an explanation or reasoning for his innocence. Innocent suspects are said to more commonly continue with plain details, whereas the guilty suspect will move from plain details to objections. The objections are an attempt by the guilty subject to gain control over the conversation as his denials begin to weaken. Once the suspect feels that the objections are not getting him anywhere he becomes quiet and begins to show sings of withdrawal from active participation in the interrogation. He is now at his lowest point and the interrogator needs to act quickly in order not to lose the psychological advantage gained by this.
5. **Procurement and retention of the suspect's attention**. Once the interrogator notices the suspect's passive signs of withdrawal, he tries to reduce the psychological distance between himself and the suspect and to regain the suspect's full attention. This may involve moving closer to the suspect, leaning forward towards the suspect, touching the suspect gently, mentioning the suspect's first name, and maintaining good eye contact with the suspect. The suspect will look defeated and depressed.
6. **Handling the suspect's passive mood**. The interrogator exhibits signs of understanding and sympathy and urges the suspect to tell the truth. Attempts are then made to place the suspect in a more remorseful mood by having him become more aware of the stress he is placing upon the victim by not confessing. The interrogator appeals to the suspect's sense of decency and honor, and religion, if possible.
7. **Presenting an alternative question**. The suspect is presented with two possible alternatives for the commission of the crime. Both alternatives are incriminating, but they are worded in such a way that one alternative acts as a face-saving device while the other implies some repulsive or callous motivation. In other words, the suspect is given the opportunity to provide an explanation or an excuse for the crime, which makes self-incriminating admission much easier to achieve. The timing of presenting the alternative question is critical, attempting to catch the suspect by surprise and making him more likely to confess.
8. **Having the suspect relate various details of the offense**. This relates to the suspect having accepted one of the alternatives given him and consequently providing a first self-incriminating admission. The interrogator should be alone with the suspect to avoid any distraction.
9. **Converting an oral confession into a written confession**. Inbau, Reid, and Buckley recommend, for nonemotional suspects, frightening the suspect into a confession by exaggerating the strength of the evidence against him and the seriousness of the offense. The "minimization" approach is recommended for remorseful suspects. The interrogator tricks the suspect into a confession by offering sympathy, providing face-saving excuses, partly blaming the victim or circumstances for the alleged crime, and minimizing the seriousness of the charges.

Gudjonsson (1992) presents a wonderfully useful outline for interviewing criminals. The same techniques are useful for interviewing offenders for forensic assessment, interrogations, or as part of an intake for entering a treatment program. It is imperative that the interviewer have an understanding of interview techniques, but also a comfortability with interviewing violent offenders.

There is a fine line between bullying and coercing a confession and working the process toward a true confession. The interviewer should be well trained and have supervised experience conducting forensic interviews. Most graduate programs do not teach effective forensic assessment. So, it is important to seek out necessary training to become proficient in interviewing offenders.

36 Why Use Sarcasm and Confrontation

CATCHING THE OFFENDER

In my previous description of the perpetrator persona (Chapter 14), recall that narcissism is what the offender acts on. Most batterers and sex offenders relate to sarcasm and confrontation. When you attempt to be empathetic, they take advantage of you. They used psychological force and, in some cases, physical force, to get their victims into vulnerable situations. They thrive on taking advantage of the empathy and respect that victims and others show them. But the best, most genuine respectful behavior I know of is meeting the offender face on, interacting with them on their level. They understand control, sarcasm, and confrontation.

You will be more likely to catch the offender off guard when interacting with sarcasm and confrontation. Catching them off guard is important as it allows for more slippage to occur, that is, the offender unknowingly revealing valuable information. However, using sarcasm and confrontation simply because you want to do so is inappropriate and potentially abusive. I am referring to utilizing sarcasm and confrontation at key points during an interview or interaction with a perpetrator. Interjecting sarcasm at the right time to catch the perpetrator off guard. Sarcasm can be used to highlight how unbelievable the perpetrator's story is. The goal is to help to clarify the offense details and, hopefully, to either get a confession or have the offender invest in telling the truth so that effective intervention may occur.

Batterers and sex offenders may let information slip out without being aware. When you throw them off guard, the perpetrator persona loses it, and often reveals valuable information. When offenders are in control of the interview, they reveal only what they feel will not significantly incriminate them, but enough to satisfy others and then back off.

Examine inconsistencies in offenders' stories. The more they focus on lying and misleading, the more cognitive energy they must invest in their lies. The more energy they invest in their lies, the less likely they will recall the previous lies or misleading information. In short, they have more difficulty following their own trail of deceit.

Pay attention to what they say and how they react to you following the sarcasm or confrontation. Their statements and behavior often give much away. They do not expect sarcasm, and carefully worded sarcasm plays on the offender's emotional reaction and memories, often revealing important relevant information.

Why an empathetic, gentle, trusting approach is ineffective and inappropriate when working with physical and sexual offenders? Well, most psychologists and social workers were taught to be respectful and empathetic toward their clients. Carl Rogers (founder of humanistic pyschology and one of the most influential psychologist in American history) led the way for this approach. At that time, many clients were self-referred and their issues were noncriminal. In fact, most clients suffered from depression, anxiety, or general relationship difficulties. Those were the days!

Today, however, our clients present with more severe pathology. Many are court-ordered into counseling, treatment, or therapy as a result of criminal convictions. The most common involve convictions for drug and alcohol abuse, domestic assault, child abuse, and sexual offending. These clients are not highly motivated for engaging in the therapeutic environment, and many lack the investment necessary to complete a meaningful, structured treatment or counseling program (whatever those are). Many enter into therapy only to satisfy court orders, and if they could, they would simply skate by in the program without making any changes. Many receive some financial assistance for therapy.

The main problem as I see it is that we professionals have been given power to effect changes in the lives of offenders. This power scares me. Not only are we expected to "treat" the identified primary problem, but also we are expected to assume responsibility for the client's behavior in the community. If the client reoffends, we may be held liable, at least in the eyes of the community. And the balancing act begins. For you see, we have been taught to use the Carl Roger's approach in relating to our clients, but that approach, while appearing appropriate to the uneducated, lacks the necessary substance to address the significant, aggressive issues offenders present with.

So I offer this. Assume a therapeutic approach somewhere between that of Columbo and Sipowitz (*NYPD Blue*) (see Chapter 35). You see, sarcasm and confrontation really are the most effective methods for addressing offender issues. Now I've really gone and done it. I said it ... yes, confrontation and sarcasm. Many therapists would never acknowledge to engaging in the use of sarcasm or confrontation. In fact, these are almost curse words to the profession. Yet I assert that most every therapist uses both sarcasm and confrontation in their work with offenders or they would not have survived in the field. Offenders present as individuals with low self-esteem, depression, and often lack social skills and many have literacy problems. But the problems at hand, that is, the presenting problem, is offending. They committed a physical or sexual assault, rape, child molestation, date rape, or child abuse.

Assaultive behavior is a result of narcissism, not low self-esteem or depression. I find it difficult to imagine striking out at someone out of depression, but could imagine striking out in anger. But anger requires the thoughts of the potential victim deserving to be hit, or that I have the right to strike out and vent. That is narcissism. The focus is not about low self-esteem or depression, but of narcissism.

Narcissism

Listen to the offender as they explain their offense. Narcissism will be evident. A narcissist can easily comply with treatment expectations. In fact, he may be a "model" client. Yet he rarely internalizes what he learns and, therefore, rarely changes. He will be the first to reoffend. And, of course, he has an elaborate explanation for why he reoffended. However, in the therapy process, the narcissist will often cause the therapist to have an upset stomach, a feeling that something is not right. But the therapist will often be unable to identify the problem. The client may appear to be making too much progress in a very short amount of time. This again may be a narcissist.

Other narcissistic clients will engage in power struggles with their therapist and probation or parole officers, anyone who has power or authority over them. They will quickly accuse others of interfering with their lives, and will blame anyone they can. Some of their complaints may appear, on the surface at least, genuine. Yet, upon closer examination, their complaints are unfounded. They power struggle simply because they can. And they know that sometimes they will succeed at splitting professionals, thereby further entrenching themselves in the delusion that they really are innocent of the offense behavior.

Narcissistic clients may insist on addressing their own childhood victimization issues or current relationship issues before addressing their offense behavior. Some may even become depressed or obsessive-compulsive and engage in acting-out behaviors. These behaviors are simply distractions. They may complain about intense urges to reoffend, but usually their urges involve other paraphiliac behavior, such as use of prostitutes, promiscuity, window peeping, and exposing themselves. Yet, even if you focus on their immediate crisis, they will tend to invest very little toward change. That's when you know you caught them in their game. They distracted you from the real reason they came to therapy, which is the offense. Any progress they make on these other issues may appear slow and they may request a break from therapy. You see, this is when they are most vulnerable and when their perpetrator persona is showing. This is the best time to insist that they refocus their energy on their offense behavior and not allow for a break or to focus on other issues.

When confronted, they may blame their own victimization for "causing" them to offend in the first place. It is cyclical. The more you focus on their own victimization issues prior to addressing their offense behavior, the more they become trapped in their own victim mode. You will waste several months with them and, in the end, they are more likely to terminate from therapy. Finally, they rarely follow though on what you encouraged them to do, such as taking psychotropic medication as prescribed.

Many attempt to refocus the issues back onto their own victimization throughout therapy. Don't be distracted. When batterers or sex offenders decompensate because of the distraction, they may:

- Exaggerate symptoms
- Complain of excessive somatic complaints
- Have onset of ailments very quickly, yet the symptoms fluctuate, getting worse when attempts to address offense interfere with their alleged pathology or medical problem
- Refuse to comply with medications

They also attempt to run out their probation or parole time before making any changes. This is a significant sign of narcissism. Think about it. Many offenders still fear the courts and find the consequences unpleasant and to some degree painful. But the narcissist or psychopath focuses on his own pain and suffering and does not comply because he wants to challenge the power. He finds it condescending to take orders from others and does not believe that he did anything wrong that would require any intervention. In fact, he will blame his victim and other circumstances without accepting much, if any, responsibility for his own behavior.

Power struggles should be redirected on the offenders. The therapy program is to address their aggressive and illegal behaviors, not to address the therapist's personality. And interviews are meant to obtain as much information as possible from the offender, guided by the interviewer. So, the power struggles need to be addressed immediately and the interviewer needs to be in the driver's seat.

Use of Sarcasm and Confrontation

Sarcasm and confrontation are therapeutic approaches used to meet the offender where he is at, and to get at the perpetrator persona rather than the social persona. The perpetrator persona is a narcissistic, self-centered, remorseless, and, at times, psychopathic persona.

It is important to not resort to the use of sarcasm and confrontation out of anger. Perpetrators are notorious for pushing buttons and power struggling. They may attempt to provoke the therapist into reacting to their ploy. However, a grounded therapist recognizes the importance of taking a deep breath and thinking before acting. Like you have never told your clients to do the same thing, right?

It is important that the therapist or interviewer be aware of his own needs when interacting with offenders. Avoid power struggles and situations where you may be tempted to "one up" the perpetrator. Remember that you are the professional, and it is your responsibility to direct the interview or therapy process in the direction you decide. If the offender is attempting to demonstrate how much smarter he is than you, call him on it. But avoid trying to put him down or out-do him, or the offender may win a situation that never should have had a winner to begin with.

Have a point in mind, a purpose for the sarcastic comment or confrontation. Know what the answer is likely to be, and anticipate how the offender may react. It is probably not a good idea to provoke a potentially violent or psychotic individual. But you could do so if you think before you act. Remember that timing is everything.

Last, but most importantly, you must be comfortable using sarcasm and confrontation or you should not use them. Not all therapists, probation and parole officers, and police officers are comfortable using sarcasm and confrontation. Stick to what your good at doing. Have a colleague assist when necessary. Remember, it is not only all right to recognize your limits of competence, but it is your ethical obligation.

Why I Believe Professionals are Reluctant to Use Sarcasm and Confrontation

Schools do not teach how to use sarcasm and confrontation to work with offenders. In fact, most teach that sarcasm and confrontation should be avoided. But forensic work does not follow the same guidelines or techniques as clinical and counseling require. Most schools do not teach how to work with offenders at all. Again, offenders require specialized training. Offenders are not the volunteer clients, they always have a reason to be cooperative, and it is generally speaking not because they really want to change.

Professors tend not to be experienced clinicians; they have often never worked with offenders. In fact, many of the professors teaching in forensic psychology have very limited experience. Those with experience may have engaged in the general forensic field, such as conducting parenting or custody or competency assessment, but few if any forensic assessments on batterers or sex offenders. I recall the professor who was interviewed for one of the news programs when the sniper was active in the Maryland and Virginia area. He arrogantly sat in a chair with his hands behind his head, and leaned back almost lying in his chair — a totally arrogant stance. He commented about the sniper's probable profile. Amazingly, it followed what the textbooks would suggest. However, his profile was wrong. Experienced profilers and experienced forensic psychologists know better, and are aware that few cases follow the "textbook" examples.

Many professionals are simply not comfortable using sarcasm and confrontation. Forensic psychologists and law enforcement personnel need to be comfortable using sarcasm and confrontation or they should not use it. Not everyone is capable of using sarcasm and not every professional is able to effectively interview violent offenders.

Examples

One way that I have effectively used sarcasm was when forensically assessing a predatory sexual offender. He claimed that he sexually molested his 11-year-old daughter because he thought it was his wife. He was very well defended. I had him describe his wife and daughter, from their bodies to their perfume to their genital development, as well as how his wife responded to him sexually. I diagramed this on a white board as the offender described the victim and his wife. After drawing the body and genital size differences, the offender began to weaken in his defense and eventually admitted that he was, in fact, aware that he had sex with his daughter, though denied it was abuse! Sarcasm was useful in this exercise.

37 Guidelines for Interviewing Sexual and Domestic Abusers

RECOMMENDED INTERVIEWING STRATEGIES

The following are recommended strategies for interviewing violent offenders. Whether the interview is to obtain a confession, to gather more information as part of a therapeutic assessment, or for treatment purposes, I have found the following strategies and guidelines effective.

1. Obtain and review all available information about the alleged or actual offense. This involves understanding the offense from both the victim's and offender's perspective. The necessary information includes the following:
   - *Police reports:* These contain very useful baseline information. The police often have interviewed the victim immediately following an offense. The victim's statements may change over time, but you will find that the overall facts remain consistent from interview to interview with the victim.
   - If the perpetrators are interviewed soon after the crime, they are usually caught off guard and make many interesting statements. They offer excuses for their conduct and often place the majority of blame onto the victim. At times, perpetrators admit the offense but offer justifications for their behavior. This can be used to confront denial at a later time.
   - *Criminal history:* This is the perpetrator's record of criminal activity. This will be addressed in more detail elsewhere in this book. But it is imperative that a thorough criminal history check is obtained. Also, for every documented police contact or conviction (regardless of whether the conviction is by plea agreement or trial conviction), it is often very fruitful to obtain copies of the criminal complaint and police reports for every crime that suggests any relation to violent behavior. I strongly recommend that you obtain the records related to any crime involving burglary, trespassing, robbery, disorderly conduct, domestic assault, assault, DWI, DUI, and any sexual-related crimes. Remember that most perpetrators receive plea agreements and, therefore, the criminal record often appears less relevant and less violent before you gather the records.
   - *Victim statements*: These can include statements made to the police, advocates, emergency room staff, and medical staff. It is never the role of a psychologist to determine guilt or innocence. However, the veracity of the victim's and perpetrator's claim is often more apparent when all of the records are obtained and reviewed.
2. Be comfortable conducting interviews with violent criminals and psychopaths. If perpetrators suspect that you know very little about the offense, are inexperienced, or personally reactive to the case, they will successfully play you. Some criminals have personality traits that make them very good at reading the interviewer and they may be adept at telling an apparent true story of the offense that has nothing to do with reality. Perpetrators who have personality traits involving narcissistic personality disorder, antisocial personality disorder, and/or psychopathy are keen and adept at distorting the truth and often appear believable. It is imperative that anyone conducting interviews or assessments of

perpetrators understand the type of crime being investigated and understand the personality traits associated with the type of crime that occurred.

3. You need to control the interview. The perpetrator should not be allowed to guide or lead the interview. This is a mistake many novices make. Stick to your outline and when the perpetrator attempts to change the direction of the interview, keep him on track. It is often a sign that you are getting close to the truth when the perptrator attempts to change the topic or side-track you.
4. Do not disclose everything you know about the perpetrator or the crime or situation in question. You have the information and, therefore, hold the upper hand. Disclosing all that you know discourages the perpetrator from telling you what he knows.
5. Perpetrators, like any of us, do not want to incriminate themselves. They are going to tell you what they want you to believe and what you already know. Be cautious in what you share with them so that you do not tip your hand and reveal your lack of information or the extent of information known. You want to retain the advantage of information that can substantiate the perpetrator's disclosure.

   For example, if the perpetrator was convicted of assault 20 years ago and that assault was actually a plea agreement involving a rape as well as a burglary charge being dismissed as part of the plea agreement, the perpetrator may erroneously assume that no one actually has the correct and full details. He assumes that no one is aware that the assault conviction actually involved burglary and rape. Many Pre-Sentence Investigation investigators fail to obtain information to fully understand what prior convictions are involved. They assume that the criminal history accurately represents the perpetrator's criminal activity. But we know this is not the case. (This will be discussed in more detail elsewhere in the book.) The perpetrator may eventually provide previously unknown details of the current or previous crimes and may even confess to crimes that were uncharged.
6. At some point in the interview, minimization and denial should be confronted. But timing is everything. It is often beneficial to allow perpetrators to tell their story in their own way. Often they give more facts than they intended, especially if they believe that they have the upper hand and are fooling the interviewer.
7. At some point in the interview, again timing is imperative, point out and confront inconsistencies and blatant lies. Confront the perpetrator on inconsistencies and implausibilities. His reaction is often very useful in profiling and assessment.
8. Do not accept partial admissions as being "all of the details of what happened," or as being "the truth." It is never the job of a psychologist, therapist, or social worker to second-guess police reports or criminal convictions. Leave that to the attorneys. We do not second-guess victim statements. Inaccuracies will inevitably occur in everyone's statements. However, there is almost always adequate truth to the statements made by victims to define what actually occurred. Perpetrators are very adept at focusing on minor discrepancies (e.g., "I did not hit her, I slapped her," the timing of the offense, all of what was actually said or who initiated it.) Encourage the perpetrator to reveal and discuss the details of the situation or offense. The goal of obtaining a confession is beneficial not only for criminal prosecution, but necessary to make effective and accurate intervention recommendations (e.g., incarceration, types of treatment, living arrangements).
9. Always question perpetrators about how they believe their alleged or actual victim was impacted by their offense behavior. Document the perpetrators' degree of remorse, empathy for the victim, and whether they accept responsibility for their offense behavior. A perpetrator may demonstrate some degree of remorse, but not accept any responsibility for the offense. Others lack any significant sense of empathy and remorse. Remember that violent behavior stems from narcissism and a lack of empathy.

10. Be aware of the goal of the interview. If the goal is to get a confession, then keep that goal in mind throughout the interview. Do not get into treatment issues, stick to the confession. If the goal is to make recommendations to the court for therapeutic interventions, keep that in mind. The direction of questioning can often get off track and result in frustration when you realize after the interview has ended that you have questions remaining. If the documented reports appear consistent with information provided by the perpetrator, move on to other areas of relevance.
11. Reinterview the perpetrator as many times as you need to obtain whatever information is necessary to complete your task. Some therapists, psychologists, and law enforcement personnel may feel uncomfortable calling the perpetrator back for more questioning, but remember that after the perpetrator has left and the report is being written, questions may arise and deficiencies in the interview data become more apparent. It is not only acceptable to reinterview the perpetrator, but it may be unethical not to do so if certain information is missing that impacts the quality and/or thoroughness of the report. It is not uncommon to have ended interviews with more questions than answers. This is generally a result of the perpetrator either taking control of the interview without your paying attention to it or a lack of focus on your part (possibly due to fatigue or other distractions). One other cause is often overlooked by inexperienced interviewers and by those untrained in psychopathy. The more comfortable the perpetrator is with hurting others, the more they possess narcissistic, antisocial, or psychopathic traits, the more likely questions will remain unanswered at the end of the interview. You may become aware of these traits as you begin to prepare the report and realize that you are unclear about some areas of questioning. Rather than second-guess your skills or whether you may not have paid attention to the perpetrator's response, reinterview the perpetrator and get the answers you need.
12. It is absolutely all right to lie, to tell partial truths, or use totally deceitful statements in order to facilitate the interview. Manipulation of information is almost always useful in encouraging perpetrators to tell the truth and to reveal more details. In fact, when the perpetrator believes that you have more incriminating information, he feels more compelled to "set the records straight" and provide statements about previously unknown crimes and even about unknown facts concerning the current situation or crime.
13. Understand and become comfortable with your style of interviewing. Not everyone is good at or comfortable with conducting interrogation or therapeutic interviews, and if you are not comfortable doing so, you should only conduct interviews with supervision. Interview styles vary widely. On a continuum, I believe that at one end is what I call the Colombo approach (see Chapter 29). This involves a gentle interview approach that allows the perpetrator to feel more relaxed and encourages the perpetrator to believe that you are taking his side and that you believe him. At the other end of the continuum is what I call the Sipowitz approach (taken from the *NYPD Blue* television show), which is direct, confrontive, and sarcastic. You must be careful, however, that the use of confrontation and sarcasm not become abusive or threatening. Another widely accepted method is what I refer to the Plausibility approach. Utilizing this approach, you offer perpetrators an out, telling them that you can understand how they could have committed the crime, but for understandable and very human reasons. The goal, of course, is to gain an admission for either prosecution or therapeutic recommendations. Not my favorite is when one combines all three approaches at different times during the interview. There are times when the Sipowitz approach is not necessary because the perpetrator confesses and is very cooperative.
14. When questioning the perpetrator, review all important aspects of the crime. Go through the crime, step-by-step, fact-by-fact. Utilize the police reports, criminal complaint, victim statements, and any other information to provide you the facts of what might have

occurred. Focus on what was said, who initiated what, what occurred and the timing of each behavior, and how the crime came to be. Pay attention to how the perpetrator isolated and/or chose his victim, the degree of violence used, and all aspects of the offense behavior.

If you are a psychologist or therapist, then you are ethically required to accurately provide a diagnosis if diagnostic criteria are met. Understand the appropriate diagnostic categories and accurately diagnose every disorder that the perpetrator presents with. Some professionals believe that it is ethical to only diagnose the bare minimum and not to diagnose disorders that overlap in problematic behavior or etiology. But if the perpetrator meets the diagnostic criteria, then he is given the diagnosis. For example, an offender may meet the diagnostic criteria for Antisocial Personality Disorder, and yet still meet the criteria for Adjustment Disorder with Disturbance of Conduct. For example, if the perpetrator is in the process of separation or divorce, they would be expected to be experiencing additional and rational pain, discomfort, stress, and grief, and, therefore, his acting out behavior may at times be more related to the temporary and current stressors than his long-standing pattern of antisocial behavior. The comorbidity of mental health disorders is far more common than once thought. (See Chapter 41 on diagnosing for more information.)

15. I always find it effective to give the impression that I know much more about the perpetrator's criminal history and offense details than I do. I may play dumb at the beginning of the interview, but if the perpetrator is defensive and minimizing, then at some point I allude to knowing other pertinent facts, even if I know nothing else. This strategy has resulted in numerous confessions of additional criminal activities and a far more rich sense and understanding of the perpetrator's thinking and violent behavior.

    One of my strategies is to ruffle and look at some papers I have, but not let the perpetrator see what is on the papers. I will then "hem and haw" and look at the papers and then look at the perpetrator, and then back to the papers. The perpetrator usually takes the bait and asks what it is that I am looking at. I then indicate that I have questions about another incident of rape or abuse. I might ask, "I know about some other things that have happened. Can you tell me what they might be?" I purposively keep my comments very vague, and will become more focused and clear with specific questioning based on the perpetrator's response.

    One example is a child molester who was being assessed for consideration of lifetime civil commitment as a sexual predator. I had suspicions that he had other child victims, but no evidence. As I ruffled papers, I indicated that I was aware of some other touching that had occurred, touching involving him and two or three other children, both male and female. After his face turned pale, I knew I was on to something. As I began to question him further, he stated, "I guess I better tell you about my girlfriend's kids." He confessed to sexually molesting his current girlfriend's three children, two boys and one girl. No one had any idea of additional victims.

16. Never fully trust the veracity of the perpetrator's claims. Nor should you fully trust statements made by family, friends, and lovers of the perpetrator, all of whom may have adequate reasons to minimize and withhold information. Whatever facts are obtained, you should repeatedly question the source to ensure accuracy and pay attention to the emotional reaction of the person telling you the information. Ask yourself, "Does the person's effect match the type of information he is telling me?" For example, if he is smiling at you while telling you about an abusive incident, but minimizes the details or impact, his smile is probably covering up significant fear and anxiety. The emotional reaction should match the information provided before you believe the person. However, a narcissist or psychopathic person can look you in the eye and lie to your face without

so much as a squint. Be careful of information you obtain that contradicts any factual information.

17. Remember that it is *never* the job of a psychologist, psychiatrist, therapist, or social worker to make statements about the guilt or innocence of anyone. That is up to the courts. At best, we can assert that the perpetrator admitted a crime, or denies the crime, and they either did or did not appear credible. Mental health professionals have the task to present all of the relevant factors and facts, regardless of whether it is in the offender's best interest or not; no favors can occur, all information should be objectively presented even when it differs from what others offer (e.g., defense attorneys, prosecutors, the victim, or offender).
18. A skillful interviewer and assessor manipulates the perpetrator up and down during the interview. The purpose of this is to test, to some degree, the perpetrator's ability to control his anger, to observe the perpetrator's anxiety or anger level, and to assess his ability to calm down. However, this never includes being verbally abusive towards a perpetrator. The skillful use of sarcasm is a true art, and only seasoned and trained assessors should use sarcasm.
19. Lastly, be confident about your ability to conduct the assessment, interview, or interrogation. Not everyone is capable or comfortable conducting an interview, assessment, or interrogation. You should have extensive training and have observed others engaging in these services before you begin to engage in them. If you are fresh out of graduate school, you still need specialized training in assessing or interrogating the specific type of perpetrators you will be working with. A Ph.D. does not make you competent to conduct these services unless you have formal graduate training in forensic psychology. Even then, you should work with and observe skilled professionals before conducting the interview on your own. Be aware that in court, any mistake you make could result in a perpetrator being acquitted, may result in the continued abuse or rape of children, identified victims, as well as other potential victim in the future. In addition, you can be held accountable by the state licensing board and civilly liable as well. Most employers, including state, county, and federal agencies, cover you if someone files a lawsuit against you in your position. However, when the complainant files a complaint with your licensing board, the government agencies and private agencies generally do not protect you. Most do not offer "board" coverage to protect you against ethical complaints, and none that I am aware of offer any coverage to protect you against a lawsuit against you personally. If the complainant is successful in court, you may have to pay thousands of dollars in damages. There is no place for professional arrogance in the forensic field.

Remember, perpetrators that admit to what the authorities already know have not given us much; they simply are admitting what they know we know. And this is fine as long as the information disclosed is adequately representative of the details necessary to complete an assessment or obtain a criminal conviction. However, even when perpetrators admit to details already known, that does not in any way indicate that they have been totally truthful, nor does it suggest that they will be truthful in the future in other matters. Always be wary of disclosures made by perpetrators. This is especially true when the perpetrator appears to be benefiting from treatment or other therapeutic interventions. Psychopathic and narcissistic perpetrators tend to do fairly well in treatment and often appear to be the "star-pupils." However, research has demonstrated time and again that psychopaths and narcissists simply become better criminals, not treated.

Strategies for the Professional

Here are some of the strategies I use with sexual offenders and physical abusers. These strategies often result in perpetrators revealing so much more information than when I use other strategies. Below are some samples of what I might say to the perpetrator.

1. *"Before we begin, I want to let you know that I have talked with others about your case."* I do not expand on who I talked with or if I have really talked with anyone.
2. *"I spoke with the Minneapolis Police Department about you. Can you tell me what that was about?"* I provide only minimal information about what I know. If necessary, and the disclosure of the information will not place anyone in danger or jeopardize a criminal case, I may read sections of the entire report to refresh the perpetrator's memory and to prove that I have, in fact, obtained the information.
3. *"Several other police departments have information about you, although not all of these resulted in the police talking with you or in criminal charges. Can you tell me a little about these police contacts?"* This is especially useful when the perpetrator has resided in different counties or states. Police often have records on every call, regardless of whether an arrest was made. It is not uncommon for domestic abusers to have many police calls to their residences that did not result in a criminal conviction or arrest, but the information may be used to establish a pattern of violence.
4. *"I know something about some other victims. Can you tell me about this?"* This is a good statement that has resulted in numerous perpetrators revealing information about previously unknown crimes.
5. At times, I may read excerpts from the criminal complaint or police reports in order to confront the perpetrator and to let him know that I know the facts.

38 Forensic Considerations for Assessing Violence

INTRODUCTION

I have engaged in the practice of forensic psychology and in assessing violent offenders for many years and I have noticed that many psychologists conduct forensic assessments for assault, violence, and sexual offenses that draw conclusions on only a 1-hour interview with the offender and perhaps an administration of a Minnesota Multiphasic Personality Inventory (MMPI-2). The psychologists often fail to interview the victim and/or spouse, fail to administer a battery of tests, and fail to gather available criminal records. As a result, many violent offenders are labeled as first-time offenders when, in fact, their criminal history is rich with violent offenses. What I propose is that forensic assessments for violence (e.g., domestic abuse, sex offenses, child abuse) follow a more detailed outline and testing protocol, and involve gathering criminal records from the courts and police departments to verify criminal history rather than simply relying on a criminal history check.

I have seen many sex offenders' domestic abuse/anger assessments, and other forensic assessments that have routinely relied heavily on self-report of the criminal or only on the criminal complaint for the current offense. Most psychologists do not even bother to obtain a criminal history check, and those who do almost always fail to gather official documentation pertaining to other relevant offenses. The result is a poor assessment, one that most often significantly minimizes the offender's history, patterns, and, therefore, minimizes the actual risk for violent reoffense. Without a thorough review of all pertinent information, examiners mistakenly indicate that someone has committed his first act of assault or his first sexual offense because the examiner did not bother to check into the criminal history or take the time to gather relevant records. I find that the following procedures allow for the most accurate, detailed assessments for violent offenders.

PROCEDURES FOR ACCURATE ASSESSMENTS

It is important to understand that the predictive accuracy of unguided clinical assessments is typically only slightly better than chance levels (Hanson & Bussiere, 1998). What this means is that regardless of how experienced the clinician or how expert the clinician, clinical opinion or assessment without use of actuarial tools and testing is not adequately accurate and appears careless.

I recommend the following structured outline for conducting forensic or intake assessments on violent offenders. This includes those charged or convicted for domestic abuse/assault, assault, and any sexual offense. Even if the offender pled guilty to a lesser offense, if the criminal complaint or police reports suggest that the offense was assault or sex offense related, or if the charges began as assault or sex offense related, then I strongly recommend adhering to the outline I present below. Remember that a plea agreement does not in any way insinuate that the offender is not guilty of all of the alleged behavior, but rather the court is accepting a plea and in return offers a lesser punishment. However, the facts presented that resulted in the original charges were still assumed to be true.

Definition

I use the term *forensic* to refer to any type of assessment or investigation used to report to the courts or to answer questions from the court. I extend this to include work done that is reported back to supervising agents as well. Forensic assessment can include sex offender assessment and a batterer assessment (e.g., domestic abuse assessment, child abuse assessment) and typically includes an assessment for the purpose of civil commitment, profiling for law enforcement, and assessing the type of treatment intervention required as well as identifying the predicted level of risk for engaging in future acts of violence.

Why Interview the Victims or Partners of Violent Offenders?

Bowen et al. (2005) reported that the predictive accuracy of partner expectation of offender reoffending significantly improved prediction. Weisz, Tolman, and Saunders (2000) found that women's expectations regarding how likely it was that their partners would become violent during a dispute in the future was the single best predictor of severe violence. Heckert and Gondolf (2004) also found that the women's perception of risk significantly improved the prediction of repeat reassult over prediction with men and women's characteristics alone. Therefore, it is imperative to include the partner/victim's report in assessments to provide more reliable assessments of reoffending and for risk for violent reoffense.

Assessment to Determine Guilt or Innocence

There are clinicians who accept referrals (usually from defense attorneys) and are asked to do either a short and biased assessment or to assess the offender to determine guilt or innocence. This, what I call "*paid-for professional whore,*" is the clinician who is given referrals from defense attorneys with the expected outcome that the offender (1) is at minimal risk for engaging in future acts of harm, (2) that the offender is not in need of any substantial therapeutic treatment intervention, and (3) that the offender should not be recommended to serve any additional time of incarceration. These are simply unethical clinicians. The responsibility of anyone conducting a forensic assessment (again, forensic meaning any assessment or services being done for the court) is to examine all of the available relevant evidence, interview all pertinent individuals, administer appropriate testing and screening tools, and then complete a thorough report that outlines all relevant facts about the offender (e.g., personality characteristics, diagnosis, risk for harm) and make responsible therapeutic recommendations. Any short cuts for any reason are simply not acceptable.

The clinician who conducts an assessment to determine the guilt or innocence of an offender is simply playing God. It is the job of the court, not a clinician to determine the guilt or innocence of a defendant. How would a clinician know the true facts? The offender has more than ample reason to distort and lie about the truth. And even psychological testing and risk screening tools have limitations — they do not prove that an offender did, in fact, engage in the alleged criminal behavior. The Association for the Treatment of Sexual Abusers (1993) prohibits an assessment being conducted for the use of an assessment to confirm or deny whether an event or crime took place. Even having a response of sexual arousal to children on the plethysmograph never proves that the individual has ever or ever will act on that attraction with sexually deviant behavior (Fruend & Blanchard, 1989; Murphy & Barbaree, 1994).

As Melton et al. (1987) indicated, clinicians should avoid being used as "lie detectors." They support leaving guilt or innocence up to the trier of fact, that is, the courts. Interestingly, the research literature does not support that clinicians have the ability to accurately differentiate sex offenders from nonsex offenders in the general population (Becker & Kaplan, 1990; Becker & Quinsey, 1993; Hanson & Bussiere, 1996; Murphy & Peters, 1992). As McGrath and Purdy (1999) commented, psychological testing, personality characteristics, background history, and psychophysiological responses have never proven to accurately identify sex offenders. And again, there are no personality tests that can differentiate batterers or sex offenders from the general population. McGrath and Purdy commented further that the majority of courts in the U.S. prohibit psychological testimony that is used to determine if an individual was more or less likely to have committed a sexual offense based on a profile.

Common Questions from Professionals

Why conduct a through assessment on batterers and sex offenders? Why not just put them into treatment? These are common questions I hear from professionals. It is important to understand that the predictive accuracy of unguided clinical assessments is typically only slightly better than chance levels (Hanson and Bussiere, 1998). What this means is that regardless of how experienced the clinician or how much the clinician is an expert, clinical opinion or assessment without use of actuarial tools and testing is not adequately accurate and appears careless.

It is important to understand that unaided clinical judgment has been found to be of unacceptably low reliability. Unaided clinical judgments from clinicians about an offender's level of risk generally have been found to be less reliable than actuarial measures (Craig et al., 2004; Hanson & Bussiere, 1996; McNeil, Sandberg, & Binder, 1998; Hood et al., 2002). In addition, utilizing structured and systematic risk assessment approaches further increase predictive accuracy (Bonta, Law, & Hanson, 1996; Janus & Meehl, 1997; Monahan, 1996). Perhaps one of the major criticisms of clinical judgment used alone is that there is an apparent tendency and bias to over-predict risk (Hagen, 1997; Hood et al., 2002) which then falsely predicts recidivism.

Understanding that there are different typologies of batterers and different typologies of sexual offenders is a significant overlooked fact, even by many professionals involved in the violence abuse and sex offender fields. It is important to understand that evidence suggests that impulsivity is a stable trait over an individual's life span, interwoven with other factors (Rosenberg et al., 1988). To adequately assess impulsivity, clinical and/or forensic interviews need to be conducted to gather information from others. In addition, life-style impulsivity can be successfully identified utilizing a criminal history check and psychological testing. The MMPI-2 (Butcher et al., 1989) and Millon Clinical Multiaxial Inventory III (MCMI-III) (Millon, 1983) are very adept at identifying impulsivity and related characterological traits. Life-style impulsivity has been found to be a powerful predictor of recidivism among sexual offenders (Prentky et al., 1991) as well as for predicting the frequency of offending (Hall, 1988; Hall & Proctor, 1987; Prentky & Knight, 1986; Rice, Harris, & Quinsey, 1990; Rosenberg et al. (1988). In addition, Lisak and Roth (1988) found that impulsivity was related to self-reported sexual aggression among college men.

Harris, Rice, and Quinsey (1998) found that the number of prior sex offenses (the tendency for offenders to recommit the same type of offenses rather than switching) and the frequency of the offense behavior are positively correlated to reoffense (Hanson, Steffy, & Gaunther, 1993; Quinsey, 1984, 1986; Quinsey et al., 1995). In addition, sexual offenders (and batterers) have been reported to endorse "rape myths," traditional vs. enlightened sex roles, interpersonal violence, hostility to women, adversarial sexual beliefs, sensation seeking, and generally right-winged socio-political ideas (Malamuth, 1996; Malamuth, Heavy, & Linz, 1993; Quinsey, 1984, 1986; Walker, Rowe, & Quinsey, 1993). Harris, Rice, and Quinsey (1998) have also found that sexual aggression is related to general antisociality (juvenile delinquency, generally violent conduct), social isolation, poor intimacy, and psychopathy and sexual deviance.

Batterers present with similar issues as discussed above. In fact, the majority of batterers would be subscribing to similar beliefs as the sexual offenders. So, the above data appear to apply to batterers as well as sexual offenders. Remember that the majority of psychological and verbal abuse utilized by batterers involves sexually orientated words and phrases that are indicative of deviant sexual fantasies and ideation.

The necessity of conducting a thorough assessment on batterers and sex offenders appears supported by numerous studies and research. The use of psychological tests and actuarial tools offer objective measures of the offender's personality structure and traits.

Forensic Examiner's Biases

Factors unrelated to alleged abuse can impact the examiner's opinion during a forensic assessment. Nuttall and Jackson (1994) found that professionals who had been the victim of childhood physical or sexual abuse were more likely to believe allegations of child sexual abuse. Kendall-Tackett and Watson (1991) found that professionals who believed that children do not lie about sexual abuse were more likely to be convinced that alleged sexual abuse actually occurred. Physicians who had a tolerance for corporal punishment were less likely to report child physical abuse (Morris, Johnson, & Clasen, 1985). Herzberger and Tennen (1985) reported that females were more likely than males to view harsh discipline as inappropriate. In addition, female professionals demonstrated stronger emotional reactions to child physical and sexual abuse allegations (Hazzard & Rupp, 1986).

In addition, Boat and Everson (1988) indicated that the profession of the evaluator was related to the type of information viewed as important. This is, of course, an unavoidable situation, though it presents with the probability of missing relevant information. Professionals were also more likely to report abuse when children came from poorer families (Zellman, 1992). Psychologists were more likely to attribute more responsibility to the offender in an adult–child sexual abuse situation when the offender was male (Wagner, Aucoin, & Johnson, 1993); yet sexually abused male children were held more responsible than the offender for the sexual contact than were female children (Waterman & Foss-Goodman, 1984; Muller, Caldwell, & Hunter, 1993; Duke & Kean, 1983).

Limitations of Testing

It is imperative to understand that not all batterers or sex offenders who score in a reoffense direction on a screening tool or test will, in fact, ever reoffend. In fact, individuals who produce deviant sexual arousal on the phallometry do not always act on their sexually deviant interests with another person. What this means is that despite scoring high on a scale or phallometric measure, or any measures, does not indicate that an individual has or ever will act on the deviant interests. Some have argued that the MMPI-2 has a "sex offender" profile when it does not. Nonoffenders can produce the same or similar patterns. So the clinician must beware of the limitations of testing and screening tools (covered in Chapter 49, Research Concerns and Assessment Tools).

It is imperative that the clinician informs the referral source and comment on the assessment report any limitation or concerns about the sources of information used and the reliability of the conclusions and recommendations. The American Psychological Association (1992) indicates that clinicians have an ethical obligation to inform referral sources and other consumers of evaluations about the limits of their accuracy. In addition to commenting about the concerns I have about the data, interviews, or testing results used in my assessments, I add the following comment to all of my forensic assessments:

> This report was based on a review of records, a clinical interview with the subject and/or with spouse or family members. Psychological testing was performed. It should be noted that the validity of the conclusions drawn in the report is subject to the limitations of scientific procedures and psychological descriptions and the impossibility of making absolute predictions.

Competence

Conducting forensic assessments requires adequate training in the specialized field being assessed (e.g., batterers, sex offenders, competence, insanity). Training must include both formal (advanced degree) and informal education as well as ongoing continuing education credits in the area of specialization. Simply attending conference trainings on topics unrelated to the area of specialty is not adequate. Continuing education must include training specific to the area of specialty being practiced. In addition, it is imperative that the clinician has had adequate and documented training and supervised experience in evaluating the population being worked with (Coleman & Dwyer, 1990). Simply having worked with a handful of batterers or sex offenders does not make a clinician competent to assess or treat an offender. Supervised experience typically takes 6 to 18 months to adequately understand the population.

Informed Consent

It is imperative that offenders are given an informed consent form to sign prior to conducting any part of a forensic assessment. The important components of informed consent include client understanding of: (1) the purpose of the assessment, (2) the nature of the assessment and the procedures to be used, (3) the confidentiality of the assessment result; (4) risks involved, (5) the advantages and disadvantages of completing the assessment, (6) fees, and (7) how the assessment results will be used and who will have access to the final report. The informed consent I use is the following.

I use the same form for any forensic assessment, changing only the title of the assessment. In addition, I use a similar informed consent form with anyone interviewed as part of the offender's assessment, including the victim and/or children. I use two forms for fees. The specific one for forensic assessments appears on the following pages (fees given are examples).

INFORMED CONSENT

To Participate in a Domestic Abuse Assessment

Scott A. Johnson, M.A., L.P., DABPS, DAACCE, ACFE

You are being assessed to determine what recommendation appears appropriate to address the problems that resulted in this referral. You are aware that the results of this assessment will be shared with the Court to comply with the Court order.

The assessment consists of (1) a review of all records made available at the time of this assessment; (2) obtaining relevant collateral information from any other source as deemed necessary (which may include law enforcement, probation offices, child and adult protection, friends, family, etc., whether by person, phone and/or mail); and (3) a clinical interview (if you choose to participate). Where appropriate, releases of information will be obtained.

You have the right not to participate in this clinical interview. Should you choose to exercise this right, your refusal may adversely affect the outcome of this assessment, but will not bias this examiner's opinion of the facts. However, this may be the only opportunity that will be provided to allow your participation and input into this process. Should you choose to participate in the clinical interview, anything you say may be used against you in that it might support criminal charges being brought against you; the loss of custody of you children; other legal action, sanctions, or restrictions; action being taken by child or adult protection agencies; or possible divorce or separation from a spouse or a relationship ending. On the other hand, information you possess may support the above entities not taking drastic or legal action against you, and may result in a better understanding of your version of the actual or alleged offense or identified problem or concerns, as well as a better understanding of your strengths and skills. Your participation allows for more accurate, appropriate, and effective recommendations being made to the referring party. You may choose to participate in the interview, but reserve the right to refuse to answer certain questions if you wish. You may also end the clinical interview at any time. It is understood that information obtained and statements made during the interview are not private and may be shared in the examiner's report to the referring person or agency. A copy of this assessment report will also be sent to the referring person or agency and a copy will also be placed in your file.

This process does involve psychological and/or other testing. The results from the testing will be incorporated into the final assessment report. The clinical interview may last up to 3 hours, but will likely last about an hour to one and a half hours. During the clinical interview, you will be asked to provide your view of pertinent questions and provide evidence or other argument on your behalf. The questions asked will be of a general and personal nature.

Under Mandated Reporting Laws, if you reveal that you/another person have abused or neglected a minor or vulnerable adult in the past 3 years, and reveal the identity of that person, that information would be reported to law enforcement. Also, under the Duty to Warn law, if you report that you intend to seriously harm yourself or another person, appropriate action will occur to ensure the safety of yourself or the intended victim, which may include notifying law enforcement. In addition, if you are pregnant, or provide information concerning a pregnant person that reveals that they/you are using certain drugs or abusing alcohol, this information would be reported to the appropriate authorities.

I, (print name)________________________________, (DOB)________________, have decided:

To participate (sign name)____________________________________ Date:_________________

Not to participate (sign name)__________________________________Date:_________________

Witnessed by: Scott A. Johnson, MA, LP, DABPS, DAACCE, ACFE Date:____________________

In addition, I use a specific Release of Information Consent Form for forensic assessments because, in Minnesota, clinicians are required to notify the victim and supervising agent of any imminent risk, whether the offender has completed or terminated the treatment program. A normal release of information would expire immediately when the individual or offender terminated services, therefore, not allowing for anyone to be notified. As a result, the supervising agent and/or victim are notified the same day should the offender terminate or be terminated from the program. This is one form of the statement used on my release of information forms and offenders initial that they agree:

> It should be noted that when Court Ordered for assessment or treatment, the **ABOVE PARTY** may be notified by phone and/or letter of sessions attended; missed; unexcused absences; failure to comply with expectations; completion of services; and/or termination/suspension of services. The above party will also receive a copy of the assessment report or the suspension or termination report. **I UNDERSTAND THAT EVEN IF I TERMINATE SERVICES THE ABOVE PARTY WILL BE NOTIFIED AS EXPLAINED ABOVE.** In cases where abuse or assault has occurred, the **VICTIM** will also be notified as explained above. My initials indicate that I understand and agree with these terms.

This simply helps eliminate any problems with notification. The offender's rights are respected by adding this type of statement because they are informed in advance of how information will be disclosed and that the releases of information do not expire if they simply choose to terminate their participation. For safety reasons, this is necessary.

FORENSIC ASSESSMENT FEES

Mending Path Relationship Center, Inc.

Scott A. Johnson, M.A., L.P., DABPS, DAACCE, ACFE

Assessments include Psychological Assessments, Domestic Abuse/Anger Management/Assault Assessments, Sexual Offender/Psychosexual Assessments, Parenting Assessments, Custody Assessments/Study, and Divorce Mediation. If you are Court-ordered to complete an assessment, this is referred to as a "forensic assessment," meaning that it is being conducted for the Court. It should be noted that insurance *may* not cover all or part of the assessment. Informed Consent is explained and obtained when any forensic assessment occurs. The assessment process and fees are described below.

All Fees Are Payable at the Time of Service or in Advance as Indicated

A diagnostic interview involves a 1 to $1\frac{1}{2}$ hour interview. The purpose of this interview is to review the offense history, to gather and review of relevant history, to assess mental health and safety concerns, and to begin to formulate intervention recommendations. Records are obtained from relevant sources.	Cost: $200 per hour
Psychological testing includes paper and pencil tests as well as risk assessment tools that are completed by the psychologist. This typically takes $1\frac{1}{2}$ to 3 hours to complete.	Cost: $100 per hour
Interview with victim. The victim will be interviewed whenever possible and when the victim is willing. This provides an opportunity to obtain the victim's version of the offense and other pertinent information, such as the abuse history. If the victim is unwilling or presents with limitations (e.g., young age, intellectual limitations, mental illness) then the examiner will rely on official records and/or interviews with client, parent, guardian, or other reliable sources.	Cost: $200 per hour This is paid in advance
Interview with spouse or partner. The partner of the client (offender) will be interviewed whenever possible. This allows for accurate information to be obtained pertaining to the client's (offender's) home life. This may occur even when the partner is not the victim.	Cost: $200 per hour This is paid in advance
A second diagnostic meeting. During this meeting the test results are discussed; additional information is gathered or clarified as necessary; recommendations are discussed.	Cost: $200 per hour
Final assessment report is completed. This final assessment report is completed and sent to the referral source, which may include the Court, a probation or parole officer, attorney, or another psychologist. The client is provided a copy of the report, and a copy remains in the client's file. The final assessment report is typically 12 or more pages in length and includes all of the information obtained from the Courts as well as the testing and interview data. This usually takes 2 hours to complete.	Cost: $200 per hour This is paid in advance
Expected additional expenses. This includes consultation with other professionals, travel time, etc.	Cost: $200 per hour. This is paid in advance
TOTAL COST OF FORENSIC ASSESSMENT	$

I understand and agree to the above charges.

Client:____ Date:__________________

Witness: Scott A. Johnson, MA, LP, DABPS, DAACCE, ACFE Date:__________________

Recommended Outline for Forensic Assessments of Sexual and Nonsexual Violence

The following is the outline I recommend to adequately and effectively assess an offender's history (see Johnson, 2005). The following areas appear to have significant relevance when assessing patterns of deviant and violent behavior.

- Review of all relevant documents and information.
- Interview with the offender.
- Interview with the victim when appropriate.
- Interview with spouse/partner.
- Abuse/assault/sex-related violent criminal history.
- Nonassault/nonsexual criminal history (including juvenile and adult history).
- Additional information related to sexual offending/abusive behavior/child abuse.
- Treatment history (including participation in *Abuse* or *Sexual Offender* treatment programs; participation in *Chemical Dependency* treatment programs; other therapeutic interventions (e.g., marital or family therapy, childhood interventions, out of home placements, any outpatient or inpatient placement to address mental health issues).
- History of suicidal ideation and suicidal attempts, also whether the offender is currently experiencing suicidal ideation or has a current plan to injure or kill him/herself.
- History of homicidal ideation; also whether the offender is currently experiencing homicidal ideation or has current plans to kill or seriously harm anyone at this time.
- History of experiencing psychotic symptoms.
- History of taking or currently using psychotropic medications.
- Diagnosis (including prior and current).
- Chemical use and abuse history and screening.
- Medical history that includes any serious conditions, which may impact reality testing or the offender's ability to reoffend with violence.
- Background information (including educational history, discipline history in school settings, residence history, and family of origin information).
- Employment history.
- Relationship/marital history.
- Development history (e.g., any delays in attaining expected developmental tasks).
- History of engaging in paraphilic/sexually deviant/sexual acting out behavior (specifically ask about each area of paraphiliac behavior, which tends to result in more disclosure — attendance at strip shows, use of sex lines, use of prostitutes, and frequency and type of pornography use [use of pornography is significantly correlated with violent behavior]).
- Identify sexual orientation.
- History of experiencing sexual/nonsexual jealousy.
- Personal abuse history (including whether having witnessed or experienced physical or sexual abuse in the household or experienced neglect).
- Perpetrated abuse history (summarize the offender's history of abuse/assault, use of weapons, committing assault that resulted in serious injury, etc.).

- Supervision history, including whether offender has violated any conditions of probation or parole.
- History of failing to follow a Restraining Order or Order for Protection.
- Psychological testing (which should include at least the MMPI-2 and/or MCMI-III, STAXI-2, SARA, HCR-20. For sexual offenders at least the MSI and STATIC-99, and possibly the ABEL Assessment and/or Penile Plethysmography.)
- Summary of all testing completed on offender.
- Clinical interview summary.
- Mental Status Exam.

In addition to the above, I include a summary of factors that have been empirically related to violent reoffense. This also allows for a more thorough review of the assessment data before formulating the recommendations and assigning a risk level for violent recidivism. These include:

Factors Considered for Risk Assessment

- The offender's score on actuarial risk prediction tools.
- The offender's criminal history.
- The offender's history of violent offenses.
- The situations and circumstances of the abuse/sex offenses (e.g., relationship between victim and abuser; the ages and vulnerability of the victims; the type of offense committed (predatory, opportunistic, or elements of both); the degree of force used (nonviolent, violent, sadistic, kidnapping, bondage, torture, use of weapons, threats, or killing of victim); the duration of each offense situation; and the length of offender's offense history.
- History of participation in recommended treatment.
- Use and abuse of drugs and alcohol, and whether the offender continues to use them despite treatment interventions.
- Information indicating the presence of additional victims and offenses not prosecuted (obtained from self report, treatment staff, police reports, the Pre-Sentence Investigation report (PSI), and other collateral sources).
- Attitude toward their offense behavior, treatment, and risk towards the community.
- Demonstration of remorse or lack of it.
- Admits, denies, or minimizes offense behavior.
- Presents with traits of psychopathy (lacks remorse, appears shallow, lacks guilt, narcissistic; important to utilize Hare Psychopathy Checklist).
- Presence or absence of mental illness and mental disorders, and whether the offender has followed recommended treatments (such as taking medication, aftercare, support groups, etc.).
- Presence of mental health diagnoses.
- Testing and assessment reports, including psychological, intellectual, and court ordered assessments and evaluations.
- Assessed the level of risk for reoffense, taking into account both the gravity and degree of future offense behavior.
- Information obtained from the clinical interview (e.g., that the offender provided adequate information to allow for an accurate assessment).
- Offense Specific Dynamics, which include: Were the victims known or strangers? Did the offender use gratuitous violence? Does the offense behavior appear sadistic in nature? Does the offense behavior appear premeditated or opportunistic? Was the offense behavior bold in nature (e.g., occurred in public place, high risk of getting caught)?

The Assessment Process

The assessment process should include the following.

1. **A diagnostic interview,** which involves a $1^1/_2$- to 2-hour interview. The purpose of this interview is to review the offense history, to gather and review relevant history, to assess mental health and safety concerns, and to begin to formulate intervention recommendations. Records are obtained from relevant sources, including the probation officer, the courts, police departments, etc.
2. **Psychological testing** is imperative. This should include psychological tests as well as risk assessment tools. Recommended tests and tools for violence risk assessment include: the Minnesota Multiphasic Personality Inventory (MMPI-2) (Butcher et al., 1989) and/or the Millon Clinical Multiaxial Inventory (MCMI-III) (Millon, 1983); the Spousal Assault Risk Assessment (SARA) (Kropp et al., 1999); STAXI-2 (Spielberger, 1999); Conflict Tactics Scale (Straus, 1979); and the HCR-20 (Webster et al., 1997). For sexual offenders, I add the Multiphasic Sex Inventory (MSI) (Nichols & Molinder, 1984) and STATIC-99 (Hanson, 1999), and when possible, the Abel Assessment for Sexual Interest (Abel, 1995) or Penile Plethysmograph, referred to as PPG (Freund, 1967). If the offender's victims include children, then the Parenting Stress Index (PSI) (Abidin, 1983) and Child Abuse Potential Inventory (CAPI) (Milner, 1986) are added. Administering only one test appears grossly inadequate and unethical. A battery of tests offers validity information across tests and more effectively identifies the offender's needs and appropriate risk level. See the reference section for descriptions of each test/tool.
3. **Interview the victim** whenever possible and when the victim is willing. This provides an opportunity to obtain the victim's version of the offense and other pertinent information, such as the abuse history. However, if the victim is not related to the offender, it is not acceptable to pressure his/her cooperation. However, if the victim is family or the partner/spouse of the offender, then they should be strongly encouraged to participate.
4. **Interview with the spouse/partner** should occur whenever possible. This allows for accurate information to be obtained pertaining to the offender's home life, including the abuse history, drug/alcohol use history, and other areas of concerns. This should occur even when the spouse/partner is not the victim of the current offense. I find this interview imperative. Most offenders who assaulted or raped nonrelatives also assault and act out with partners/spouses. I inform offenders that I will not be able to complete the court-ordered assessment without their spouses/partners participation. I always interview the spouse/partner. There is no other way to obtain relevant facts about the assaultive history without this interview, and no other way to establish rapport so as to offer information about whether abuse continues to occur. In addition, I find that many victims are not provided with adequate information about victim resources and may not have anyone else to talk with about the abuse. This allows for a prime opportunity to help the victim/spouse/partner obtain therapeutic services that might otherwise be missed. Only about two clients a year choose to complete the assessment elsewhere because they do not want their spouse/partner interviewed. But this is generally an indication that the abuse is more severe and ongoing, therefore, warranting the offender's resistance.
5. **Interview the children** whenever possible. This is perhaps the most difficult interview to arrange. Many perpetrators and victims will refuse to allow their children to be interviewed. The reasons for their resistance is typically that they do not want their children to be "traumatized" by the interview process or by discussing the violence that occurred within the home between the parents. However, the children are already victimized and need an opportunity to share their experience and to have their experience validated. In addition, the children have valuable assessment information about the family

environment, the frequency and type of violence that is occurring, as well as information about their parent's substance use. The children also provide information about safety concerns within the home. Children provide valuable information and should be interviewed whenever possible.

6. **Conduct a second or third diagnostic meeting** during which time the test results are discussed, additional information is gathered or clarified as necessary, and the recommendations are discussed.

Importance of Gathering Official Records

It is imperative for examiner's to obtain and utilize all pertinent information concerning a violent offender. This certainly includes interviews with the offender and the offender's partner/spouse, and when possible and appropriate, an interview with the victim. The pre-sentence investigation report, the criminal complaint, and the police reports provide much information about the offender, and are imperative to review to complete a forensic assessment. Oh yes, even police reports that did not result in criminal charges, and charges that were dismissed for any reason (other than acquittal by a jury), provide crucial information about an offender's criminal history. It may be necessary to contact law enforcement wherever the offender has resided, not just where the current offense occurred. Treatment records are also excellent sources of information. It appears unethical to conduct a forensic assessment without these sources of information.

Another problem is that plea agreements occur so often that they are more the norm than the exception to the rule. But if an offender pleads guilty to a lesser offense, that offense still remains as described in the original charges unless the judge orders the criminal complaint be rewritten or if the offender was acquitted by jury. Therefore, what appears on the criminal history check is often misleading. For example, if the offender was originally charged with domestic assault, terroristic threats, and disorderly conduct, but pleads guilty to the disorderly conduct, the only charge that will show up on the criminal history check is the disorderly conduct, and that is what assessors typically have access to. The problem is that the disorderly conduct is assault related, and remains assault related regardless of the final plea agreement. Therefore, it is important to gather more information concerning any charge or conviction that appears related to violence. Commonly related charges include any type of assault or sexual offense; robbery, burglary, disorderly conduct, and trespass.

When I conduct forensic assessments pertaining to sexual predators and sexual psychopaths, it was very common to find that most (approximately 70%) had more than two prior sexual offenses that were pled down to lesser charges, such as disorderly conduct. All of a sudden, what looks like a sexual offender who is being assessed to address his/her first sexual offense is actually being assessed to address a documented pattern of sexually violent behavior, and this is simply the first time they were convicted directly of a sex offense. The same holds true for offenders who commit domestic assault, assault, and child abuse.

Case Examples Justifying Gathering as Much Official Information as Possible

Case Number 1

This was an adult incarcerated sexual offender, approximately in his early 20s. He was currently incarcerated for attempted criminal sexual contact. According to his criminal history check, he had a sex-related offense 3 years prior for window peeping, two convictions for robbery, two convictions for burglary, three auto-related felonies, as well as a conviction for escape. Although clearly a career criminal (his felony criminal history began when he was approximately 14), he had only one known serious sexual offense. After ordering copies of the criminal complaints and police reports for the above convictions, however, a new picture emerged. One of his prior attempted burglaries involved window peeping on a female and attempting to enter the house when the victim was home and alone. While waiting to make his plea, he sought sex offender treatment, suggesting that this would have been a rape burglary rather than simply a burglary. His other three related burglaries and robberies involved the same situation. He and some friends broke into the home of an elderly couple, beat the man, and he attempted to rape the woman. Had I simply based my assessment on the criminal history, none of these other details would have been known, and a violent, serial rapist would have been set free.

Case Number 2

This involved an adult male in his late 20s. He had numerous convictions for sexual offenses and for assault. Upon further records checks, it was learned that all of his assault convictions involved assaulting female strangers to obtain sexual contact by force. He also had a conviction for window peeping. Without reading the criminal complaint or police report, one would assume this was a fairly innocuous offense. However, the police report indicated that he had been window peeping on this same female on numerous occasions and that he had stolen her underwear from the apartment dryer, more suggestive of stalking than simply window peeping. Of interest was a theft charge. I obtained the police report for that offense and found that it involved confronting a female and requesting sex, assaulting her, and when questioned, he admitted that he intended to rape her. How misleading the "theft" conviction was. The same was true for a burglary conviction, which turned out to be related to one of the attempted rape convictions. As a result of this individual's violent sexual history, I sent teletypes out to all of the local police departments where he had ever resided since adolescence. It turned out that he had numerous police contacts involving window peeping, possibly stalking of one female, and at least one incident that may have been an attempted rape, all occurring when he was an adolescent, and none resulting in arrest. After a review of his extensive history of attending sex offender treatment, it was discovered that he failed to complete every program as a result of his acting out with violent and sexually violent behavior while in the programs. In fact, disciplinary records indicated that he even engaged in violent sexual and non-sexual behavior toward female staff, and that he regularly entertained sexually violent thoughts of female staff. Although his documented criminal sexual behavior history indicated that he was perhaps untreatable and very dangerous, the additional collateral information was priceless in profiling him. It was only after obtaining the collateral information that his thoughts and plans of raping a female staff were revealed.

Case Number 3

This case also involved another adult male in his late 20s. He had two known sex offense convictions as well as several adult and juvenile convictions for disorderly conduct and burglary. After obtaining the police reports and/or criminal complaints, it was learned that all of his assault convictions were pled from violent sexual offenses. The burglary involved striking the female victim with a club and attempting to rape her. It turns out that all of his numerous criminal convictions were sex related, and that he was becoming far more physically violent with each sex offense.

Case Number 4

This involved a male in his early 20s. He was currently incarcerated for two separate counts of False Imprisonment, each involving two different females. Upon reading the criminal complaint, it turns out that both of these incidents involved kidnapping and physically and sexually assaulting females. The victims were strangers. After reading the criminal complaint and police reports, it turned out that there was a third victim who managed to escape. The police reports and criminal history indicated that he also had convictions for aggravated battery and disorderly conduct as well as for violation of order for protection. After obtaining the records on each of his convictions, it turned out that all involved seriously assaulting his significant other. When interviewed, he also admitted that he was convicted of aggravated criminal sexual abuse, which was nowhere to be found in the records. He indicated that this involved sexually molesting one sister and attempting to rape another.

Discussion

I have noticed far too many psychologists conduct forensic assessments for assault, violence, and sexual offenses that draw conclusions on only a 1-hour interview with the offender and perhaps an administration of an MMPI-2. What I find to be more effective, accurate, and professional is outlined above. Psychologists have an ethical obligation to utilize all available resources when assessing an offender, especially when risk is to be estimated. The public, and especially the original victim, is at grave risk from being assaulted or raped again if the psychologist underestimates the offender's actual risk, and this will occur far too often when psychologists fail to utilize the above (or similar) outlines and fail to gather the available information and reports. Take the time to obtain relevant police reports, criminal histories, psychological records and treatment reports and incorporate the information into the assessment. Let's set a new standard and collectively establish a "recommended" and "acceptable" outline for forensic assessments addressing violence.

SUMMARY

- Official sources of information provide much more detail than the offender provides.
- The information contained in the police report or criminal complaint is usually enough to warrant criminal charges being filed or resulting in a conviction or plea agreement, then it must be accurate and factual to a significant degree to assume truth.
- Criminal history checks are very often deceiving; a disorderly conduct conviction could have been the result of a plea agreement and the original charge may have been assault, criminal sexual conduct, robbery, or a more serious crime than "disorderly conduct" would suggest.
- Often times police reports, criminal complaints, and Pre-Sentence Investigation reports contain valuable and rich information concerning the offender's history, including relevant information pertaining to offender's criminal history, family and relationships, employment history, mental health history, and may also suggest the use of alias's and/or other police reports or criminal charges that did not result in formal convictions or that were eventually dismissed or sealed. All history has some degree of relevance. You cannot accurately assess or treat an offender or accurately assess psychopathy or assess the true risk for violent recidivism without a detailed history.
- It is often fruitful to contact the local police departments and ask for any information known about the individual. At times, local law enforcement have several police reports indicating sexual and nonsexual violence that do not result in arrest or criminal charges. This information is very useful as it suggests a much more significant violent history regardless of whether there was an arrest or conviction.
- Prior psychological assessments and psychological testing offers rich information to compare the offender's current and persistent personality traits and behavior. Often juvenile treatment records provide rich information concerning an offender's history of engaging in violent behavior. Remember that out-of-home placements are typically the result of either abuse occurring in the home or the child engaging in assaultive or sexually assaultive behavior, both of which are indicators for higher risk of engaging in future acts of violence.
- The children of the perpetrator and victim will provide valuable information not only about the frequency and type of violence that is occurring, but also about current and ongoing safety concerns, substance use, and other risk assessment information.

- Conducting a thorough assessment offers an invaluable opportunity to assess the offender's true personality traits and features, as well as a better understanding of the offender's deviant thinking.
- Only through a thorough assessment and interview process can the assessment be complete; it is important to assess all aspects of the offender's life to adequately assess the degree of pathology and deviant thinking and deviant and violent behavior. This includes physical and sexual assaults, rapes, child abuse and molestation, and pet abuse.

39 Where to Gather Relevant Information

SOURCES FOR GATHERING ASSESSMENT INFORMATION

Conducting forensic assessments on batterers and sex offenders is a difficult, complicated, and an exhausting time-consuming procedure. Doing a thorough assessment requires a willingness to invest the amount of time and energy necessary to gather all pertinent information. I will use the terms *abuse*, *sexual assault*, *rape*, *violence*, and *aggression*, interchangeably. Information should be gathered from the sources discussed below, though other sources may be included.

OBTAINING RELEVANT RECORDS

It can be difficult and time consuming to gather all records pertinent to an offender. However, records help establish a prior history of abuse and sex offense-related behavior as well as other impulsive and problematic behaviors. Diagnosing always is made more effective when records are gathered, especially to determine premorbid functioning and to help in diagnosing personality disorders and psychopathy. Do not be intimidated by the record-gathering process.

You will find most agencies are very cooperative and understanding when they know that you are gathering the records for a forensic assessment or for intake for treatment. Obtain a Release of Information form from the offender for each and every entity for which you are requesting information. The release of information/consent form should be a two-way consent form authorizing both the clinician and the entity to release verbal and in writing any and all records pertaining to the offender. (See the sample consent form at the end of this chapter.) I often call the agency and find out to whom I should send the release of consent form. I am amazed at how much information people share even before they receive the consent form. They can tell you whatever they want and you can use the information, but you must have a consent form if you reveal any personal information about the offender. Any citizen can contact the police or military and they will likely release records to anyone requesting them because the records are most often public information. However, always obtain a signed consent form. Records may be obtained from the following sources.

Batterers and Sex Offenders — The Offenders

Offenders have much information to offer to help to assess their level of risk, treatment intervention (including adjunct treatment, such as substance abuse and medication), and the degree to which offenders accept responsibility for their offense behavior. Important areas in which to question offenders include (but are not limited to) the following:

- How they explain their offense and how they claim to have committed the offense (e.g., how did they make the decision to offend).
- Degree of empathy and remorse. For example, are they able to identify how their victim was or may have been injured, or the potential impact on the children?
- Degree of denial or acceptance of committing the offense and of their responsibility for having committed the offense. Examine whether their responsibility for committing the offense is external (e.g., blaming the circumstance, use of drugs or alcohol, psychological

problems, such as depression, or the victim for their choice to commit the offense) or internal (e.g., accepting that they made the choice to offend).

- How they explain any differences between their version of the offense and the victim's version as well as explaining any forensic evidence that disagrees with their claims. This is an important area to explore. Offenders can make up rather elaborate excuses for how their victims or the evidence is misleading or false.

Victim, Partner, and Children

The victim, current partner of the offender (e.g., significant other, spouse, intimate partner), and the children should be interviewed whenever possible. Some offenders will refuse to allow contact and others will attempt to sabotage the interview process by interfering in some manner. It is imperative that collateral contact be made with the victim, current partner, and the children because they all have valuable information about the offender's history as well as about current and future risk of harm. It is not the offender's option to complete a forensic assessment (any court-ordered or probation-ordered assessment) or treatment of any kind without collateral contact being made — ever. The offender must be held to the same standard of assessment and therapeutic intervention as any other offender would be, without exception.

It is understandable why some clinicians feel awkward requiring collateral contact and involvement of collateral individuals. However, one of the main differences between forensic assessment and clinical or counseling assessment is that the offender is mandated to compete the assessment and to follow the recommendation being made. The offender is not truly a voluntary client, but rather is coerced via court order and risk for negative consequences (e.g., incarceration, loss of custody, divorce) to cooperate.

There is also a significant difference between court-ordered (forensic) treatment and voluntary treatment. Forensic-based treatment (e.g., treatment for abuse or anger management, sex offender treatment) requires ongoing risk assessment. The offender's risk status is truly a dynamic factor that requires monitoring not only when the offender is present in treatment, but monitoring all of the time via contact with the victim, current partner, children, and the supervising agent. The offender's risk level can change in a matter of seconds.

The clinician must be trained in forensics, trained in assessing and treating batterers or sex offenders, and trained in conducting risk assessment on violent offenders. In addition, the clinician working with a forensic population needs supervision from a clinician who has years of experience working with the same population. Lastly, the clinician requires ongoing consultation and continuing education and training in their specialty. No clinician should ever work with offenders without all of these aspects addressed. Clinicians who fail to follow these recommendations fall victim to their own arrogance. They believe that somehow they are immune from being manipulated or lied to by any offender, and that they can accurately predict risk of harm and recidivism without having adequate consultation or training. These are dangerous clinicians and usually these are the clinicians that defense attorneys seek out because of their lack of appropriate background. They can be more easily manipulated by the defense counsel as well.

At any rate, it would be nearly impossible to assess the risk for current or future acts of violence against any victim if ongoing contact was not made with the offender's partner, victim, and children. How else could it be determined whether any ongoing violence was occurring? And the victim, children, and partner should have, whenever possible, contact with a clinician other than the one who is working with the offender. It is also important to have contact with the partner, victim, or children following the assessment interview to determine if the partner is in danger of retaliation from the offender as a result of cooperating with the offender's assessment.

Information obtained from the victim, partner, and the children often reveals a more engrained pattern of abuse or rape, a significantly more concerning substance–use history, as well as more personality and prior violent history not available from other sources.

Children (the Sponges and Barometers of the Household)

The children or stepchildren of the offender have a lot of valuable information to offer as well. The children are typically much more aware of the abuse that occurs in a home or relationship even when they are not present when the violence occurs. They are like *sponges*, absorbing the violent energy and then they are *barometers,* acting out when they have reached a sufficient degree of impact from the violence they observe, feel, or of which they are actually the target. Children and adolescents tend to be more forthcoming, open and honest than professionals give them credit for. The children provide information that the offender and victim are unable or unwilling to notice. The information provided by the children often reveals a more severe pattern of abuse or rape than what the victim reveals, as well as valuable information concerning the offender's and victim's substance-use history.

Areas to Question

Important areas to question the victim, partner, and children about include (but are not limited to) the following areas:

- Current threats or harm, intimidation, or blackmail from the offender.
- Degree of fear they experience for future acts of violence from the offender, which helps to rate the offender's risk for future acts of violence.
- Frequency and duration of assaults. For example, how often are the assaults, including psychological and verbal abuse, occurring; how long does the average incidents last; how long did the most recent incident last?
- History of when the abuse first began (typically months or years before the current intervention), which includes any acts of violence that have occurred regardless of whether the incidents were reported.
- Identify the specific types of abuse that have occurred, not just those involved in the current offense (e.g., use the Johnson Force Assessment [Johnson, 2001b]). This helps provide a much more robust and accurate history of the violence that has occurred over the history of the entire relationship.
- Assess the victim's/partner's/children's view of the offender's substance use/abuse; any psychological problems or interventions or treatments the offender has experienced.
- Victim's/partner's/children's knowledge of any prior abuse or sexual violence (or allegations) the offender has or might have engaged in with others. Include any type of violence, including child abuse or neglect, date rape, pet abuse, aggressive driving, or other examples.
- Any information about violation of a restraining order, contact order, or probation violation.

Official Sources of Information

Collateral information should include official sources of information. Official sources are presumed to be reliable because the individuals documenting the records are trained and often licensed personnel (e.g., police officers/sheriff's deputies, clerks of court, judges, prosecutors, county attorneys). Offenders and victims often are more open and descriptive when initially interviewed by law enforcement than at a later time. Offenders are often caught off guard when initially questioned and often provide either blatant lies, admissions of guilt without realizing what they are saying, or giving well-rehearsed preplanned explanations. Official sources of information include (but are not limited to):

- **Law Enforcement**. Police reports, police interview reports/notes, arrest reports, and complaints provide valuable information about the offender's violent behavior as well as the offender's history or engagement in any criminal activity. Any reports that law enforcement has, including allegations or reports that never resulted in an arrest, should be considered when assessing an offender. Patterns emerge only when professionals take the time to diligently seek out, obtain, review, and utilize information from any official source.

  Police departments often have records on offenders that date back many years. I have often found that many offenders have prior police contacts that did not all result in arrest or prosecution, yet still provide robust information about the offender's modus operandi and violent and nonviolent criminal history. Call the police department and request any and all records related to the specific offender. The offender should sign a release of information. However, records are considered public information. I often request information from the local police departments in every city and state in which the offender has resided. The rationale is that many offenders have unstable lifestyles that result in many moves. Even juvenile records provide valuable information.

  When the offender has lived in another country, contact Interpol (the international police agency). The agency will at times have information on the offender that other agencies do not. It may be necessary to contact the local police department in the city and country the offender resided in as well.
- **Court Documents and Records**. This includes filings for any type of orders for protection, harassment, restraining orders, etc., though not all information is to be considered 100% factual unless a trier of fact was involved (e.g., judge, and a conviction or acquittal occurred). The information is still to be used to help establish or identify any pattern that emerges.
- **Military Records**. These are often public records as well. It is important to know the type of discharge the offender received and if there was any disciplinary or legal action taken. I have found the military to be highly cooperative when gathering information.
- **Evidence**. This includes tape recordings, photographs, hard copies of e-mails, letters, any tangible item. I have had offenders produce tape recordings of specific incidents where they claim that their victim was actually the perpetrator of the crime; yet, in all of such cases, the tape actually proved the offender wrong, proving that the offender was, in fact, the perpetrator.

  For those offenders who are incarcerated, I have found that searching their cells provides valuable information on some of the more predatory offenders. It is not uncommon to find pornography, department store catalogs and newspaper advertisements used for masturbatory purposes; the pictures used most often involved children or adolescents. In addition, some predatory offenders often have notebooks in which they write about their schemes and violent intentions. All of these types of evidence help to better understand the potential risk that harm offenders pose. If offenders are being supervised or incarcerated, knowing that their cell could be searched at any time without warning, yet the continue to engage in their deviant behavior, they are telling everyone around them that they are not concerned about their deviant behavior or of possible consequences, that is, they are not concerned about continuing to engage in violent behavior.
- **Treatment Records**. Any type of assessment report or treatment records provide valuable insight into the offender's psychological state, mental disorders, and degrees of cooperation or lack thereof. Treatment records, such as assignments (e.g., logs, offense histories offenders are expected to write), often provide more detail into the offender's violent behavior and frequently reveal additional violent incidents as well as identifying victims who have not reported the crime or children who were unable to the same. Of course, all licensed clinicians in the U.S. are mandated to report incidences of abuse and neglect

as well as of imminent risk for harm to self or others. So, no clinician can or should ever fail to report additional victims per their state's guidelines and ethical guidelines.

Mental health and treatment records provide valuable information about the offender's mental status and history of problematic behavior. I often request a summary of the treatment rather than all of the case notes because the case notes (1) are usually handwritten and difficult to read and (2) I am more interested in a summary of contact than actual content of each sessions. Most often clinicians and programs will have an intake summary, progress notes, and a discharge summary. I also request copies of all testing completed, including raw scores. These often suffice. I request more detailed records or a verbal consult with the clinician to address any information that appears unclear. It is interesting how many offenders will portray themselves as having completed previous treatment programs with flying colors and then information received from the programs contradicts the offender's story.

- **Supervision and Incarceration**. Records are kept pertaining to an offender's incarceration history, that is, whatever they do while incarcerated in jail, workhouse, or prison is documented. These records provide factual evidence of the offender's ability to control his deviant and violent urges and behavior, and whether he is complaint with therapeutic recommendations. Many prisons offer anger management and sex offender treatment to the inmates. These treatment records should always be reviewed as part of any forensic assessment. Other records of interest include any disciplinary action taken or warnings given; job performance; participation in religious or other activities; and of course, all contacts attempted or made via visitors, mail, etc. Always obtain copies of any rule infractions as these infractions are basically criminal behavior that continues despite incarceration.

  The same is true with any contact the offender attempts to or actually makes with another person. Batterers and sex offenders, who continue to engage in any form of abuse or violent behavior, including psychological force, are continuing to offend despite being incarcerated. In addition, any violations of a "no contact" order while incarcerated is a very serious stalking and harassing incident. Sex offenders with child or adolescent victims who attempt to engage in contact with minors while incarcerated are very serious predatory sexual offenders who will almost always reoffend upon release. Offenders who continue to engage in criminal or criminal type behavior, or who continue to engage in any form of abuse or violent behavior while incarcerated, are dangerous offenders who will almost always continue their violent behavior upon release.

  Incarcerated offenders have guards around them and have every aspect of their life open to scrutiny and observation, when they choose to continue to engage in violent or criminal behavior while incarcerated, they are proving themselves to be dangerous offenders. If offenders are unable or unwilling to cease their violent behavior while incarcerated, why would they choose to or be able to cease their violent behavior once released, at a time when they would have far less constraints.

  Supervising agents (e.g., probation and parole officers) also have valuable information. Whether an offender has cooperated with expectations of supervision, followed through with therapeutic recommendations, or complied with medications, indicates the offender's degree of cooperation and of potential risk. Offenders who fail to follow recommendations of supervisors or even those who postpone following through with recommendations may pose more risk of harm than those that comply. Offenders who have a stake in their community and their own life usually follow supervision without any question. Unfortunately, so would psychopaths!

- **Schools**. Learning institutions can be cooperative when record gathering. If the offender indicated problematic behavior or disciplinary action when attending any grade of school or college, records should be obtained. Schools do not always maintain disciplinary records, though problematic students tend to be remembered by the staff.

- **Employers.** Places of employment have records that help establish work history and patterns of problematic behavior. I have found most employers are cooperative as well. Many offenders experience problematic behavior with anger and harassing behavior at work, which is useful information.

Safety Issues

I have said this before and will say it again. It is important to have contact with the offender's victim, partners, and/or children following their involvement in the offender's assessment interview to determine if they are in danger of retaliation from the offender as a result of cooperating with the offender's assessment. Some offenders will take revenge; however, many will not. But the issue of the safety of anyone should never be ignored or minimized by the clinician or law enforcement. Offenders can present with two extremes of dangerousness for revenge. At one end of the continuum are those offenders who give clear warning of intent to harm, those that make overt and direct threats, and who may by impulsive. However, at the other end of the continuum are those offenders with psychopathic traits who tend to give little warning and who appear calm and cooperative just before harming others. It is better to over-react than under-react when safety issues arise.

EXAMPLE OF TWO-WAY CONSENT FORM

AUTHORIZATION FOR MUTUAL RELEASE OF INFORMATION

I, DOB____hereby authorize:

Mending Path Relationship Center, Inc. and

Scott A. Johnson, M.A., L.P. DABPS, DAACCE, ACFE

PO Box 538

Chanhassen, MN. 55317

(612) 269-3628

TO RELEASE TO and RECEIVE FROM:

____

____

____

____

Phone:______________________________

The following information:
_____Intake/Evaluation/Assessment Summary
_XX_Court/Probation Documents/Police Reports/Child Protection Reports/Court Orders/Victim Statements/Related Information
_XX_ Records related to disciplinary action, charges, suspensions, corrective action plans, concerns, and related material
_____Admission/Discharge Summary
_____Case/Progress Notes
_____Psychological Testing (Specify below)
Summary
_____Raw Scores and Profiles
_X__Verbal Consultation
_X__Other (Specify) _XX_Concerns __ Updates _XX__Safety issues
_____Participation in sessions
_____Interview information to be included in spouse/partner/abuser (name)____________ Assessment Report

This information will be used for the purpose of:
_____Enhancing Support Network/Crisis
_____Support Continuity of Care
_____Updating Referent
_____Updating Primary Care Physician
_X__Complying with Court Order
_____Insurance Coverage
_____Participation in Sessions/Updates/Mediation
XXX Other (specify): to complete forensic assessment or intake

It should be noted that when Court Ordered for assessment or treatment, the ABOVE PARTY may be notified by phone and/or letter of sessions attended, missed, unexcused absences, failure to comply with expectations, completion of services, and/or termination/suspension of services. The above party will also receive a copy of the assessment report or the suspension or termination report. I UNDERSTAND THAT EVEN IF I TERMINATE SERVICES, THE ABOVE PARTY WILL BE NOTIFIED AS EXPLAINED ABOVE. In cases where abuse or assault has occurred, the VICTIM will also be notified as explained above. My initials indicate that I understand and agree with these terms.

This authorization specifically includes all information obtained and recorded during or after my involvement with or treatment by you and/or your organization. I authorize you to discuss the above specified aspects of my treatment and/or evaluation/contacts with the person specified above. I do not re-authorize the use of this information by any third party.

I understand that my records are protected by law and cannot be disclosed without this written consent unless specified by law. I also understand that I may revoke this authorization at any time except to the degree that action has been taken prior to my revoking of it. I further understand that my revocation of this authorization may adversely affect my eligibility for continued services with Mending Path Relationship Center, Inc. and with Scott A. Johnson, M.A., L.P.

This consent automatically expires without my expressed revocation upon fulfillment of the above-stated purpose. I further understand that I have the right to review the information in my file pursuant to Minnesota law.

Client's Signature:____________________________________Date:________________________

Parent or Guardian (if applicable):_______________________Date:________________________

Witness: Scott A. Johnson, MA, LP______________________Date:________________________

40 Treatment for Batterers and Sex Offenders

Treating batterers and sex offenders is complex. Research has not demonstrated consistently that treatment intervention is all that effective in reducing recidivism. This may be because the therapists are not adequately trained, or at the very least, not trained with the same information and consistent information and techniques. It may also be that the programs are failing as well, with inconsistencies and financial struggles that are not offering the appropriate treatment modality or structured treatment goals, and perhaps also having inadequately trained and inexperienced staff. Of course third reason why treatment programs may not be as effective as we would prefer is that the offenders have personality disorders and traits that make them untreatable to some degree. If they are not internally motivated for change, and their personality, need for control, and deviant thoughts and behavior are strong and habituated, then perhaps change is not as likely. Treatment professionals are not yet ready to give up on the notion of successfully treating offenders.

Of course, my colleague Dr. Listiak (Listiak, 2005) points out an interesting dilemma: perhaps it is not the offender that is failing but rather the professionals failing the clients. This is an interesting perspective, perhaps treatment professionals do not yet have adequate programs. There is certainly room for improvement when considering the fact that the two fields, domestic abuse and sexual offenders, have not shared their findings, successes and failures with each other. So, perhaps the professionals have yet to develop an effective treatment approach. But that having been said, the effectiveness of what we have to offer relies on the offenders' choosing to cease their violent behavior.

Treating batterers and sex offenders is a difficult task and requires several things that regular therapy does not. First, therapists must receive formal training specifically on assessing and treating batterers and sex offenders. Simply having a graduate degree or having been in the field for years should never be accepted as competence. Training should include classroom, ongoing conference training, ongoing consultation, and formal face-to-face supervision until the therapist has a thorough understanding of offender dynamics. Remember that the vast majority of batterers and sex offenders who present for treatment are not voluntary in the normal and usual sense; they are court ordered and choosing to comply with the court order to avoid consequences. In addition, ongoing risk assessment of the offender is an inherent part of the treatment process. Poorly trained therapists are not competent to balance therapy and risk assessment a the same time.

Second, programs and therapists need to follow clearly established policies about holding the offender 100% accountable for all treatment and supervision rules and recommendations. This means that under no circumstances are rule infractions to be ignored, dismissed, or casually explained away by the offender. This is especially true of any order for protection (e.g., restraining order): any infraction should be considered as a serious violation and direct assault and the offender should receive serious consequences. Violating orders for protection demonstrate more serious psychopathology and significantly more risk and needs to be viewed that way. Contact with victims outside of therapist supervised sessions is not allowed until adequate treatment progress has been made and safety issues (including safety from verbal abuse) have been addressed.

Third, the offender does not dictate the type of treatment modality, attendance, or length of program. Offenders often have to miss work to attend therapy and are often financially stressed. However, they are to attend the program as any other offender would with no exception. Sometimes offenders have to sell belongings (e.g., boats, other recreational belongings) to pay for treatment.

Remember that any rule that the offender is allowed to break or bend is another form of the offender taking control of the situation and doing things on their terms. Offenders are good at finding and exploiting therapist weaknesses.

I have seen far too many therapists allow the batterer to guide treatment decisions. I have heard far too many therapists comment that they offer the type of treatment that their client (abusers) are willing to complete rather than offering a treatment that is appropriate for the batterer. It is my professional opinion that therapists who allow their batterers, sex offenders, and even substance abusers to dictate the type of treatment are not competent to be practicing and are engaging in unethical conduct.

Consider a medical example. Doctors do not fail to treat medical conditions because patients refuse a specific treatment. They may try lesser interventions first, but they follow a standard treatment plan for most medical conditions. And physicians may well refuse to treat a patient who fails to follow medically necessary interventions. I have also seen psychiatrists who refuse to treat patients who refuse to comply with recommended treatment and medication regimes. It is not unethical to offer treatment, I believe it is unethical to offer substandard treatment interventions solely based on what a patient is willing to comply with. The belief that "any" or "some" treatment intervention is better than no intervention is not empirically supported.

Batterers, sex offenders, and even substance abusers are adept at conning others into offering the least intense treatment intervention. It is the professionals who are obligated to follow standardized and acceptable treatment interventions and if a patient refuses, it is ethical to refuse to treat that individual. Remember that substandard treatment is not better than no intervention at all, in fact, substandard intervention usually worsens the problem. Minimizing the patient's symptoms and offering a lesser treatment program than warranted only serves to worsen the abuser's cognitive distortions about how minimal their problem is, when in fact abuse, sex offending and substance abuse generally worsens without appropriate and effective intervention.

Whenever possible, the therapist and supervising agent should have ongoing contact with the offender's victim, spouse/partner to ensure that the offender is complying with restrictions and that any ongoing assault or abuse is being reported. Simply trusting the word of the offender is not adequate or effective. It is my experience that when talking with the victim, spouse or partner, that the ongoing psychological and sexual aggression that is occurring is reported. The victim and partner/spouse will report ongoing abuse and sexual violence if they trust that the therapist or supervising agent will take the report seriously and take action to hold the offender responsible. It is imperative that the supervising agent be involved in the entire assessment and treatment process, be updated about progress as well as failure to comply, or be told of any risk concerns. The importance of involving the supervising agent and having the supervising agent and court hold the offender immediately accountable (e.g., revoking probation and implementing incarceration for noncompliance) is well documented (e.g., Feder & Dugan, 2002; Berk, Campbell, Klap, & Western, 1992; Sherman, 1992).

THEORIES OF VIOLENCE AND TREATMENT MODELS

Abusers tend to be a group for whom it is extremely difficult to induce long-term change (Lawson, Dawson, Kieffer, Perez, Burke, & Kier, 2001; Dutton, 1998; Gondolf, 1997). This is not all that difficult to understand given that many batterers have personality disorders that are highly resilient to change. Individual differences also play a role in why it is important to understand why treatment is effective for some in decreasing recidivism but not for others. Let's examine the theories of violence using a treatment model outline.

Ventilation Model

This model is based on the premise that suppressed anger needs to be expressed and vented (Hamberger & Hastings, 1993). This is based on the frustration–aggression hypothesis (Philipchalk, 1995). Research does not support that the expression of anger reduces the likelihood of physical violence (e.g., Berkowitz, 1973). In fact, it is quite clear that batterers are very capable of expressing their anger and rage (Pence, 1989).

Insight-Orientated Model

Robertson (1999) describes this psycho-dynamic therapy as being based on the premise that violence is symptomatic of underlying internal conflicts or other intra-psychic problems (e.g., unresolved issues from childhood; personality disorders; failure to develop appropriate attachments; fear of intimacy; poor self concept; and obsessive-compulsive behavior (Adams, 1988; American Psychological Association Presidential Task Force on Violence and the Family, 1996; Carden 1994; Hamberger & Hastings, 1993). Insight-orientated therapy has been strongly criticized as being inappropriate for batterers, mostly because only a small percentage of battering can be attributed to the psychopathology of the batterer (Tifft, 1993). In addition, though batterers may have depression and anxiety or low self-esteem, these are often a consequence of the battering rather than the cause (Adams, 1988).

The psychodynamic theory explains battering as being a result of a negative parent-child relationship (van der Kolk, 1988; Zeanah & Zeanah, 1989; Bowlby, 1988). Basically, attachment needs were not attended to and the child never developed adequate security, sense of protection, or capacity for intimacy. When parents are absent physically or emotionally, the child's development may be thwarted.

Systems or Interactional Model

Robertson (1999) describes this approach as having its origins in family systems therapy (Adams, 1988). The battering is seen as an interactive, dynamic interpersonal transaction (Zelas, 1995), a series of coercive exchanges leading up to aggression or forced submission (Deschner, 1984). Giles-Sims (1983) viewed domestic abuse between partners as being mutually caused elements based on the couple's need to maintain homeostasis and predictability. The systems/interactional approach views battering as a couple's problem, viewing battering couples to be held accountable for the abuse, not one person. This approach focuses on how each person (batterer and victim) plays a role in the violence. From this perspective, the woman's nagging, refusal to have sex, being over-involved with the children make her culpable for her being abused (D. Adams, 1988; Eisikovits & Edleson, 1989; Hamberger & Hastings; 1993; Pressman & Sheps, 1994; Tifft, 1993). This could seriously jeopardize the safety of the woman. This is a dangerous and less effective treatment model.

Feminist-Insight Model

Robertson (1999) describes the feminist-insight model as based on the assumption that battering serves to control women partners and that battering is pro-social in that it is often socially condoned. Men are believed to have greater power, are more likely to terrorize, injure and kill their partners, and their violence is often condoned (Adams, 1988; P.J. Adams, 1999; Hamberger & Hastings, 1993; Pence & Peymar, 1993). Intervention is primarily addressed via education rather than therapy.

The feminist theory believes that battering is a normative response to male socialization that supports the domination of women (Bograd, 1988; Kaufman, 1992; Walker, 1984). Research has supported that approximately eight times more males who hold patriarchal ideologies and who support violent attitudes toward women batter (Kantor & Strauss, 1989; Stith & Farley, 1993). However, Lawson et al. (2001) point out that the feminist approach does not account for violence that occurs within same-sex relationships nor does it account for the unprovoked abusive behavior of female batterers (Gelles, 1995). This theory also does not account for the large number of males that reside in a patriarchal society who do not engage in violence against their partners (Dutton, 1995; Feldman & Ridley, 1995).

Feminist theory views battering as a social and political problem (Straus & Steinmetz, 1974). They assume that to understand the abuse of power and control it is necessary to understand the social construct of gender as it influences the perpetuation of the dominant male role, the subordinate female role, and the domestic violence that results when such influences come into conflict. Feminist theory includes an emphasis on helping victims identify the impact of the violence on personal behavior, self-empowerment, problem solving, the effective use of social and legal institutions, and the use of therapeutic support groups (Register, 1993).

Cognitive-Behavioral Model

Robertson (1999) describes the cognitive-behavioral model as being based on social learning (Bandura, 1977). Battering is considered to have been learned through observation of role models (especially parents) and via trial and error learning experiences. Intervention includes educational efforts to teach men the damaging, negative and self-defeating ramifications of battering. The subsequent consequences of battering are examined, including loss of love, trust, and loss of the relationship. Men are taught how to identify and understand the physical, emotional and cognitive cues of their violence. Cognitive therapies are utilized that help to challenge the rigid and irrational thoughts that support and encourage battering.

From a social learning theory perspective, which falls under the cognitive-behavioral camp, violence is learned through direct observation as a means to resolve conflict and as a means to obtain what one wants, as well as to maintain control in relationships (Bandura, 1979; Dutton, 1998; Lawson et al, 2001). Research has indicated that between 60% and 80% of male batterers experienced violence in their family homes (Gelles, 1987; Hotaling & Sugarman, 1986; Straus, 1990). However, not all males who observed interparental violence became batterers themselves (only approximately 30%) (Kaufman & Zigler, 1987).

Seligman (1975) and Gelles (1979) found that the theory of "learned helplessness" emphasized that women who grew up in violent homes tended to be more tolerant and accepting of violence as adults. Painter and Dutton (1985) referred to "traumatic bonding" as the strong emotional ties that develop between two people when one partner intermittently harasses, beats, threatens, abuses, or intimidates the other. This is also referred to as the Stockholm Syndrome (Kuleshnyk, 1984), which occurs when victims and/or hostages befriend their abusers or hostage takers, and will adamantly defend their hostage takers.

The primary advantage of the cognitive-behavioral model is that it recognizes the functional value of battering to the batterer (e.g., releasing tension, avoiding unpleasant situations, enforcing victim compliance) and places responsibility for the violence on the man alone (Hamberger & Hastings, 1993). Though critics have complained about how gender power issues do not appear to be incorporated into the cognitive-behavioral approach, this is a misperception. Following a treatment outline that addresses issues such as problem solving, understanding the abuse cycle, assertiveness and conflict resolution, the approach effectively helps educate and challenge batterers about their battering behavior and cognitions that condone the violence. This model is easily adapted to include additional treatment approaches and areas of concern as necessary, therefore being a very flexible and practical approach. It is important to understand that although some programs adopt a purely educational model utilizing a cognitive-behavioral approach that demonstrate a belief that men are somehow powerless over their angry and violent behavior (Adams, n.d.; Adams, 1988; Gondolf & Russell, 1986), the effective use of this model necessitates a theoretical belief that men are never, ever powerless over their anger or violent behavior. Batterers engage in deliberate and purposeful behavior (Gondolf & Russell, 1986), injuring their victim in areas of the body not likely to be noticed by others, and batterers are able to postpone battering until they are alone with the victim, necessitating total control over their battering behavior (Tonne, 1992). The cognitive-behavioral approach is the most widely used model to treat battering and appears to be by far the most effective.

SUMMARY

In summary, the Cognitive-Behavioral Model is perhaps the most widely used and the most effective. However, individual differences and needs should be addressed as part of any effective batterer program.

GROUP TREATMENT INTERVENTION

Batterers

It has long since been accepted that batterers should receive group based treatment intervention versus individual treatment intervention. I find this to be true for several reasons. First, for the safety of treatment staff, a group session allows for more immediate safety when confronting or challenging the batterer. Second, the group setting allows for more effective and in depth connection with other batterers and increases the likelihood that the batterers will be better able to identify their own problematic thoughts and behavior. Group treatment allows opportunities for batterers to understand that they are not alone, that it is expected that all past and subsequent acts of violence and abuse be reported, thereby decreasing their sense of isolation. The other group members offer a source of modeling which is extremely powerful as a therapeutic technique and modality. Further, group members can challenge their peers when group norms are not met, as well as support and affirm them in their efforts to change (Eisikovits & Edleson, 1989; Hamberger & Hastings, 1993; Pence & Paymar, 1993; Pressman & Sheps, 1994).

It is important to understand that some of the critics of group intervention express concern that group formats tend to follow structured curriculums and may appear to lack spontaneity and opportunities for the batterer to be adequately challenged or encouraged to strive for more indepth personal growth or increased self-awareness (e.g., Jennings, 1987). However, groups that are run well make time and opportunity to explore more personally relevant issues and allow for members to remain in the group beyond the structured number of group sessions.

Group treatment is believed to be more appropriate than individuals or couples' counseling because it provides peer support for becoming non-abusive (Crowell & Burgess, 1996) and, unlike couples' counseling, avoids the risk of blaming the victims (e.g., Healey et al., 1998). Groups were also less expensive than individual or couples' counseling.

Tolman and Edleson (1995) found that the standard of batterer treatment includes group treatment, anger management training, skill building, and resocialization, and they found that socialization theory and cognitive-behavioral orientations were incorporated into these programs.

Sexual Offenders

Schwartz (1995) indicated that sex offenders may covertly sexually assault their individual therapist and exercise power and control in the therapeutic relationship without the therapist even being aware of it. Sexual offenders need a group setting to help with learning new socialization skills, which a group setting offers the opportunity to do. Schwartz strongly supports and recommends a group setting for treating sexual offenders. The acceptable modality for treating sexual offenders is group therapy.

COUPLES & MARITAL THERAPY

Overall, the body of research supporting use of couples' therapy for domestic abuse intervention are sparse and often unsophisticated (Stith, Rosen, & McCollum, 2003).

Stith, Rosen, McCollum, and Thomsen (2004) indicated that they found treating domestic abuse within a multi-couple format was effective in reducing physical violence within the intimate partner relationship. Their research and program, however, only treated atypical batterers, very low-level risk of violence men and women. However, their study presented major concerns and limitations. The primary problem with their study was that they included only men and women who were highly motivated to participate in the multi-family group format; all couples were excluded in which injuries occurred during assaults that required medical attention; and the batterers were low-level violence men only and could not have substance abuse problems. Their sample size was of course too small to allow for any generalization and their population voluntary. This does not appear to represent the vast majority of men or women referred for domestic violence programs. However, for those rare circumstances where there is no documented or alleged moderate to severe abuse, ever occurring, not even once, and no indication of any significant psychological abuse (e.g., threats, harassment, stalking), then couples' therapy appears to hold some promise. However, the literature does not yet support the use of couples' therapy, there are too few studies and the studies that do exist have significant flaws in their methodology or populations.

Jacobsen and Gottman (1998) suggested that there is a place for couples' therapy in treating physical aggression. They argued however that couples' therapy is not appropriate in treating cases of battering that includes control, subjugation, and intimidation. This of course eliminates the vast majority of men and women referred for batterer treatment. They further indicated that encouraging couples' counseling places the battered woman at increased risk for violence that is both more severe and more frequent. Regardless of what the therapist says about the unacceptability of violence, the fact that the therapist is treating both of them (abuser and victim) implicates them both in the perpetration of the violence (Jacobson & Gottman, 1998).

Some studies indicate that approximately 50% to 70% of battered women remain in a relationship with their batterer (e.g., Feazelle, Mayers, & Deschner, 1984). It is clear that some services must be provided to address couples and relationship issues, though how this is done must be carefully structured and care taken to ensure victim safety.

Couple and marital therapy should never, ever be offered for any batterer or sex offender, even when the victim insists. This is to allow adequate time to assess the history of abuse that has occurred. Even when there has been only one serious incident of physical abuse, when asked appropriate questions, most victims will admit that there has been moderate to significant psychological violence and throwing of objects that led up to the "one incident" that resulted in the referral for treatment, in which case the problem is moderate to severe, not minimal. The primary problem of battering (and at times with sex offenders) has little or nothing to do with the relationship, but rather with the batterer's own psychopathology and coping skills. It is unconscionable and unethical to engage in marital or relationship therapy until the batterer and victim have completed their own structured treatment intervention. Anecdotally, it appears that engaging in marital or relationship treatment intervention, even when both the batterer and victim are in the process of completing their own treatment programs, only serves to strengthen the batterer's minimization of the problem and appears to increase the likelihood of future violence.

It is common that the victim or batterer or sex offender will insist on couples, marital, or family therapy in their attempts to avoid group therapy. It is never advisable to engage in couples, marital or family therapy, or sessions with the victim and offender, until (1) the offender has either completed their own group treatment program; or (2) when the therapist has seen demonstrated progress on the offender's part in treatment; the offender has been in treatment at least 3 months so that adequate time has passed to begin to see if any changes are lasting; and (3) the victim, partner, or children are ready for the session. In addition, it is the therapist who makes the decision to have or not have couples, family, or marital therapy. It is never up to the victim or offender to dictate or demand that couples, marital, or family therapy occur. The therapist should not give-in to demands of the offender or victim. Offenders and victims have their own reasons for trying to return to the homeostasis of abuse that they were accustomed to. They tend to have little patience and often want and then even demand couples, marital, or family therapy prematurely. The therapist is ultimately responsible for making the final decision for any joint session. The therapist's main job is to ensure that any contact or sessions that occur with the victim and offender are not allowing any form of abuse to occur.

For sexual offenders, couples, marital and family therapy may prove to be a valuable adjunct much sooner than with batterers. This is especially true if the victim was not a family member. If the victim was a family member, however, then the conjoint or family sessions should follow the above recommendations made for batterers.

When treating an incest offender, it is especially important to understand all of the abusive and inappropriate boundaries and behavior that have occurred, including abuse towards the spouse, other children, and pets. All forms of abuse must be identified and to do so requires interview of all family members, not just the victim. The therapist must understand all of the direct and obvious seductive, controlling, and abusive behaviors, statements and gestures the offender engages in, as well as the far less obvious, indirect and subtle behaviors, gestures and statements. The victim/s will be able to provide this information far more accurately than the offender, generally speaking because the offender mislabels their own behavior to be far less inappropriate and deviant than the victim. I recommend that the following guidelines I wrote for engaging in sessions involving the victim/s of incest with the offender. These sessions are often referred to as reunification sessions. I will speak more about this later.

DENIERS GROUP

Denial occurs in the vast majority of batterers and sex offenders at some point in the intervention process. Maletsky (1991) found that over 85% of this sexual offenders referred to his program denied all or part of their crime upon intake. When minimization was included as a form of denial, the percent of deniers increased to 98%. Most offenders, however, admit some if not all of their offense behavior, at least to sufficient degree to offer meaningful intervention. Barbaree (1991) identified three forms of minimization: (1) minimizing the harm to the victim; (2) minimizing the extent of previous offenses; and (3) minimizing individual responsibility for the offenses.

Happel and Auffrey (1995) suggested that denial has at least 12 steps, portraying denial as a process. Their steps include: (1) denial of the crime itself; (2) denial of responsibility for the crime; (3) denial of intent and premeditation; (4) denial of deviant arousal and fantasies; (5) denial of the frequency of the deviant acts; (6) denial of the intrusiveness of the offense behavior; (7) denial of injury to victims; (8) denial of sexual gratification from the offense behaviors; (9) denial of various types of grooming behavior; (10) denial of risk management activities; (11) denial of the difficulty of change and need for help; and (12) denial of relapse potential and possible recidivism. I believe that offenders (batterers and sex offenders alike) experience difficulty with several of these types of denial at different times during intervention. I have had offenders (both batterers and sex offenders) demonstrate several types of denial at different times while in a treatment group and upon completing the program, they presented with far less to no denial.

There are some programs that offer sex offenders (I am not aware of specific deniers groups for batterers) specific deniers group or individual intervention to address denial. It is expected that the offenders eventually admit their offense behaviors and become motivated for therapeutic intervention. Though this appears to be a compassionate and realistic option, I have to question the point. When working in one of the sex offender treatment programs I co-facilitated a deniers group. Anecdotally, many of the men were in the group longer than the 6 months maximum. In short, only approximately 40% to 60% of the men ever moved from the deniers group into a regular treatment program group. Approximately 10% of those who moved into a regular treatment group actually successfully completed the treatment program (that is, met all treatment requirements). Now this was not a formal study by any means, but I did not find that a deniers group was economical nor did the majority of offenders appear to benefit from the group, and I therefore question the necessity of such a group. Perhaps the criminal justice system is a better intervention than treatment for offenders who are unable or unwilling to admit to their offense behavior and unable or unwilling to comply with treatment and intervention recommendations. Happel and Auffrey (1995) justified excluding deniers from treatment citing the importance of limiting the time and money spent on offenders who are unwilling to admit their deviance or to invest in the group. Hanson and Bussiere (1996) found that sex offenders who were unmotivated (most often deniers) to attend treatment and who failed to complete treatment were at greater risk for general recidivism than those who completed treatment.

It is unclear how batterers benefit from deniers-based programs, but I would hypothesize that batterers would respond similarly to sexual offenders. There needs to be more research on batterers and denial and treatment.

What appears to be the more effective intervention for deniers is to place them into ongoing treatment groups rather than to have a group full of deniers only. The offender who is denying their offense is allowed to participate in the group for a specific amount of time, and if the denial does not significantly decrease, the offender is then terminated and referred back to the court. Schwartz (1995) found that the denial of many of the sex offenders placed into group with non-deniers decreased as the offender was expected to respond to the support and peer pressure of the group to admit culpability. The offender found support and non-judgment from the therapist for admitting their offense behavior. Confronting denial may include various techniques, including the plethysmography, polygraph, having to repeatedly tell the details of the offense until it matches official records, and review of official records of offense details and victim accounts. Marshall (1994) found that most of the deniers in his group admitted most or all of their offense behavior once participating in the treatment group. Schlank and Shaw (1997) indicated that sex offenders appear able to benefit from treatment even when they are in denial and that many will admit their offense behavior as they participate in treatment.

SUMMARY

It should be noted that the degree of psychopathology never lessens the offender's responsibility for their violent behavior. All of the above models utilize a group approach. Under very rare circumstances should a batterer be treated on an individual basis. For much of the same reasons mentioned above by Schwartz, batterers too will be likely to con the therapist on an individual basis. In addition, when treating offenders on an individual basis, the therapist becomes too much of an ally and collaborator in the batterer's minimization and ongoing abuse. Couples and marital and family therapy should only be utilized after the offender has demonstrated significant progress in their treatment program, and the victims have also completed any necessary therapy and safety has clearly been assessed, including ongoing verbal and psychological abuse.

TREATMENT OUTLINES

There are suggested treatment outlines for batterers that I recommend that all involve similar strategies and theories. The Duluth Model (Pence & Paymar, 1993) is the unchallenged treatment of choice for most communities to address battering (Healey, Smith, & O'Sullivan, 1998). It has demonstrated significant effectiveness in ending and modifying battering behavior. A similar approach is that by Decker (1999). His treatment outline and program are very well grounded and similar to the Duluth Model (see his work for more information).

My own (Johnson, 2001) program utilizes similar strategies to the Duluth Model and also incorporates aspects of sex offender treatment. In fact, I use this same and/or a slightly modified version for both batterers and sex offenders. My outline for treating batterers and sex offenders follows a basic cognitive-behavioral approach. The same outline can be used with both batterers and sex offenders, modifying slightly as needed. My outline is included here in summary form. The batterer receives a detailed outline that specifies the curriculum for each group session, and each group session lasts for 2 hours. There are numerous assignments that are completed for each group. Offenders not completing assignments as expected continue in the program until all assignments are complete. It is not uncommon for some offenders to even repeat the course to complete their treatment. Interestingly, I have not had any offender complain about continuing in the program longer than expected. Their motivation and investment in their treatment becomes evident.

In my program, each batterer purchases a workbook that contains all of the assignments and treatment materials required. Each offender is aware of what is expected each week. Couples, marital, and family sessions occur as necessary, solely at the discretion of the therapist, not the offender. I also will not complete the forensic assessment or treat the offender if their victim and/or partner/spouse is not interviewed or participating as required. The victim and/or partner/spouse have valuable information about the offender and the abuse or sexually deviant behavior.

During an initial session with the offender and victim/spouse, I openly discuss the offense and explain that the offender is 100% responsible for all of the abusive or sexually deviant behavior, irregardless of any relationship difficulties. I also clearly state that the victim is to utilize 911 if the offender violates any of the treatment rules, violates any order for protection, or becomes abusive, intimidating, or threatening. Absolutely no form of psychological or verbal abuse is tolerated and any physical or sexual abuse that occurs after treatment begins generally results in termination and recommendations being made to the supervising agent for a period of incarceration.

Robertson (1999, pp. 11) suggests that effective batterer programs should:

1. Incorporate an explicitly feminist analysis of battering (e.g., the batterer maintains power and control over his partner);
2. Prioritize the safety and autonomy of women over the confidentiality of the batterer;
3. Have a primarily educational approach versus therapeutic approach in which the cultural and social context of battering is addressed;
4. Within that framework, incorporate cognitive-behavioral techniques to help men learn non-violent behaviors;
5. Emphasize the need for participants to take responsibility for their own behavior;
6. Monitor participants, particularly their use of violence;
7. Have well-developed links with battered women's organizations to whom they are held accountable; and
8. Are well integrated with the criminal justice system to ensure adequate consequences for their use of violence.

These are all important components to ensure success of treating the batterer as well as ensuring the safety of the victim. These suggestions are similar to what others have said as well (e.g., Pence & Paymar, 1993; Decker, 1999; Johnson, 2001).

EFFICACY OF TREATMENT

BATTERERS

Several studies raised serious concerns about whether the most common batterer program models are effective (e.g., Dunford, 2000; Feder & Dugan, 2002; Davis et al., 2000), while the latest literature reviews on the subject have failed to reach a consensus (e.g., Babcock, Green, & Robie,

2004; Bennett & Williams, 2004; Feder & Wilson, 2005). Batterer programs overall have demonstrated a small but positive effect (Babcock, Green, & Robie, 2004; Bennett & Williams, 2004). It is unclear at what percentage of "treatment success" and percent of batterers who cease their violent behavior indicates that the current efforts exerted to supervise and treat batterers is worth the investment. Many studies suggest that at least approximately 50% of the batterers continue to batter, even after demonstrating some time without battering. But that philosophical discussion is for another time and book.

It is important to be aware of the limitations of many of the studies. The source from which data on reoffense are gathered is often flawed and lengthy follow-up is not generally engaged in, so the true percentage of batterers who have ceased their physically abusive behavior is unknown. Some studies did not indicate whether ongoing psychological abuse constituted treatment failure or reoffense, and some studies did not count ongoing psychological abuse (including threats, harassment, violation of no contact orders, or stalking) as treatment failure. So at best, some of the studies have questionable outcomes given the poor definitions.

Support that Treatment Intervention Is Effective

Some studies have found that batterers do change, that they can cease their abusive behavior (again, definitions of whether psychological abuse constituted reoffense were at times ambiguous). Tolman and Edleson (1995) found that the majority of men ceased their physically violent behavior for some time following intervention. Successful treatment outcomes typically ranged from 50% top 80% within follow-up time frames. However, it is unclear how durable these changes are.

Davis et al. (2000) did find that men in their 26-week program did significantly better than men in their short-term program or men in their control group (untreated). The victims of the men completing the long-term program reported fewer incidents of new abuse. Dunford (2000) found that there were no significant differences in outcomes between the control (untreated) and treatment group of men, finding that mandated counseling was ineffective for a navy population.

Men who choose to complete treatment tend to be more educated, more often employed, and less likely to have a criminal records than men who drop out (Babcock & Steiner, 1999). Cessation of violence was consistent with age, legal marriage, and unions of longer duration, all suggesting more stable relationships (Aldarondo & Kaufman-Kantor, 1997).

Tolman and Bennett (1990) concluded that the majority of men stopped their physically violent behavior within the measured period of time. But they added that it was difficult to determine whether the cessation of physically abusive behavior was solely the result of psychological treatment alone or whether other factors contributed. Other contributing factors for treatment success may include victim action (separation or treatment of it), police contact, prosecution, disapproval of others, and other naturally occurring processes (Tolman & Bennett, 1990).

The most successful program is one that incorporates sanctions from the criminal justice system (e.g., mandated attendance, immediate consequences for failing to follow-through with treatment recommendations) (Carlin, 1988; Hamberger & Hastings, 1993; Pence & Paymar, 1993; Tolman & Edleson, 1995). In addition, the inclusion and availability of integration of services for battered women to ensure adequate safety for battered women are essential (e.g., Currie, 1988; Edleson, 1995).

The best prognosis was found when the violence was less severe and the history of violence less prolonged, and where power imbalances between the couple were minimal (Bowker, 1983). The role of isolation is paramount for batterers maintaining power and control over their victims. When victim isolation was decreased, the likelihood that husbands would continue to abuse was lessened. The most vulnerable wife is one who stands alone against an equally isolated husband (Baumgartner, 1993).

Rosenbaum, Gearan, and Ondovic (2001) found that court-ordered men who completed treatment had the lowest recidivism rate. Grusznski and Carillo (1988) described completers as more educated, with higher levels of employment, more likely to have been exposed to interparental violence, and less likely to have been victimized themselves, as compared to dropouts. Rosenbaum et al. (2001) also found that partner aggressive men who were mandated to treatment had higher completion rates than self-referred men.

Women batterers were more likely to complete treatment when they were court-ordered into treatment and when supervised by the court than when attending treatment on a voluntary basis (Rosenbaum et al., 2001).

Support that Treatment Intervention is Less or Ineffective

Some have found that once violence against women has begun, the violence continues and increases in frequency and intensity over time until the woman is seriously injured or killed, or leaves her violent partner (Pagelow, 1981; Walker, 1984; Ferraro, 1988; Dobash, Dobash, & Cavanagh, 1985).

Gondolf (2000) reported that out of 618 court-referred male batterers, approximately 45% had reoffended with physical assault within 30 months. It did not appear to matter what type of batterer program the batterers were involved in and the length of services was not a factor. Feld & Straus (1989) found that 57% of batterers continued to use abuse.

Several studies have found that some batterers cease their violent behavior. Feld and Straus (1989) found that a significant number of abusers ceased their violent acts over time: 33% of "serious" abusers in one year used no violence in year two; 10% used minor violence, and 57% continued to use severe violence. Woffordt, Mihalic, and Menard (1994) found that approximately half of marital violence was suspended over a there-year period of time. It is unclear what the new recidivism rates would be, that is, after the 3-year period measured.

Jacobson et al. (1995) suggest that Gottman's Type 1 batterers are not likely to respond to treatment. They argue that battering is primarily a public health problem, not a problem for psychotherapist. This is an important statement of fact. Therapists should be careful not to significantly alter their program to fit the patient's denial, lack of cooperation, minimization or other resistance. Offenders should be offered consistent treatment interventions. Type 1 offenders tend to be more personality disordered and therefore more difficult to treat. These offenders also likely lack necessary motivation to complete treatment and to enact behavioral changes.

Jacobson et al. (1995) indicated that current treatment reviews suggest that recidivism rates following completion of existing treatments may not be lower than those that would have occurred in the absence of treatment programs and those that complete treatment programs are not more likely than dropouts to cease violence in the future.

Gottman et al. (1995) indicated that treatment programs based on the assumption that batterers lack impulse control may be a mismatch for Type 1 men. In fact, Type 1 men are over controlled versus under controlled. This point helps reiterate the need for identifying the specific type of offender and matching the offender with the right treatment or interventions. This helps to illustrate how important it is to accurately typologize the batterer into an approximate category of offender.

Lawson et al. (2001) indicated that group treatment is more effective than no treatment (Dutton, 1986; Gondolf, 1997). However, men who completed treatment programs were only slightly less likely to reoffend than men who completed no treatment, dropped out of treatment, or refused treatment (Rosenfeld, 1992). Men who completed treatment often reverted back to battering within 6-12 months (Hamberger & Hastings, 1993). Recidivism rates ranged from 24% to 40% and were similar for both voluntary and court-ordered men.

Waltz et al. (2000) as well as others have indicated that it is important to consider and address personality disorders and Axis II traits. Short-term, psychoeducational, or exclusively sociocultural based treatment programs focusing on anger management appear unlikely to adequately address the batterers' needs and therefore may only be minimally effective. Batterers who are Family-Only batterers may benefit from anger management and relationship problems due to the fact that they tend to have little if any Axis II pathology. Personality traits become an important consideration when placing batterers and sex offenders into specific groups, taking into account the personalities of all of the group members. For example, it would be very difficult to have a group composed of primarily offenders with borderline personality disorder.

Hare (1995) indicated that approximately 25 to 35% of men in domestic abuse programs are psychopaths. He also indicated that psychopaths are 2 times more likely to reoffend following release from prison and 4 times more likely to reoffend after release from an intense therapeutic community program. Psychopaths do not recognize they have a problem are not likely to benefit from treatment programs. In fact, because the brains of psychopaths are wired differently, they do not respond to emotional words or cues. In short, psychopaths simply become better at offending with treatment intervention.

Female victims who had less education were more likely to experience violence; however, female victims of abuse were less likely to remain in the relationship than women with more education. The reason for this finding is that women of increased educational status may experience more shame about having been a victim, and may also fear losing much more in terms of social and familial status as well as financial status, which is supported by the findings of Aldarondo & Kaufman-Kantor (1997).

Women who were isolated were more likely to experience ongoing violence. Isolation from friends and family and limited or prevented access to income raised the probability of continued violence, as did behavioral demonstrations of beliefs in the right to degrade female partners through name-calling and put-downs. It was also found that violence occurring early in a relationship increases chances of the violence continuing (Aldarondo & Kaufman-Kantor, 1997). Attempts to isolate the victim and to limit access to financial resources also predicted future violence.

Lawson, Dawson, Kieffer, Perez, Burke, and Kier (2001) found a significant decrease in the frequency of violence in their study of thirteen male batterers but not a significant positive treatment effect. Only four remained violence free after a year. Nine continued to engage in physical and psychological abuse despite completing treatment.

The likelihood that violence will continue was correlated with the frequency of assaults, the youth of the offender, living in a common-law relationship, and living in a union of short duration (Johnson, 2003). The subgroup of offenders most likely to reoffend is characterized by antisocial personality traits and would fall into the category of the generally violent/antisocial subtype identified by Holtzworth-Munroe and Stuart (1994) (Bowen, Gilchrist, & Beech, 2005). In addition, rehabilitation of such offenders has limited impact and sometimes results in counterproductive outcomes (e.g., Hemphill, Hare, & Wong, 1998).

Another study found that dropouts were younger and had lower levels of employment, and that they had more police contacts for substance-related offenses and other nonviolent offenses, and they presented with higher degrees of psychopathy, and were more likely mandated to treatment (Hamberger & Hastings, 1989). Rosenbaum et al. (2001) also found that batterers who were self-referred failed to complete treatment at a higher rate than those who were court-ordered, and that having witnessed father-to-mother aggression, and having a history of aggression in previous relationships of their own, were more characteristic of drop outs than treatment completers. Ford & Regoli (1993) found no significant differences in the rates of rearrest between men ordered into treatment and those who were not.

Women are more likely to drop out of treatment without being mandated and when they are not supervised to complete treatment (Rosenbaum et al., 2001).

CHARACTERISTICS OF BATTERERS MOST LIKELY TO REOFFEND — STATIC RISK FACTORS FOR VIOLENT RECIDIVISM FOR BATTERERS

Static (stable) risk factors for violent reoffense with batterers include young age, violence in the family of origin, excessive substance abuse or dependency, antisocial behavior including generalized aggression and non-violent offending, prior assault history, severe personality disorder and/or psychological problems, neglect or abuse as a child, program drop-out, and personality pathology (Aldarondo & Sugarman, 1996; DeMaris & Jackson, 1987; Hamberger & Hastings, 1990; Hart & Ogloff, 1997; Hanson & Wallace-Capretta, 2000; Hilton, Harris, & Rice, 2001; Holtzworth-Munroe, Bates, Smutzler, & Sandin, 1997; Saunders, 1995; Stith, Smith, Penn, Ward, & Tritt, 2004b; Tolman & Bennett, 1990).

Perhaps one of the most powerful predictors of future violent reoffense was having engaged in previous domestic violent behavior (Babcock & Steiner, 1999; Chen, Bersani, Myers, & Denton, 1989; Dutton et al., 1997). The general rule of thumb in forensic assessments is that the past is the best predictor of future behavior.

DYNAMIC RISK FACTORS FOR VIOLENT RECIDIVISM FOR BATTERERS

Dynamic risk factors for violent reoffense with batterers include anger (Beasley & Stoltenberg, 1992; Dutton, Saunders, Starzomski, & Bartholomew, 1994; Eckhardt, Barbour, & Davidson, 1998; Eckhardt & Jamison, 2002), patriarchal and pro-domestic violence attitudes (Sanders, Lynch, Grayson, & Linz, 1987), interpersonal dependency (Barnett, Martinez, & Bluestein, 1995; Kane, Stanger, & Riciardelli, 2001; Murphey, Meyer, & O'Leary, 1994), and external locus of control (Gondolf, 1985; Rouse, 1984; Theodore, 1992).

Bowen, Gilchrist, and Beech (2005) found that batterers who reoffended reported significantly greater levels of interpersonal dependency than those who did not reoffend. They also found that a high level of emotional reliance upon intimate partners is one of the defining characteristics of what Dutton (1998) identified as the abusive personality, which overlaps with the dysphoric/borderline offender category identified by Holtzworth-Munroe and Stuart (1994).

Bowen et al. (2005) reported that the predictive accuracy of partner expectation of the offender reoffending significantly improved prediction. Weisz, Tolman, and Saunders (2000) found that women's expectations regarding how likely it was that their partner would become violent during a dispute in the future was the single best predictor of severe violence. Heckert and Gondolf (2004) also found that the women's perceptions of risk significantly improved the prediction of repeat reassult over prediction with men and women's characteristics alone. Therefore it is imperative to include the partner/victim's report in assessments to provide more reliable assessments of reoffending and for risk for violent reoffense. This is especially true when assessing or treating the batterer. There should be an ongoing contact between the treatment provider and the victim and current partner and children of the offender to ensure that no further violence is occurring. There is no alternative to assessing the offender's escalation and abuse potential other than being in contact with those around the offender.

TESTING THE EFFECTIVENESS OF BATTERER PROGRAMS AND JUDICIAL MONITORING

Most programs incorporate a cognitive-behavioral approach and techniques including skills-building, anger management, and learning new alternative behaviors for addressing conflict (Healey et al., 1998). Unfortunately many programs attempt to offer a "one size fits all" approach which fails to take into account the diversity of batterers (Healey et al., 1998). As I cannot say enough, every batterer should by placed into the appropriate type or category of offender. Typologizing needs to be a routine activity before placing an offender into any treatment intervention.

Judicial Monitoring

Research has shown that the effectiveness of a specialized domestic violence probation supervision produced significantly lower rates of reoffending and arrest-free periods compared with probationers receiving traditional supervision (Klein, Wilson, Crowe, & DeMichele, 2005). Taxman (2002) suggested that a new model of supervision be utilized to surveil and control the offender with evidenced based therapeutic practices that engage the offender in a process of change. Approaches based on surveillance or supervision alone have been found to be relatively ineffective. Others have demonstrated that judicial monitoring has been shown to lead to a reduction of reoffense for batterers (Labriola, Rempel, & Davis, 2005). Men who were mandated into counseling were significantly less likely to recidivate than men who were not (10% versus 31%, respectively) (Palmer, Brown, & Barrera, 1992).

However, not all studies have produced similar findings. Offenders who were not sentenced to monitoring demonstrated no significant differences in probability of re-arrest for any crime, for domestic abuse, or for criminal contempt, which signals domestic violence with the same victim. Labriola et al. (2005) also found no significant differences between the two groups (sentenced to monitoring or not) in survival time (the number of crime-free days prior to first re-arrest). Their findings did suggest that a possible positive impact of judicial monitoring in reducing the total number of domestic violence re-arrests; however, considered in light of other outcome measures for which monitoring did not make a difference, the general finding was that judicial monitoring failed to produce the hypothesized results.

Labriola, Rempel, and Davis (2005) failed to demonstrate any impact of attendance at a batterer program on recidivism; both official re-arrest records and victim reports showed nearly identical rates of re-abuse between offenders assigned to a batterer program with judicial monitoring and those assigned to judicial monitoring only. They also found no effect of judicial monitoring, that is, no effect of those who were and those were not sentenced to monitoring. This was in contrast to the findings of the Brooklyn batter field test that court control/supervision of the batterer was critical to reducing recidivism (Davis, Taylor, & Maxwell, 2000).

THE IMPORTANCE OF STAKE-IN CONFORMITY

It certainly makes sense that batterers who have a stake in their future and community would invest more in complying with court ordered mandates and might work harder at being law abiding. Having a *stake-in conformity* has been defined as including being employed and length of employment; older age; being married; residing with an intimate partner; and residential stability and owning their own home. In short, batterers having an stake-in conformity have far more to lose from an arrest or incarceration and are more likely to comply with court orders.

Feder and Dugan (2002) found that the men involved in a court-mandated treatment program did not cease their violent and abusive behavior and that there was still evidence of physical abuse 6-12 months postsentencing. In addition, approximately one-third of the men failed to attend the batterers' program, and 100% of these men were violated for one of more conditions of probation (71% were specifically violated for failing to complete the batterers' program). They suggested that men who were mandated into treatment and who were held accountable by their supervising agent for completing treatment, when the man had a stake in his future (stake-in-conformity), that is, the man did not want to be rearrested or to go to jail, had a better prognosis. Regardless of whether the men were in treatment, men who were supervised and had supervising agents willing to revoke probation, tended to not reoffend. Similar findings were obtained from the Minneapolis Spouse Abuse Experiment and the Spouse Assault Replication Programs (Sherman, 1992). Subsequent meta-analysis of these studies found that men with stake-in-conformity were less likely to reoffend because of the intervention (in this case, arrest) (Berk, Campbell, Klap, & Western, 1992;

Sherman, 1992). Specifically, when an offender had a high stake in conformity, an arrest was more likely to deter him from further abuse.

The Broward Experiment (Feder & Dugan, 2002) failed to find any significant outcome differences between batterers assigned to a six-month batterer program and probation and others assigned only to probation. However, they suggest that batterers who were employed, married, owned their own homes, or otherwise had a stake in the community were less likely to reoffend. This offers support that batterers who have a stake (investment) in their community, a stake in their future, a stake in conformity, tended to reoffend at a significantly lower rate than batterers without a stake, regardless of treatment intervention.

The Brooklyn Experiment (Davis et al., 2000) showed that men who attended treatment sessions for a longer period (26 weeks compared with 8 weeks) committed fewer new violent acts. However, batterers were more likely to complete the shorter program. Interestingly, there was no effect of actual program attendance, or of assignment to a group, suggesting that the monitoring function of the programs over a longer time period may be more responsible for reducing subsequent violence than actual treatment intervention.

The Ontario Experiment (Palmer, Brown, & Barerra, 1992) found that men assigned to the program had consistently lower rearrest rates than men in the control (no treatment) group, and 70% of the men completed the program.

Pence and McDonnell (1999) suggest that the Duluth Model is not simply a curriculum based batterer treatment program, but requires and involves full coordinated community responses and interagency cooperation and involvement. Mazur and Aldrich (2003) and Sack (2002) suggest that batterer intervention programs function largely as tools for monitoring rather than for changing their personality or behavior. This of course requires that the supervising agent and the court be active in their responsibilities to hold the offender accountable and to violate conditions of probation when the offender fails to comply with all expectations of supervision. Anecdotal experience and numerous studies (e.g., Babcok & Steiner, 1999; Harrell, 1991; Palmer, Brown, & Barerra, 1992) have found that courts rarely penalize offenders with noncompliance with program mandates. Gondolf (2002) reported that this system failure enabled a small subgroup of men in his study to chronically reoffend. The most effective approach appears to require aggressive participation by the supervising agent and the court.

SEX OFFENDERS

The efficacy of treatment in reducing sexual reoffense continues to be in question. Perhaps the problem is that defining "reoffense" remains ambiguous. Reoffense for any criminal activity (sexual and nonsexual) increases recidivism rates far more at times than counting only sexual reoffense. Numerous studies (see literature review by Craig, Browne, Stringer, & Beech, 2004) have demonstrated success, again with the caveat of problems measuring true recidivism rate and what constitutes treatment success.

Sex offenders, like batterers, are at risk for reoffense for their lifespan. Several authors have found that sex offenders may be at risk for reoffense for over 20 years, that is, over the course of their entire lifespan (Furby et al., 1989; Gibbons, Soothill, and Way, 1981; Hanson et al., 1993). Hanson, Scott, and Steffy (1995) found that the general recidivism for sexual offenders ranged from 50 to 80%.

PREDICTORS OF SEXUAL RECIDIVISM

Rice, Quinsey, and Harris (1991) found in one study in which 37% of treated and 31% of untreated sex offenders reoffended within 7 years. They found no significant difference between the treated and untreated group. This is despite the fact that most of those who attended treatment completed treatment. Hanson, Steffy, and Gauthier (1993) found similar results. Rapists as a whole tend to reoffend slightly at a higher rate than other types of sex offenders.

Almost all of the predictors of sexual recidivism were historical or extremely stable variables. The predictors of general and sexual violent recidivism were similar to general recidivism among nonsexual offenders (e.g., age, marital status, juvenile delinquency, sexual deviance, and antisocial personality disorder) (Rice, Quinsey, & Harris, 1991; Champion, 1994; Gottfredson, & Tonry, 1987). For child molesters, Hanson et al. (1995) found that 35% of their sample reoffended within 10 to 31 years. Marques, Day, Nelson, and West (1994) found that 14% of untreated child molesters and 14% of rapists reoffended within 4 years, therefore, difficult to demonstrate a treatment impact.

Pithers et al. (1988) found that for sexual offenders, anger was more of an issue for rapists than other types of sexual offenders. He further found that factors including anger, cognitive distortion, low victim empathy, and offense planning appear to differentiate recidivists from non-recidivists. Those sex offenders that reoffend tended to show an increase in anger and subjective distress just prior to reoffending; these were acute symptoms, not stable risk factors (Hanson & Harris, 1998). This may well pertain to batterers as well. Hanson and Bussiere (1996) found that nonsexual violence was higher for rapists than nonrapists. Recidivism was found to increase with the number of previous sexual offenses, the selection of victim, and the selection of unrelated victims (Frisbie & Dondis, 1965). Recidivism (not necessarily just sexual) increased with follow-up time; there was no evidence that clinical treatment reduced risk for any type of offender, and no indication that recidivism rates varied by subtypes of offender (Harris et al., 1998).

Hanson and Bussiere (1996) conducted a meta-analysis on sex offender research. They found that the strongest predictors for sexual reoffense was deviant sexual preferences (e.g., sexual attraction to children) measured by plethysmography, early onset of sexual offending, age (which was negatively affected), never having been married, general deviant sexual practices, a history of prior sex offenses, and to a lesser extent, a history of diverse sexual crimes. Specifically, they found that deviant sexual preferences were related to sexual recidivism for both rapists and child molesters, but was strongest for the child molester. Having a male victim and having a victim who was a stranger were less important predictors, but were significantly related to sexual recidivism. They also found that sexual recidivism was associated with sexual deviance, and to a lesser extent, general criminological factors. The predictors of general and nonsexual recidivism were similar to general recidivism among nonsexual criminals (e.g., age, marital status, juvenile delinquency, antisocial personality disorder) (Champion, 1994).

Perhaps the single best predictor of sexual recidivism was deviant sexual interest to children as measured by phallometry (Hanson & Bussiere, 1998). Among sex offenders, nonsexual violent recidivism was predicted by many of the same variables that predicted recidivism among nonsexual offenders (Andrews & Bonita, 1994).

Hanson and Bussiere (1996) also found that the predictors of general recidivism were associated with a history of juvenile delinquency, prior offenses of any type, antisocial personality disorder, being single, youthfulness, and race. Dynamic factors included alcohol abuse problems or if the offenders were intoxicated at the time of the offense, those who deny their offenses or were unmotivated for treatment. They also found that incest offenders had the lowest rate for recidivism.

Hanson and Bussiere (1996) further found that sex offenders were more likely to report deviant sexual fantasies when upset (McKibben, Proulx, & Lusignan, 1994). To the extent to which sexual offenders are distressed does not predict recidivism, but sexual offenders appear to react deviantly when distressed.

Hanson and Harris (1998) indicated that negative emotional states were common precursors to reoffending for both rapists and child molesters (Pithers, Beal, Armstrong, & Petty, 1989; Pithers, Kashima, Cummings, Beal, & Buell, 1988). Other identified risk factors included cognitive distortions, low victim empathy, and social skills deficits. In addition, negative mood may be an acute risk factor for sexual reoffense (Proulx, McKibben, & Lusignan,1996; McKibben et al., 1994). In their studies, they found that deviant sexual fantasies tended to follow episodes in which the offenders felt stressed or upset.

Hanson and Harris (1998) further found that sexual recidivists tended to have poor social supports, attitudes that supported sexual assault, chronic antisocial lifestyles, and poor self-management styles. They also tended to have long histories of sexually deviant behaviors, prior treatment failures, low intelligence, and poor childhood environments. Supervising agents found that the recidivists cooperated poorly with supervision as indicated by being disengaged, manipulative, or absent. The recidivists tended to show an increase in anger and subjective distress just prior to reoffending. Therefore, the psychological symptoms appeared acute in nature rather than stable factors. Interestingly, both the nonoffenders and recidivists had equivalent levels of life stress and negative affect.

Hanson and Harris (1998) further found that the strongest predictor of violent recidivism (including sexual offenses) was what Quinsey et al. (1997) identified as "Dynamic antisociality." This included complaints about staff, no remorse, ignores previous violent acts, unrealistic discharge plans, negative attitudes, poor self-management, and lack of cooperation with supervision. This appears to apply to batterers as well.

In summary, the stable dynamic factors tended to be the most robust predictor for recidivism. Additional strong predictors of sexual recidivism included sexual deviancy (as measured on the penile plethysmography (PPG) or Abel Assessment for Sexual Interest (Abel, 1995) as well as a criminal lifestyle.

Hanson and Morton-Bourgon (2004) updated the previous findings of Hanson et al. (1996, 1998). In summary, they found that an antisocial orientation (e.g., unstable lifestyle, history of rule violations) was a significant predictor of both violent non-sexual recidivism and general recidivism as was deviant sexual interests. They identified new predictors of sexual violence, including sexual preoccupation, conflicts in intimate relationships, hostility, and emotional identification with children. They found that the general sexual recidivism rate is associated with two broad factors: a) deviant sexual interests, and b) antisocial orientation/lifestyle instability (Hanson & Bussiere, 1998; Quinsey, Lalumiere, Rice, & Harris, 1995; Roberts, Doren & Thornton, 2002). Sexual recidivism increased when deviant interests are present (Hanson & Bussiere, 1998). They found that crimes, including sexual crimes, are often associated with an antisocial lifestyle and lifestyle instability (crime-prone personality). These individuals tend to engage in a range of impulsive, reckless behavior (e.g., excessive drinking, frequent moves, fights, and unsafe work practices) (Caspi, Moffit, Silva, Stouthamer-Loeber, Krueger, & Schmutte, 1994; Gottfredson & Hirschi, 1990). Hostility and lifestyle instability are associated with sexual recidivism for both child molesters and rapists (Prentky, Knight, Lee, & Cerce, 1995; Rice, Quinsey & Harris, 1991).

Hanson and Morton-Bourgon (2004) commented that the newly identified dynamic factors for recidivism may be amenable to change (conflicts in relationships and hostility). They found that sexual recidivism rate to be approximately 14% over a 5-year period and that the sex offender would likely reoffend with a nonsexual versus sexual offense. Again, a caveat — these reoffense rates likely underestimate the actual reoffense rates because many sex offenses remain undetected. The sexual offenders with identifiable deviant sexual activities were the most likely to continue sexual offending. The evidence was strongest for sexual interest in children and for general paraphilias (e.g., voyeurism, exhibitionism). Phallometric assessments of sexual interest in rape, however was not significantly related to sexual recidivism.

Of significant interest is that sexual preoccupation (high rates of sexual interests and activities) was a risk predictor for sexual, violent, and general recidivism (Hanson & Morton-Bourgon, 2004). High rates of masturbation, pornography use, and impersonal sex were also found to be common among sex offenders (Kafka, 1997). The possible connections between sexual preoccupations and sexual offending may include a general lack of self-control, specific problems controlling sexual impulses, and a tendency to overvalue sex.

Hanson and Morton-Bourgon (2004) found that all forms of recidivism were predicted by an unstable, antisocial lifestyle, characterized by rule infractions, poor employment history, and reckless, impulsive behavior (Hanson-Morton-Bourgon, 2004; Bonta, Law, & Hanson, 1998; Gendreau, Little, & Goggin, 1996; Quinsey et al., 1995). They further found that antisocial orientation (e.g., noncompliance with supervision, probation violation) was an important predictor of violent and general recidivism and was one of the most significant predictors of sexual recidivism. They also found that lack of self-control may lead to a variety of criminal behavior (e.g., Gottfredson et al., 1990).

Hanson and Morton-Bourgon (2004) also reinforced that the lack of an intimate partner and conflicts in the offender's intimate relationship were associated with increased risk for sexual recidivism (also Hanson & Bussiere, 1998). Another intimacy factor for extrafamilial child molesters was an emotional identification with children, that is, feeling closer to children than adults, and those who have children as friends. They may even feel as though they are children themselves (Wilson, 1999). Intimacy deficits and persistent sexual offending are associated (Hanson & Motron-Bourgon, 2004; Malamuth, 2003; Ward, Hudson, Marshall, & Siegert, 1995). Hanson et al. (2004) indicated that child molesters may turn to a child for sexual contact because they lack the skills to relate to adults and probably feel lonely. They indicated that although low social skills and loneliness are common among sex offenders, neither of these factors was associated with increased risk for sexual recidivism. They hypothesize that social inadequacy and rejection are less important to offending than are the strategies used to address such problems.

Similarly, general psychological problems (e.g., depression, anxiety) are common among sex offenders (Raymond, Coleman, Ohlerking, Christensen, & Miner, 1999) but have little or no relationship to recidivism. Sex offenders are more likely, however, to respond to stress and depression through sexual acts and fantasies (Cortoni & Marshall, 2001). Although chronic negative mood is unrelated to recidivism, acute deterioration of mood is associated with increases in deviant fantasies among sex offenders (McKibben, Proulx, & Lusignan, 1994) and may indicate that the offender is at the gravest risk for reoffense (Hanson & Harris, 2000). It is the manner with which the sex offender copes with the problems that leads to recidivism, not the problems. The personal histories of sexual offenders frequently included physical, psychological and sexual abuse (Lee, Jackson, Pattison, & Ward, 2002; Smallbone & Dads, 1998). Hanson and Morton-Bourgon (2004) however, did not find that aversive childhood environments had any significant relationship to sexual reoffense. Attitudes tolerant of sexual assault showed a small relationship with sexual recidivism.

To reduce recidivism, intervention must address enduring characteristics which have been referred to as criminogenic needs (Andrews & Bonita, 2003), stable dynamic risk factors (Hanson & Harris, 2000) or causal psychological risk factors (Beech & Ward, in press). Hanson reported in his most recent study (Hanson & Morton-Bourgon, 2004) that a number of potentially dynamic risk factors were significantly related to sexual recidivism, which include sexual preoccupations, conflicts in intimate relationships, hostility, emotional identification with children, and attitudes tolerant of sexual assault. What they were not certain about is whether changes on these factors are associated with reductions in recidivism. They also found a smaller significance of having deviant sexual interest in children as measured by phallometry, though still is a predictor. Employment instability also significantly predicted sexual recidivism.

Doren (1998) found that much of the research on sexual offense recidivism has serious shortcomings that limit the application of the findings. Reconviction rates are often used as the source of reoffense, but reconviction rates do not represent the true reoffense rates (e.g., Furby et al., 1989). Remember that most sexual offenses are never reported or are reported years following the offense. Unofficial data of new sexual offending (e.g., arrests, probation or parole reports, self-report, and victim report) increased the measure of sexual recidivism 170% (Barabree & Marshall, 1988). The major flaw in research tends to be the far too short period of time at risk (e.g., how many months is the offender able to access new victims) and definitions of reoffense. Therefore all current professional research appears to potentially underestimate the accurate base rate for sexual recidivism because the studies end too soon. Many of the studies identified subjects wrongly as nonrecidivists when in fact they did reoffend. Doren referenced Prentky et al. (in press) who found that 52% of the child molesters in their study reoffended within the 25-year follow-up period, but even this tends to underestimate the true base rate for sexual reoffense for child molesters. Hanson et al. (1995) suggested that the accurate base line for extrafamilial child molesters to be approximately 44 to 52%, again, a conservative percentage. Other research produced similar findings. Extrafamilial child molesters reoffend at a conservative estimate of approximately 52%.

Prenkty et al. (in press) found that 39% of the rapists in their study reoffended with a new sexual charge over 25 years; but the corresponding reconviction rate was only 24%. This highlights how reconviction rates grossly underestimate reoffense rates. The reoffense rates assume that all rapists who reoffend are caught, which is far from reality. Prentky et al. (1997) indicated that rapists tend to reoffend at a rate of at least 39% over a 25 year period; this likely underestimates the true rate of rapist-recidivism by as much as 62%.

Rice and Harris (1997) and Quinsey, Rice, and Harris (1995) found that mixed offenders (those with both child and adult victims) exhibit nearly total recidivism for violent offenses, including sexual offenses.

Hanson, Broom, and Stephenson (2004) found that the program they examined did not appear to be effective in reducing recidivism. The recidivism rates between the treated and untreated sex offenders were similar and no significant treatment effect was obtained.

Craig, Browne, Stringer, and Beech (2004) indicated that several static factors have repeatedly been demonstrated to predict sexual recidivism: past criminal history (Rice et al., 1990; Proulx et al., 1997), prior sex offenses (Quinsey et al., 1995; Hanson et al., 1995; Hanson et al., 1993; Hanson & Bussiere, 1998), psychopathy or personality disorder (Rice et al., 1990; Hanson & Harris, 1998; Seto & Barbaree, 1999), age and time spent in custody (Broadhurst & Mailer, 1992; Browne et al., 1998), and paraphilias and deviant sexual interests (Quinsey et al., 1995; Proulx et al., 1997; Hanson & Harris, 1998; Hanson & Bussiere, 1998.

It was reported by Langstrom and Grann (2000) that the factors predicting general recidivism included early conduct disorder, previous convictions, psychopathy and the use of death threats or weapons at the index sex offense; the predictors of sexual recidivism included previous sex offenses, poor social skills, male victims, and two or more victims involved in the index offense.

Dynamic factors are those amenable to change. These include unemployment, substance abuse, criminal lifestyles or negative social influences, and impulsivity (poor self-control/management), which have been associated with sexual recidivism (Hanson & Harris, 1998, 2000; Serin, Mailloux, & Malcolm, 2001). Affective disorders (e.g., subjective distress and anger, anxiety or depressive disorders), poor social skills, poor victim empathy, cognitive distortions, and high levels of deviancy have all been positively associated with dynamic risk in sexual offenses (Hanson & Harris, 1998, 2000). Discontinuation of treatment and rejecting treatment and length of treatment programs were also considered dynamic factors for risk of sexual recidivism (Fisher et al., 2000). Delinquent behavior during treatment and poor treatment compliance and cooperation were also related to increased risk of sexual recidivism (Browne et al., 1998; Hanson & Harris, 1998).

FACTORS *NOT* RELATED TO SEXUAL RECIDIVISM

Hanson and Bussiere (1996) identified several factors that *do not appear* to be related to sexual recidivism, which include anxiety, depression, low self-esteem; substance abuse problems, having a history of being sexually abused, and general psychological problems (including anxiety, low self-esteem, and depression). Increasing the subjective well-being of the sex offender was found to be unlikely to reduce their recidivism rate (Hanson et al., 1993). They also found that there was no evidence to suggest that treatment reduced sexual recidivism. Hanson and Morton-Bourgon (2004) identified several additional factors in their subsequent study that were not related to sexual recidivism. These included denial of offense, low victim empathy, and low motivation for treatment.

In general, they found that evaluations of treatment progress showed little if any relationship to recidivism.

PSYCHOPATHIC SEX OFFENDERS

Adult psychopathic rapists and psychopathic child molesters who have been treated are more likely to recidivate violent (80% at 6 years post-release) than treated nonpsychopathic rapists and child molesters (20% at 6 years post-release) (Quinsey, Lalumiere, Rice, & Harris, 1995). Serin, Mallioux, and Malcolm (2001) found that sex offenders with high psychopathy scores showed the highest recidivism rate. Seto and Barbaree (1999) found that an interaction between psychopathy and treatment behavior significantly predicted both sexual and nonsexual recidivism. Interestingly, men who were more psychopathic and were rated as showing greater treatment progress and change by their therapists were more than five times as likely to commit a new serious offense.

Quinsey, Harris, Rice, and Cormier (1998b) found that psychopathy was a significant and powerful predictor for sexual reoffending along with a measured sexual deviance on the penile plethysmograph.

SUMMARY

By far the most effective treatment intervention for batterers and sexual offenders is group therapy with a cognitive-behavioral orientation. Psychopharmacology is also indicated for many offenders. Though many batterers and sex offenders have psychopathology, the mental disorders never, ever excuses the offender of having 100% responsibility for their behavior. The idea that a high-risk sex offender can be turned into a low risk sex offender thorough treatment has no significant empirical evidence (Rice et al., 1998). The need for intensive community supervision is imperative for violent offenders.

Although the research on gay and lesbian relationship violence is scarce, most data suggest that the percentage of gays and lesbians that experience interpersonal violence is similar to that of heterosexuals. In addition, the typologies of batterers did not appear to differ based on sexual orientation.

It is important to recognize that there will always be offender outliers who will not present with consistent characteristics or motivation for treatment. Some may present with high-risk reoffense factors and yet not reoffend even without treatment.

I assert that there is little significant difference between the batterer and sexual offender when is comes to risk prediction, personality traits, and the importance of all of the above identified risk factors. Both batterers and sex offenders utilize pornography and sexually deviant fantasies, both engage in psychological, physical and sexual abusive behavior, and both share similar patterns of recidivism. The question of the true impact of treatment intervention remains unclear and will require more defined and longitudinal research.

RELATIONSHIP VIOLENCE/DOMESTIC ABUSE/SEX OFFENDER PROGRAM GOALS

	Target Problem Area Addressed	Goals
1.	Program Participation	Attend all therapy sessions as schedule. Actively participate in all therapy sessions. Complete all assignments given on time. Keep financial accounts.
2.	Physically, Psychologically, and/or Sexually Abusive Behavior	Admit all incidents of neglect, psychological, physical, and/or sexually abusive behavior toward others. Admit all incidents of illegal behavior (share your criminal behavior history). Demonstrate a detailed understanding of your own abuse cycle. Learn and regularly utilize a method of monitoring violent fantasies and behavior. Demonstrate an understanding of the types and areas of abuse. Share own relationship and sexual history, including abuse and sexual victimization. Demonstrate an understanding of the role of cognitions in your abusive and offense behavior. Demonstrate an understanding of your motives underlying your abusive/offense behavior. Eliminate the pairing of anger and frustration with violence and sexual behavior, and understand the role of emotions in your offense and violent behavior. Develop and demonstrate empathy toward your victim's experience. Develop and demonstrate an understanding of the role of shame and guilt in offense/abusive behavior.
3.	Relationship Skills: Communication, Intimacy, and Sexuality	Examine family of origin and primary relationship and resolve significant conflicts that you are a part of. Demonstrate appropriate coping skills/problem-solving skills relevant to anger, stress, deviant urges, and deviant thought, and to resolving conflicts with others. Demonstrate comfort with appropriate problem resolution skills in relationships with significant other, co-workers, family members, and others. Cease abusive behavior toward others. Maintain appropriate boundaries with others, including children. Develop and demonstrate an understanding of healthy, appropriate relationship and sexuality needs and skills. Demonstrate appropriate parenting skills.
4.	Appropriate Sexuality	Demonstrate comfort with own sexuality through use of masturbation, dating, and/or responsible sexual contact with romantic partner. Learn and regularly utilize a method of monitoring appropriate and deviant sexual fantasies and behavior. Reinforce only appropriate sexual fantasies and behavior, and demonstrate effective methods to control deviance.
5.	Responsible Behavior	Abide by all conditions of probation/supervised release and treatment. Demonstrate honesty in therapy sessions and in the community. Demonstrate a willingness to make amends to victims as necessary and as deemed appropriate. Utilize consistent feedback from therapist, group members, probation/supervising agent prior to acting on any major life decisions, including, but not limited to, employment, sexual contact with others, changing relationship status, financial commitments, etc. Demonstrate use of conscience in relationships. Demonstrate ability and willingness to consistently control deviant and inappropriate behavior. Demonstrate good judgment in everyday decisions and situations. Maintain financial support of self and others (including child and spousal support, rent/mortgage payments, car payments, treatment expenses, etc.). Develop and implement a plan to repay the victim and community.

RELATIONSHIP VIOLENCE/DOMESTIC ABUSE/SEX OFFENDER PROGRAM GOALS (CONTINUED)

	Target Problem Area Addressed	Goals
6.	Relapse Prevention	Develop and maintain an adult support network that includes both men and women. Demonstrate appropriate interventions for deviant and unhealthy cognitions. Demonstrate resolution to own history of abuse and sexual victimization. Develop and employ a reoffense prevention plan. Develop and utilize appropriate stress management plan. Demonstrate an understanding of the impact and consequences of your assaultive/abusive behavior on your victims and others. Follow any recommendations for aftercare or other supportive adjunctive or post-treatment interventions or services.
7.	Other Relevant Problem Areas	Resolve marital/primary relationship problems, including couples or marital therapy when required. Resolve problems with substance abuse and mental health issues as required, including assessment and treatment, support groups, following any prescribed behavior, avoiding any inappropriate behavior, and taking medication as prescribed. Maintain appropriate health and hygiene. Develop and engage in personal interests, hobbies, and recreational activities.

Additional goals may be added at any time to address specific individual needs and circumstances.

Source: Scott A. Johnson, M.A., L.P., DABPS, DAACCE, ACFE , 2001/2002©, Services for Relationship Violence, Inc. Mending Path Relationship Center, Inc.

RELATIONSHIP VIOLENCE AND DOMESTIC ABUSE TREATMENT PROGRAM

MENDING PATH RELATIONSHIP CENTER, INC.

(FORMERLY SERVICES FOR RELATIONSHIP VIOLENCE, INC.)

Scott A. Johnson, M.A., L.P., DABPS, DAACCE, ACFE

Following is the weekly group topic outline. This outline may change at anytime without notice. (The numbers in parentheses indicate which of the 45 goals are addressed within each weekly group. Other goals may be addressed as well, but these are the key goals addressed in the specified week.)

Week	Topic
1	Introduction (5, 6, 8, and 10).
2	Problem Solving and Problem Resolution Skills: Discussion of healthy and unhealthy problem-solving techniques (14, 17, 18, 20, 22, 30, 31, 35, 36, 38, and 39).
3	Discussion and examination of physical force and hands-on/touch-offenses and discussion of physical assault and rape (5, 8).
4	Discussion and examination of psychological force, psychological abuse, and the impact on the victim (8).
5	Discussion and examination of psychological force and psychological abuse continued; begin presentations of Offense Grid and examination of offender dynamics and reasons why offense occurred (10, 12, 13).
6	Discussion of abuse-supporting thoughts and a discussion of the continuum of abuse and offense cycle (9, 11).
7	Discussion of cognitive factors involved in offense behavior, including defense mechanisms, cognitive distortions, and trigger events/stressors (9).
8	Continuation of discussion of impact of cognitive distortions and application of rational emotive therapy concepts to better understand offense (12, 14, 16).
9	Offense Cycle and Boundaries Explored: A detailed discussion of the offense cycle and the importance of maintaining appropriate boundaries (11, 12, 20).
10	Empathy Explored: Understanding the importance of developing a clear knowledge of emotions and behavior indicative of empathy; examining the victim's experience of the offender's lack of empathy (15, 16, 28, 33, 34, 40).
11	Discussion of intimacy and emotions: Understanding the importance of clearly identifying specific emotions experienced and what intimacy needs are being fulfilled (13, 14, 15, 16, 21, 23, and 24).
12	Discussion of how abusive behavior and thinking are learned (10, 15, 16, 37).
13	Discussion of the characteristics indicative of healthy and abusive relationships (21).
14	Relationship problems explored: Discussion of problems experienced in relationships and how to effectively address and overcome the problems (21, 22, 23, and 30).
15	Communication explored: Discussion of healthy and unhealthy communication patterns (30, 32).
16	Impact of abuse on victims: Examining the effects of abuse on all impacted, including relatives, children, and others close to both the offender and/or victims (17, 18).
17	Reoffense Prevention Plan: A discussion to help in the development of a realistic reoffense prevention plan (24, 25, 34, 38).

Individual sessions: Prepare for goals 5, 6, 8, 9, 10, and 37; meet with partner. Ongoing goals: 7, 8, 13, 14, 17–20, 22, 23, 25–39, and 40–44.

Source: Scott A. Johnson, M.A., L.P., DABPS, DAACCE, ACFE , 2001/2002©, Services for Relationship Violence, Inc. Mending Path Relationship Center, Inc.

41 Diagnosing the Abuser and Sex Offender

COMORBIDITY

It is imperative that the offender be assessed and diagnosed with all diagnosis that meets the offender criteria. In the *DSM-IV-TR* (APA, 2000), it often states in each diagnostic category that if all of the criteria are met for a specific diagnosis, then all should be given.

It is important to understand that mental health disorders may occur as solo diagnosis. However, the comorbidity of mental health disorders is far more common than once thought. It is not uncommon to find a mental health patient presenting with multiple diagnoses, and offenders as well. It is my professional opinion that when conducting a forensic assessment or intake assessment that any and all diagnoses that a batterer or sex offender meets the *DSM-IV-TR* diagnostic criteria for should be adequately given. It is not the clinician's job to give the lesser of the diagnoses, but rather to provide a neutral and thorough diagnostic assessment. I have seen far too many clinical and forensic assessments on batterers and sex offenders that either lack appropriate documentation of relevant history and testing necessary to adequately and appropriately diagnose an offender or where the clinician fails to give all relevant diagnosis because they do not want to "over diagnose" or "make the individual look bad." I have heard clinicians, including those conducting forensic assessments, comment that they did not want to give all diagnoses that the individual met criteria for because they did not want to portray the individual in a negative manner. We as clinicians do not create or cause the symptoms to be present. We are observers and gather information from official sources, testing, and interviews. Any symptoms that the offender presents with are his and his alone, we do not have the right to exacerbate symptoms to give extra diagnoses any more than we have the right to minimize or ignore symptoms to avoid diagnosing.

Ethically, it is really not an option. Any clinical or forensic assessment should include all diagnosis that the offender meets *DSV-IV-TR* criteria for, without exception. Even if it means that the batterer or sex offender has three personality disorders, substance abuse, and two or more paraphilias, if they meet the diagnostic criteria for each diagnosis, then each and every diagnosis should be made and put into the offender's report, treatment intake, forensic assessment, and treatment notes. An accurate picture of the offender's personality and disorders is important to ensure that adequate safety measures are in place to protect the public, the victims, and the offender. In addition, it is difficult to recommend the most accurate and appropriate treatment intervention without a clear understanding of the diagnoses, given that the diagnoses describe the mental disorders of the offender. Also, appropriate diagnosis allows for more accurate offender typologizing to occur.

It has now been accepted that in certain segments of the psychiatric community, comorbidity of mental illness is the rule rather than the exception (Nemeroff, 2002; West, Zarin, & Pincus, 1997). Evidence suggests that patients that have schizophrenia or schizoaffective disorder have a significant occurrence of comorbid alcoholism or substance abuse (Buckley, 1998; Kamali et al., 2001). Additional evidence has shown that the presence of comorbid substance abuse among psychotic patients may also increase the risk of accompanying physical–medical disorders (e.g., cardiac problems, skin infections, respiratory conditions (Dickey et al., 2000), which may further complicate the assessment or treatment process. A high percentage of adults diagnosed as having Attention Deficit Hyperactivity Disorders (ADHD) had a history of at least one lifetime comorbid disorder, and a small percentage of these adults still reported a present comorbid disorder (Spencer et al., 2002). Similarly, adolescents and adults with Obsessive-Compulsive Disorder indicated the presence of a variety of comorbid conditions, e.g., depression, anxiety, and Post-Traumatic Stress Disorder (PTSD) (Gershuny et al., 2002; Geller et al., 1995; Pinccinelli et al., 1995).

It has also been found that in addition to accompanying Obsessive-Compulsive Disorder, major depression had its own set of reported comorbid disorders, including Conduct Disorders in depressed children (Knapp et al., 2002) and Panic Disorder with Agoraphobia in depressed adults (Posternak & Zimmerman, 2002). Similarly, patients diagnosed with PTSD demonstrate moderate levels of both comorbid anxiety and major depression (Davidson et al., 2001). Patients diagnosed with alcohol dependency often present with comorbid major depression (Salloum et al., 1998; Prescott, Aggen, & Kendler, 2000). However, there is no causal relationship between these disorders. Moderate levels of comorbid Panic Disorder, Obsessive-Compulsive Disorder, and Anxiety have been reported with patients with Bi-Polar I (Pini et al., 1997). Patients with Bi-Polar II disorder tended to have comorbid personality disorders, substance abuse, and anxiety (MacQueen & Young, 2001; Vieta et al., 2000).

Common Diagnosis for Batterers and Sex Offenders

Common diagnoses for abusers include the Adjustment Disorders (see caveat below), Mood Disorders, Sadism, Intermittent Explosive Disorder, Antisocial Personality Disorder, other Personality Disorders, Delusional and Psychotic Disorders, and Substance Abuse and Dependencies. Of course, other diagnoses may apply as well. Some of the more common diagnoses are discussed below.

Personality Disorders

Three of the more commonly occurring Personality Disorders seen with batterers and sex offenders are Antisocial Personality Disorder, Narcissistic Personality Disorder, Borderline Personality Disorder, and Dependent Personality Disorder. Of course, it is also common for batterers and sex offenders to have a Personality Disorder, NOS — meaning *not otherwise specified* — and is used when the individual presents with symptoms of specific personality disorders, but not enough in any single category to warrant a specific personality disorder diagnosis.

"Personality traits are enduring patterns of perceiving, relating to, and thinking about the environment and oneself that are exhibited in a wide range of social and personal context." (American Psychiatric Association, 2000). The *DSM-IV-TR* (APA, 2000) goes on to state that "only when the personality traits are inflexible and maladaptive and cause significant functional impairment or subjective distress do they constitute personality disorders." Batterers were found to be diagnosed as having personality disorders 35 to 85% of the time (Dutton & Starzomski, 1993; Hamberger & Hastings, 1986; O'Leary, 1993). Personality disorders involve behaviors that can impair an offender's interaction with others, how the offender understands his offense behavior, and how likely the offender is to demonstrate resilient change.

Antisocial Personality Disorder (APD) involves "… a pervasive pattern of disregard for, and violation of, the rights of others…" (APA, 2000). This pattern generally begins in childhood or adolescence. However, at times the only source available to verify childhood Conduct Disorder, Oppositional Defiant Disorder, or the presence of any antisocial behavior is the offender. Of course, when only the offender is present to offer history information, the veracity of his claims is always suspect. I find it appropriate to diagnose Antisocial Personality Disorder without information indicating childhood or adolescent problems if the offender currently clearly meets the diagnostic criteria for for this disorder and I will label it *Antisocial Personality Disorder, Adult Onset*. The *DSM-IV-TR* (APA, 2000) lists the following diagnostic criteria for Antisocial Personality Disorder: a pervasive pattern of disregard for and violation of the rights of others since age 15, as indicated by at least three of the following: (1) failure to conform to social norms with respect for lawful behaviors; (2) deceitfulness (e.g., lying, use of aliases, conning others for personal profit); (3) impulsivity or failure to plan ahead; (4) irritability and aggressiveness as indicated by repeated physical fights or assaults; (5) reckless disregard for the safety of self or others (e.g., reckless driving); (6) consistent irresponsibility (e.g., repeated failures to sustain consistent work or honor financial obligations; and (7) lack of remorse (e.g., being indifferent to or rationalizing having hurt, mistreated, or stolen from another). Gottman et al. (1995) stated that Type 1 batterers are likely to have this diagnosis. In the antisocial male, there is a 25% prevalence of alcoholism (Brown, 2005). There appears to be support that individuals with alcohol and/or drug abuse or dependency also meet the criteria for Antisocial Personality Disorder.

Narcissistic Personality Disorder involves a "...pervasive pattern of grandiosity (in fantasy or behavior), need for admiration, and lack of empathy..." (APA, 2000). According to the *DSM-IV-TR*, this would be evident by at least five of the following: (1) a grandiose sense of self-importance (e.g., exaggerates achievements, talents, expects to be recognized as superior without expected achievements); (2) preoccupied with fantasies of unlimited success, power, brilliance, beauty, ideal love; (3) believes that he is "special" and unique and can only be understood by or associate with others who are equally "special;" (4) requires excessive admiration; (5) has a sense of entitlement; (6) is interpersonally exploitive; (7) lacks empathy (e.g., unwilling to recognize or identify with the feelings and needs of others; (8) is often envious of others or believes that others are envious of him; and (9) shows arrogant, haughty behavior or attitudes. Gottman and colleagues' Type 1 batterers are likely to present with this disorder. In general, individuals with Narcissistic Personality Disorders are polydrug users with a preference for stimulants (Brown, 2005).

Dependent Personality Disorder involves a "... pervasive and excessive need to be taken care of that leads to submission and clinging behavior and fears of separation..." (APA, 2000). The *DSM-IV-TR* states that this would be evident by at least five of the following: (1) difficulty making everyday decisions without an excessive amount of advice and reassurance from others; (2) needs others to assume responsibility for most major areas of his life; (3) has difficulty expressing disagreement with others because of fear of loss of support or approval; (4) has difficulty initiating projects or doing things on his own (e.g., lacks self-confidence in own judgment, not simply a lack of motivation or energy); (5) goes to excessive lengths to obtain nurturance and support from others, even to the point of doing unpleasant things; (6) feels uncomfortable or helpless when alone because of exaggerated fears of being unable to care for himself; (7) urgently seeks another relationship as a source of care and support when a close relationship ends; and (8) is unrealistically preoccupied with fears of being left to take care of himself. Gottman et al. (1995) found that many Type 2 batterers present with Dependent Personality Disorder or with dependent features.

Borderline Personality Disorder involves "...a pervasive pattern of instability or interpersonal relationships, self-image, and marked impulsivity (APA, 2000) per the *DMS-IV-TR* (APA, 2000). According to the *DMS-IV-TR*, this would be evident by the presence of at least five of the following: (1) frantic efforts to avoid real or imagined abandonment; (2) a pattern of unstable and intense interpersonal relationships characterized by alternating between extremes of idealization and devaluation; (3) identity disturbance; (4) impulsivity in at least two areas that are potentially self-damaging (e.g., spending, sex, substance abuse, reckless driving, binge eating); (5) recurrent suicidal behavior, gestures, or threats, or self-mutilating behavior; (6) affective instability due to a marked reactivity of mood; (7) chronic feelings of emptiness; (8) inappropriate, intense anger or difficulty controlling anger (e.g., frequent temper outbursts, constant anger, recurrent physical assaults); and (9) transient, stress-related paranoid ideation or severe dissociative symptoms. Brown (2005) found that individuals with Borderline Personality Disorder often use alcohol and drugs in self destructive and unpredictable ways.

Personality Disorder Not Otherwise Specified (NOS) involves possessing features of several of the Personality Disorders, though not enough of any specific disorder to warrant a specific diagnosis. However, the traits must still cause clinically significant distress or impairment in one or more areas of the individual's life. In addition, an individual may be diagnosed as having a

Personality Disorder NOS that is not specified in the *DSM-IV-TR* (e.g., Sadistic Personality Disorder, Sadistic Traits, Passive-Aggressive Personality Disorder).

Sadism and Sadistic Personality Disorder

If the perpetrator meets the diagnostic criteria for sadism, it is wise and ethically appropriate to diagnose it. Though not adopted for the *DSM-IV* or *DSM-IV-TR*, Sadistic Personality Disorder is a genuine mental health diagnosis. It appears appropriate to use Personality Disorder, NOS, with sadistic Features/Traits. Many batterers and sex offenders are sadistic. The primary difference between Sadism (generally referred to as Sexual Sadism in the *DSM-IV-TR*) and Sadistic Personality Disorder is that the latter would exclude engaging in sexually sadistic behavior. I often see forensic and treatment assessments written about abusers and sexual offenders who clearly are sadistic, but the clinician failed to diagnose the disorder. I believe that most clinicians simply do not adequately understand sadism.

Sadism involves causing pain, suffering, or humiliation to the victim, and experiencing arousal. "Experiencing arousal" is inferred when the perpetrator continues to engage in the painful or humiliating behavior even after being aware of the victim's suffering or humiliation, and continuing to engage in the physical abuse, verbal abuse, or sexual contact.

Perpetrators are not sadistic when they hit their victim to gain compliance, and then cease the physically assaultive behavior once they gain compliance or cooperation from the victim. If the perpetrator engages in gratuitous violence, this is not necessarily sadistic behavior either. Gratuitous violence is utilizing more assaultive behavior than was necessary to gain victim compliance; however the violence itself is not arousing to the perpetrator. If the victim was repeatedly punched in the face and the offender ceased raping her, for example, that would not be a sadistic perpetrator. Had the fact that the victim had a bloody face decreased the perpetrator's arousal and ceased the assault, then the behavior was not sadistic in nature. A sadistic perpetrator would continue to rape despite the victim's bloody face, and would likely continue to physically assault the victim during and after the rape. A sadistic abuser would likely continue to physically assault the victim even after the primary assault is finished.

"Continued assault" does not necessarily mean continued punching or slapping. During the course of a physical or sexual assault, a sadistic perpetrator would likely change the type of assaultive behavior, engaging in various demeaning or injurious acts. For example, during the sexual assault, the sadistic rapist might begin to pinch the nipples of his victim, insert objects into the victim's orifices, and later may urinate on the victim. Humiliating behavior really has a very global definition. Any physical behavior or verbal comment made can be potentially humiliating to a victim. An example would be an abuser who repeatedly punches (or repeatedly slaps, hits, kicks, etc) his victim even after the victim is compliant or no longer able to physically resist.

Summary of Personality Disorders

In summary, numerous studies have demonstrated the resiliency of personality-disordered traits in offenders (Meyer & Deitsch, 1996; Carden, 1994; Dutton, 1998; Hart, Dutton, & Newlove, 1993). Personality disorders often have their etiology in childhood and often present as chronic, pervasive, enduring, and inflexible patterns and traits of perceiving and relating to the world. Personality disorders are likely to be life-long disorders with a very poor prognosis for change. Many personality-disordered people are rigid, suspicious, and oppositional individuals who are not likely to engage in a therapeutic process voluntarily nor are they likely to significantly change their problem cognitions or behavior. It is recommended that an individual is off all drugs, including alcohol, at least 1 year to accurately assess for personality disorders and other possible impairments (Brown, 2005). Without allowing for ample time for the effects of the alcohol and/or drugs to clear the system and for the brain to return to premorbid functioning, personality disorders may be erroneously under- or over-diagnosed

PSYCHOPATHY

Psychopathy involves two sets of factors (Hare, 1991). Factor 1 includes affective and interpersonal traits, such as callousness, lack of empathy, shallow affect, and manipulativeness. Factor 1 has been referred to as *aggressive narcissism* by Meloy (1992). Cleckley (1941/1988) refers to these as the primary personality features of psychopaths. The second factor includes behaviors indicative of instability and antisocial lifestyle, such as poor anger control, parasitic lifestyle, impulsivity, and irresponsible behavior. In summary, Factor 2 relates to chronic antisocial behavior (Hare et al., 1990). Meloy (2002) indicated that three factors have been identified that further delineated the primary psychopath as: an arrogant, interpersonal style, a deceitful interpersonal style, and a deficient affective experience (Cooke & Michie, 2001).

Several authors (Vassileva et al., 2005; Harpur, Hare, & Hakstian, 1989; Kossen et al., 1997) have indicated that Factor 1 correlates more highly with narcissism, with some emotional anomalies, and with the distinctive interpersonal style of the psychopathy more than Factor 2. By contrast, Factor 2 is more strongly correlated to Antisocial Personality Disorder and to alcohol and substance use disorders than is Factor 1 (Harpur et al., 1989; Smith & Newman, 1990).

Vassileva et al. (2005) indicated that "true" or primary psychopaths are usually characterized by a lack of anxiety, whereas secondary psychopaths may appear psychopathic, yet experience substantial anxiety and negative effect which may contribute to their impulsive antisocial behavior (Blackburn, 1975; Karpman, 1941; Lykken, 1995). Vassileva et al. (2005) found that secondary psychopaths were characterized by severe alcohol and drug dependence, together with elevated scores on the social deviance factor of psychopathy (PCL-R Factor 2). They also found that this group of psychopaths presented with significantly elevated levels of anxiety. They also found that primary psychopaths were characterized by elevated scores on the personality features of psychopathy (PCL-R Factor 1), in conjunction with interpersonal and affective behaviors indicative of psychopathy. Further, Factor 1 traits tended to predict violent recidivism better than Factor 2 traits, and that primary psychopaths tended to have more convictions for crimes of assault that the secondary psychopaths.

Hare (1995b) estimates that 25 to 35% of men in domestic abuser programs are psychopathic. Gottman et al. (1995) stated that Type 1 batterers are more likely to be psychopathic. Hare (1982) found that psychopathic individuals show a heart rate decrease when anticipating aversive stimuli. Approximately 50 to 75% of prison inmates meet the diagnostic criteria for Antisocial Personality Disorder, only 20 to 30% will meet the diagnostic criteria for Psychopathy as defined by Hare (1991) (Hare, Hart, & Harpur, 1991; Harpur & Hare, 1994).

Forth and Kroner (1994) labeled approximately 26% of convicted rapists, 35% of serial rapists and murderers, 18% of mixed sex offenders, and 5% of incest offenders as psychopaths. Prentky and Knight (1991) found higher percentages of psychopaths among smaller samples of rapists (45%) and child molesters (30%). Meloy (2000) found higher rates of psychopaths among sexual homicides.

It is imperative to accurately assess and diagnose psychopathy as the literature is ripe with studies (e.g., Quinsey et al., 1995; Serin, Mallioux, & Malcolm, 2001; Seto & Barbaree, 1999; Quinsey et al., 1998b) that find that psychopaths recidivate at much higher rates than nonpsychopaths and with much more violence.

Meloy (2002) indicated that "sexual predation, and predatory violence in general, are served by a number of characteristics of the psychopath: (1) low levels of anxiety and autonomic hyporeactivity, (2) chronic emotional detachment that subserves a lack of empathy, (3) sensation seeking, (4) fearlessness, (5) manipulativeness, (6) deception, (7) criminal versatility, (8) lack of guilt or remorse, and (9) "shallow affect." Meloy referenced Hazelwood and Warren (2000), who indicated that sexual offenders involved impulsive or ritualistic types. They believe that impulsive sexual offenders would more likely be psychopathic given their impulsivity, risk taking, cruelty, minimal use of fantasy, and generally antisocial history. Meloy indicated, however, that the ritualistic offender (which includes the sexual sadist) is just as likely to be psychopathic, but in a more refined, deliberate, erotically channeled, and fantasy-motivated manner.

It is my professional opinion that most clinicians also do not adequately understand that psychopathy is not the same as Antisocial Personality Disorder. It appears that the most appropriate training to diagnose psychopathy would come from Dr. Robert Hare (1991, 1995), and utilizing his tool, the PCL-R (the Psychopathy Check List-Revised) (Hare, 1991). Psychopaths are incapable of experiencing genuine emotion and are, therefore, exceptionally more dangerous. They tend to fool most clinicians regardless of how experienced or well trained the clinician may be. However, clinicians trained in assessing psychopathy are far less likely to be fooled.

Intermittent Explosive Disorder

Intermittent Explosive Disorder involves (per the *DSM-IV-TR*):

1. Several discrete episodes of failure to resist aggressive impulses that results in serious assaultive acts or destruction of property.
2. The degree of aggressiveness or destructiveness is grossly out of proportion to any precipitating psychosocial stressors.
3. The aggressive episodes are not better accounted for by other diagnosis.

However, this can and frequently exists with other diagnosis. Note that destruction of property is included here and is often overlooked by inexperienced and poorly trained clinicians. So even a batterer or sex offender who has physically abused on only a handful of occasions may be appropriately diagnosed as having Intermittent Explosive Disorder. In addition, batterers who throw objects or destroy objects also qualify for diagnosis of Intermittent Explosive Disorder. It is not necessary to ascertain that the offenders "intended to destroy" the property damaged, simply that the offender did, in fact, damage property.

Mood Disorders

Depressive Disorder involves a period of at least 2 weeks during which there is either a depressed mood or the loss of interest or pleasure in nearly all activities, and the presence of at least four other symptoms listed in the *DSM-IV-TR*. Mood Disorders involve a disturbance in mood. There are several types of mood disorders. Major Depressive Disorder involves having at least five or more major depressive episode:

1. Depressed mood most of the day.
2. Significantly diminished interest or pleasure in most/all activities.
3. Significant weight loss/gain or loss of/increase in appetite when not dieting.
4. Insomnia or hypersomnia.
5. Psychomotor agitation or retardation.
6. Fatigue or loss of energy.
7. Feelings of worthlessness or excessive or inappropriate guilt.
8. Diminished ability to think or concentrate.
9. Recurrent thoughts of death or suicidal ideation, with or without a plan.

The symptoms must cause significant impairment in social, occupational, or other important areas of the individual's life, and not be a result of other conditions. It is not common for people with depressive disorders to engage in violent behavior. However, underlying depression impacts self-esteem and a sense of self-worth, which may make offending with battering or a sexual offense more likely with offenders who are already predisposed to engaging in violent behavior.

Bi-Polar Disorder (otherwise known as Mania or Manic Depression) is described by *DSM-IV-TR* as involving "… a distinct period in which there is an abnormally and persistently elevated, expansive, or irritable mood" and these additional symptoms must be present:

1. Inflated self-esteem or grandiosity.
2. Decreased need for sleep.
3. More talkative or pressured speech.
4. Flight of ideas or racing thoughts.
5. Distractibility.
6. Increase in goal-directed activity.
7. Excessive involvement in pleasurable activities that have a high potential for painful consequences.

During periods of Manic episodes, individuals may become significantly obsessed, ritualistic, engage in activities that result in high degrees of pleasure (e.g., spending, sexual promiscuity), and violent behavior. The violent behavior could be either an impulsive and random act or a highly planned and obsessively planned attack. Never should the diagnosis of any type of Bi-Polar Disorder lessen the degree of responsibility on the offender. Offenders with a history of Bi-Polar Disorders who fail to take prescription medication as prescribed and who engage in violent behavior should be held criminally responsible as well. The past is the best predictor of future behavior, and failure to comply with medication suggests that individuals will at some point fail to take their medication again, and if they engaged in violent behavior once, they are far more likely to do so again.

I do not believe that batterers and sex offenders with diagnosis of depressive or anxiety disorders or substance abuse/dependency actually abuse because of these disorders. There is no empirical support that indicates otherwise. For example, the vast majority of individuals with depressive disorders and/or anxiety disorders do not engage in violence. The same can be said for those with chemical abuse/dependency issues. There are thousands of individuals who present with anxiety disorders, depressive disorders, and substance abuse/dependency who will not become batterers or sex offenders.

Paraphilias

Paraphilias are "recurrent, intense sexually arousing fantasies, sexual urges, or behaviors generally involving (1) nonhuman objects, (2) the suffering or humiliation of oneself or one's partner, or (3) children or other nonconsenting persons that occur over a period of at least 6 months" (APA, 2000). The *DSM-IV-TR* also indicates that the diagnosis of Pedophilia, Voyeurism, Exhibitionism, and Frotteurism "…can be made if the person has acted on these urges or the urges or sexual fantasies cause marked distress or interpersonal difficulties." It is assumed that a person who has acted on paraphilic urges or fantasies will probably have been experiencing the urges or fantasies for at least 6 months. Only the offender could tell you how long they experienced the urges or fantasies, and offenders are not the most forthcoming individuals with sharing honest information. However, regardless of how long the individual has experienced the deviant urges or fantasies, the fact that they have acted upon them is enough to warrant the paraphilic diagnosis.

"For Sexual Sadism, the diagnosis is made if the person has acted on these urges with a nonconsenting partner or the urges, sexual fantasies, or behaviors cause significant distress or interpersonal difficulty" (*DSM-IV-TR*). So, the fact that an offender engaged in any of the above paraphilic behaviors is sufficient evidence to warrant giving the offender the specific diagnosis.

Important Information About Paraphilias

Sex offenders tend to present with multiple paraphilias (Abel et al., 1987; Abel, Mittleman, and Becker, 1985). Abel and colleagues demonstrated that nonincarcerated sex offenders: (1) were well educated, (2) report a substantially higher number of victims than once thought, and (3) sexually molest young boys at a significantly higher rate than young girls. Most sex offenders have multiple paraphilias (Abel, Mittleman, & Becker, 1985). Therefore, while assessing and treating offenders, it is important to assess for the presence of all paraphilic behavior. Without identifying the extent of the offender's pathology, treatment cannot be expected to adequately address the offender's needs nor could the supervising agent adequately supervise. The full extent of risk and pathology must be clearly identified and this is only possible when a thorough assessment occurs.

Arrest records are grossly inadequate sources of the types of sexually deviant and sexually violent behavior sex offenders engage in (Abel et al., 1987; Frisbie & Dondis, 1965). As an example, Abel et al. (1987) found that the average number of paraphilias per sex offender in their study of 561 sex offenders was at least two. In addition, they found that some individuals committed very large numbers of paraphilic acts, but not always with the same victim. They also found that pedophiles involved with children outside of the home will occasionally return to the same victim, but will most often find new victims. Abel and Osborn (1992) also found that of 453 pedophiles, the offenders had molested 67,112 victims, totaling 106,916 acts of child molestation. Interestingly, sex offenders molesting boys were far more likely to have large numbers of victims and larger numbers of completed paraphilic acts than sex offenders molesting a female victim. Interestingly, out to the 453 pedophiles, 97 offenders completed 28,997 nontouch child molestations acts (e.g., exhibitionism, voyeurism). Approximately 99% of the victims of nontouch sex offenses were girls. Of the 453 pedophiles, 371 had completed actual touching of their victim for a total of 77,919 actual child molestations (60% of the victim were boys). Less than approximately 21% of pedophilic acts are committed against related child victims.

I (Johnson, 1997, 1998) conducted research involving incarcerated adult male sexual offenders. Interestingly, 92 offenders volunteered for the study with no remuneration. They identified having approximately 823 victims, 773 were female and 50 were male. Eight offenders had numerous victims and engaged in voyeurism as well as touch offenses and when taking out the number of victims involved in nontouch offenses, there were 219 female victims and 35 male victims involved in touch offenses. None of the offenders had only one victim.

Abel et al. (1987) found that of their 561 sex offenders involved in the study, the offenders completed a total of 291,737 paraphilic acts. This number is shocking when taking into account that most treatment professionals fail to understand the importance of conducting a thorough assessment to determine the full extent of the offender's sexual interest and criminal and paraphilic history. Child molesters tended to have the most victims and to have completed the most paraphilic acts. Therefore, it is unlikely that a sex offender would have only one paraphilia if an appropriate and through assessment were completed. Abel et al. (1987, p. 687) stated that "... individuals with paraphilias have a general deficit of control, leading them to carry out various categories of paraphilic behavior and with various categories of victims."

Pedophilic Sex Offenders

In their study of 45 male pedophiles in treatment, Raymond et al. (1999) found that 93% of their subjects met criteria for an axis I disorder in addition to Pedophilia. In the lifetime of the offenders, 67% had met diagnostic criteria for mood disorder, 64% met diagnostic criteria for an anxiety disorder, 60% substance abuse disorder, 53% for another paraphilia diagnosis, and 24% for a sexual dysfunction. They also found that personality disorders were common as well, including Obsessive-Compulsive, Antisocial, Avoidant, Narcissistic, and Paranoid. The authors also concluded that the lack of progress for offenders is often attributed to poor motivation, inattentiveness, resistance, and denial. These factors for lack of progress might well be explained by the comorbidity of other mental disorder diagnosis other than the primary diagnosis related to the offender's need for treatment. Therefore, it is imperative, as I have said before, to make a complete and comprehensive diagnosis regardless of whether the diagnoses given are all the primary focus of treatment.

Adjustment Disorders

It is important to understand how overused and misused Adjustment Disorders are. The *DSM-IV-TR* defines an adjustment disorder as a "... psychological response to an identifiable stressor or stressors that results in the development of clinically significant emotional or behavioral symptoms." Understand that Adjustment Disorders involve a significant or extreme reaction to some specific event. Abuse, rape, or molesting a child never, ever, is diagnostic of an Adjustment Disorder. The fact that the offender becomes depressed and anxious following an assaultive or violent event that he perpetrated does not cause a problematic reaction on his part. The fact that an offender may actually become or appear to become depressed or remorseful is not a reaction to the offense, but rather to the consequence for his offense behavior. I believe that diagnosing an offender with Adjustment Disorder only serves to minimize the seriousness of the offender's *choice* to engage in verbally or physically violent behavior or sexually violent behavior. If the offender is reacting to and "adjusting" to having injured his partner or being prosecuted, I will also hesitate to use this diagnosis, primarily because he needs to address his perpetration issues. It is not about how victimized the abuser or sex offender perceives himself to be, but about being held accountable for his violent behavior. The victim and children may appropriately be diagnosed as having an Adjustment Disorder, but not the offender.

Substance Abuse and Dependency

DSM-IV-TR defines Substance Abuse as "... a maladaptive pattern of substance use manifested by recurrent and significant adverse consequences related to the repeated use of substances." In addition, at least one of the following occurred within a 12-month period per the *DSM-IV-TR*: (1) recurrent substance use resulting in a repeated failure to fulfill major role obligations at work, school, or home; (2) repeated use in situations in which it is physically hazardous (e.g., driving under the influence); (3) recurrent substance-related legal problems (e.g., arrests for DUI, DWI, Assault, Disorderly Conduct, any criminal activity); and (4) continued substance use despite having persistent or recurrent social or interpersonal problems caused by or exacerbated by the effects of the substance (e.g., arguments, fights or assaults).

Substance Dependence involves "a maladaptive pattern of substance use, leading to clinically significant impairment or distress," and is manifested by three (or more) of the following, occurring anytime in the same 12-month period: (1) tolerance; (2) withdrawal; (3) the substance is often taken in larger amounts or over a longer period of time than intended; (4) there is a persistent desire or unsuccessful attempts to cut down or control the substance use; (5) a great deal of time is spent in activities to obtain the substance (e.g., numerous visits to the doctor), use the substance, or recover from the effects of the substance; (6) important social, occupational, or recreational activities are given up or reduced because of the substance; and (7) the substance use is continued despite knowledge of having a persistent or recurrent physical or psychological problem that is likely to have been caused by or exacerbated by the substance (e.g., use despite medical problems, relational problems).

It is clear that many batterers and sex offenders likely experience substance abuse or dependency problems at some point in their lives. Alcohol and drug use are often associated with abuse (Lawson, 2003; Leonard & Blaine, 1992; Hotaling & Sugarman, 1986). Kantor and Straus (1989) found that heavy and binge drinkers were two to three times more likely than moderate drinkers to abuse their female partners. Some studies have shown that individuals with alcohol and/or drug abuse or dependency also meet the criteria for Antisocial Personality Disorder (Brown, 2005). Understand that alcohol and drugs can contribute to impulsive behaviors and aggression. In addition, sex offenders are often under the influence of alcohol in at least 50 to 75% of case studies (Koss, 1988; Barbaree et al., 1993; Murdoch, Pihl, & Ross, 1990). There is no brand of alcohol or drug that *makes* someone or *causes* someone to abuse or rape or molest.

As discussed in Chapter 14 (The Role of Alcohol and Drugs in Abuse), substance abuse or dependency never *causes* an individual to engage in violent behavior. The individual still makes a conscious choice about the behavior they choose to engage in, whether sober, intoxicated, or high. Being "under the influence" never causes or justifies abuse or sexual offenses. The vast majority of people who abuse or are dependent upon alcohol and/or drugs will never engage in violent behavior. Therefore, no drugs or amount of alcohol causes someone to engage in violent behavior. The predisposition was present long before the use of the chosen substances.

Dissociative Disorders

Dissociative Disorders disruptions in the functions of consciousness, memory, identity or perception. *DSM-IV-TR* indicates that there are five types: Dissociative Amnesia (inability to recall important personal information); Dissociative Fugue (sudden and unexpected travel in addition to inability to recall personal identity or assuming new identity); Dissociative Identity Disorder (presence of two or more distinct identities or personalities, formerly called Multiple Personality Disorder); Depersonalization Disorder (persistent or recurring feelings of being detached from own mental processes or body); and Dissociative Disorder Not Otherwise Specified.

Schizophrenia

Schizophrenia is characterized by a mixture of characteristic signs and symptoms that are present for at least 1 month and some symptoms lasting longer than 6 months. *DSM-IV-TR* indicates that at least two or more symptoms are present, each lasting at least 1 month: (1) delusions, (2) hallucinations, (3) disorganized speech, (4) grossly disorganized or catatonic behavior, and (5) negative symptoms. In addition, the person has experienced some degree of dysfunction in the areas of social and/or occupational functioning. The symptoms have lasted at least 6 months, though they may vary.

The Stockholm Syndrome

The Stockholm Syndrome (Kuleshnyk, 1984) is at times not as well understood by clinicians. However, it does help explain why victims are at times particularly resilient about returning to and at times protecting their perpetrators. Though this syndrome was developed to help explain why hostages befriended and protected their captors, I find it applies to victims of battering, sexual offenses, and child abuse.

The Stockholm Syndrome involves: (1) feelings by captives towards their captor, (2) negative feelings by captives toward the police or authorities trying to arrange for their release or safety or intervention, and (3) positive feelings by the captured toward their captor. This may make it difficult for victims of battering and of sexual offenses to leave the offender or to report the violence to others or to cooperate with prosecution. The victim may fear losing the only significant relationship they have and may also fear the loss of their identity of having been in a relationship. Graham and Rawlings (1987) and Graham, Rawlings, and Rimini (1988) applied the Stockholm Syndrome to battered women.

Predisposition for Disorders

Many individuals (including offenders) have a predisposition for developing and having certain mental health and medical disorders. Predisposition means that an individual is at risk for developing a disorder; predisposition may be for a positive or negative outcome (e.g., whether or not the individual develops a specific disorder). For example, as a result of genetic/biological factors, certain individuals are more prone to develop heart disease, diabetes, and muscular sclerosis. When the certain genes are triggered, or certain environmental factors are present, the disorders activate or surface. Social and familial factors also significantly impact the development and onset of mental health disorders as well. Negative factors can predispose an individual for substance abuse or violent behaviors. Being raised in an abusive or emotionally neglectful home and having substance abusing parents and relatives can also increase the likelihood of developing disorders. Positive factors predisposing an individual in not developing mental health disorders include growing up with parents who are emotionally available and caring and certainly the lack of a history of family substance abuse.

When offenders have a predisposition to be self-centered, lack frustration tolerance, and believe that it is appropriate to utilize violence to get what they want when they want it, they have become predisposed to engage in violent behavior.

In general, people who develop mental health disorders have an increased predisposition for developing other disorders. However, with exception to moderate to severe psychotic disorders, the mental health disorders do not *cause* an offender to offend. The offender makes a conscious decision to harm another person, plain and simple as that. The mental health disorder may contribute to the ease by which the offender makes a decision to batter or commit a sex offense, but the disorder does not *cause* the offender to offend. For example, someone with Antisocial Personality Disorder does not just sexually assault an adult or molest a child. But an individual with Antisocial Personality Disorder who already possesses a predisposition for violence is far more likely to offend with violent behavior because he has already given himself permission to be violent. Remember what my colleague Veronika Vallerie (1995) asked, "What kind of beer makes me want to molest a child?" It is not the beer or the mental disorder, but rather the offender who makes a conscious choice to offend.

SUMMARY

Again, I strongly recommend that all batterers and sex offenders be diagnosed with all appropriate and relevant diagnoses. The purpose of an assessment or intake evaluation is to ascertain the extent of the presenting and identified problem. To accurately and appropriately give all diagnoses that are criteria-based is the ethical thing to do. I become frustrated with clinicians who comment that "I only give the lesser of the diagnosis that the offender meets criteria for," "I don't want to make them look bad." The purpose of assessments is to correctly identify the presenting problem even when the diagnoses appear to portray the individual in a negative way. Remember that it is not the clinician or the *DSM-IV-TR* that is portraying the individual in a negative or positive manner, but rather it is the history and symptoms and problematic behavior of the batterer or sex offender that dictate the diagnosis.

I do understand that defense attorneys routinely attack the clinician for making their client appear to be a monster or animal, arguing that the clinician is over-diagnosing. The clinician must always follow ethical guidelines and give the appropriate diagnosis. The defense attorney and offender never, ever should guide the assessment, diagnostic, or treatment process. The attorneys never should be allowed to guide sound clinical diagnostic procedures.

The above diagnoses apply to the batterer and the sexual offender. However, sex offenders often present with paraphilias more often than the batterer. But many batterers utilize pornography on a regular basis and may be more prone to becoming sadistic in their abusive behavior. For many offenders, they present with two to four diagnoses.

It is recommended that an individual be off all drugs, including alcohol at least 1 year to accurately assess for personality disorders and other possible impairments (Brown, 2005). Remember that alcohol and drug use impair the brain and, therefore, impact behavior and decision making.

Especially with chronic alcohol and/or drug users, it takes months, not weeks, to fully detoxify the brain. This is one reason why individuals convicted for any type of violent behavior (psychological, physical, or sexual), no matter the degree, should have at least a 5-year probation period to allow for a more accurate diagnostic assessment.

42 The Johnson Force Assessment (JFA) for Batterers and Sexual Offenders

The Johnson Force Assessment (JFA) tool was first developed as a research tool to identify the types of force that physical batterers and sex offenders used in the commission of their offenses. It is important to understand and identify the specific types of comments used by offenders, and the specific types of behaviors they use to groom or abuse their victims. Numerous researchers (e.g., Briere & Malamuth, 1993; Finkelhor, 1979; Goodchilds & Zellman,1984; Groth & Birnbaum, 1978; Kaufman et al., 1993; Koss, 1987; Lang, and Frenzel, 1988; Langevin, 1985; Makepeace, 1981/1986; Malamuth, 1981; Miller, 1988; Miller & Marshall, 1987; Muehlenhard, 1988; Rogers & Renshaw, 1993; Peters, 1976) have identified specific strategies used by offenders as well as having identified offense prevalence and patterns. It is important to understand all of the subtle and overt grooming and abusive behavior an offender engages in. No behavior or comment, however seemingly innocuous, can be ignored.

Two separate research studies have been completed. The first involved adult male sex offenders who were on probation from their sex offenses and who were involved in outpatient sex offender treatment programs in Minnesota (Johnson, 1995, 1997). The second study (Johnson, 1998) involved adult male sex offenders who were incarcerated in Minnesota prisons and who were also involved in sex offender treatment. Unpublished data was gathered on batterers.

The JFA is a thorough screening tool to use when the offender first enters treatment, and then to use as needed during treatment and prior to completing treatment. Information obtained ongoing offers the most information about whether the offender has ceased his use of force. For example, each offender could complete a JFA every week to identify the types of force he engaged in during that week. The offender's significant other/partner and support people could also complete a JFA at the time of intake and on a weekly or monthly basis to identify ongoing abuse. Interestingly, offenders did not engage in retaliatory behavior from their partner's disclosures on the JFA. It should be noted that couples and marital sessions did not formally occur until after the offender had made significant progress in the batterers program. This was less an issue with sex offenders. The results of the ongoing JFA could then be discussed during group and couples or family sessions. I have used the JFA with offenders, the offender's significant other and family, and with victims.

The JFA is easy to use and involves only clinical interpretation. Of course, offenders have been known to be less than honest in admitting the use of force, especially on an ongoing basis. However, once a therapeutic relationship has been established, the clinician should be able to determine the degree of honesty or deceit of each offender.

I have used the clinical version with offenders in an outpatient sex offender treatment program. For some, I had their family and partner complete the JFA on a weekly basis. The results were discussed during a monthly meeting. The offender, partner, and family were instructed not to discuss the JFAs until their monthly meeting. It was found that offenders gained insight from the use of the JFA.

Directions

Directions are simple. Inform the offender that several types of force occurred during the offense. It is also expected that the offender utilizes similar types of force in his daily life. Everyone uses some types of force in their lives, whether as part of their job (to pressure someone to purchase something) or when we want someone to do something. For example, we have all pressured others to engage in some activity that we want to do, even when the others do not want to do the same thing. However, in most situations, the type of force we use would not be considered abusive, and the worst harm likely experienced by the other person would involve seeing a movie they did not want to see, or eating some other food than what they preferred. Although the use of force is not always illegal, any use of force becomes part of how we as individuals cope with life's stressors. The clinician's observations about how the offender utilizes force in a group, marital, family, or individual sessions also provides information about what types of force that offender uses.

For the victim and other support persons (family or significant other), it should be explained that their input is very important, and that by providing honest responses, they are encouraging the offender to pay attention to forces they might otherwise miss. For example, in the daily interaction between the offender and his partner, the offender may be overly controlling and insensitive to the other's needs.

Interpretation of the JFA

Interpretation is easy. Simply look at the completed form and discuss any force indicated, regardless of how often the behavior or statement was made. The more types of force identified and the more often the force occurs, the more serious the problem, and, possibly, the higher the chance for reoffense. The JFA has not undergone empirical testing and as a result there is no formal method to say that any certain number of types of force identified places an offender at a specific risk for violent reoffense. However, any ongoing force/aggression is serious and demonstrates that the offender has not ceased his abusive and offending attitude or behavior and, therefore, remains at risk for continued violence because he is, in fact, continuing to engage in the violence while involved in treatment.

SUMMARY

It is important to assess the occurrence of all forms of force; again, force includes all types of violence that batterers and sex offenders engage in. In addition to my own clinical and forensic experience and research, others have found that identifying types of force used is very helpful. In fact, identifying types of force and coercion used in offenses is not only useful for identifying violent crime but for treatment as well. How could reoffense be prevented without understanding how the offender grooms his victim or to clearly understand the specific types of behavior and comments the offender engages in? Contact the author for information on obtaining a usable version of this tool.

ABUSE CHECKLIST: VICTIM SECTION

Please review the following information. Be open and honest.

Personal Victimization History

Have You Ever Been the Victim of:	When a Child (Age 13 or Younger)	When an Adolescent (Age 14 to 17)	As an Adult	Recent/ Ongoing
Verbal/psychological abuse				
Physical abuse				
Sexual abuse				
Neglect				
Sexual assault or rape				

COERCED: To be talked into, tricked, pressured, etc.

FORCED: To being physically forced or threatened into engaging in sexual behavior or some other type of behavior.

Specific Examples of Abuse

	Type of Abuse (Check all that apply)	Occurred When a Child (13 or Younger)	Occurred When a Teenager (Age 14 to 17)	Occurred as an Adult	Relationship of Abuser(s) (Include All Abusers)	Indicate Whether Abuse is Recent or Ongoing
	Verbal and Psychological Abuse					
1.	Name calling ("bitch," "asshole," "slut")					
2.	Swearing/cussing at you ("fuck you," "go to hell")					
3.	Being called sexually derogatory names					
4.	Being belittled					
5.	Being verbally insulted					
6	Sulking to manipulate you or get his/her way					
7.	Sneering at you					
8.	Yelling and screaming at you					
9.	Being followed around house					
10.	Having your whereabouts monitored or having to account for your time					
11.	Controlling your money					
12.	Controlling who you see or have as friends					
13.	Frequently interrupting you					
14.	Coercing or forcing you into humiliating or harmful behavior					
15.	Staring or glancing at you to intimidate					
16.	Withholding affection from you					
17.	Treating you as if you were inferior or a servant					
18.	Acting jealous					
19.	Being criticized					

20.	Being falsely accused of unfaithfulness or other things					
21.	Being demanding of you					
22.	Standing near you to intimidate					
23.	Directly intimidating you					
24.	Destroying your property					
25.	Destroying shared property					
26.	Being disciplined verbally					
	Threats					
27.	Threatening to harm you					
28.	Threatening to hit you					
29.	Threatening to spread rumors about you					
30.	Threatening to harm your children					
31.	Threatening to tell others your secrets or private or sensitive information					
32.	Attempting to or actually harming your children					
33.	Threatening to harm your pets					
34.	Attempting to or actually harming your pets					
35.	Threatening to commit or attempting suicide					
36.	Threatening to kill you					
37.	Threatening to harm others					
38.	Driving recklessly or dangerously to scare you or to prove a point					
	Physical Abuse					
39.	Pushing you					
40.	Slapping you					
41.	Punching you					
42.	Kicking you					
43.	Scratching you					
44.	Pinching you					
45.	Pulling your hair					
46.	Burning you (with lighters, cigarettes, etc.)					
47.	Restraining you (using body weight to hold you down, restraining your hands, etc.)					
48.	Ripping your clothing					
49.	Attempting to kill you					
50.	Forcing you to take medication or drugs to restrain or harm you, to get "high," or to cause you to become unconscious					
51.	Providing you with drugs without your knowledge or permission, spiking your food or drinks, slipping you drugs					
52.	Using a weapon against you (knife, gun, bat, belt, etc.)					
53.	Grabbing you					
54.	Shoving you or throwing you bodily					

55.	Choking, strangling, or smothering					
56.	Attempting to or actually physically harming your children or physically abusing them in any way					
57.	Attempting to kill you					
58.	Being physically disciplined by your partner					
59.	As a child or teen, being physically abused in any way (includes discipline that left marks, bruises, or created fear					
	Sexual Abuse					
60.	Forcing any sexual contact					
61.	Forcing or insisting unwanted sexual contact					
62.	Being forced or coerced into any sexual contact					
63.	Being talked into or pressured into sexual behavior					
64.	Being physically forced into any sexual contact					
65.	Demanding sex					
66.	Allowing sexual contact out of fear					
67.	Engaging in sexual contact with you without asking					
68.	Unwanted touching of genitals					
69.	Pushing your sexual boundaries					
70.	Expecting or demanding sexual contact from you					
71.	Engaging in sexual contact with you when you are impaired, intoxicated, "high," or sleeping					
72.	Being pressured or forced to go farther sexually than what you are comfortable with					
73.	Being blackmailed or tricked into sexual contact					
74.	Intentionally getting you drunk or "high" to take advantage of you sexually					
75.	Being drugged or having drinks spiked to take sex from you					
76.	Spiking your drinks or food to force sexual contact					
77.	Talking sexually about you in front of others; telling others about you sexually or about your sexual contact together					
78.	Watching/viewing pornography in your presence when you do not approve or forcing you to watch/view pornographic material (movies, magazines, internet, etc.)					

79.	Forcing you to attend strip shows or attending strip shows					
80.	Expecting, coercing, or forcing you to view pornography					
81.	Expecting, demanding, or forcing you to act out what is seen in pornography					
82.	Being aware that your partner views any type of pornography (includes magazines, videos, Internet, sex lines, etc.)					
83.	Expecting or demanding sex when in the presence of others or when in public places (parks, lakes, cars, etc.)					
84.	Forcing or talking you into having sex with others or forcing you to watch them have sex with others					
85.	Having others watch you engage in sex					
86.	Demanding you to perform specific sexual acts or to dress or talk in certain sexual ways					
87.	Your partner using prostitutes or having a sexual affair					
88.	Forcing sex when you are asleep, when you are ill, when having sex is adverse to your health, refusing to respect your limitations or "no's," when you are intoxicated or when you don't feel safe refusing					
89.	Groping or touching your body in public or anywhere when you do not want to be touched; demeaning you					
90.	Not having your sexual needs met, not asking you what you would like or want sexually; focus only on partner's demands					
91.	Not being allowed privacy when dressing, showering, or when in states of partial or full nudity					
92.	Being treated like a sex object					
93.	Being called sexual names					
94.	Sex being used as a punishment or a way to make up					
95.	Being forced or coerced into having your picture taken when naked, partially naked, or during sex					
96.	Having your sexual activities video taped					
97.	Making insulting comments about your genitals, body, or the way you are sexual; comparing you to others sexually					

98.	Forcing or insisting on engaging in humiliating or painful sexual contact					
99.	Causing pain or humiliation to you when sexual (slapping your buttocks, pinching your breasts, inserting objects into you, etc.)					
100.	Being forced or talked into using sexual devices or objects during sex					
101.	Being forced or talked into masturbating					
102.	As a child or teen, engaging in sexual contact with any adult					
103.	Any sexual contact with a relative (parent, grandparent, uncle, aunt, sibling, cousin, etc.)					
104.	As a child or teen, being sexually molested or abused by anyone					
105.	As a child or teen, being sexually molested by a stranger					
	Neglect					
106.	Failing to keep commitments					
107.	Not participating equally in expected activities					
108.	Failing to express intimacy often enough					
109.	Focusing only on their own needs and concerns					

Offender's and Abuser's Questions

	Age First Engaged in:	Age Last Engaged in:	Number of Victims (if applicable):	Ever Been in Trouble for Engaging in the Behavior:
Voyeurism (window peeping, looking at someone naked, bathing, dressing, or in a state of dressing without their permission or knowing)				
Exhibitionism (exposing your genitals to unsuspecting people)				
Frottage (intentionally rubbing against someone's genitals, i.e., breasts, while passing, without consent)				
Transvestism/Crossdressing (dressing in the opposite sex's clothing for sexual gratification)				
Telephone Scatology (making sexually explicit telephone calls)				
Pedophilia (engaging in sexual contact with a child or being sexually aroused by children)				
Sexual Sadism (becoming sexually aroused to the pain, suffering, or humiliation of another)				
Sexual Masochism (becoming sexually aroused to the pain, suffering, or humiliation of yourself during sex)				
Masturbated in Public (masturbating in public places, bathrooms, etc.)				
Fetishism (sexual arousal to objects, i.e., underwear)				

Have you ever engaged in any of the following behavior and, if so, please provide the following information:

Have You Ever Abused Anyone in the Following Ways?	When a Child (age 13 or younger)	When an Adolescent (age 14 to 17)	As an Adult	Recent/Ongoing
Verbal/psychological abuse				
Physical abuse				
Sexual abuse				
Neglect				
Sexual assault or rape				

COERCED:To be talked into, tricked, pressured, etc.

FORCED:To being physically forced or threatened into engaging in sexual behavior or some other type of behavior.

If you have ever abused anyone, please complete the following information: please be honest.

	Type of Abuse (Check all that apply)	Occurred When a Child (13 or younger)	Occurred When a Teenager (age 14-17)	Occurred as an Adult	Relationship of Victim(s) (Include all victims)	Indicate Whether Abuse is Recent or Ongoing
1.	**Verbal and Psychological Abuse**					
2.	Name calling ("bitch," "asshole," "slut")					
3.	Swearing or cussing ("fuck you," "go to hell")					
4.	Calling someone sexually derogatory names					
5.	Belittling					
6.	Being verbally insulting					
7.	Sulking to manipulate getting your way					
8.	Sneering at them					
9.	Yelling and screaming					
10.	Following them around the house					
11.	Monitoring their whereabouts or having them account for their time					
12.	Controlling the money					
13.	Controlling who they see or have as friends					
14.	Frequently interrupting					
15.	Coercing or forcing humiliating or harmful behavior					
16.	Staring or glancing to intimidate					
17.	Withholding affection to get what you want or to punish					
18.	Treating them as if they were inferior or a servant					
19.	Acting jealous					
20.	Criticizing					
21.	Falsely accusing them of unfaithfulness or other things					
22.	Being demanding					

23.	Standing near them to intimidate					
24.	Directly intimidating them					
25.	Destroying their property					
26.	Destroying shared property					
27.	Disciplining your partner					
	Threats					
28.	Threatening to harm					
29.	Threatening to hit					
30.	Threatening to spread rumors					
31.	Threatening to harm children					
32.	Threatening to tell others their secrets or private or sensitive information					
33.	Attempting to or actually harming children					
34.	Threatening to harm pets					
35.	Attempting to or actually harming pets					
36.	Threatening to commit or attempting suicide					
37.	Threatening to kill					
38.	Threatening to harm others					
39.	Driving recklessly or dangerously to scare or to prove a point					
	Physical Abuse					
40.	Pushing					
41.	Slapping					
42.	Punching					
43.	Kicking					
44.	Scratching					
45.	Pinching					
46.	Pulling hair					
47.	Burning them (with lighters, cigarettes, etc.)					
48.	Restraining them (using body weight to hold them down, restraining their hands, etc.)					
49.	Ripping their clothing					
50.	Attempting to kill					
51.	Forcing them to take medication or drugs to restrain or harm, to get "high," or to make them unconscious					
52.	Providing drugs without their knowledge or permission, spiking their food or drinks, slipping drugs					
53.	Using a weapon against them (knife, gun, bat, belt, etc.)					
54.	Grabbing					
55.	Shoving or throwing them bodily					
56.	Choking, strangling, or smothering					

57.	Attempting to or actually physically harming children or physically abusing them in some way					
58.	Attempting to kill					
59.	Using physical discipline on children that resulted in any type of mark being left (includes discipline that left marks, bruises, or created fear)					
60.	Physically disciplining your partner					
	Sexual Abuse					
61.	Forcing any sexual contact					
62.	Forcing or insisting on unwanted sexual contact					
63.	Forcing or coercing into any sexual contact					
64.	Talking into or pressuring into sexual behavior					
65.	Physically forcing any sexual contact					
66.	Demanding sex					
67.	Accepting sexual contact when your partner is experiencing fear					
68.	Engaging in sexual contact without asking					
69.	Engaging in unwanted touching of genitals					
70.	Pushing sexual boundaries					
71.	Expecting or demanding sexual contact from them					
72.	Engaging in sexual contact with them when they are impaired, intoxicated, "high," or sleeping					
73.	Pressuring or forcing to go farther sexually than what they are comfortable with					
74.	Blackmailing or tricking into sexual contact					
75.	Intentionally getting them drunk or "high" to take advantage of sexually					
76.	Drugging or spiking drinks to take sex from them					
77.	Spiking drinks or food to force sexual contact					
78.	Talking sexually about them in front of them or others, telling others about them sexually or about your sexual contact together					
79.	Watching/viewing pornography in their presence when they do not approve, or forcing them to watch/view pornographic material (movies, magazines, Internet, etc.)					
80.	Forcing them to attend strip shows or attending strip shows					

81.	Expecting, coercing or forcing them to view pornography					
82.	Expecting, demanding or forcing them to act out what is seen in pornography					
83.	Knowing that your partner is aware that you view pornography (includes magazines, videos, Internet, sex lines, etc.)					
84.	Expecting or demanding sex when in the presence of others or when in public places (parks, lakes, cars, etc.)					
85.	Forcing or talking them into having sex with others, or forcing them to watch you have sex with others					
86.	Having others watch you both engage in sex					
87.	Demanding they perform specific sexual acts or to dress or talk in certain sexual ways					
88.	Using prostitutes or having sexual affair(s)					
89.	Forcing sex when they are asleep, when they are ill, when having sex is adverse to their health, refusing to respect their limitations or "nos," when they are intoxicated, or when they are not safe refusing					
90.	Groping or touching their body in public or anywhere when they do not want to be touched; to demean them					
91.	Not having their sexual needs met, not asking them what they would like or want sexually; focus only on your own demands/needs					
92.	Not allowing privacy when dressing, showering, or when in states of partial or full nudity					
93.	Treated like a sex object					
94.	Calling them sexual names					
95.	Using sex as a punishment or a way to make up					
96.	Forcing or coercing into having their picture taken when naked, partially naked, or during sex					
97.	Having your/their sexual activity video taped					
98.	Making insulting comments about their genitals, body, or the way they are sexual; comparing them to others sexually					

99.	Forcing or insisting on engaging in humiliating or painful sexual contact					
100.	Causing pain or humiliation to them when sexual (slapping their buttocks, pinching their breasts, inserting objects into them, etc.)					
101.	Forcing or talking them into using sexual devices or objects during sex					
102.	Forcing or talking them into masturbating					
103.	As a child or teen, engaging in sexual contact with any adult					
104.	Any sexual contact with a relative (parent, grandparent, uncle, aunt, sibling, cousin, etc.)					
105.	When a child or teen, sexually molested or abused by anyone					
106.	As a child or teen, being sexually molested by a stranger					
	Neglect					
107.	Failing to keep commitments					
108.	Not participating equally in expected activities					
109.	Failing to express intimacy often enough					
110	Focusing only on your own needs and concerns					

Source: Scott A. Johnson, M.A., L.P., DABPS, DAACCE, ACFE, 2001/2005©, Mending Path Relationship Center, (612) 269–3628.

43 Assessing Sexual Interest and Arousal

MEASURING OFFENDERS' SEXUAL INTEREST

Actuarial measures designed to predict general (any, including sexual) recidivism were strong predictors of general recidivism among sexual offenders. Assessing the sex offender's sexual interest is of utmost importance. It is unrealistic and naïve to believe that a sex offender could successfully be treated (e.g., significantly decreasing the risk for recidivism) without first assessing (1) the focus and scope of the offender's sexual interest, (2) the degree and strength of interest to deviant fantasy, and (3) to measure changes in sexual interest over the course of treatment. There are several means of assessing the sex offender's sexual interest.

MEANS OF ASSESSING SEXUAL INTEREST

The first means of assessing sexual interest involves examining the behavior of the offender. The behavior involved in any documented deviant or criminal behavior, whether allegations or proven fact, help define the offender's deviant and violent interest. Included here is when the offender acknowledges having a substance abuse sexual interest. If the offender has engaged in voyeurism, forced his wife into sexual contact, and is verbally abusive towards others with sexually derogatory words and names, then this offender can clearly be assessed as a rapist and voyeur. However, it is highly unlikely that the offender's known sexual and violent criminal history adequately reveals the full scope of the deviant sexual interest.

In addition, the offender is the least reliable reporter of fact. Consider that they often have little, if any, true gain in being completely honest. Second, the offender's deviant sexual interest has become "normal" to him and, therefore, he may not accurately report all of his deviant sexual and violent fantasy and interest. Lastly, he may not be able to adequately label or identify all of the content that sexually arouses him.

Another method involves examining the offender's criminal history. However, this is one of the most flawed because many offenders receive plea agreements, which means that the actual conviction that shows up on a criminal history check most often does not accurately represent the offender's true crime. For example, as I have often found conducting forensic assessments, offenders can plea to disorderly conduct when originally charged with assault, domestic assault, child abuse, or sexual offenses. So, the criminal history check is often flawed. The only way to combat this problem is to obtain the Criminal Complaint and police reports for each and every known arrest and conviction. In addition, Gebhard et al. (1965) found that pedophiles have often been found guilty of less than three paraphilic acts per offender; yet Abel et al. (1987) found that the sex offenders involved in their study engaged in 23 to 282 completed paraphilic acts per offender.

Another method is plethysmography. This involves placing a electronic sensor ring around the penis or a photosensor into the vagina of women and measuring the most minute and unconscious response to sexual and violent stimuli. This method has demonstrated significant reliability and validity (Rosen & Beck, 1988; Pithers & Laws, 1995; Harris & Rice, 1996; Murphy & Barbaree, 1994). Numerous studies (Harris et al., 1992; Lalumiere & Quinsey, 1994). Quinsey and Lalumiere, (1996) have shown that when properly administered, phallometric erectile measures have acceptable reliability, convergent validity, and divergent validity. There are some concerns of plethysmography, which include ethical concerns about the use of nude stimuli materials that include naked children as well as concerns about the invasiveness of having something attached to the penis or inserted into the vagina. One drawback is that the plethysmography testing takes several hours to adequately complete and to get the interpretation.

A related method is the Abel Assessment of Sexual Interest (Abel, 1995). This method utilizes visual reaction time to measure sexual interest. There is sufficient support that this method is accurate and valid (Abel et al., 1998; Harris et al., 1996; Langevin, 1985). This procedure is far less intrusive and requires less time to administer and interpret the results. The Abel Assessment for Sexual Interest utilizes multiple measures that make it difficult to manipulate or fake (Abel, 1995; Abel et al., 1994). In addition, the stimulus material utilizes all clothed models, thereby eliminating the ethical issue of using nude models or child nude models.

Johnson and Listiak (1999) compared the Abel Assessment for Sexual Interest with Penile Plethysmography on a prison-based population of male sexual offenders. The results were that the Abel Assessment for Sexual Interest was comparable to Penile Plethysmography and the Abel Assessment even surpassed plethysmography on several of the categories of assessment. (See Johnson and Listiak, 1999, for more details.)

The last method to be discussed in the assessment of sexual interest is the use of polygraphy. The use of polygraph has now become standard protocol in assessing sexual offenders during and following treatment, as well as used by supervising agents to determine whether the offender has (1) fully disclosed all victims and incidents, (2) engaged in any recent risk behavior, (3) complied with supervision and treatment expectations, and (4) whether the offender has reoffended.

Section IV

Treatment Issues

44 Medication Interventions

Psychopharmacological interventions are becoming more necessary and effective with batterers and sex offenders. Although no medication will stop offending behavior, medication can lessen adjunct issues that can lessen the offender's irritability, depression, paranoia, and sexual drive.

The psychopharmacological approach appears to demonstrate some effectiveness for sexual offenders. Land (1995) indicated that sex offender treatment could be augmented by the use of the antiandrogens and mood stabilizers. The antiandrogens include medroxyprogesterone acetate (MPA or Depo-Provera®). This acts as a chemical castration for the offender, interfering with the frequency or intensity of erection and ejaculation. However, no research has demonstrated that the use of MPA alone significantly reduces sexual reoffense. The men using MPA are still often capable of achieving erections, engaging in intercourse, and ejaculating.

Land indicated some mood stabilizers and antidepressants include fluoxetine and clomipramine. Fluoxetine is an antidepressant and antiobsessional agent. This medication decreased sexual fantasies, urges, and sexual contacts, therefore, the offender's urge to engage in deviant or violent sexual behavior is less frequent and less intense. Clomipramine is used to treat obsessive-compulsive disorder and sexual disorders. It is a tricyclic antidepressant. This medication helps decrease obsessional sexual preoccupation and compulsive sexual behavior.

McGrath and Purdy (1999) mention that cognitive-behavioral and hormonal treatment have shown significant treatment effects with sex offenders (see also Alexander, 1994; Hall, 1995; Marshall et al., 1991). The hormonal drugs help give offenders more control over their deviant sexual interests by reducing the levels of the sex drive.

For batterers, the primary psychopharmacological interventions include the antidepressants, mood stabilizers, and the new generation of atypical antipsychotics. Cesta (2005) indicated the following in reference to the use of medication to address anger management:

There are many medications that have been shown to modulate anger. These include antidepressants, mood stabilizer, and the new generation of medications called atypical antipsychotics. Discussing the quality of one's anger management difficulties with a psychiatrist can lead to a choice of medications or combinations of medications, which may decrease the unwanted symptom of anger, violence, and uncontrollable outbursts. Medications such as Depakote®, Tegretol®, and lithium are the classic mood stabilizing medications and have great promise and some success of decreasing anger and the violence associated with its extreme presentation. Even common antidepressants can modulate anger in certain patients and things such as Prozac®, Zoloft®, Paxil®, and others can work to great benefit in a patient with anger and untreated depression. Some of the newer medications, such as Zyprexa®, Risperdal®, Seroquel®, and similar drugs, have a broad range of capability, and anger management is one of the categories of symptoms they can be very useful in controlling. In summary, anger is a symptom like any other for any disease and a qualified physician can certainly determine what might be effective treatment in combination with therapy and behavioral changes.

WHY USE MEDICATION

I have found that mental health professionals are resistant to utilizing medications to address anger management, sexual offender behavior, and substance abuse. The old school thinking would support that the psychopharmacological approach was not effective and unnecessary. However, the recent research demonstrates that psychopharmacological intervention is an important adjunct to therapeutic intervention. One without the other is likely far less effective. From a psychological perspective, it is naïve to believe that an offender's anxiety, obsessions, and depression would be effectively addressed only with psychoeducational or therapeutic intervention when anxiety, obsession, and depression have clearly established biological roots. I challenge professionals to try both a psychopharmacological and cognitive-behavioral approach to treating batterers and sexual offenders.

SUMMARY

The use of medication alone has not been empirically shown to significantly decrease sexual reoffense. Cognitive behavioral-based therapies utilizing group therapy in conjunction with psychopharmacological intervention appear to be the most promising.

45 Treatment Excuses

"BUT I WANTED TREATMENT"

Some of the most common excuses I hear from abusers and sex offenders in regards to why they fail to complete treatment are so obviously wrong and twisted. The irrational and distorted thinking processes that perpetrators experience are simply amazing and I find their excuses interesting. If their children gave them the same or similar reasons for not attending school, the perpetrators would have likely confronted their children and not accepted their reasoning.

Abusers and sex offenders require specialized treatment programs, including domestic abuse or anger management treatment, sex offender treatment, and, often in addition to the specialized treatment, many require completing a structured substance abuse treatment program prior to beginning any other treatment. They need to stop abusing drugs and/or alcohol before entering any other treatment.

When abusers and sex offenders fail to complete specialized treatment, they are telling you that they do not believe that they need the treatment. They make many excuses, and at times their explanation appears to have some possible legitimacy or plausibility. But remember that most abusers and sex offenders are good at conning others. They utilize con, lies, justification, minimization, and, of course, blame to avoid taking responsibility for their crimes. Let's examine some of the most common excuses that I have heard in my years of working with physical abusers and sexual offenders.

"There Was Not Enough Time to Complete Treatment"

There are occasions when time simply does not allow an offender to complete a treatment program during a specific period of time. For example, if they are incarcerated in prison, the length of their sentence may prevent them from being accepted into the program. Many programs simply cannot afford to accept people into their programs knowing that they do not have sufficient time to complete the program. If they accepted these offenders, the quality of the programming would be grossly affected, as would the benefits for the other offenders in treatment. For those on probation, they may only receive financial assistance to help pay for the cost of treatment while they are on probation. Once probation ends, so does the funding. Domestic abuse/anger management and sex offender treatment can be costly, approximately $1500. However, most insurance companies cover the majority of the cost, and some counties help pay for their probationer's treatment costs.

Sex offenders are far more likely to go to prison than abusers. For the majority of sex offenders, their prison sentence or probation period allows for the participation and completion of treatment. However, if the offender is on probation and violates the conditions of his probation, he may be sent to jail or prison, but at times, the length of his sentence limits and/or precludes participation in an appropriate treatment program. Probation may end before treatment is completed. In Minnesota, when sex offenders leave prison, they are assigned an additional term of 5 to 10 years of parole, during which time they have ample time to comply with treatment directives. When offenders break the rules while in prison, they receive discipline reports and at times, segregation. They may not be allowed to participate in treatment while being punished for their rules infractions. However, offenders who chose to engage in the inappropriate behavior are fully aware that they would likely receive consequences, and that the consequences would likely limit their ability to participate in treatment. So, in reality, they chose not to complete treatment by engaging in inappropriate behavior that resulted in their being disciplined. This is a choice, not something that simply happened.

In my experience, the vast majority of abusers who fail to complete treatment simply waited until their probationary period was nearly complete. They either fail to realize or simply do not care that the assessment process can take 2 to 3 weeks to adequately and ethically complete. And the recommended treatment is almost always a 16- to 24-week group for a batterer and 12 to 24 months for a sex offender. So scheduling an assessment or treatment when their probationary period is about to end simply will not work. They either have their probation period vacated and go to jail or, in many cases, their probationary period is extended an additional year to allow for treatment. Despite perpetrators being allowed more time to complete their required therapeutic intervention, an extended probationary period is a good thing, keeping perpetrators on probation for an extended period of time.

Refusing or Postponing the Participation of Friends, Family and Significant Others in Treatment

Offenders are masters of finding multiple excuses for avoiding full participation in treatment. They may claim that their work schedule conflicts with treatment times, that members of their support network, family, spouse, or significant other are unable or unwilling to participate. Again, these excuses are unacceptable. If offenders believed that they were in need of treatment and if they believed that they had problems with their anger, sexual behavior, violence, etc., they would prioritize treatment. Further, if they were honest with their support people and loved ones, it is highly likely (based on anecdotal experience) that their support people would participate in the treatment program as requested. It is my experience that when offenders refuse to involve others in their treatment, it is because either they have not accepted full responsibility for their offense to their support network or that they have not explained to the others that they really want and need them to participate. When reluctant support people have attended one or two sessions with the offender, they almost always are hooked and return for future sessions. If offenders have not been honest with their support network about the offense behavior, how can the support network prioritize participating in the treatment program? It is difficult and unlikely that others would choose to do so unless the offenders convince them that they did, in fact, commit the offense and that they need the support of these others in treatment.

Claims of Not Being Able to Afford Treatment

There certainly are some offenders who are unable to pay for their treatment, but every offender can afford at least to pay for some portion of their treatment. They had no problem finding money for drugs or alcohol, buying cigarettes, purchasing junk food and pop, for engaging in recreational activities or vacations, or purchasing new clothing. I have worked with several offenders who claimed to have no money to pay even a portion of their treatment fees, yet they purchased boats, new car or trucks, or were unable to provide verifiable explanations for where and how their money is spent. Perpetrators often come into the assessment with new tennis shoes and clothing, many of which are expensive brand names. They cannot afford these brands, but they have no sense of priorities and no or minimal financial and budgeting skills. I have had two offenders sell their boats to pay for treatment. After all, it was either pay for treatment and remain in the community on probation, or be terminated from treatment and have their prison sentence executed. And the probation officers and the courts upheld my requiring the sale of the recreational crafts. This is one good example of how establishing a working relationship with the probation and parole officers and the courts is helpful. Educate the supervising agents and the courts about your program and your program expectations so that they can support your goals. I find it amazing how almost all offenders manage to pay for treatment when their excuses for not paying are not accepted. The bottom line is that the offender must make treatment the most important activity and priority in their lives. Their job, family, and freedom all rely on successful treatment completion. When they refuse to cooperate with treatment expectations, they are telling you loud and clear that they care little about their offense or their freedom, and they should receive swift and significant consequences.

"I Found God, and He Has Forgiven Me; Now I Don't Need Treatment; Why Don't You Just Let Me Get on with My Life?"

Claims of religious "healing" and having been "forgiven" are common from offenders. While it is certainly not the expertise of a psychologist, therapist, or law enforcement official to question the veracity of the religious beliefs of the offender, it is within our duties to challenge the offenders on how their newly found faith relieves them of responsibility for completing treatment to address the abuse or sexual offense behavior. I have yet to hear a religious leader claim that God supports an offender in not accepting full responsibility for offense behavior or that the offenders, once they confess their offense behavior to God, are no longer expected to complete treatment. In fact, I have been told by several religious leaders of various faiths that they teach that God would expect the offender to complete treatment. To accept full responsibility involves and requires completing any and all required or recommended therapeutic interventions. It is important to accept that an offender's claim of being "forgiven" by God for the offense behavior may be genuine. I mean, who are we to question what occurs between any offender and God. But being "forgiven" does not take away the expectation of addressing the offense behavior with what ever means have been established.

I have seen many offenders who have claimed that they found God while in prison. However, I have also found that many who have found God (following convictions for battering or sex offenses) do not engage in behavior indicative of their new found faith, such as changing their foul language or attending religious services. They continue to engage in deviant and inappropriate behavior. In short, regardless of whether their new found faith is genuine, they are still to be held to the same standard and expectations as any other batterer or sex offender, follow the same restrictions, and serve the same probationary or parole sentences.

I believe that an active spiritual life is important for living a life style that promotes caring for and respecting others. The lack of spirituality may make it easier to justify criminal behavior.

"I Pled Guilty Only to These Certain Facts, Not the Other Facts; Read My Plea Agreement"

Offenders may plead guilty to only certain behaviors contained within the criminal complaint, and the judge may accept their partial plea. However, unless the criminal complaint is amended, the offender is considered guilty of all behavior he is accused of. If the criminal complaint is amended, then the offender should only be held accountable for that which the amended complaint accuses. If acquitted by jury, the same holds true. We simply cannot hold offenders accountable for behavior that they were not found guilty of, even if we believe that they did engage in the behavior, *unless there continue to be allegations by the victim or others or the offender confesses.*

An example of this would be an offender who pleads guilty to engaging in sexual intercourse with a 14-year-old girl or boy and claims that it was consensual. They deny using any force other than their age difference. The criminal complaint indicates that the offender actually utilized physical force and threatened the victim. The fact that the offender only pled guilty to part of the offense behavior and the fact that the judge accepts the plea, does not erase the fact that the offender used force and threats — unless, of course, if the judge orders the complaint amended. If a perpetrator denies punching or restraining his wife, but admits to pushing her out of the way so he could leave, and he pleads guilty only to that admission, he is still to be held accountable for the punching and restraining.

A "plea" is often accepted because the perpetrator is at least admitting some degree of the offense as being true. The courts then rely on mental health professionals to address the full scope of the criminal behavior. The perpetrator may receive a decrease in incarceration time, a decreased fine, and a decreased probationary period. But the perpetrator is still guilty of all of the facts contained within the Criminal Complaint.

"I Am Not Sexually Attracted to Children!"

Like we never hear this, right? It is difficult for many sex offenders to admit their offense behavior, especially when their victims are children. What kind of monster molests a child or their own children? Only a perverted sorry excuse for a human being would be capable of this. But are all child molesters sexually attracted to children? Surprisingly, the answer is "no."

Not all child molesters have a primary sexual attraction to children, and not all even have a significant attraction to children. For some, they offend more out of anger and revenge than for sexual reasons. Remembering, of course, that rape and sexual assault, child molestation, and incest are crimes of power and control. Some child molesters would not likely show any significant sexual arousal or attraction toward children on the Abel Assessment or the plethysmography. These offenders tend to have only one or two victims, and they offended on only a couple of occasions or, perhaps, only once. For the offender who has offended on numerous occasions, and especially those who have multiple victims, they tend to be the ones with a significant sexual arousal and interest in children.

Think of it this way. Why have you never engaged in sexual contact with a dog? (Probably few if any of you have.) Or why not be sexual with someone of the same gender? The answer is simple — because most people do not have a sexual attraction or arousal to dogs or someone of the same gender. When a person engages in sexual contact with children more than once, likely they have some degree of sexual attraction and arousal towards the victim or towards that specific age and gender, or, for some, they are sexually indiscriminate, and they care little about what age or gender the victims may be.

"My Problem is Alcohol or Drugs, Not Sex or Violence"

Many abuse and sex offenses are alcohol and/or drug related. In fact, statistically, approximately 70% of abuse and sex offenses occur when the offender is under the influence of drugs or alcohol. However, it appears that in most sex offenses, the role that alcohol or drug use plays is peripheral. It is a reasonable assumption that simply being under the influence of alcohol or drugs does not cause a person to become abusive or to commit a sexual offense. For some individuals, the use of alcohol or drugs serves as a disinhibitor. It may become easier to abuse or to commit a sexual offense or any crime, for that matter, while under the influence. Caution is warranted when substance use is claimed as an excuse.

It may well be that an individual appears to offend only when under the influence of some substance. But consider the complexities of a sexual offense. First, the offender must have chosen a victim, whether by opportunity or predatory pursuit. Next, the victim must be isolated to prevent escape and intervention from others, and to ensure that the victim will do as they are told. Then the actual assault occurs. The victim's and offender's clothes must be removed or partially removed. The offender then engages in various sexual acts, and forces the victims to perform certain sexual behavior as well. At some point the offender decides that he or she has been satisfied and that the offense is over. The victim may be returned to a certain location, left at the scene, or brought to another location.

All of the offense behavior involved numerous decisions and interactions. The offender must respond not only to the victim's resistance or cooperation, but to the behavior of others who may interrupt the offense. The offender decides what does and does not feel good, and what behavior to engage in. The offender decides what type and how much force, violence, or torture to use on the victim. The offender focuses on what will bring him pleasure, and must react to the victim's behavior. All of this requires that the offender pay attention to and react to the offense situation as a whole, not just certain aspects. The same or similar scenarios apply to batterers.

If someone was under the influence of drugs or alcohol to any degree, whether somewhat under the influence or totally intoxicated or high, he remains under conscious control of his behavior. Think about it. Any type of sexually assaultive behavior requires significant degrees of thought. Therefore, regardless of whether the offender was under the influence of drugs or alcohol, he still was aware of what he did during the offense. Understand that "blackouts" do not occur *during* the offense, but have to do with memory recall *after* an event or offense. However, in my clinical experience, even those with significant histories of chronic drug or alcohol abuse recall at least 90% or better of their offense behavior. It is imperative that the offender is held 100% responsible for his offense behavior regardless of his choice to use alcohol or drugs, and irrespective of whether he chose to recall and admit the offense details.

In reference to the claim that "alcohol caused me to offend," my colleague Dr. Valliere puts it this way. She asks "what kind of beer makes me want to have sex with children?" The same question can be asked about abuse, rape, or torture. The answer is quite simple: there is no beer, other alcoholic beverage, or drug that makes someone rape or abuse or molest children. It is that simple. The offender had previous thoughts about abuse or sexual offending prior to the use of substances. The use of the alcohol and/or drugs simply made it easier to stay focused on the violent act and easier to give oneself permission to abuse or sexually offend. Without the predisposition of violent thoughts or sexually abusive behavior, it is highly improbable that anyone would engage in the violent acts.

Do All Abusers and Sex Offenders Need Treatment?

It is my professional opinion that all abusers and sex offenders require some degree of intervention, whether a psychoeducational group or class, or long-term intensive treatment. In addition, they may require additional types of treatment or medical intervention as well, and some may benefit from the use of psychotropic medication. Many batterers and sex offenders are in need of substance abuse treatment. Some need medical and psychiatric attention to address issues such as depression, attention deficit-hyperactivity disorder (ADHD), and possible treatment for sexual dysfunction. Many also are in desperate need of therapy to address issues such as intimacy, their own childhood abuse or neglect, relationships, and socialization.

Be cautious of any therapist or psychologist who indicates that a batterer or sex offender is not in need of any significant intervention. Of course, some offenders require little intervention. For example, a first-time offender who utilized no physical force, who sexually molested a relative, or who committed statutory rape (engaged in sexual contact with an adolescent minor who consented) may only require completion of a psychoeducational program. Such a program educates the offender on boundaries, sexuality, and relationships as well as other topical areas. These offenders tend to function fairly well aside of their recent abusive or sexually assaultive behavior. They tend not to present with personality disorders or other mental health disorders, and tend not to have abusive backgrounds or difficulty with employment or in relationships in general.

When an offender has more than one offense, more than one victim, or utilizes physical force, or uses a weapon or threats, they likely require far more structured therapeutic intervention, such as a structured moderate to long-term anger management or sex offender treatment program. These offenders have established a pattern of battering or sexually assaultive behavior, and/or have little regard for the safety and well being of the victim (and likely no regard for the rights or safety of others as well). It is these offenders that tend to be at higher risk for recidivism.

It is also important to understand that clinical interview and assessment, as well as psychological tests, require at least some cooperation of the offender to render an accurate profile or picture of the offender's personality and needs. If the offender is cooperative, the next concern is the degree to which he is willing and capable of providing honest answers and the degree to which his is willing to reveal information that may not yet be known, such as the presence of additional victims or additional behavior engaged in during a specific offense, or of prior instances of battering.

I have seen reports that rely too heavily on the offender's version and apparently the report writer minimizes the offender's needs and unfortunately the offender's risk for reoffense. One common issue that I often find erroneously under diagnosed is that of whether an offender has a diagnosable mental disorder, including substance abuse. Some of the most widely used assessment tools rely on information that the offender provides. If offenders claim that they have not been abusing drugs or alcohol, or that they were not under the influence of drugs or alcohol at the times of the offenses, or that they were not intoxicated or high when the offenses occurred, but were under the influence, some substance abuse assessors may well indicate that the offenders are not chemically dependent or in need of substance abuse treatment. Unfortunately, many of these offenders are later diagnosed as having a substance abuse disorder and in need of substance abuse treatment, but this occurs after sentencing at a time when the offenders may not have enough time on probation or a long enough sentence in prison to complete any mandated treatment intervention.

On the other hand, I have read reports where the assessor indicated that the offenders' primary problem is their substance abuse, and that sex offense and violence issues are not as important to address as the substance abuse. This type of conclusion significantly underestimates the offenders' motives and needs, as well as the offenders' choice not only to use and abuse chemicals, but to engage in physically or sexually abusive behavior. It is absurd to believe that the person had no thoughts of engaging in sexual or other violent behavior prior to using alcohol or drugs. Some people actually believe that offenders only think of violence when under the influence. The bottom line is this. When an offender commits a physically violent act or sexual offense when under the influence of substances, the only role that the chemicals played was one of disinhibition; but disinhibition only to the behavior already thought about, not to new behavior that the offender did not think about prior to committing the offense. Therefore, it is naive to believe that substance abuse is the primary problem of a batterer or sexual offender.

In some cases, offenders fail to cooperate with an assessor. They may refuse to answer any questions at all or they may choose to answer only certain questions, or they may provide false information. The belief that an offender would be less than truthful may seem like a novel thought to some, but one must carefully consider what the offender has to gain or what is at stake. For example, if an offender who committed a violent rape only admits to physically assaulting the victim, he may actually be able to plead only to assault. As a result, the sexual assault charges may be dismissed as part of the plea, and the county attorney may accept this to avoid a lengthy trial. After all, the person is still convicted of the offense. But, as I discussed elsewhere in this book, regardless of what the offender pleads guilty to, he actually plead guilty to everything in the criminal complaint, unless the judge amends the criminal complaint, which rarely happens. So, for the offender to admit to you that he raped the victim, when he was allowed to only plead to the physical assault would obviously not be in his best interest. A batterer who pleads guilty to Disorderly Conduct is another example. Disorderly Conduct is a domestic abuse-related offense and remains that; the facts and offense details do not get erased because of a specific plea agreement. Therefore, to rely heavily on the offender's admissions to complete an assessment is unprofessional and unacceptable. It is imperative to give the most weight to official sources, such as the criminal complaint, the Pre-Sentence Investigation report, victim statements, police reports, court transcripts, and, of course, other collateral sources.

An offender may not admit that he engaged in all of the offense behavior for many reasons, the most common of which will be discussed elsewhere in this book. The offender may simply lie to avoid being held accountable for his actions, which may be heinous. The offender may not yet be ready to admit to himself or his family that he actually engaged in behavior that is socially reprehensible. In fact, some offenders may view people who committed such offenses as monsters and perverts, and it would be devastating to their very identity to admit that they, too, may be the monster and pervert. Others are simply psychopathic and narcissists. They admit to some aspects of their offense behavior while denying other aspects. They do this to enhance their ego and perhaps to gain the thrill of fooling the professional. It is important to also understand that offenders who are truly antisocial personalities or psychopaths may provide some true facts, and, by doing so, they hope that you will not attempt to verify the veracity of the statements they have made. Always assume the offenders are lying. Ask them to repeat answers to important questions several times during the assessments. Listen for inconsistencies and facts that may be implausible.

When diagnosing an offender, you must rely, at the time, more on the factual information and offender's history rather than on self-report. For example, pedophilia requires that the offender has had sexual fantasies involving some type of sexual behavior with children that has occurred over a 6-month period, or that he has engaged in sexual contact with children. Many offenders fail to report their sexual fantasies. It is important to diagnose even if the offender fails to self-report necessary behavior. For example, the offender has sexually molested one or more children over the past 3 to 12 months, and the sexual contact occurred twice per week. The duration alone suggests that the offender has had sexual fantasies of the victims, as it is accepted as fact that sexual behavior follows fantasies. One must first have the fantasy before acting. As for substance abuse, offenders may report only sporadic abuse of substances. However, it appears reasonable that if they were under the influence of substances at the time of the offense, then they likely can be diagnosed as having a substance abuse problem. In addition, the diagnosis of pedophilia can be made on the fact that the offender molested a child. The molestation or rape of a child is proof of fantasies involving the victim or other children. The degree, frequency, or intensity of the fantasy is irrelevant.

The Offender's Attitude Interferes with Therapeutic Intervention

There are times when an offender's personality or general attitude interferes with or prevents effective treatment intervention or assessment completion. A batterer or sex offender may wait until the very last day of his probationary period to seek the court-ordered assessment, despite the fact that the assessment may take 2 or more weeks to complete. The offender may refuse to answer questions or may provide only partial answers. Some offenders are openly sarcastic and controlling during the interview. The therapist or examiner should avoid getting caught in power struggles. However, the professional must be in control of the interview process at all times. The use of sarcasm, humor, and empathy, at the right time, often helps the offender become more cooperative.

Offender's who refuse to cooperate or who become aggressive or threatening during an interview or treatment session should be suspended or terminated. Although they have untreated problems that involve violence, their aggressive, violent or threatening behavior should never be minimized or tolerated. In addition, their victim/spouse/partner should be warned if they leave a session in an angry and escalated mood, being aware of current risk of harm to their victim/spouse/partner. Their supervising agent should always be updated as soon as possible about threat or violent behavior. If a direct threat is ever made towards the therapist, victim, or spouse/partner, the police and supervising agent need to be notified immediately. Never underestimate the capacity of a violent individual to seriously harm or kill someone. The fact that he is becoming threatening to a mental health professional or others is a clear sign that imminent violence may occur. Safety always wins out over confidentiality and over feelings of embarrassment or confusion. When in doubt, think protection first, processing the incident later.

Some offenders are not going to cooperate with an assessment or with treatment. In such circumstances, the assessment may have to be completed without their participation, as long as adequate records are available for review. If the victim and/or spouse/partner are available, they should be interviewed regardless of whether the offender cooperates. The victim and spouse/partner have valuable information to provide. I believe that these types of offenders should be incarcerated because they truly require external control, and they lack empathy and remorse, and have no problem threatening or harming others.

Whenever an offender communicates a threat or demonstrates escalating or violent behavior, it must be taken serious. Anytime a professional ignores or minimizes such behavior, the professional has successfully taught the offender that they will not take necessary action to protect others or themselves, and, therefore, it is the professional who aids and abets the offender in the next violent act.

46 Therapist Concerns

PLEA AGREEMENTS

When assessing and working with batterers and sex offenders, it is important to understand the limitations of the therapist, social worker, and psychologist. For purposes of this book, I will refer to clinicians and mental health professionals who assess and treatment sex offenders as therapists, psychologists, social workers, counselors, any mental health professional. I have several concerns about therapists. For this book, I will use the term *official version* to refer to the court-accepted version of an offense. This is usually the criminal complaint. All of the offense details, however, may not be included in the criminal complaint and, therefore, it is important to examine police reports, arrest reports, and victim's statements to ascertain the true details of any given offense.

Plea agreements certainly save time and money on the part of the state, country, and prosecuting city. However, offenders often claim that they pled guilty only to specific parts of the official version. The fact, though, is that when they pled guilty to an offense, they were found guilty of everything contained in that official version, regardless of whether they admitted to all of the details. The state is simply allowing them to receive less punishment by way of a plea agreement.

The same is true when additional charges are dismissed as part of a plea agreement. This is common. For example, the offender may have been charged with Criminal Sexual Conduct 1st Degree, Criminal Sexual Conduct 2nd Degree, Burglary, and Assault. The offender pleads to the Burglary and Assault, and the court dismisses the Criminal Sexual Conduct charges as part of the plea agreement. As a result some therapists might not hold the offender accountable for the sexual assault. That is dangerous and erroneous in that the offender was not found innocent of sexual assault, but simply not given a criminal sentence for the sexual component of the offense. Both judges and prosecuting attorneys agree that when the additional charges are dismissed as part of a plea agreement, the offense details in the official version are considered to be true and accurate. What that means is that despite the offender pleading guilty to everything but the sexual assault, he is, in fact, guilty of the sexual assault and can and should be held fully responsible for all documented behavior involved in the offense.

Consider the batterer who is charged with Domestic Assault, Criminal Sexual Conduct, and Terroristic Threats. The criminal complaint outlines that the offender physically assaulted the victim, was sexually violent, and threatened to kill the victim. The final result was a conviction by plea for Disorderly Conduct. Many therapists might disregard the offense details for consideration that the pled offense was for Disorderly Conduct and not for the other violent behaviors involved in the offense. Regardless of the final plea, it is important to hold the offender accountable for everything in the criminal complaint and criminal charges.

The plea agreement basically means that the court simply did not issue an additional sentence for each of the charges. The plea agreement saves time and money for the state and county as well as for the offender. Avoiding a trial is a realistic necessity whenever possible. In addition, the plea agreement saves the victim from having to testify in court, which can be very traumatizing.

There are only three exceptions that I am aware of to not hold the offender responsible for everything contained in the official records. The first is when the court amends the official version of the offense. When this occurs, the official version will be rewritten, and will include only those details that the court finds the offender guilty of. This does not happen often. The second is when the charges are dismissed separate and independent of a plea agreement. The Pre-Sentence Investigation report and sentencing sheet or transcripts will clarify what charges were dismissed as part of a plea agreement. The offender should be held 100% accountable for all behaviors described in the official version regardless of what they pled guilty to.

The third is when the offender is found "not guilty" by jury. The unfortunate reality of a jury verdict of "not guilty" is that the offender may well be guilty of the offense, but the jury was unable to find him guilty. On the other hand, the offender may well be innocent of the charges. This may be a result of the prosecuting attorney not charging the offender with all of the possible charges or counts that his offense warranted. An example of this comes to mind. An offender committed a violent rape. The prosecuting attorney charged him with 12 counts, which included burglary, assault, several other charges, and two involving Criminal Sexual Conduct 1st Degree. The problem the jury indicated was that the offender did not appear to have committed Criminal Sexual Conduct 1st Degree, but rather 2nd, 3rd, and possibly 4th Degree. Because the attorney failed to add these charges to the offender's complaint, the jury was forced to find him "not guilty" of rape. My point here is simple. I strongly encourage attorneys to charge all possible offenses to offenders, so that juries have the option to find them guilty of a lesser offense, thereby holding offenders accountable for their actions. The offender, incidentally, admitted to me that he coerced the victim into sexual contact, but indicated that he did not physically assault her until after the rape. Another offender slipped through the system. I am not intending to criticize county or prosecuting attorneys. They work under a great deal of political, time, and financial constraints. They do a good job. I am just making recommendations.

One additional troubling plea is the Alford Plea. What this plea basically allows is for an offender to plead guilty to something, but to deny the offense details. This is really a last ditch effort to procure a plea and conviction without having to prove that an offense occurred. For example, this plea may be used when the victim refuses to testify, when it is questionable whether the victim is believable or stable enough to provide good testimony, or when there simply is not enough evidence to win a case without the offender's admission. This is rarely used with sex offenders anymore in Minnesota. However, even an Alford Plea is an admission of guilt. The offender should still be held accountable for all of the offense behavior described in the official version.

THERAPIST CONCERNS AND OBLIGATIONS

Now let's turn our attention to the issues involving concerns about therapists. Therapists who work with batterers and sex offenders have several obligations, which will be discussed below. These are concerns that I have, and may or may not be representative of concerns of others.

The therapist has an obligation to hold the offender 100% responsible for their offense behavior. It would be nice if batterers and sex offenders, as well as any other court-ordered client, would be honest about his behavior and motivated for treatment and change. However, this is not the case for many types of clients who are ordered by the court to undergo assessment and treatment. Many offenders tend to lack genuine motivation to change, and often have little interest in accepting responsibility for their offense behavior. It is the job of the mental health professional to address these problems, and many effectively do so. However, there are some limitations that must be acknowledged and respected.

NONTREATABLE OFFENDERS

It should be accepted that there are offenders who are simply not treatable. Some of these offenders may not yet be sufficiently motivated to change and some may be in the precontemplative stage. However, there are those who simply will not respond to treatment either because they are unwilling or unable to engage in the therapeutic process or unable or unwilling to change their violent behavior. Offenders who may be unmotivated or unwilling to engage in treatment often include recidivistic batterers, child molesters, serial rapists, sadists (sexual and nonsexual), and psychopaths.

In fact, there is no recognized treatment that has had any positive effects on treating serial rapists, sadists, and psychopaths; the literature has been scarce and the studies available rely on a low number of subjects. Many have attempted to treat these individuals, but to my knowledge, none have had success. Of course, success unfortunately is defined only in terms of having no additional convictions for subsequent sexual crimes. However, it appears that these types of offenders may have multiple victims for every known victim. It may take years to successfully identify and apprehend these offenders when they reoffend once released from jail or prison. Additionally, many offenses may go unreported. Consider for example that most victims of domestic abuse report that the abuse has occurred on numerous occasions prior to the reported offense. The same is true with sexual offenses.

Some sexual offenders and batterers can be successfully treated, but many cannot. It appears to be an appropriate definition of "successful" treatment to say that the offender completed the program requirements. It would be great if the definition of successful treatment would be no further reoffense, no verbal or physically abusive behavior, and no sexually abusive behavior. It is important to recognize that despite legal and therapeutic intervention, some simply will not cease their abusive or sexually abusive behavior. However, anecdotally I have found that approximately 80% of the batterers fail to cease their verbally and physically assaultive behavior and reoffend within 6 to 24 months. For the 20% that do cease their physically violent behavior, most will continue to be verbally abusive, manipulative, and intimidating. Therefore, from my own experience, most abusers will not cease their abusive behavior. The literature reports mixed results, yet no study reported a total cessation of all abusive behavior for the majority of their sample.

Sexual offenders, depending on the type of offender (typology), have differing reoffense rates. Most sexual offenders reoffend at a rate between 30 to 80% over the lifetime of the sex offender. (This is covered in more detail in Chapter 40.) Much of the respected research on sexual offenders suggests that there was no consistent significant difference between reoffense rates for sex offenders who completed treatment vs. those with no treatment.

Concerns about Medication

Some offenders do well when on medication, but medication alone has never proved to eliminate or significantly decrease sex-offending behavior in the long run. This may be due to the fact that the majority of batterers and child molesters, like other types of sex offenders, stop taking their medication after a time. It is almost impossible to mandate that someone take medication every day or week for the rest of his life. This would require vast financial and personnel resources that few if any state, county or city would be able to afford. It is important for therapists to acknowledge the fact that medication is marginally effective primarily due to the low compliance rate of offenders once supervision has decreased or ceased.

Understand that medication may be specific to sex offense behavior. For example, some medications (e.g., antiandrogens and mood stabilizers) help decrease obsessive thoughts and may help decrease sexual arousal. However, they have not significantly demonstrated success in reducing sexual offending behavior when used alone. Even castration did not demonstrate significant change in reoffense rates. Medications for batterers (e.g., antidepressants, mood stabilizer, and the new generation of medications called atypical antipsychotics) help decrease anxiety, depression, and irritability. The batterer, like the sexual offender, may not achieve significant success without the combination of medication and therapy.

One additional concern arises for the child molester, serial rapist, sexual sadist, and sexual psychopath. It appears that these types of offenders may be in one of three categories. (A more thorough discussion of typologies of offenders can be found in Chapter 17.) The first is the low energy, depressive offenders who lack an apparent self-identity and appear to have low self-esteem. They typically resist therapeutic interventions by developing somatic complaints, family crisis, and by making shallow and oft times obvious false disclosures. In short, they attempt to avoid discussing their offense behavior and resist making any significant changes in their risk and social behavior. These types of offenders may make superficial and insignificant changes to avoid being terminated from the treatment program. They tend to evoke or seek empathy from others, and may portray themselves as learning disabled or lacking skills, when, in fact, they are of normal intellectual functioning and have managed to establish some sense of a social life. They may be in a program for a significantly longer period of time than other more motivated offenders. They often are prematurely terminated from treatment after several years of progressless participation. Make no mistake, these offenders are highly narcissistic and will reoffend when the opportunity presents itself.

Treatment of the Moderately Motivated Offender

The somewhat-to-moderately motivated offenders admit some or most of their offense behavior, and tend to believe the victim's version over their own. They eventually admit to most, if not all, of this offense behavior as they increase their participation in treatment over time. This is the offender who appears to experience what may be considered "normal" achievements and difficulties. They actively participate in treatment, often are model clients or at least productive and invested treatment participants. They may somewhat resist some expectations placed upon them, but over time they become compliant and increasingly more invested. They appear genuine and tend to make stable progress over time. These are the most likely offenders to involve significant others and support people as part of their treatment plan.

Treatment of the Aggressive and Narcissistic Offender

Aggressive and narcissistic offenders make it clear from the outset that they have no investment in treatment. They blatantly challenge authority and program rules, and often demand exceptions. They cause concern because they at times appear ready to act on aggressive impulses. They rarely comply with treatment expectations that require them to admit offense details and that require them to involve others in their treatment program, such as significant others. They tend to have a transient lifestyle, with many short-term relationships and marriages, have several children that they refuse to pay child support for, and their employment is sporadic. If they were in the military, they likely would receive a dishonorable discharge or an equivalent type of discharge. They likely have been terminated from jobs and it may be unclear how they make their money. There are some who hold down stable jobs, who remain in relationships and marriages long-term, and who assume financial support for their children. There really are two extremes of this type of offender. These are the obsessional followers and stalkers (Meloy, 1998). They have a sense of entitlement that appears brazen.

Therapist Must Hold Offender Responsible for Behavior

It is not the role of the therapist to "protect" the offender/client from consequences of the offense behavior. The therapist must hold the offender 100% responsible for all of his behavior and not over-protect him. There are some therapists who attempt to use ethics and therapeutic relationships as justifications for not holding their clients fully accountable and, at times, protecting their clients from possible legal consequences. Common examples will be discussed. Some therapists will not follow federal mandated reporting laws that require the therapist to report to law enforcement or child/adult protective services anyone who appears to be maltreating a child or vulnerable adult. They often claim that to report every incident damages the therapeutic relationship they have with the offender and, therefore, to follow the law means to decrease the chances that the offender will report other important information. Some therapists take it a step farther by claiming that the entire group would be negatively affected, and may result in the group members not fully and truthfully disclosing their offense history.

These therapists scare and anger me. The very behavior that resulted in a referral for treatment may be occurring in other venues with other victims. The therapist needs to hold the offender accountable not only for reporting inappropriate and violent behavior, but to be ceasing such behavior. Every time offenders get away with abusive or inappropriate behavior, they learn how to be better offenders and con artists. They learn that the therapist will accept such behavior, within limits. The offender may generalize that if the therapist condones such behavior, then the courts will as well. Basically the result is that the offender becomes a better sex offender, and becomes more confident in offense-based behavior. The offenders simply refine their offense behavior. These are the therapists who should lose their licenses or agree to not work with court-ordered clients. Strong boundaries are imperative for the clinician working with batterers and sex offenders. Ethical guidelines help shape appropriate boundaries the same as organizations that set guidelines for working with offenders (e.g., ATSA, APA)

Another example is when the therapist attempts to misrepresent the true risk that the offender poses in the community. Rather than admit that their client remains at risk, they minimize the risk to the public. This occurs in several ways. Some therapists refuse to share pertinent information with other professionals or law enforcement. Some therapists poorly document the offender's treatment, assignments, or refuse to document the offender's offense history. This is in and of itself an ethical violation.

The end results are severely damaging to the public. How could an appropriate and accurate risk assessment be completed on an offender when the therapist fails to clearly document relevant information. Another way this occurs is when the therapist refuses to keep copies of the required assignments in the offender's clinical file. Many return the assignments back to the offenders to do with as they please. These are common examples of how some therapists over-protect offenders. A strong therapeutic alliance between the offender and clinician can improve treatment success rates.

Therapists Should Not Reinterpret Offense Details

As stated above, it is important that therapists not reinterpret offense details and not attempt to prove what did or did not occur in any specific offense. Many therapists try to interpret offense details and end up re-trying the case from the offender's perspective. It is as though the therapist believes that they are the better trier of fact than the courts. This is very dangerous because no therapist is qualified to decide what did or did not occur in the offense. The only information that the therapist has to go on to challenge the official offense version is their client's word and version. We know that offenders never lie; well, all right, they do. They have little to benefit by being totally honest and forthcoming. It is important that therapists not play the role of the prosecuting and defense lawyer, police officer or judge. A polygraph can go a long way to help clarify some offense details.

An example of this would be the therapist who makes the comment that "everything in the official version, criminal complaint, or arrest report is not true." No one would dispute that the official version of an offense may contain errors, and no one would expect that a victim would be able to offer 100% of the truth as based on possible eye witnesses. The victim experienced significant emotional, spiritual, sexual and physical trauma, and may have even feared for her life. However, it is not, and has never been the job of a therapist to reinterpret an offense. The court documents are supposed to represent the truth, and the offender was found guilty of those details based on a jury verdict of guilty or a plea of guilt by the offender.

So, who is right? Is it possible that an offender may have committed the offense, but perhaps did not actually do everything the victim claimed? Certainly. But my point is this. When offenders are found guilty, they are found guilty of every detail in the official version of the offense, regardless of what they admit in court. Questions concerning this section should be addressed to law enforcement and court personnel in your area. I have found the Attorney General's Office and County

Attorney's Offices extremely helpful in addressing these concerns. After all, they really are the legal experts and are the most appropriate people to ask. Again, the use of polygraph would be beneficial.

Therapist is Responsible for Administering and Interpreting Tests

It is the responsibility of the therapist to appropriately administer and interpret psychological and psychophysiological tests, and to make use of the results with the offender in treatment. Therapists must be competent to accomplish this task. They must obtain training and supervision to maintain competence. Ongoing education, such as conferences and consultation, is required. It is also important to recognize that no test determines the guilt or innocence of anyone. Even the polygraph does not determine guilt or innocence.

I believe that it is important to understand and appreciate the limitations of psychological testing. In addition, it is important to do a battery of tests, not just one or two. This will help avoid the offender being able to produce false testing results. Offenders I have worked with or assessed had a Minnesota Multiphasic Personality Inventory (MMPI-II) (Butcher et al., 1989) or Mellon Clinical Multiaxial Inventory (MCMI-III) (Millon, 1983) profile indicative of either no pathology or, at the other extreme, profiles indicative of personality disorders, depression, or paranoia. Interestingly, many offenders produce subscale scores indicating that they have no problems or concerns in their primary relationship or family, have low anger and anxiety subscale scores, and many with lower scores on substance abuse scales than official data supports.

Sometimes it appears that unless the offender is at the extremes for either having no personality disorder or has several disorders, the testing tells little about the offender that was not already either known or could have reasonably been assumed. However, offenders producing profiles falling within the "normal" range should be dealt with cautiously. It may well be that they are not presenting with any serious psychopathology, in which case, therapeutic intervention may be more successful because of the lack of pathology and perhaps increased motivation. On the other hand, they may be in denial, not wanting to change, or even innocent of the charged offense. Much more detailed assessment would need to occur to ascertain the facts. Again, polygraph would be useful in this situation. However, most offenders who produce profiles falling within the "normal" range, who have offense histories, are indicative of an established pattern of verbal, physical, or sexual violence, may well have produced a valid profile, one that is validly demonstrating that these offenders are psychopathic and narcissistic. That is, they have few if any concerns and they are not emotionally reactive. Therefore, they are less likely to produce overly pathological profiles.

Understand that the psychopathic offender experiences little if any remorse, concern for others, or even concern for consequences. So, producing normal profiles on testing may well be accurate; offenders are not concerned about the harm they caused, do not likely believe that they caused harm or if they did believe that they caused harm, they do not feel guilty about it and likely justify their behavior. (See the work mentioned by Meloy in Chapter 3 in this book for more details about psychopathic offenders.)

For the sexual offender, psychophysiological and psychosexual testing, specifically the Plethysmograph (PPG) and Abel Assessment for Sexual Interest (referred elsewhere in this book as the Abel Assessment), offer details concerning an offender's sexual preference, interest, and arousal. These tests, along with the polygraph, are state-of-the-art and should be used with each and every sexual offender. It certainly makes sense to use the tests that offer the most accurate assessment of the problem being addressed.

I have two concerns regarding use of plethysmography or the Abel Assessment. The first is that the fact an offender has, or appears to have, a sexual interest or sexual arousal to deviant stimuli in no way indicates that they have ever acted on these interests or arousals. The fact that an offender is sexually interested and sexually aroused to female children is in no way an indication that they have or ever will behave sexually towards female children, or any children for that matter.

What the results indicate is that the person has some (or significant) predisposition for engaging in such behavior, based solely on his response on the tests. If he has actually engaged in sexual contact with female children, for example, and his results on the plethysmograph and/or Abel Assessment indicate a sexual interest and sexual arousal to female children, you can confidently report that this offender is sexually attracted to female children. However, you cannot, based on the plethysmography or Abel Assessment results, state that the offender did, in fact, commit or will ever commit a sexual offense. Offenders that have a demonstrated sexual response to deviant stimuli on the plethysmograph or Abel Assessment may or may not ever act out their identified deviant sexual interests. Research (e.g., Abel et al., 1998; Hanson & Bussiere, 1996, 1998, 2004; Quinsey et al., 1995; Proulx et al., 1997; Hanson & Harris, 1998) has demonstrated that having deviant sexual interest as being highly predictive of subsequent sexual acting out. Therefore, while having a deviant sexual response on measures of sexual interest does not prove that an offender has or ever will sexually offend, it is highly probable that they have or will do so.

The second concern is that both of the above procedures are most accurate if used as a measure of increased or decreased sexual interest or sexual arousal by administering the tests several times throughout an offender's treatment. For example, conduct a test at the beginning, in the middle, and near the end of treatment. This is the most powerful and accurate reading and measure of the offender's interest and arousal to sexually deviant stimuli. When administering either test on only one occasion, the opportunity to obtain a baseline reading is lost. For example, let's assume that an offender was only tested prior to completing a treatment program. The most accepted belief in the sex offender field is that the offender has the ability to decrease or increase deviant or appropriate sexual interest and sexual arousal, but that it would be nearly impossible to totally eliminate these deviant patterns. Given my example, the offender's test results will indicate that they have some degree of deviant sexual interest and arousal. No one would ever be able to say with confidence that the offender's pattern has improved or not. Perhaps it worsened in treatment. Without a baseline reading, one could never make a confident or competent statement concerning the offender's sexually deviant interest or arousal.

However, some therapists attempt to do so. They indicate that the offender successfully completed all of the program requirements and has appeared to demonstrate healthy sexual interests and behaviors. Even the offender may indicate that he now experiences deviant sexual interests or arousal far less than before treatment, or that he now never has deviant interests or arousal. Both therapists and offenders often over estimate their own ability to accurately identify the risk of reoffense and the degree of deviant sexual interests and arousal. Without a baseline test, the truth will never be known. If it were truly that easy to do without plethysmography or the Abel Assessment, then therapists would be more accurate in their prediction of reoffense. The bottom line is this: When using the plethysmograph or Abel Assessment, it is strongly recommended that the tests be run several times throughout the offender's treatment in order to obtain accurate and useful information. Sex offenders may or may not eliminate their deviant sexual interests or arousal, and even if they are successful at eliminating or significantly decreasing their deviant arousal and interest, it does not mean that they will no longer pose a risk for engaging in sexually violent reoffense. In the same manner, offenders that continue to have some degree of deviant sexual interest will not necessarily reoffend, though they remain at increased risk for sexually violent reoffense. There are additional factors to be considered, such as change in dynamic risk factors.

Ongoing Education Important for Therapists

It is the responsibility of the therapist to receive ongoing education pertinent to the area that they practice in. The most widely attended and recognized conference for therapists who work with sex offenders is the annual conference sponsored by the Association for the Treatment of Sexual Abusers (ATSA). This conference offers state-of-the-art training pertaining to current research and clinical information taught by those currently working with sexual offenders. How can anyone stay on top of the ever-changing field of sexual offender treatment and risk assessment without participating in annual training? The answer is that they cannot. Therapists who fail to attend this conference at least every other year really miss out on a lot. I would be skeptical of anyone who works with sexual offenders who does not maintain continuing education specifically concerning the assessment or treatment of sexual offenders. Contact ATSA for further details. For therapists working with batterers, there is no one specific conference to recommend. The bottom line is that clinicians working with batterers or sex offenders need ongoing training to maintain their competence.

Limits of Confidentiality and Therapist's Obligation

It is the responsibility and obligation of the therapist to inform the offender of the limits of confidentiality. "The member must inform the client of the purposes, goals, techniques, rules of procedure, and limitations that may affect the relationship at or before the time that the counseling relationship is entered" (American Personnel and Guidance Association, 1981). While it is important to protect the client from possible harm, the court-ordered batterers and sex offenders surrender the right to confidentiality, in so far as that any and all information they reveal concerning the existence of additional offenses and victims become known. Often this occurs as a result of completing a sexual history, offense autobiography, or journal of all inappropriate, deviant, or illegal sexual behaviors. In addition, some states have expectations of programs that work with batterers that require the therapist to notify the supervising agent and the victim of when the batter completes treatment, if the offender fails to complete treatment, and of any risk concerns. This should be made clear to the offender. I have included a copy of the release of information consent form that I use with offenders with a special section to address this (Chapter 39). It is imperative to inform clients of the limits of confidentiality.

Under the area of welfare of the consumer, APA (1981, 2002) ethics state that "... psychologists clarify the nature and direction of their loyalties and responsibilities and keep all parties informed of their commitments. Psychologists fully inform consumers as to the purpose and nature of an evaluative, treatment, educational, or training procedure, and they freely acknowledge that clients, students, or participants in research have freedom of choice with regard to participation." Under this principle, *section b* of the ethics statement states, "When a psychologist agrees to provide services to a client at the request of a third party, the psychologist assumes the responsibility of clarifying the nature of the relationship to all parties concerned." In summary, the offender has the right to know and understand the limits of confidentiality and to understand the service being rendered.

Difference between Voluntary and Involuntary Clients

Let me end with a brief note about treating involuntary clients. All clinicians are initially trained with the assumption that clients are voluntary, that the client is participating in the therapeutic intervention with motivation for change and an eagerness to actively participate. However, court-ordered clients are not the "norm" of clients. They have consequences for not complying with recommendations and require a far more active clinician. The clinician must have an in-depth understanding of the characteristics of the offender as well as a thorough understanding of risk assessment. Although it is desired to establish and maintain a working and effective therapeutic relationship between client and clinician the power imbalance must exist to a larger degree with batterers and sex offenders. Involuntary clients are under order of the courts, social services, child protection, ethical boards, or employers, to complete a recommended treatment intervention or face more severe consequences. Even if the offender appears motivated for treatment, the clinician must not lose sight of the fact that the offender is not voluntary. The offender can choose to comply with the order for treatment, but they are still involuntary clients. Remember that the offender was referred for therapeutic intervention as a result of having engaged in harmful, violent, or reckless behavior.

On the one hand, following mandated reporting laws is simple, the clinician does what ethical standards and the law dictate. It is never the clinician's job to protect the client/offender from receiving consequences as result of the offender's behavior. Some clinicians, however, are not suited for working with violent or involuntary clients. For example, the clinician must follow any and all mandated reporting laws and ethical standards without exception. I recall one clinician who did not want to report that her client, a female child abuser, had once again physically abused her child, while in treatment. The clinician stated that she did not want to make the mandated report because of the consequences the client/batterer would receive, and that she feared the therapeutic relationship would be irreparably damaged. After all, it was the client/batterer who reported the current abuse to the clinician.

I have found it very interesting from an anecdotal standpoint that even when the offender is read an informed consent or has the mandated reporting laws explained, many batterers and sex offenders will still choose to make disclosures. I have not read in the literature or experienced in my own practice any significant loss of a therapeutic relationship as a result from making a mandated report. Though some clients/offenders may terminate treatment following a mandated report being made, many will find another program to complete treatment and some will have more severe legal consequences. However, the offenders were informed of the limits of confidentiality and the clinician must abide by the reporting laws. The bottom line is that any failure to make a mandated report or to take action in any case where an offender has reoffended while in treatment, only serves to (1) give the offender the message that the clinician is likely intimidated by the offender, after all, the clinician failed to follow the law; and (2) the relationship between the clinician and offender has now become an *aiding and abetting in the commission of a crime* relationship. The therapists should be held legally and ethically responsible for their failure to follow laws and ethical standards. And last, but perhaps the most important aspect of failing to make a mandated report, is that the victims are not being protected to the degree they could be. And worse yet, the abuse is likely to continue as a result of the reinforcement the offender received, that is, the unspoken support of the clinician by failing to make the mandated report.

47 Confidentiality in Treatment

Confidentiality in treatment is an important and complex concept. On one hand, treatment participants have the right not to incriminate themselves by disclosing the existence of additional abuse or sexual abuse victims. On the other hand, full disclosure is imperative for full investment in treatment to occur. In addition, it is imperative that the additional victims, especially children and vulnerable adults, receive the treatment they desperately deserve and need. It is also important that the offenders be held fully responsible and accountable for their offense behavior, including offenses that they have not been charged or convicted for.

There are differing views pertaining to what guarantees are afforded to physical and sexual offenders while involved in treatment. Two of the most common assignments in treatment are the autobiography and criminal history, often presented as one goal. The offender is expected to document significant events in his life, including history of having been abused as well as documenting additional crimes, including those that he was not arrested or convicted for.

The dilemma is this. When offenders make disclosures that they have neglected, physically, or sexually abused a vulnerable adult or minor, that information must be reported to law enforcement or child or adult protection within 24 hours under law, and ethical standards require such compliance with reporting laws. Many therapists and mental health professionals fear that, if the offender is aware of the mandate reporting laws, he will withhold such information for fear of prosecution, or that in some way the information would be used against him. Ethical standards and laws require that anyone who receives psychological treatment, including offenders, must be made aware of the limits of confidentiality, including the mandated reporting laws. Contrary to this fear, the vast majority of offenders will eventually disclose information concerning unreported or unadjudicated criminal offenses involving sexual or physical abuse, neglect, or rape. This author has found that most offenders, once established in treatment, will make such disclosures. There will, of course, always be offenders who simply will not make any such disclosures, and it is often this handful of offenders who are used to support therapist concerns about the reporting laws. The bottom line is this. If the offenders are mandated into treatment following being convicted of physical or sexual abuse, or even if they are a self-referral, they must make a decision to fully comply with treatment expectations. There will always be those who refuse to do so.

It is difficult to imagine how an offender would benefit from any type of treatment intervention without fully investing in the treatment process. Full disclosure is a requirement, a necessity to gain anything useful from the treatment experience. If an offender is allowed to "complete" or "graduate" from abuse or sex offender treatment without making all disclosures of his history of engaging in abuse or sex offense behavior, then he has successfully conned and manipulated his therapist, and will have learned that he is even smarter and cleverer than the mental health professionals. Unfortunately, he is at higher risk for reoffense because he was allowed to avoid taking responsibility for his full criminal abuse history.

Another concern voiced by therapists is that if they report the offender's disclosure, they risk losing the offenders trust; the therapeutic relationship that took so long to establish could be damaged beyond repair. The unfortunate thing is that this belief demonstrates therapist immaturity and a general lack of competence. In addition, this demonstrates a therapist who has his/her own issues to address, and it is my professional opinion that these issues create an impaired professional. To work with a predominately court-referred clientele requires different skills than working with general psychological problems. Very few offenders would quit treatment at the point of the therapist making a mandated report. The offenders are usually court ordered, which means if they cease participation in treatment, they will go to jail or prison. In addition, if the offender reveals information that is reportable, remember that the offender, like any recipient of any psychological service, must have been made aware of the limits of confidentiality prior to beginning treatment or psychological services. Therefore, any disclosure made while in treatment occurred with full knowledge that the information would have to be reported. Therefore, even if offenders choose to leave treatment, they were unamenable for treatment anyway. They were not ready to accept full responsibility for their criminal behavior.

When offenders disclose that they have reoffended, that is, they have physically or sexually abused or raped again while in treatment, some therapists again fail to report the information to law enforcement because they do not want to lose the offender or his trust. But think about it. This is really a "no brainer." The offender reoffended while involved in a treatment program that addresses offense behavior. Offenders are not allowed to have "slips" or "relapses" in the same way an alcoholic may. They are not allowed to reoffend while in treatment and be given special handling by treatment staff, which only serves to support the offender's belief (and obviously, in this case, the therapist's as well) that the offender's criminal behavior is out of the offender's control, and that he did not mean to offend but was merely his own victim of circumstance. Offenders who reoffend with the same or similar type of behavior, while involved in a treatment program or receiving individual counseling or therapy, are demonstrating that they are indeed unamenable for treatment and are incapable of ceasing their offending, abusive behavior without more stringent consequences and in need of highly structured treatment settings, including residential treatment, state mental hospital programs, and even prison. While there certainly are physical and sexual offenders who lack internal control over their criminal and abusive offense behavior, understand that they are still held accountable for their behavior, and are likely candidates for civil commitment to highly structured treatment programs that remove the offender from the public for many, many years and, for the majority, for life. Therapists, who refuse to make mandated reports when the offender makes disclosure, abet further abuse. It is my belief that such therapists and programs that allow such disregard for the laws should be held criminally liable for the abuse they aided in covering up.

Offenders can make useful disclosures without creating a mandated reporting situation. They can report the physical or sexual abuse that occurred and provide an accurate description of their behavior, but not provide the victim's true identity. For example, an offender can report an additional five child victims, the victim's rough ages, the relationship between the offender and victim in general terms (e.g., family or nonfamily), and the approximate time frame of the abuse. Without specific details that would identify the specific identity of the victim or victims, no report could be made to the authorities. But the offender could then choose to make full disclosures without fear of prosecution and still be held accountable for the extent of his abusive or sexually assaultive behavior.

I have personally found that there are offenders who, after careful discussion, decide to report additional victims. They are aware that the information must be reported to the appropriate authorities. These tend to be the lower intellectual functioning individuals or those who simply are motivated to be held fully accountable for their criminal behavior. Interestingly, I also find that the more narcissistic or psychopathic the offender, the more he is likely to disclose further abuse because he feels immune from prosecution. The arrogance and need of recognition for their behavior often result in their revealing more offense-related information than expected.

I find it best to make all mandated reports directly to law enforcement, rather than the social service agencies, such as adult or child protection. The reason is that I have found that social service agencies tend to be far more understaffed, over-worked, and, as a result, are less likely to take necessary action. Law enforcement agencies, on the other hand, are far better trained to intervene and make necessary decisions for arrest and criminal prosecution. At any rate, make the report to whatever agency you trust will give the case the most time and energy.

The therapist should be careful not to assume the role of an intermediator between the offender and the police department or County Attorney's Office. That is a role that should be reserved for the offender's family, friends, and attorney. In fact, assuming such a role appears to be a conflict of interest and results in an impaired professional. It is unacceptable for the therapist to provide therapeutic services and also serve as an advocate or legal advisor. The therapist is biased in that they have an investment in the offender's treatment as well as in the number of offenders who complete their treatment program. In addition, it is exceptionally difficult for a therapist to provide a neutral risk assessment of the offender when engaged in treating the offender.

Pertaining to providing forensic assessment and treatment, I find that it is never an acceptable practice to go from the role of treatment provider to forensic examiner with the same patient. A forensic examiner providing any type of assessment for the courts must be neutral and unbiased. A previous therapeutic relationship significantly decreases the ability of the professional to provide an unbiased examination of the offender. Also, previous attitudes of either liking or disliking the offender are going to exist if a prior treatment relationship existed, again, making it nearly impossible to provide an unbiased examination. However, moving from the role of forensic examiner to treatment provider may be acceptable when the forensic examination has a clear end, and the offender was offered other treatment options other than the examiner. If the offender wants to receive treatment from the examiner, that is fine, as long as the offender is appropriate for the professional's program. However, offenders must to given the names of other providers and treatment programs so that they have a choice of providers, not simply the examiner.

When the offender reports previously unreported abuse or sexual offenses, I have found that at times, the County Attorney's Office was willing to offer the offender less or no further time served in jail or prison for the offender's complete cooperation. That would usually involve the offender pleading guilty to the new charges, and cooperating with continued treatment programming. Everybody wins in this situation. Offenders are held 100% accountable for their offense behavior, the County Attorney's Office does not have to spend significant financial and staffing resources building their case to go to trial (because the offender admits the offense), law enforcement does not have to complete a lengthy investigation because the offender provided most of the facts, and, most importantly, the new victims were identified and could then be offered therapeutic services. Remembering that at times, when the offender makes such disclosures, the victims have not yet even reported the abuse. Everybody wins.

Another issue that frequently arises is what information the treatment staff chooses to document. I believe it is unethical to not document information provided by an offender involving additional acts of abuse or sexual assault/rape/molestation. If the offender reveals information on additional violent crimes or additional victims, then that information should be clearly documented. This information is 100% related to his current treatment issue and needs to be documented. In addition, not documenting the information gives the offender support in deviant ways. Perhaps the most damaging is that by failing to document additional acts of violence or victim identifies, the offender is far more likely to minimize his offense behavior and more likely to justify his behavior as being insignificant. In addition, it is difficult to provide the necessary ongoing risk assessment when pertinent facts of additional victims or offenses are not clearly documented. Remember, that if it is not documented, it did not happen. For a therapist not to document relevant risk information is simply unconscionable. When other offenses are found out and the courts come to the therapist inquiring about statements the offender indicated that he made to the professional, the professional needs to rely on documentation. Offenders will do whatever they can to get the professional in trouble or to discredit the professional. Accurate documentation is the best protector a mental health professional has.

Further, professionals failing to document such information often are the same professionals failing to make mandated reports. The role of a therapist or psychologist is never, ever to protect the client/offender from experiencing consequences for his abusive or sexually abusive actions.

Therapists and psychologists who protect their clients are, in my professional opinion, impaired and unable to provide safe and competent forensic or therapeutic services for offenders.

Informed Consent

When I conduct forensic assessments on abusers or sex offenders, I read verbatim my Informed Consent form. It is intimidating, yet outlines most all of the possible outcomes resulting from any disclosure the offender makes. It is imperative that any professional conducting a forensic assessment (this is also referred to as a "court-ordered assessment") provide a written informed consent to the offender. The reason I read it verbatim to each and every offender is that I account for literacy issues that have yet to be identified (because this occurs on the initial visit with the offender) and then the offender's argument that he did not understand "informed consent" or how the information will be used is weak at best, and I have never had an offender win an appeal based on that argument. The offender has a copy of the Informed Consent form and he follows along as I read it to him. I also have a similar version for the victim and for the spouse or partner because it is important that they are also aware of the limitations of confidentiality for the statements or other information that they provide. At the end of this chapter is a copy of my informed consent form. For abusers and sex offenders, it would be the same form. The same form is also used with the victim, spouse and partners, with minor modifications.

In my experience of assessing over 850 sexual offenders for civil commitment consideration while employed as the Civil Review Coordinator for the Minnesota Department of Corrections, only approximately 10 offenders refused to sign the Informed Consent and refused to participate in the assessment interview, despite being fully aware of the possible outcome, which included being referred for civil commitment and possibly spending the rest of their lives in a state hospital. In my experience with batterers, I have only had approximately three that refused to sign the Informed Consent form on the first assessment interview, but they did sign it on the second visit. However, I did not discuss the offense details without the Informed Consent form signed.

For offenders who are involved in a treatment program and making satisfactory or better progress, the disclosure of additional victims probably will not impact their completion date from the treatment program. And, if the disclosure of additional victims results in the County Attorney's Office deciding that additional punishment is not warranted, then the offender may receive minimal additional sentencing or an extended probationary period rather than additional incarceration. So, at times, it is in the offender's best interest to reveal additional victims rather than waiting until the victim reports a crime. Many children and adult victims report the crimes of abuse and molestation and statutory rape at some point in time. So, the offender disclosing the offense before the victim reports it is often in the offender's best interest and more often results in plea agreements with the County Attorney's Office.

I was one of the experts on a Governor's Task Force in Minnesota to address the issue of civil commitment for sexual predators. One of the topics included *use immunity*. *Use immunity* would offer offenders immunity for any information on new sex crimes while participating in sex offender treatment. The final outcome was that *use immunity* did not work and allowed offenders to escape any prosecution or additional restrictions necessary to protect the community. Even one of the most prominent defense attorney concurred, indicating that it would be difficult to define what crimes would be safe from prosecution (*use immunity*) and what crimes would not be included in *use immunity*.

It was decided that offering *use immunity* for disclosures made in treatment pertaining to the presence of additional victims does not appear to be a comfortable or practical option. *Use immunity* would prevent law enforcement and the County Attorney's Office from using disclosed information to investigate and/or prosecute an offender for any crime that he voluntarily discloses while in the course of sex offender treatment. The primary problem here is one of accountability. How can the offender be held accountable and responsible for his criminal behavior if he cannot be prosecuted and held to pay some consequence for the sexually assaultive or abusive behavior? The answer is that he cannot with *use immunity*. Imagine that a sexual offender reveals that he sexually molested several other children and these children have not been identified and no report of child molestation exists. The offender reports these crimes to his therapist and, with *use immunity*, the offender can never be prosecuted for those crimes. So the offenders can continue to offend as long as they continue to report their behavior. Another reason why *use immunity* does not work is that the good of the community should be the utmost issue at hand, and the victims deserve to be identified and offered necessary assistance and resources.

Perhaps the most compelling reason against *use immunity* is that the offender (1) made a conscious decision to engage in the offense behavior, and (2) made a conscious decision to disclose information pertaining to additional violent offenses. Therefore, any information disclosed was offered freely and may be used for or against the offender. In addition, the offender needs to be held 100% accountable for his offense behavior history, not for just one offense. Think of this example. An offender is arrested and believes that the authorities will become aware of additional violent criminal acts, or that the offender was not arrested, but is concerned that authorities may find out about the offenses. So, the offender enters treatment and discloses all of the details for his violent criminal behavior. If he had *use immunity*, he could not be held criminally accountable for his offense behavior. And if he is in treatment and discloses that he murdered one of his victims, and that the offense was never detected, would he then be untouchable for that offense? For these and a variety of other reasons, *use immunity* was not implemented and it is unlikely that it ever will be.

INFORMED CONSENT

To Participate in a Domestic Abuse Assessment

Mending Path Relationship Center, Inc.

Scott A. Johnson, M.A., L.P., DABPS, DAACCE, ACFE

You are being assessed to determine what recommendation appears appropriate to address the problems that resulted in this referral. You are aware that the results of this assessment will be shared with the court to comply with the court order.

The assessment consists of (1) a review of all records made available at the time of this assessment; (2) obtaining relevant collateral information from any other source as deemed necessary (which may include law enforcement, probation offices, child and adult protection, friends, family, etc., whether by person, phone, and/or mail); and (3) a clinical interview (if you choose to participate). Where appropriate, releases of information will be obtained.

You have the right not to participate in this clinical interview. Should you choose to exercise this right, your refusal may adversely affect the outcome of this assessment, but will not bias this examiner's opinion of the facts. However, this may be the only opportunity that will be provided to allow your participation and input into this process. Should you choose to participate in the clinical interview, anything you say may be used against you in that it might support criminal charges being brought against you; the loss of custody of your children; other legal action, sanctions, or restrictions; action being taken by child or adult protection agencies; or possible divorce or separation from a spouse or a relationship ending. On the other hand, information you possess may support the above entities not taking drastic or legal action against you, and may result in a better understanding of your version of the actual or alleged offense or identified problem or concerns, as well as a better understanding of your strengths and skills. Your participation allows for more accurate, appropriate, and effective recommendations being made to the referring party. You may choose to participate in the interview, but reserve the right to refuse to answer certain questions if you wish. You may also end the clinical interview at any time. It is understood that information obtained and statements made during the interview are not private and may be shared in the examiner's report to the referring person or agency. A copy of this assessment report will also be sent to the referring person or agency and a copy will also be placed in your file.

This process does involve psychological and/or other testing. The results from the testing will be incorporated into the final assessment report. The clinical interview may last up to 3 hours, but will likely last approximately 1 to $1^1/_2$ hours. During the clinical interview, you will be asked to provide your view of pertinent questions and provide evidence or other argument on your behalf. The questions asked will be of a general and personal nature.

Under Mandated Reporting Laws, if you reveal that you/another person have abused or neglected a minor or vulnerable adult in the past 3 years, and reveal the identity of that person, that information will be reported to law enforcement. Also, under the Duty to Warn law, if you report that you intend to seriously harm yourself or another person, appropriate action will occur to ensure the safety of yourself or the intended victim, which may include notifying law enforcement. In addition, if you are pregnant, or provide information concerning a pregnant person that reveals that they/you are using certain drugs or abusing alcohol, this information would be reported to the appropriate authorities.

I, (print name)________________________________, (DOB)____________, have decided:

To participate (sign name)________________________________ Date: ________________

Not to participate (sign name)______________________________Date: ________________

Witnessed by: Scott A. Johnson, MA, LP, DABPS, DAACCE, ACFE

Date:_______________________________

48 Preparing for the Victim–Perpetrator Session

REUNITING PERPETRATORS WITH THEIR VICTIMS

Some clinicians attempt to reunite the perpetrator with the victim and or their family as soon as possible. Research has indicated that the time the perpetrator is absent from the family may cause the family unit harm, especially for young children. However, the research does not indicate that any of the effects are permanent. In fact, most research seems to indicate that the children continue to care for their parent despite the parent's absence, especially when the custodial parent supports the memory of the missing parent to the children.

The effects of reuniting the perpetrator with their family too soon can cause significant harm. What I am referring to is the effects of psychological force and abuse. When the perpetrator has made sufficient progress in therapy, and the victims have also made strides in their own therapy and healing process, the reunification process supports growth and healing. If one has not made progress in therapy, or simply is not ready for a meeting with the other, it should not occur at that time.

Imagine the effects of a child or adolescent who was the victim of sexual abuse, and the abuser was the father. If the father does not understand how he made the decision to offend, has not identified and begun to resolve the underlying issues surrounding the offense, and does not have a reoffense prevention plan in place, the victim is likely to sense that he is not sincere and may even feel threatened by him. This may be difficult for others to understand or to even be aware of. I suggest that all professionals involved in the reunification process become educated about psychological force, and to examine carefully the readiness of both the victim and perpetrator prior to arranging a session where they will be together.

I will share the criteria I use to determine whether a batterer or sex offender is ready to engage in the reunification process. Each clinician may adapt this outline to fit his or her particular orientation and comfort. Again, I offer the following as an outline, not an empirically supported body of work. I will use the term perpetrator to include batterers and sex offenders.

How to Assess Perpetrators' Readiness and How to Prepare Them

1. Voice tone. How does the perpetrator sound? Is his language rough or is it gentle and safe? This is very important to assess. Remember that the perpetrator utilized subtle as well as overt coercion and control and the victim will be highly tuned into detecting it far before the clinician or other caregiver. Listen carefully to the victims when they describe how perpetrators abused them and what was said. Have the victim describe the voice tone.
2. Lack of blaming on the part of the perpetrators. They need to accept and take responsibility for 100% of the offense behaviors. No one else was to blame, in any way. They cannot blame stress, alcohol, or drugs, or any other issue for their choice to offend. Do they really believe that they are, in fact, 100% responsible for their choice to engage in sexual offense behaviors? If so, then continue.
3. Understand perpetrators' offense cycles and how they made the decision to offend. This is imperative long before any reunification between victim and perpetrator occurs. Again,

it is nearly impossible to protect the victim from the perpetrator's abusive behavior and comments if the clinician and perpetrator do not understand the perpetrator's individualized cycle and decision to offend. Even during a therapeutic session, the perpetrator could further psychologically harm the victim by not being able to explain his cycle or how he gave himself permission to offend.

4. Understand and accept that there isn't a cure for batterers or sex offenders, they could reoffend at any time, any place, against anyone. Statements such as: "I know I will never do it again," are indicative of cognitive distortions that support the perpetrator truly believing he did not have the power to reoffend, which is an extremely dangerous belief for an offender to hold. I would guess that he held this belief just prior to committing his sexual offense, except just the opposite; he was unable to control himself, he was out of control and had to act out. There are two sides to every story and to every issue. If an offender believes he will never reoffend, then he believes that he was powerless in some way and had to offend in the first place. You cannot have one of these beliefs without the other.
5. Demonstrate genuine empathy for their victims. In addition, they should be able to identify all of the people who have been in some way affected by their offense — family members, children, spouse, friends, etc. Genuine empathy includes, but is not limited to, identifying the effects on the victim at the time the offense occurred, current effects, and potential future effects. Be explicit and graphic when discussing the effects and discuss potential extremes in reactions, such as prostitution, suicide, etc. Empathy also includes the perpetrator assuming full responsibility for his offense behavior and not blaming the victim or anything or anyone else for his choice to offend.
6. Be able to identify and discuss their grooming process. Many perpetrators deny that they groomed their victims. However, they did not just one day decide to sexualize and engage in sexual fantasies of their victims, and then suddenly make the decision to sexually assault them. The batterer may have begun with subtle psychological abuse, putting the victims down, using sarcasm, to groom their victims for more overt and serious forms of control and abuse. Isolation of the victim (e.g., limiting access to financial resources, to friends and family) is also part of the grooming process for both batterers and sex offenders. The grooming for most offenders began weeks or even months prior to the offense occurring. The offender, despite making any significant progress in therapy, will harm his victim if he is unable to discuss with his victim how he groomed her. Further, it would be nearly impossible for the perpetrator to provide a feasible reoffense prevention plan if he is unable to identify and discuss his grooming behaviors openly and in detail.
7. Sessions with victims and perpetrators should **NEVER** occur too soon. Both require time to demonstrate their healing and that takes time — months not days. Do not rush the process or you will end up in the role of colluding with the perpetrator in minimizing the hurt, pain, anger, and confusion the victim is likely experiencing. The victim's therapist must participate at least to the degree that he/she assesses the victim to be prepared for and psychologically ready for the session. The perpetrator is not the one who dictates when the sessions occur, ever.
8. Do not be intimidated by the perpetrator. They can use depression or intimidation to manipulate you into a session with their victim. Stand your ground and do not allow the perpetrator to guide the reunification process. It is not uncommon for perpetrators to threaten to sue because the clinician would not allow them to move back home or to meet with their victims at the time they requested. Yet, I am not aware of any clinician having been successfully sued for that reason. Perpetrators drop their threats after their attorney reminds them of the court order which states that they are to follow any and all recommendations of the treatment program.

9. Any pressure, threat, or intimidation you feel from the perpetrator is a clear indicator that he is not ready to meet with his victim. In fact, if he is threatening you, the person who holds power over his freedom in some way, he will likely threaten his victim if given the opportunity to do so. Therefore, the reunification process must be delayed.
10. It is imperative that both the perpetrator and victim each have a different therapist. One can imagine how difficult it would be for the victim to trust the therapist who works with the person who assaulted her. It also becomes a difficult task for therapists to respect and maintain appropriate boundaries with information entrusted to them by the perpetrator and victim. Some information may be disclosed during the meeting, which the perpetrator or victim should not have access to.
11. Have the perpetrator write a letter to his victim. This letter can then be discussed and revised in his therapy group and with his therapist. This allows an opportunity for the letter to be revised months prior to the session occurring. It is imperative that the language of the letter be carefully listened to and examined. It is easy for a perpetrator to blame others or even his victim for his offense behaviors in very subtle ways. Carefully listen to the overall attitude of the letter. Does the message being given accomplish the goal for why it was written or does it only confuse the issues?
12. The tone of the perpetrator while the letter is read offers significant data on how well he believes he is responsible for the offense. In addition, effect should be appropriate for the setting. Guilt and shame certainly would be expected and would be appropriate, but when in excess, it is indicative that the perpetrator is not yet ready to meet with his victim. Victims are like sponges, absorbing without limit the subtle and direct emotional tones and the underlying messages of the perpetrator's language, his choice of words and his attitude. Be aware that the words and phrases used may have a unique meaning to the victim and/or her family. Scrutinize every aspect of the letter, settling only for a letter that successfully tells the victim that the perpetrator is accepting full responsibility for his choice of offending, and that he is able to explain the process by which he chose to offend as well as his reoffense prevention plan.
13. When the letter appears to be well written, send a copy to the victim's therapist to review. This allows him/her to prepare the victim for the content and explicitness of language. DO NOT HAVE THE VICTIM'S THERAPIST READ THE LETTER TO THE VICTIM, OR HAVE THE PERPETRATOR MAIL THE LETTER TO THE VICTIM. The perpetrator needs to read it directly to the victim and the victim may have a letter as well to read to the perpetrator. The point in using a letter is to be able to limit the content of the letter, prepare the perpetrator for the meeting, and to offer some sense of safety for the victim.
14. When the perpetrator is taking responsibility for his offense, it is imperative to mention how he sexually abused his victim. To simply apologize for "sexually abusing" her does not even come close to admitting what happened during the offense. Further, many victims feel cheated by such a general statement that really does not identify the specific behaviors they endured. The perpetrator should acknowledge the exact behaviors he engaged in. For example, a letter may read like this: "I apologize for what I did to you. It was wrong for me to fondle your breasts and vagina, and even more wrong of me to insert my finger into your vagina... ." The point is to own the exact behaviors that occurred.
15. Some may be concerned about the explicit language in the letter. Many parents certainly do not want the victim to hear this language, and even a few (thankfully, only a few) therapists resist this as well. But exactly who are we protecting by withholding the explicit language, the victim? Remember that it was the victim who experienced and survived the sexual abuse first hand. The victim knows what the perpetrator did to her, she needs to have the perpetrator admit to the behaviors, not minimize the touching

under the blasé category of "sexual abuse." The perpetrator is let off the hook when he is not expected to inform his victim of exactly the behaviors he forced her to do. The language can be modified based on the developmental level of the child. However, I caution that most are capable of hearing the truth up front, uncensored because they were there.

16. There are at least two possible serious effects of not including explicit language in the perpetrator's letter. The first is that the victim never heard the perpetrator admit to the behaviors that actually happened. The second is that a direct effect of not being explicit with the sexual behaviors the victim was subjected to engage in places an immense sense of shame on the victim. They are taught a direct lesson, and this lesson is that it is too uncomfortable and unbearable for adults to discuss how they were sexually abused. I cannot even begin to imagine how victims could learn to live with themselves, having been sexually abused, and not being allowed or encouraged to openly discuss their experiences. The victims learn to not discuss the details with others, not for their own protection, but to protect the perpetrator and other family members. This is a revictimization.

Effects of Engaging in a Meeting between the Offender and Victim Too Soon

1. The perpetrator believes that he has made more progress than he actually has. The effect is that the potential risk for reoffense is minimized.
2. Both perpetrator and victim may question if professionals are concerned about potential risk or if either are truly ready because the meeting is occurring.
3. Both may believe that the perpetrator possesses less risk because the meeting would not be occurring otherwise.
4. The focus of treatment for both the victim and perpetrator may become the issue of reunification, when, in fact, they have not yet healed their wounds.
5. It is difficult to assess the ramifications of the sexual abuse, especially the types and intensity of psychological abuse. It may take days before the victim begins to react to the meeting.
6. The family needs time to deal with both the shock of the offense and the difficulties of having the perpetrator living away or having no contact. It generally takes 3 to 6 months for everyone to settle down and for the problem areas of the family system and the degree of the perpetrator's abuse to be correctly assessed and appreciated.
7. The therapist should control the pace and timing of future contacts between the victim and perpetrator, being careful and attentive to the needs and issues of each. Again, both the victim and perpetrator may attempt to rush the reunification process. The therapist should be aware that victims often want contact with the perpetrator soon after the assault occurred, especially when the perpetrator is a relative. Be careful to recognize that victims are faced with the dilemma of both loving and hating their abuser, and that they may request or even demand contact long before they are even aware of how they were affected by the assault.
8. If the first session goes well, schedule more sessions. Make sure to process the sessions in-depth during individual sessions soon after the joint session. Do not rush contact outside of the therapy session. I recommend sessions continue within the therapist's office only for 2 to 6 months before even considering contact elsewhere. The reason for this is that it takes time for the perpetrator and victim to process their emotional reactions of the meetings. Each needs time to think about the previous session. Problems may arise during or after the sessions. The victim may experience fear of the perpetrator, which may include nightmares, depression, sexual acting out, or may be having flashbacks of the sexual assault.

9. Again, it may take several days or even a couple of weeks for victims to begin to experience negative effects, and may take time before they tell their therapist. They may believe that they are then hindering the reunification process if they tell anyone. Victims should be educated about all of the possible effects of sessions with their perpetrator, and strongly encouraged to tell their therapist. It is also advisable to have the perpetrator give the victim permission during the session to tell the therapist of any problems that arise, even if it delays future meetings.

During the Victim/Perpetrator Session

1. The therapist should set boundaries and establish ground rules. Include the victim in this process. The clinician is in control of the session and can end the sessions at any time or give the perpetrator a timeout, if needed.
2. Encourage victims to bring anyone they may need for support, including but not limited to other family members, friends, clergy, etc. Again, both should have their therapists present. If one does not have a therapist, he/she should begin to work with one as soon as possible, but PRIOR to the session. As stated before, there are drawbacks to utilizing only a single therapist.
3. Seat the perpetrator and victim across from each other. This allows for a sense of physical safety, and makes it easier for the therapists to observe the interaction more clearly.
4. Pay attention to and observe body movements and postures, language, and voice tones. Intervene immediately if anyone crosses any verbal, physical, or sexual boundaries. Included here is if the perpetrator uses language that appears in any way blaming others for his behavior, and any language, voice tones, gestures or behaviors that appears abusive, threatening, or intimidating. For the victim, if she appears overly trusting, she backs down, or appears to blame others or herself for the assault, intervention is warranted.
5. Limit, if not avoid, touches and hugs. I prefer not to allow any type of touching to occur too early in the healing process. In most circumstances, neither the perpetrator nor the victim is ready for touches from the other during the first several sessions. A hug, as innocent as it appears, can actually undermine months of progress in therapy. Touching too soon often gives a false sense of security and comfort in times of stress. Further, touches may result in subtle abuse and intimidation, or the occurrence of flashbacks of the physical or sexual assault. If this were to occur, especially early in the process (within the first three sessions), significant harm may result and therapy progress could be undone.
6. Stick to the agenda as much as possible. I understand that the therapist should allow the victim to set the pace, and if what is being discussed shifts, it is important to ascertain if the perpetrator and victim are both ready for any deviation for the agenda. One example comes to mind. The victim asks the perpetrator questions concerning the use of alcohol in the assault process, and the perpetrator continues to believe that alcohol caused the assault. The perpetrator blaming his offense behavior on alcohol could adversely affect the victim and and this could increase the victim's anger. Although the victim becoming angry would be an appropriate response, she is, to some degree, revictimized by the perpetrator not taking responsibility for the assault behaviors. One other example is if the victim asks the perpetrator how he came to make the decision to commit the assault. The perpetrator may continue to hold beliefs that the victim provoked the assault and may blame the victim. Needless to say, the victim could suffer setbacks in her therapy progress. Hopefully, the therapists of both the victim and perpetrator have prepared them, and have ascertained that both have genuinely made progress enough to participate in this type of session.

7. Remember that timing is everything. Pace the session to accomplish whatever time will allow. End the session early if either the victim or perpetrator appears to have had enough or if the therapeutic benefit appears to have diminished. Don't rush the process. Whatever is not accomplished during one session can be addressed during the next.
8. Allow for the victim to arrive early. This will prevent unsupervised and unwanted contact from occurring in the waiting room. Have the perpetrator leave shortly before the victim, so as to allow the victim to process the session. If safety appears to be a concern, have the victim leave first so that the perpetrator has less opportunity to have unsupervised or unwanted contact with the victim. Both therapists must take responsibility to address upfront any concerns that arise, especially safety concerns.
9. Demonstrate genuine and appropriate attitudes during the session. If the victim or perpetrator observes one of the therapists expressing an overly positive or negative attitude, erroneous assumptions and messages may be perceived. Give feedback and make comments openly and freely.

49 Research Concerns and Assessment Tools

BACK TO THE BASICS OF PREDICTION

It is important to differentiate tests' or screening tools' ability to identify if an individual belongs to a certain "high risk" or "recidivism" group vs. predicting if a specific individual will, in fact, reoffend. As in the case of placing a specific batterer or sex offender into the high-risk group for violent or general recidivism, all an individual needs to do is to appear to possess certain traits, behavior, or history. For example, it has been demonstrated that sex offenders who have prior sex offense convictions and have a deviant sexual interest that includes sexual arousal involving children are at high risk for sexual recidivism and are likely to not respond favorably to sex offender treatment efforts or legal interventions (e.g., Hanson & Bussiere, 1996; Frisbie & Dondis, 1965). However, there is no test or screening tool available to date that can predict if this specific sex offender *will* reoffend. A screening tool or test may support the placement of this individual into a group of men who are at high risk for sexual reoffense. But there isn't a test or screening tool available to correctly identify, with acceptable certainty, that offenders of the high-risk group will indeed reoffend, and it is unlikely that there ever will be such a screening tool. If the score of the actuarial tool places the offender at a 70% rate for reoffense, it is still unknown whether the offender is, in fact, in the 70% of the offenders who *will* reoffend or in the 30% of offenders with the same score that *will not* reoffend.

The ethical and professionally correct statement is that a given individual is at high risk for violent or sexual recidivism, and that he fits into a group of batterers or sex offenders who have sexually and/or violently reoffended. Again, this is based on the offender possessing a number of identified risk factors that strongly correlate with sexual reoffense.

Generalizability

It is important to understand how the test or risk assessment tool being used fits for the population of the offender for which it is being used. I have seen tools that state in the manual that you can use the tool on females as well, despite the fact that the tool was not standardized on any female population. We can certainly assert that we may expect offenders to share many traits or risk factors; however, we are unable to speak with certainty or even significant probability until appropriate research has been conducted.

Definitions

It is imperative that research clearly defines the things being measured. The most difficult issues I see that create problems include the lack of appropriately defined recidivism, not clearly defining the type of treatment the offender is receiving, failing to adequately differentiate offenders into specific typologies, and inadequate time at risk for the offender. It is imperative that recidivism is clearly defined. Many studies rely on rearrest and conviction rates to determine reoffense. Anyone not arrested and convicted of a new assault or sex-related offense is not counted as a recidivist. But his is exceptionally problematic. As research has demonstrated for many, many years, offenders have committed many offenses for which they did not get caught and will not ever be. I recall a general rule of thumb being that for every offense or incident of offense known, there are generally other offenses and victims that will not be charged or identified. Also, the victim, offender's current and prior spouses and partners, and supervising agent will need to be contacted yearly to identify whether reoffense has occurred. Even when victims were uncooperative when they first reported the offense, they still may provide accurate reoffense data at a later time. And given that most offenses do not result in immediate disclosure or criminal charges, it is imperative to gather as much information from as many available sources as possible to determine the true recidivism rate. Lastly, most offenses result in a plea bargain. Simply performing a criminal history check is insufficient to identify new arrests or convictions for assault or sex-related behavior. And I also believe that any violent behavior counts as recidivism.

Batterers and sex offenders may cease their hitting or raping or molesting behavior for a time. But most will continue to engage in coercive, seductive, manipulating, and verbally abusive behavior and language, which still is abuse and, therefore, either a reoffense or ongoing abuse that never ceased. Many studies fail to specify whether these batterers who cease physically abusing also ceased their psychological abuse as well. Numerous studies have indicated that many, if not most, batterers will reoffend at some point (e.g., Tolman & Edleson, 1995; Gondolf, 2000; Hamberger & Hastings, 1993). In addition, the under-reporting of sexual offenses and limited ability to clear all cases makes the use of rearrest rates significantly lower than the actual reoffense rates (Janus & Meehl, 1997). Barbaree and Marshall (1988) reported that unofficial sources show 2.4 times more reoffense than official records.

The types of treatment for batterers and sex offenders used to vary like brands of soups. That has changed. Standards are now available for treating both batterers and sex offenders. Programs that follow the most acceptably effective treatment modality (group) and follow cognitive-behavioral and psychopharmacological approaches are the most effective. But labeling a program "cognitive-behavioral" does not mean that the program follows the same principles or techniques that any other program follows. So, to be useful, the program must identify what specifically it is that they do, what assignments are used, and how groups are structured. How program material is implemented also varies across treatment programs taking into account program philosophies, beliefs, training and background of clinicians, and adherence to standardized treatment protocol.

If the offenders being studied are not placed into relevant and appropriate typologies, then how will anyone know the true recidivism rate for the offender or for a specific treatment program? It is clear that not all typologies of batterers or sex offenders reoffend at the same rate. So, programs that offer reporting on typologies should be meaningful and useful. For example, most research suggests that incest offenders tend to reoffend at a significantly lower rate than child molesters. That one example is even too simplified. Some incest offenders have both male and female victims and subsequently reoffend at a higher rate than those with only a female victim. Other incest offenders have both familial and nonfamilial victims and may even have adult victims. Again, these offenders would likely reoffend at a higher rate, even though all would be considered incest offenders on some level. Batterers also fit into different typologies and the characteristics of each typology of offender may impact reoffense rates. The bottom line is that, just as the medical profession assesses and treats syndromes and diseases specific to each disorder, so, too, must offenders be typologized and treated to address specific personal issues and personality characteristics.

The last primary issue (more than will be discussed below) is measuring true recidivism rates with meaningful *time at risk*. *Time at risk* is the amount of time the offender is in a situation when he could reoffend. Time incarcerated is still time at risk, but not anywhere to the same degree as time in the community. Most studies measure recidivism in a 1- to 5-year time frame. This is hardly adequate. Over the course of a lifetime, the offender has many opportunities to reoffend. And no research has successfully proven the first 5 years at risk is only when most offenders will reoffend. Over the course of 20 to 40 years, the rates of recidivism continue to increase. At some point, most typologies of offenders reoffend at over a 50% rate; many higher. But few studies follow their offenders 20 to 40 years. Despite the recidivism rate declining in percentage over time, offenders do continue to reoffend. Despite a lesser percentage of offenders reoffending later in life, many will continue to reoffend.

Published Work

The fact that an instrument or tool has been published in a peer-reviewed journal does not guarantee that it should be utilized, is ready to be used, or that it is appropriate for your population. A careful review should be conducted to examine the areas mentioned below.

First, and perhaps most important, the statistical validity and reliability should be clearly explained and understood. If either is low, then the tool or instrument should not be used for anything other than a general screening. Even if no other screening tool exists for the purpose at hand, it is unethical and unprofessional to utilize an instrument or tool that has yet to be standardized on your specific population. I have read correspondents from one of the authors of the Minnesota Sex Offender Screening Tool-Revised (MnSOST-R) in which he encouraged the use of the MnSOST-R with probationers despite the fact that it had not yet been standardized on that population. The most concerning issue is that this author was encouraging the misuse of the tool. The fact that no tool exists to use with a specific population does not in any way make it ethical or appropriate to apply assessment screening tools or protocols to an untested population. Ever.

Second, the population of the subjects included in the study should be clearly identified. This should include the race, gender, age, socioeconomic background, marital history, sex offense history, general criminal history, treatment history, and mental health and chemical dependency history. These factors are important to consider for several reasons. The most important of which is that the tool or instrument should only be used with populations that are very similar. For example, if the tool or instrument was standardized on incarcerated adult male sex offenders, whose mean ages ranged from 25 to 40, middle socioeconomic, and primarily Caucasian, then it would be inappropriate to utilize the tool or instrument with a population primarily consisting of men who are African American, of lower socioeconomic status, and in another age category.

Another overlooked area for researchers raises serious concern about the credibility of the research pertaining to the treatment success for batterers and sexual offenders and recidivism data. Most studies group all male batterers or sex offenders and label them together into a single category, which is as "batterers" or "sexual offenders." However, as research has demonstrated, not all types of offenders recidivate at the same rate, nor do all types of batterers or sexual offenders respond well to treatment intervention. I believe it is extremely important for researchers to separate their subjects on the basis of various typologies. So, when a study reports that 30% of its sample reoffended, in my opinion, it means little unless there is a breakdown of the different types of the offenders in the study. For example, I want to know how many same-sex child molesters reoffended, how many stranger rapists, etc., not simply how many "sex offenders" reoffended.

Researchers must take responsibility to accurately report their data and many fail to do this by lumping all sex offenders together. This usually occurs because of small sample sizes, which prevents breaking down into types. In addition to the typologies, it is also imperative to know what the victim pool of the sexual offender is. This appears rather simple to do. All available sources of information should be utilized to determine the offender's victim pool. Does the offender have only child, adolescent, or adult victims, or has he crossed age categories? Does he have male and female victims and, if so, are the male victims children while the female victims are adults? This would certainly be important information. And what about the relationship between the sexual offender and all of his known victims? Is this strictly an incest offender? Has he offended outside the immediate family or community? Does he even know any or all of his victims, or for the most part, are the victims strangers? Or perhaps he has both known and stranger victims, as well as victims that are family members. Typology and victim pool are important areas to pay attention to. Unfortunately, many of the research articles published have very low occurrences for several specific types of offenders, yet they present the data as if these areas were taken into account or by failing to mention it at all, perhaps attempting to avoid addressing the issue altogether.

Third, the limitations pertaining to the use of the tool or instrument need to be clearly identified and mentioned in detail. This includes identifying the population that is most appropriate or not appropriate with which to use the instrument or tool, the required training necessary for competent and consistent administration, scoring, and interpreting of scores or profiles. It should never be assumed that because someone is a licensed psychologist that they are, therefore, competent to use a tool or instrument. They must still receive appropriate training and supervision before using the tools or instruments. For example, if the tool or instrument provides weight to whether someone has entered, completed, or was terminated from either substance abuse, domestic abuse, or sex offender treatment, and the offenders in your jail or prison are not likely to have all of the necessary treatment offered to the majority of the offenders, then the tool or instrument should not be utilized. It would be inappropriate to have points unaccounted for and still use the tool or instrument. If the tool is not standardized for female offenders (and most are not), then it should not ever be used with female offenders outside of research context.

One common problem I have seen involves the use of the Hare Psychopathy Checklist-Revised (PCL-R) (Hare, 1991). Many psychologists administer this test and incorporate the results in sexual civil commitment assessments. If the psychologists involved in a case arrive at more than a two-point difference, it is likely that at least one or perhaps both of the psychologists have not received adequate training specific to the PCL-R. Dr. Hare and his staff conduct the only acceptable training of which I am aware. I have heard several psychologists comment that following their training with Dr. Hare, they recognize that they had been incorrectly scoring items. Competence in testing and assessment requires — no, demands — that a psychologist remain up to date with current training and continuing education. In addition, the clinician should obtain adequate supervision and demonstrate quality controls, following all established protocols.

Further, I assert that no psychologist should conduct forensic assessments unless they have worked both as a therapist treating the population they are assessing as well as having experience assessing the risk for reoffense. To assess for forensic purposes requires a deep, mature understanding of treatment issues as well as advanced training and experience with the specified population. Psychologists who conduct forensic evaluations may be asked to address the issues of whether an individual is mentally ill and dangerous, sexually dangerous person, competent, may be asked to assess issues pertinent to custody, and, of course, may assess whether someone should be civilly committed. The fact that a specific psychologist is competent in one area of forensic psychology, however, does not imply that the same psychologist is competent for any other area of forensic psychology. Another area of frequent concern is when psychologists misuse research and developmental instruments or tools, misreporting "percentages" and "risk of reoffense" by misinterpreting what the actual results mean. It is important to only apply those research studies that are directly relevant to the issue at hand.

Psychologists who fail to work within their competency and training, who fail to participate in specific continuing education, and who fail to keep current with the ever-changing issues and techniques have no business engaging in the practice of any type of psychology. I have testified in many court proceedings where I have had to comment on how another psychologist's work was unacceptable or incompetent. Every time a psychologist practices outside his/her true area of competence, the resulting consequences, at the very least, undermine the entire psychological field.

Limitations of Prediction Tools

One of the primary questions asked of forensic psychologists is what degree of risk a specific abuser or sex offender poses for violent reoffense. As we all are aware, there is no crystal ball or fortune teller that can accurately predict which abuser or sex offender will reoffend. At best, we are able to utilize screening tools to identify which abuser or sex offender possess factors that have been correlated sufficiently with reoffense. I have often testified at civil commitment hearings pertaining to sexual predators. It is common to hear professionals comment with certainty about an individual's risk for sexual reoffense in terms of a percentage. For example, to say that Mr. Smith is at a 70% chance level for sexual reoffense is an example of overstating the ability of a test or screening tool. At best, a screening tool can place an individual into a group of sex offenders who reoffend at a specific percentage rate (e.g., 70%), but no one can say with any degree of certainty which of the sexual offenders within that group (e.g., of those within the 70% range) will actually reoffend and which will not. For example, if a specific offender scores within the 70% range, we can not predict with any certainty whether that specific offender will reoffend because he is in the 70% range or whether he will not reoffend because he is actually in the 30% range of those who will not eventually reoffend. No test or screening tool currently available can predict with 100% accuracy which specific individual will reoffend and which will not.

Many psychologists become defensive with the above stated limitations of screening tools and tests, but, the reality is what we have to work with, and the actuarial tools greatly enhance the prediction of future acts of violence. If we can comfortably place a specific offender into a category of offenders known to reoffend at a high rate, then we have increased the possibility for increasing necessary supervision for that specific offender or increasing sanctions to further decrease the chances of violent reoffense. However, some offenders will not reoffend with or without close supervision or attention, while others will reoffend despite close and strict supervision. In fact, some offenders will find ways to continue to offend even while incarcerated. Interestingly, I have seen sex offenders continue to offend violently while incarcerated, and a few who had contrived plans to access additional victims outside of the prison with the help of spouses, intimate partners, and business schemes. So, at any rate, it is suffice it to say that we can place an individual into a category of abuser or sex offender who are at high risk for violent reoffense with relative certainty, but we cannot predict which of these offenders within the high-risk group will actually reoffend. Some of the more reliable and accurate tools are mentioned below.

Risk Assessment Tools

Karl Hanson is one of the most well-known researchers studying the risk assessment of sexual offenders. He conducted a well-known meta analysis (Hanson & Bussiere, 1996; Hanson & Morton-Bourgon, 2004) that involved analyzing numerous studies in an attempt to identify the most robust, accurate factors that relate to violent sexual recidivism. As a result of the study, 69 factors were identified that related to violent recidivism. Dr. Hanson developed the *Rapid Risk Assessment for Sexual Offense Recidivism* (RRASOR) (Hanson, 1997). This is an easy to score tool that places a sexual offender into a risk category with significant accuracy. His latest tool is the *STATIC-99* (Hanson, 1999) that combines the RRASOR and another tool, the *Structured Anchored Clinical Judgment* (SACJ-Min) (Grubin, 1998). This tool appears to be one of the best available to predict which sex offenders are in a higher risk group for engaging in future acts of sexual reoffense. However, even this tool is unable to predict which specific individual will reoffend and which will not; again, the tool produces a score that places the offender into a risk category. The RRASOR and the STATIC-99 were found to be more accurate screening tools at predicting sexual offense recidivism (Barbaree et al., 2001) than the other sex offense screening tools, including the MnSOST-R.

The Sexual Violence Risk-20 (SVR) (Boer, Wilson, Gauthier, & Hart, 1997) is an assessment method or procedure, rather than a test or scale. It was developed primarily for use in sexual violence risk assessments conducted in criminal and civil forensic contexts. This is useful when administered as part of a battery.

Multiphasic Sex Inventory (MSI) (Nichols & Molinder, 1984) is a self-report questionnaire designed to assess a wide range of psychosexual characteristics of the sexual offender. This is most useful to assess the cognitive distortions and degrees of minimization. There is limited validity for this tool.

The Minnesota Sex Offender Screening Tool-Revised (MnSOST-R) (Epperson et al., 1998) was developed by the Minnesota Department of Corrections for purposes of identifying which sexual offenders are at higher risk for sexual reoffense. The MnSOST-R appears to have high face validity, and appears accurate in identifying which sexual offenders reoffend at a higher rate. However, the MnSOST-R remains untested and unduplicated. When I worked as the Civil Commitment Review Coordinator for the Minnesota Department of Corrections, I found that the MnSOST-R significantly underestimated risk level and found that approximately 72% of the sex offenders referred for consideration of civil commitment actually had scores significantly lower than those recommended by the MnSOST-R, which suggests that this tool is not as robust and accurate as the authors portrayed. In addition, Craig et al. (in press) indicated that Barabree et al. (2001) found that the MnSOST-R was far better at predicting general recidivism than sexual recidivism.

The *penile-plethysmograph* (PPG) *phallometry* (Freund, 1965) has certainly proven itself in possessing the ability to accurately identify the sexual preferences of sexual offenders, despite some critics. It appears to accurately identify deviant sexual interest, attraction for sexual violence, and other important dimensions as well. Numerous studies have documented the utility and accuracy of the penile-plethysmograph. The strength is in assessing pedophiles. Abel et al. (1981) and Avery-Clark and Laws (1984) were also able to separate more physically aggressive offenders using phallometry. Numerous studies have also indicated that sex offenders who have a deviant sexual interest and deviant arousal reoffend at a higher rate than sex offenders who do not have a deviant sexual interest and arousal (e.g., Hanson & Bussiere, 1996). The major impediments in utilizing the plethysmograph is that it is an invasive procedure, it relies on nude slides (including those of children), and it is expensive. It is also susceptible to faking, with some offenders able to suppress or alter their response (Abel et al., 1975; Laws & Holmes, 1978; Quinsey & Bergersen, 1976) and some sexual offenders, especially intrafamilial offenders, do not display deviant sexual arousal (reported in Milner & Murphy, 1995).

Gene Abel, M.D., developed the *Abel Assessment for Sexual Interest* (Abel, 1995), which measures sexual interest utilizing visual reaction time, self-reported arousal, and a thorough sexual interest questionnaire. Without getting into all of the specifics of how the Abel Assessment works, it has withstood the test of time and critical review. Johnson and Listiak (1999) conducted and published research demonstrating that the Abel Assessment was highly accurate for identifying most sexual offenders having deviant sexual interest and reported arousal. The limitation is that the Abel Assessment does not adequately identify the sexual offender's interest to violence in general, though Dr. Abel never purported that it would. The Abel Assessment has produced impressive results. The Abel Assessment impressively was as effective as the PPG for assessing deviant sexual interest.

Personality tests such as the *Minnesota Multiphasic Personality Inventory,* 2nd edition (MMPI-2) (Butcher et al., 1989) and the *Millon Clinical Multiaxial Inventory-III (*MCMI-III) (Millon, 1983) identify with much success the personality characteristics of any individual. This aids in correctly diagnosing individuals. However, there is no "sex offender" or "abuser" profile on either test. Offenders may present with certain characteristics fairly consistently, but individuals who are not abusers or sex offenders may also present the same or similar characteristics. There is no personality trait that has been proven to apply solely to abusers or sexual offenders; that is, no personality traits or profile that discriminates sex offenders and abusers from nonabusers and nonsex offenders. Therefore, there is no MMPI-2 or MCMI-III profile specific to batterers or sex offenders.

The Hare Psychopathy Checklist-Revised (PCL-R) (Hare, 1995) has been one of the best developments in the field of risk assessment. Dr. Robert Hare's tool helps differentiate psychopaths from others. Antisocial Personality Disorder is not the same as psychopathy, and most individual's having Antisocial Personality Disorder are not psychopaths. The tool clearly establishes specific definitions to utilize to differential and appropriately diagnose psychopathy. This is exceptionally important because psychopathic offenders are by far the most dangerous. Psychopathic offenders should not be treated. Research has consistently demonstrated that psychopaths benefit from treatment only by becoming more cunning, conning, and more dangerous. They do not cease their abusive or sexually exploitive behavior.

The Spousal Assault Risk Assessment Guide (SARA) (Kropp et al., 1999) is designed to screen for risk factors of spousal or family-related assault. This tool appears robust in identifying those abusers who are at higher risk for engaging in future acts of violence. The SARA can also be applied to those in dating relationships. However, it is not normed (standardized) and, therefore, should not be used with female abusers.

The Assessing Risk for Violence (HCR-20) (Webster et al., 1997) is a checklist for violent behavior. This tool is considered a research tool; however, it appears useful as part of a testing battery to assess the likelihood of an offender engaging in future acts of violence.

State-Trait Anger Expression Inventory-2 (STAXI-2) (Spielberger, 1999) identifies the types of anger an individual experiences and how the anger is most likely to be expressed. This can be used with both male and female abusers. This has demonstrated adequate and robust reliability and validity.

Child Abuse Potential Inventory (CAPI) (Milner, 1986) is a self-report instrument used as a screening tool for the detection of physical child abuse, not sexual abuse. This has demonstrated adequate and robust reliability and validity.

The Parenting Stress Index (PSI) (Abidin, 1983) is a screening and diagnostic assessment designed to yield a measure of the relative magnitude of stress in the parent–child relationship. The PSI is divided into two parts. The first part is the Child Domain, which measures the qualities of a child that make it difficult for parents to fulfill their parenting roles. The second part is the Parent Domain, which measures sources of stress and potential dysfunction of the parent–child system that may be related to the parent's functioning. This has demonstrated adequate and robust reliability and validity.

Summary of Testing

There are other tests that I use when assessing child abuse, but the above are the most common tests and tools I find most useful in administering an effective test battery for the purposes of completing a forensic or treatment based assessment. The more tools and tests used, the more accuracy tends to occur. A battery helps to address concerns that an offender might produce a profile on one protocol of dubious validity while producing a more accurate profile on another, similar protocol (e.g., MMPI-2 and MCMI-III). An offender might score within normal range on the STAXI-2, yet his responses on the MMPI-2 and Conflicts Tactics Scale suggest an anger problem.

COMPETENCE

It is important that a psychologist or anyone else administering psychological tests or tools be competent in the specific tool/test he/she is utilizing. Simply having an advanced degree does not constitute "qualified" or "competent." The Abel Assessment, penile-plethysmograph, and PCL-R require specialized training as outlined by the developers of the tests.

It is important to differentiate the test's or screening tool's ability to identify that an individual belongs to a certain "high risk" or "recidivism" group vs. predicting that a specific individual will, in fact, reoffend. As in the case of placing a specific sex offender into the high-risk group for sexual recidivism, all an individual needs to do is to appear to possess certain traits, behavior, or history. For example, it has been demonstrated that sex offenders who sexually molest male children, have stranger victims, have prior sex offense convictions, and have a deviant sexual interest that includes sexual arousal involving children are at high risk for sexual recidivism and are likely to not respond favorably to sex offender treatment efforts or legal interventions. However, I cannot say this enough: There is no test or screening tool available to date that can predict that this specific sex offender will reoffend. However, a screening tool and/or test may support the placement of this individual into a group of men who are at high risk for sexual reoffense. But no test is able to correctly identify with enough certainty which offenders of the high-risk group will indeed reoffend.

When people present information about a new screening or assessment tool or instrument at a conference, such as at the Association for the Treatment or Sexual Abusers (ATSA), all that is being done is to disseminate information concerning the tool. It does not constitute a peer review nor does presenting at a conference contribute anything significant to the applicability, appropriateness, or effectiveness of any tool or instrument. In fact, until the tool or instrument is published in an appropriate peer-reviewed journal, it should not be utilized. However, the fact that an instrument or tool has been published in a peer-reviewed journal does not guarantee that it should be utilized. A careful review should be conducted to examine the areas mentioned below.

Psychologists are required to adhere to the ethical standards set forth by the American Psychological Association (APA, 2002) as well as the ethical guidelines set forth in the state where they practice. Continuing education must occur frequently that involves batterers and/or sex offender specific assessment, treatment, or that is focused on forensic assessment of the batterer and/or sex offender if the psychologist engages in forensic work with batterers and/or sex offenders.

Problems with Research Problems with Efficacy of Treatment Programs

Robertson (1999) identified several problems with research on the effectiveness of batterer treatment programs poor treatment specifications (difficult to determine how and what was evaluated), unspecific definitions of "success," over-reliance on self-report data, short follow-up periods, and failure to distinguish program effects from other factors in a batterer's life (e.g., separation, arrest).

The primary problem with self-report is, of course, that batterers and sex offenders sometimes lie. When addressing self-report of reoffense, men regularly under-report their violence (e.g., Dutton, 1986; Tolman & Bennett, 1990; Tolman & Edleson, 1995). This is even more a problem when current partners and victims are not interviewed to determine the veracity of the offender's claim.

Relying on rearrest records presents significant problems. First, most programs and researchers fail to invest the necessary time and effort into gathering records. It is a time-involved process. Second, the chance of a domestic assault resulting in an arrest was found to be approximately 1.37% and the chances of conviction just 0.73% (Dutton, 1987). Consider that most batterers (e.g., Dutton, 1986; Tolman & Bennett, 1990; Tolman & Edleson, 1995) and sex offenders (e.g., Abel et al., 1987; Abel & Osborn, 1992; Abel, Mittleman, & Becker, 1985; Frisbie & Dondis, 1965) are found to have committed numerous violent acts for which they are not convicted. My own experience has demonstrated that most offenders have committed far more instances of battering and sex offenses than their arrest records ever show.

Some programs define "success" as a reduction in violence or violent acts rather than an absence of violence (e.g., Edleson, 1995). At times, this means that programs labeling batterers as having "successfully completed treatment" may mean that the batterer has not engaged in physically assaultive behavior for the identified time period. The time periods are almost always significantly too short to be meaningful and this will be addressed elsewhere. However, most criminals can cease their problematic behavior when under the microscope of probation or any other form of supervision. Substance abusers have often ceased use of substances long enough to complete treatment or to pass a drug test. Batterers and sex offenders are equally capable of ceasing the behavior that is being closely monitored. Consider that sex offenders and batterers do not engage in exactly the same behavior each and every time they offend. The batterer utilizes far more psychological violence than physical or sexual violence and the psychological abuse tends to be ongoing (e.g., Eisikovits

& Edleson, 1989). Sex offenders do not always engage in touch offenses, but rather utilize grooming behavior and hands-off offenses quite often (e.g., Abel et al., 1987; Abel & Osborn, 1992).

In addition, when considering how treatment success is defined, understand that Eisikovits and Edleson found that approximately two thirds of the labeled "nonviolent" men were reported to have been engaging in the use of threats against their partners post-treatment. So despite completing a domestic abuse treatment program and having no further reported acts of engaging in physically violent behavior, they continued to terrorize their victims. In addition, psychological violence is a precursor to physically and sexually violent behavior.

Short follow-up periods have always plagued research. Perhaps the best and most promising research data are obtained when examining a short amount of time post-treatment (Tolman & Bennett, 1990; Tolman & Edleson, 1989). However, all of the research I have examined have increasing numbers of reoffense when extending the reporting period to include the offender's life span rather than just 6 months to 3 years post-treatment. Most studies have inadequately small sample sizes (Hamberger & Hastings, 1993). Another confounding issue here that is generally under reported is attrition rates. Eisikovits and Edleson indicated that many men drop out of programs when external motivators are removed (e.g., charges dismissed, wives/victims accepting the abusers back home).

Methodological problems also complicate research. It is often difficult finding an adequate control group. Volunteers are not the "norm" for batterers, sex offenders, or victims. In addition, attrition data are often omitted from the study, making it difficult to understand the true percentage of batterers, sex offenders, and victims who drop out of studies for any reason. As a result, many programs have an over-reported success rate.

Changing the behavior of batterers is made more complicated and difficult by several factors. First, batterers are rarely self-motivated to change (American Psychological Association Presidential Task Force on Violence and the Family, 1996). Second, batterers receive immediate positive reinforcement for their use of violence (e.g., compliance, sex, money, being left alone, or things being done for them). Negative consequences, when they do occur, occur following the assault (Myers, 1995).

Through their use of violence and manipulation, batterers usually succeed in controlling their partners. There exist powerful and accepted cultural values and beliefs that support men's privilege to exert power and control over women, and to take charge and exert power within the family unit (Russell, 1988). In addition, language plays an important and significant role in maintaining the batterer's position of power. Batterers are very adept at imposing their reality on others (Adams, Towns, & Gavey, 1995). Batterers often use terms of ownership (e.g., "my wife," "my money," "my house") and often use sexually derogatory and abusive language.

Robertson (1999) points out that an interesting paradox occurs with treating batterers. While the focus of treatment may be on addressing the power exerted by the batterer on others, batterers often report feeling powerless in treatment, which may impact their ability to invest in the change process. This is a result of several factors, including court orders; fears about the loss of relationships, marriages, or their role in the family; a perceived or real appreciation that they (the batterers) are less able to identify and express feelings than their partner/victim; possible experiences of having been abuwsed themselves; and possible loss of control involving substance abuse (Adams, n.d.; Currie, 1988; Pressman & Sheps, 1994).

50 Miscellaneous Issues: A Summary

IMPORTANCE OF OBTAINING COLLATERAL INFORMATION

The importance of obtaining collateral information cannot be underscored enough. A competent forensic assessment or interrogation requires that the interviewer or examiner have all of the available records and information. Without adequate records, the offender has the advantage and will be more able to offer deceit as fact. In addition, without adequate records, the full scope of the offender's offense history and potential for violent reoffense may be grossly underestimated.

MODUS OPERANDI

The modus operandi (MO) refers to the specific characteristics of offenders. This generally focuses on three things: (1) to protect their identity (how to maintain anonymity during an offense), (2) to ensure success, and (3) to successfully escape. Batterers and incest offenders do not care about anonymity; after all, the victim is almost always someone with whom they have an ongoing relationship. But, the specific MO may also include the pattern of violent behavior, how they set the victims up, and the specific violent behavior involved in the offense, as well as how the offender manipulates the victim into remaining in the relationship and decreasing the likelihood of the victim telling anyone about the violence. Specifics about how the offender behaved and statements made before, during, and following the offense also are part of the offender's MO. (See Hazelwood [1995] for more detailed information about modus operandi.)

LIMITATIONS OF TESTING AND PROFILES

It is important to understand the limitations of psychological testing, risk actuarial tools, and profiles and typologies. It is examiners' and interviewers' responsibility to have read the appropriate testing manuals and corresponding reviews and literature for techniques and testing used. Simply asking around is not adequate. Understanding the limitations is extremely important. A good example is the population that was used to standardize a test. I have seen several testing tools that included only men in their research sample, yet indicate in their test manual that the tool can also be used with female offenders. That is not appropriate or ethical. If the sample did not include females, at best, all you can do is compare the females scores to the males, *but the scores mean nothing*.

It is also important to have appropriate training. For example, the Psychopathy Checklist-Revised (PCL-R) (Hare, 1991) requires training and supervision before using the tool. The results of not having adequate training and supervision are easy to see — the final score is either too low or very high. Whenever there is more than a three-point discrepancy between the PCL-R scores of two examiners, it is likely that one or both lack necessary training and experience. In addition, there is no "sex offender" or "batterer" profile on the Multiphasic Personality Inventory-2 (MMPI-2) (Butcher et al., 1989) or the Millon Clinical Multiaxial Inventory-III (MCMI-III) (Millon, 1983). There are certainly common homogeneous characteristics shared among sex offenders and homogeneous characteristics among batterers. However, those same characteristics are also common among nonbatterers and nonsex offenders.

Pertaining to profiles and typologies, there will always be some offenders that do not fit neatly into one category or typology, and this is normal. Some offenders are evolving, that is to say that they are becoming another typology of offender, usually becoming more severe in their violent behavior. At other times, offenders simply fits in two categories and this is their idiosyncratic typology involving two or more typologies.

Importance of Using Official Sources of Information

The importance of using official sources of information cannot be emphasized enough. Official sources include any report or assessment from a mental health or law enforcement professional. This includes police reports, criminal complaints, assessment and treatment notes, and testing. Relying only on interview data is never acceptable for forensic purposes. Any information that was used to determine criminal charges, accepted by the courts, or used for assessment or treatment purposes should be reliable even if not 100% accurate. Without official records, the examiner or other professionals involved have inadequate information to work with.

Clarifying Charges, Convictions, and Plea Agreements: Avoiding the Traps of Plea Agreements

It is important to clarify any criminal record for offenses any way related to violence. These include robbery, burglary, assault, disorderly conduct, any type of assault or sex offense, including nontouch sex offenses. This also includes harassment, stalking, restraining orders/orders for protection, and any violation of supervision or orders for protection. It is important to obtain the police reports and/or criminal complaints for any arrest or conviction. This allows for the examiner or police officer to understand the offender's criminal history. Many violent offenses are commonly pled to lesser offenses, such as disorderly conduct. Regardless of the final pled charge, the details of the offense remain. If it was an assault or sex offense pled to disorderly, then the disorderly was sex or assault related. This counts against the offender's known criminal history and known violent history. It does not matter why a plea was accepted.

Victim Selection

It is important to understand how an offender selects his victims. Pay attention to the patterns that emerge. Is the offender selecting women for relationships that are more passive and depressed, or with children? How vulnerable does the victim appear (e.g., emotionally, mental health issues, financially, single parent)? Is the victim selection opportunistic or predatory? Predatory offenders select victims based on specific criteria. Opportunistic offenders select victims based on the situation at hand and the victim pool at the time. Incest offenders generally select victims within the home, but may also select children or adolescents outside the home. How and why offenders choose their victim is integral in understanding and intervening with them. Hazelwood (1995) does an excellent job of discussing the offender's selection of his victims.

Memory Problems

When offenders complain about memory problems, compare the symptoms presented with known memory impairment and *Diagnostic and Statistical Manual*, 4th ed. *DSM-IV* criteria. If the time frame of the memory impairment is only an hour or so, then it is generally not genuine memory impairment. Offenders who have engaged in numerous acts of verbally, physically, and/or sexually violent behavior may experience genuine difficulty recalling specific details of each and every offense. After all, their offense behavior has become habituated and one offense situation is similar to the next. But overall, it is my experience that offenders should be able to recall most details of most of their offenses without any memory impairment.

Drug and alcohol use may impair memory as well. However, it in no way excuses violent or other criminal behavior. In addition, people do not just act differently when intoxicated or under the influence of drugs, they make choices. They have expectations and act accordingly. In addition, the thoughts that resulted in physically or sexually violent behavior were premorbid to the substance use. That is, you don't just one day get drunk and then sexually molest someone, or hit someone. You have the thoughts before using substances. In short, ask yourself: "What kind of beer do I need to drink to assault someone or to sexually molest a child?" There is none.

Survival Strategies

It is important to engage in self-care to prevent burnout. Working with violence takes a toll no matter how good a person is at his/her job. There are some things you can do to prevent burnout and to enjoy your job more fully.

- Know your limitations. Despite having a degree (e.g., Ph.D.) and years of experience, everyone has limitations. What makes a person effective at forensic evaluations tends to interfere with general counseling and assessment skills. A police officer who spends more time profiling or conducting interrogations may tend to not be as comfortable or effective with general police work. Forensic work requires a separation of self to some degree, the ability to pay attention to your emotional reaction without allowing the emotions to be experienced fully or expressed. This is counterproductive in many professional duties, but required for forensic work. It is imperative to know what types of crimes you are comfortable working with. If you work primarily with victims, then you probably are not as effective with the offenders, and vice versa. Many psychologists and police officers are not good with confronting denial, and many are not comfortable utilizing the skillful art of sarcasm and direct confrontation. The ability to listen to an offender describe an offense without reacting to the pain or horror is a skill that some people are blessed with, more than others. It is important to work within your interest and within your limits.
- Establish and regularly use your support network. Having colleagues to discuss cases with is imperative for mental health and well being. Even if you do not question how you handled the case or interview, discussing the case with others helps to solidify your knowledge base and offers an opportunity to vent the emotional reaction that had to be stifled during the interview or interrogation. Family and friends also provide a source of support even if you cannot share the details of the case. For example, after interviewing or interrogating a batterer or sex offender, you may find yourself experiencing anger or disgust toward the offender. You may find yourself wanting to hug or be close to your spouse and children to appreciate your family after hearing about violence. This is common and needs to be attended to. Take care of your personal needs. Appreciate that you have a healthy and supportive family and network of friends and family. This is even truer when you relate your own family to the victims of the offender you were interviewing.
- Document, document, document. The old adage is still true: "If it is not documented, then it did not happen." Whatever forensic work you do, document everything. Every call, observation, document reviewed, everything. Inevitably, you will forget to document something; no one is perfect. But if you make this standard practice to document clearly, thoroughly, and often, few important details will be missed. When I conduct a forensic assessment, I use my laptop computer and type as the offender, victim, or family members respond to questions. This avoids the trap of documenting after the interview and, therefore, omitting key information and key reactions. Do not fool yourself into believing that you will remember statements or quotes an hour or more after the interview. And

you definitely will not recall all of the emotional reactions you had to the offender's or other's statements. Offenders give cues to deceit and cues to their psychopathology and these cues need to be documented as soon as possible.

- Set clear boundaries. This includes setting clear boundaries with offenders and their family as well as with your job. It is easy for an offender or his victim to con the examiner or interviewer. Our emotional reaction is difficult to control because we are first and foremost human. It is never, ever the job of the therapist, interviewer, interrogator, or police officer to be the caretaker for the needs of others. We help educate, confront, challenge, and support, but we do not take on the burden of "helping" to address the situation at hand. What I am talking about here is setting the boundary that the professional has a job to do and that is all that needs to be done. Do not set yourself up by assuming too much responsibility for the victim or offender. Life is not always fair, and good people sometimes get the short end of the stick.
- Utilize regular consultation. Even when you are the expert, you need consultation. Discussing your case with others allows for consideration of other perspectives and helps you to defend your work. Sometimes you edit your work as a result and at other times you find support that you were effective and accurate in your work. Consulting also helps to assure you that you covered all relevant aspects of the case and showed due diligence for the court when you defended your assessment or interrogation results. In addition, be open to feedback from colleagues. A good expert realizes that we never stop learning. Every interview and interrogation teaches something different about the offender and about batterers and sex offenders in general.
- Do not personalize the work. You are not a "police officer" or "forensic psychologist," but rather, this is your profession. People whose identity is their work tend to burn out sooner and tend to be less satisfied with life, in general. When you leave the office, leave the work behind whenever possible. Change your clothes, take a shower, go for a walk, play with the dog, and talk with your family. Basically, get into the "personal life" mode and just be a person. No longer a police officer or forensic psychologist, but just another person. Relax and tend to life's other joys. You will be far more effective and refreshed in your job duties when you can separate your professional identity/life from your personal identity/life.
- Recognize danger signs of burnout. Never believe that you are immune from burnout. We are all at risk. Narcissism helps to sugar coat this reality.
- Ongoing education. A competent expert never believes that they have learned it all. Maybe you have learned enough for now, but taking refresher courses or trainings is imperative. The training should be in the areas of your expertise, not in irrelevant or side areas. Make certain that you are receiving continuing education credits (CEU) training in your specified area of expertise. Otherwise, you will be out of touch with the more recent research and practices. There is nothing worse for an expert than to be on the stand testifying in court and be discredited because of failing to keep up to date with the literature or adequate updated training. I have had to testify in court to the lack of current competence of alleged "experts" who failed to keep current with literature and those who failed to receive adequate CEUs in their area of expertise.

51 Recommended Treatment Assignments for Batterers and Sex Offenders: The Most Important Assignments

USEFUL ASSIGNMENTS FOR BATTERERS AND SEX OFFENDERS

There are many assignments that are useful for batterers and sex offenders. Of course, it makes sense that there are core assignments that are strongly recommended to be completed and other assignments that are helpful and useful, but not necessarily as important. The literature is not yet supported that any specific assignment is more or less useful than any other one. What I will discuss below are some of the core assignments that I find most effective, useful, and imperative for batterers and sex offenders to complete as part of their treatment.

OFFENSE CYCLE

As discussed in Chapter 1, the offense cycle is imperative to understand from a step-by-step perspective if the offender is to understand and learn from their offense history. In addition, an understanding of the offense cycle is important for victims, intimate partners, and other support people to understand when to intervene on behalf of the offender and also being better able to identify when the offender is escalating and when the others need to protect themselves. I will not go into further detail of the offense cycle as it is detailed in Chapter 1. It is sufficient to say that if the offender only understands his offense cycle from a three-stage perspective, he is more likely to lack a more in-depth understanding of his cycle and may be more likely to miss important cues that he is, in fact, in his cycle and moving closer and closer toward possible reoffense.

OFFENSE GRID

The *offense grid* helps to identify the specific stressors, behavior, thoughts, and feelings that occurred before, during, and after the offense. After the batterer or sex offender has completed his offense cycle, the *grid* helps to examine the offense dynamics and events on a more detailed level. In the BEFORE section, the offender first identifies all relevant or possibly relevant historical information and factors (e.g., childhood events and traumas, history of substance abuse, divorces, mental health and medical conditions). These factors help to understand the offense in perspective of possible events or conditions that may have predisposed the offender to engage in violent behavior, though nothing ever makes or causes the offender to offend, ever. Simply put, this section helps to identify possible areas to explore later in therapy after the offender has completed the primary offender treatment program.

The offender then identifies specific thoughts, feelings, and behavior that occurred prior to the actual offense. I find it helpful to have the offender begin with identifying the actual behaviors that occurred, step-by-step, and then to identify the thoughts and emotions that occurred. The offender will identify at least three thoughts and three feelings for each and every behavior identified.

Remember that no behavior occurs without some cognition or thought, and feelings and cognitions are intertwined to occur almost simultaneously. Therefore, every behavior has multiple thoughts and feelings associated with it. When the offender experiences difficulty identifying the thoughts or emotions for a specific behavior, I ask the offender, "What would you normally tell yourself or do you think you would have to tell yourself to justify engaging in that specific behavior?" it doesn't matter what order the offender identifies the thoughts or emotions, just that every behavior has several thoughts and feelings identified. This process helps the offender to understand that he is fully in control of his behavior and that he, therefore, has the power to change how he responds to his thoughts and feelings. The BEFORE section relates to Stages 1 through 6 of the offense cycle presented in Chapter 1.

The DURING section relates to everything that occurs during the offense, from the point of initial verbal, physical, or sexual contact until the offense ends. Again, every behavior is identified. Each and every behavior that occurred during the offense is identified and again for every behavior identified at least three thoughts and three feelings are identified that would support the behavior. So, for example, the behaviors identified for a DURING section might include:

Batterers:

- Derogatory name calling
- Lunging forward toward the victim
- Grabbing the victim's right arm
- Pushing the victim with the other arm
- Knocking the victim down
- Helping the victim get up
- Grabbing the victim by both arms

Sex Offenders:

- Closing the door to the bedroom
- Approaching the bed where the victim is laying
- Removing ones own pants (step-by-step: each piece of clothing removed is a separate fact/behavior)
- Removing or telling the victim to remove her clothing (step-by-step: each piece of clothing removed is a separate fact/behavior)
- Touching the victim's breasts
- Kissing or licking the victim's breasts
- Digital penetration of the victim's vagina
- Having the victim fondle the offender's genitals

So each and every behavior is identified and examined. The offender needs to understand that at any point during the offense they could have stopped. Therefore, at every point during the offense, the offender had to convince himself that what he was doing was all right, that the victim needed to or deserved to have the behavior occur, and that he was justified. So every behavior relates to the next, cvery self-talk/cognition feeds into the next, and the emotions lead to the next escalating emotion. As part of the treatment process, the offender would learn to counter the self-talk that occurs at each and every stage, therefore, increasing the opportunities for the offender to stop offending. And, with the above examples, it is important to understand that batterers and sex offenders often engage in the same and similar behavior during offenses, that batterers utilize sexually derogatory language and often engage in aggressive sexual contact with their victims and, likewise, sex offenders utilize psychological and physically aggressive behavior in the commission of their offenses. So, the grids that offenders produce often are 5 to 10 pages in length and often take offenders 1 to 4 months to complete, getting feedback as they make multiple presentations of their offense grid to their group and therapist.

The AFTER section addresses everything that has occurred up to the point of entering treatment.

An Interesting Assignment

I would often have offenders read a completed offense grid to their support network during a session (again involving the offender's current partner, family, friends, and sponsors if involved in self-help groups). This would involve reading in detail the offenses grid that was to have been completed in the offender's group and involved a thorough description of the facts, thoughts, and behaviors that occurred before, during, and following an offense episode, and included graphic examples and direct statements the offender used or thought. In short, it was a word picture of the offender's ugly and deviant offense. The offender read this to his support group.

I found an interesting result from my experiences with sex offenders who had committed incest or child molesting offenses. I required that all of the men schedule a meeting in which they were to involve their support network which included, at the minimum, their current partner, family members, including parents (regardless of the offender's age), immediate family members, at least one close friend, and sponsors if involved in any substance abuse support groups. Of course, the offenders were not happy to comply with this seemingly cruel and unrealistic expectation. But, the vast majority of the offenders complied. During this meeting the offender was to present the assignment referred to as the *offense grid* (discussed below in Recommended Treatment Assignments).

Several important things occurred. First, it was the first time that the offender had actually told his support network and loved ones the truth about the offense. Most of the offender's support people stated that the offender had never admitted most of the offense details prior to this session. That is to say, the offender had never come clean with telling people what he actually did during his offense, how he set his victim up, and how he maintained secrecy from the support people. In addition, most of the support network had never understood how the offender made his conscious decision to offend, accepting that somehow the offender may have genuinely been "out of control," under the influence of drugs or alcohol and, therefore, not responsible for his behavior, or that the offender simply made a mistake that involved no intent to harm or the offender lacked an understanding of what he was doing. It was the first time the offender had informed his support people that he, the offender, made a conscious decision to harm and violate his victim.

Another important finding was that the offender's support network began to question the offender about other incidents and other areas of concerns. They also asked questions of the clinician to better understand the offense. It was interesting that many of the support people asked how they could better support the offender in not reoffending, and many had already developed realistic supportive strategies (e.g., paying more attention to the offender's cues leading up to possible offense behavior, paying attention to the offender's use or substances, or when the offender isolates). By the end of the sessions, none of the support people had any question as to holding the offender 100% responsible for the offense behavior, no one was blaming the victim or the offender's partner.

Perhaps the most important result of reviewing the grid with support people was that the offender now had a more effective support network, a network of people that better understood the offense cycle and importance of intervening when warning signs were present. The offenders generally reporting had improved relationships with their families and partners and were better able to openly discuss their stressors and concerns with their support network. Many of the offenders reported that it was the first time they had actually been able to be honest with their loved ones about the offense and that they were relieved to find so much support. Only two offenders had partners who choose to terminate the intimate relationship and it was because the offender had become involved with women who had female children that would have been potential targets for the offender. In all, the vast majority of offenders not only maintained their support network, but they developed more effective support by involving all of their support people in the treatment process.

The Decision to Offend

The next assignment is the *Decision to Offend* (see Chapter 4). This assignment helps offenders to further understand their offense cycle and decision-making. Though it may appear that this is very similar to the *offense cycle* and *offense grid*, it is a necessary tool to help offenders accept responsibility for their behavior and for implementing changes to help reduce the risk for violent reoffense.

The Cost of the Offense

I find it exceptionally valuable and important for offenders to identify and explore the cost associated with their offense behavior. The cost involves all areas of the offenders' and victims' lives from the time the offense began and into the future. It is often difficult to identify all associated costs, but an attempt should be made to identify the most salient costs. Costs are really the impact of the offense.

I have offenders complete the Cost of Abusive Behavior form (see end of this chapter) toward the end of their treatment. It is important that they have this to help with motivating not reoffending as well as to increase the degrees of empathy the offender might demonstrate. The areas to examine include the impact of he offense on the offenders' and victim's family. This includes identifying how the offender's partner/spouse, children, family of origin, and relatives have been impacted. Another important area to explore is the impact of the offense on the offender's community. This includes examining the cost or impact to the courts, police, neighbors, churches, and the community in general.

The offender is required to contact any resource necessary to identify the cost and time involved in the offense and offense-related activities (e.g., treatment for the offender, treatment for the victim, need for separate residences). The offender needs to contact the court administrator to determine the cost of prosecuting the offense, which includes the cost for the prosecutor, clerk of court, transcriber, bailiff/deputy, and fines. The offender also needs to contact the police department to determine the costs associated with the 911 call, the hourly rate of the officers who responded to the initial call (multiplied, of course, by the number of officers responding); cost of any incarceration; and cost of supervision. I am in no way attempting to cover all relevant aspects of this assignment in this chapter. These areas are indicated in the actual assignment. The point is that the offender is expected to clearly identify as much of the related costs/impact of his offense behavior as possible.

When the offender examines the cost of treatment for self, victim, and family, the figures grow significant. Even insurance payments count in this assignment. When the offender has completed this assignment, it is common that the total financial costs range between $8000 to $18,500 for a first-time offense. This helps the offender understand the costs associated with his offense. Battering and sex offenses do not just impact the offender's and victim's lives, but the entire community.

Letter of Amends

At some point in time the offender should be making amends in whatever way is safe and in the victim's best interest. The *Letter of Amends*. When done therapeutically and with caution, this has the potential to be a healing and valuable process for the offender, victim, and family. However, this is also one of the most risky assignments for the victim and family. (See the cautions in Chapter 48.)

Perhaps the most important thing to remember here is that the victim's well being is what guides this process, not the family's or offender's wishes or needs. The victim has the right to refuse any contact with the offender, and contact includes letters. In addition, below are some important considerations:

1. Any contact the offender might have with the victim should always be discussed with the clinician first. It is the responsibility of the offender's clinician to contact the victim's clinician to coordinate services and contact should never, ever be made without the victim's clinician involved.
2. The victim must be ready for and willing to participate in reading of the *letter of amends* or participating in a session where the offender will read the letter to the victim. The victim might well be willing to read the letter, but refuse to participate in any sessions or direct contact with the offender. This should always be respected. It is not the offender's wishes, but the victim's wishes that take precedent. This is just as true even if the victims appear vindictive in their refusal to meet with the offenders. I have heard some clinicians who attempted to persuade a victim to meet with the offender. The victim should never be persuaded to meet with the person who offended against them.
3. The *letter of amends* should be thoroughly reviewed and edited with the offender's group and clinician prior to forwarding the letter to the victim. It is the offender's clinician who is responsible for ensuring that the letter is not in anyway (1) blaming or projecting of responsibility of the offender's offense behavior, (2) that the offender is not asking for forgiveness, (3) that the offender was open and specific with all behavior that occurred, taking into account the developmental level of the victim.
4. I strongly encourage the offender's clinician to forward a copy of the completed *letter of amends* to the victim's clinician to allow time for that clinician to (1) make sure that the letter appears appropriate, which offers a second professional opinion; and (2) that the victim is prepared to hear the contents. The victim's therapist may or may not review the letter in advance of a session with the victim if deemed appropriate to do so.
5. At times, once both clinicians have approved the letter, the parents of the victim (if the victim, of course, is a minor) may have an opportunity to preview the letter in advance as well, though I strongly encourage this occur in the presence of at least one of the clinicians.
6. When the letter is read to the victim, it is best to have both the offender's and victim's clinicians present, as well as the parents if the victim requests their presence. Sometimes the victim requests the presence of a close friend or current partner and this should always be granted. Ensure that the victim feels a sense of safety ands security.
7. If the victim does not want contact with the offender, there is a restraining order not allowing contact, or other safety concerns are present. It is the clinician's role to determine if a session should even be offered. It may be that the most effective way for the victim to read the *offender's letter of amends* is simply to sent it. Clinical and forensic judgment should be used.
8. It is essential that the clinicians ensure that no unprepared or unallowed contact occurs between the victim and offender. In most situations it is advisable to have the victim leave the session to allow ample time to leave the area before the offender does. This may help prevent unwanted or inappropriate contact from occurring. It is also the clinician's job to ensure that offender, victim, and family understand that just because the *letter of amends* was read (see end of chapter following the Cost of Abusive Behavior form), it does not indicate that the offender and victim have ongoing or unsupervised contact. That is an entirely different clinical and legal matter.
9. I do not recommend that the offender ask the victim for forgiveness. The fact that the offender is in treatment and complying does not mean that the victim has healed from the offense or that the victim will be happy for the offender. In fact, the offender making such a request to read a letter of amends may further harm the victim in the sense that

the victim feels like the offender is simply moving on and yet, the victim continues to experience the ramifications of the offense and may suffer the impact for life. It is the clinician's responsibility to address this issue. The offender can and will still heal and may choose not to reoffend even without a session with the victim. The letter can be sent to the victim's clinician and can or may not result in a session to allow the offender to read it. Writing the letter is a therapeutic event regardless of whether the offender shares it with the victim or not.

There are other areas of concern that may arise in regard to contact between the victim and offender, as well as with amends letters, that are not discussed here. This is an intricate process that will always require an astute clinician's full attention and involvement.

SUMMARY

I have addressed some of what I consider to be crucial and effective assignments in treating batterers and sex offenders. There is no doubt numerous assignments that other clinicians find useful or essential. I am simply offering some that I have found most useful and am aware that many other programs use similar assignments.

OFFENSE GRID

	FACTS AND BEHAVIOR	THOUGHTS	FEELINGS
Before	A: Relevant history B: Immediately preceding offense		
During			
After			

Source: Scott A. Johnson, M.A., L.P., DABPS, DAACCE, ACFE , 2000©, Mending Path Relationship Center, Inc.

DECISION-MAKING PROCESS

	Abusive Decision Making
1.	Experience stress, triggers, and provocation. You recognize that you are escalating.
2.	Fail to utilize appropriate intervening strategies. You decide not to do what you know you need to do:
	Refuse to take a deep breath
	Refuse to take a time-out
	Refuse to reframe the situation
	Blame them and the world for any injustice you experience
	Give yourself permission to be "out of control"
3.	Personalize situation. You personalize everything that is said and done to you. You are unwilling to consider the possibility that no hurtful result was intended, or that it may have simply been coincidence that it happened to you.
4.	Giving up. This is the Abstinence Violation Affect (AVE). This involves giving yourself permission to abuse by justifying that you are far too close not to do it, or that you have already begun, and not that you are "out of control" and must finish what you started. You tell yourself:
	That "I cannot handle anymore"
	That "they provoked me and asked for it"
	That "they made me angry"
	That "I am angry and am unable to control my anger"
	"Well, I have already crossed the line, what difference does it make now"
	"The hell with it"
5.	Justify your escalating and violent response with supporting self-talk. You tell yourself that you are not abusing, that you are simply responding with just cause. After all, they hurt you, so you should hurt them.
6.	Engage in violent and abusive behavior and talk. You tell yourself that it is okay to be abusive and violent.
7.	Rationalize that they had it coming and that you were "out of control." You blame them for your abusive behavior.
8.	Attempt to move on and forget the abusive situation. You apologize for your violent and abusive behavior and ask for forgiveness. You promise that it will never happen again. You expect and demand immediate forgiveness and expect that they forget the incident and move on.
9.	You return to a state of heightened alert, never reaching "normal" or relaxed again. You decide to move on and forget that it happened.
10.	You are ready for the next battle. The self-fulfilling prophecy is alive and beaming here; you expect others will harm you, provoke you, or cause some sort of problems for you, and you are ready and waiting.
	Healthy Decision Making
	Thinking of what it is that I want. (e.g., "What I want from this person," "What I want this person to do," "What I do not want to do.")
	Strategizing how to get what you want, or the best way to word your request.
	Thinking through the possible outcomes, consequences, or reactions of your requests.
	Am I willing to accept the consequences of my requests or decisions.
	If you are able to identify possible controlling or abusive outcomes, then your decision or request is likely controlling or abusive.
	Take the time to confirm decisions prior to acting on them.
	If unsure about the possible consequences or if unsure about the possibility of your behavior being controlling or abusive, ASK.
	When feeling the urge to act immediately, ask yourself if the decision really must be made at this time (i.e., Is it really imperative to act now? Is it an emergency that cannot wait?). Take the time to confirm with your partner before acting.

Source: Scott A. Johnson, M.A., L.P., DABPS, DAACCE, ACFE , 2000©, Mending Path Relationship Center, Inc.

COST OF ABUSIVE BEHAVIOR

Examining the cost of abusive behavior can be a draining process. To understand the actual costs involved, examine each of the following areas impacted by your abusive behavior. Carefully and thoroughly identify and examine the emotional, spiritual, and financial impact of your abusive behavior in all of the following areas identified below.

1. **Family.** Your family has been and will continue to be impacted by your past abusive behavior and certainly will be impacted by your future abusive behavior. How has your abusive behavior affected how others view you? Identify and examine the impact of your abusive behavior on your family in the following areas:
   - ***Your Spouse***. How has your abusive behavior affected your relationship with your partner? In what ways has your relationship changed? Examine the damage to trust in your relationship with your partner.
   - ***Your Children.*** How has your abusive behavior affected your relationship with your children? In what ways has your relationship changed with each of your children? Examine the damage to the trust in your relationship with each of your children.
   - ***Your Family of Origin.*** How has your abusive behavior affected your relationship with your family of origin (including your parents, siblings, inlaws, and other relatives)? In what ways has your relationship changed with each of your family members? Examine the damage to trust in your relationship with each of your family members.
2. **Society.** Society has been and will continue to be impacted by your past abusive behavior and certainly will be impacted by your future abusive behavior. How has your abusive behavior affected how others view you? Identify and examine the impact of your abusive behavior on society in the following areas:
   - ***Your Neighbors.*** How has your abusive behavior affected your relationship with your neighbors? In what ways has your relationship changed with each of your neighbors? Examine the damage to trust in your relationship with each of your neighbors.
   - ***Local Police.*** How has your abusive behavior affected your relationship with local law enforcement? In what ways has your abusive behavior impacted how law enforcement deals with abuse? Examine the damage to trust law enforcement officer's experience with abuse-related crimes, especially pertaining to repeat offenders.
   - ***Houses of Religion.*** How has your abusive behavior affected your relationship with your religion? In what ways has your abusive behavior impacted how religions deal with abuse? Examine the damage to trust religious officials experience with abuse-related crimes, especially pertaining to repeat offenders.
3. **Financial.** Your abuse has cost your family and society considerable amounts of money. Ongoing abuse will certainly increase the actual financial cost to your family and society, in general. Identify and examine the impact of your abusive behavior on your family and society in the following areas:
   - ***Your Family.*** Identify the actual financial cost of your abusive behavior to your family. This includes the actual money paid for fines, damaged property, attorneys, living expenses (if separated), and divorce.
   - ***Law Enforcement.*** Identify the actual financial costs incurred for all law enforcement services related to your abusive behavior. This includes the cost of all police that responded to any of your abusive incidents; the cost of the detective's time; transportation costs; incarceration expenses; and any other expenses involved in the intervention, investigation, and prosecution of your abusive behavior.
   - ***Medical.*** Identify the actual financial costs incurred for all medical services related to your abusive behavior. This includes the cost of ambulance transportation, emergency room costs, expenses for medical supplies, medications, and follow-up services.
   - ***Psychological Interventions.*** Identify the actual financial costs incurred for all psychological services related to your abusive behavior. This includes the cost of transportation to appointments and support groups, psychological assessments, crisis counseling and therapy services, support group services, and follow-up services. Include the estimated number of sessions your victims will or may attend in your calculation.
   - ***Prosecution.*** Identify the actual financial costs incurred for all legal services to prosecute you related to your abusive behavior. This includes the cost of transportation; cost for county attorneys to investigate, prepare, and prosecute your crimes, costs for court, including the costs for the judge, bailiff, court reporter, and any other court-related personnel; incarceration costs; costs for probation or parole officers; and any other pertinent costs. Include past, current, and future costs.

4.	**Self.** Your self-image and identity has been and will continue to be impacted by your past abusive behavior and certainly will be impacted by your future abusive behavior. How has your abusive behavior affected how you view yourself? Identify and examine the personal impact of your abusive behavior in the following areas:
•	***Self-Image.*** How has your abusive behavior affected your self-image? In what ways has your credibility and integrity been impacted? Examine the damage to trusting yourself relating to your anger and need to abuse and control pertaining to your partner, family, and other relationships.

The above is not meant to identify all of the possible areas impacted by your abusive behavior. Please identify and address any areas that are not identified above that you believe are relevant.

For the above, you are expected to contact appropriate people to calculate the costs. This may include contacting the court administrator, clerk of court, police or sheriff departments, hospital administrators, and victim service providers.

Source: Scott A. Johnson, M.A., L.P., DABPS, DAACCE, ACFE , 2002©, Mending Path Relationship Center, Inc.

LETTER OF AMENDS

The letter of amends provides you an opportunity to take full responsibility for your abusive behavior and to apologize for it. The letter will be written and discussed with your therapist and group. When the letter appears appropriate and complete, you will schedule a session with your victims (generally with your primary victim/partner) and therapist and share the letter. This allows the victim or victims an opportunity to safely hear your amends. In some cases, the victims may not be able to or want to participate in this session. There are some cases when it will not be possible to meet with your victims or to send the letter to them. Your own healing occurs when you have demonstrated genuine efforts to make amends with your victims. Understand that you have more than one victim. There is the victim you directed the majority of the abuse toward (primary victim). Other victims include those who witnessed or were impacted by the abuse, including children, other family members, and friends. It is important that you demonstrate genuine empathy (understanding and compassion for your victim's experience) and remorse (regret, sorrow, and guilt about your abusive and violent behavior). You may also write letters to your other victims if you wish (e.g., children, family members).

The areas to be included in your letter are outlined below.

1.	**Acknowledge your abusive behavior.** This means clearly and specifically identifying all of your abusive behavior that has occurred throughout the course of the relationship. You will identify specific statements, words, and behaviors you have used to abuse and harm your victims. This will include the verbal and psychological, physical, and sexual abuse. It will also include how you have neglected your victims and family, used your children against your victims, and financial irresponsibility's.
2.	**Take full responsibility for your abusive behavior.** This involves admitting your *choice* to abuse. In addition, you will admit and accept full responsibility for how you set up situations or your victims in order to justify abusing them. Accepting full responsibility includes:
3.	Clearly and in detail identifying and accepting responsibility for your *plan* to abuse.
4.	Accepting full responsibility for all of your abusive behavior (verbal/psychological, physical, and sexual).
5.	Accepting full responsibility for how your abusive behavior *impacted* and *affected* your victims.
6.	**Identify *actual* and *potential* effects of the abusive behavior.** This involves making a list that identifies specific examples of how your victims have actually been harmed and impacted, as well as identifying how they may potentially be impacted or harmed. This includes short- and long-term effects. Demonstrate an understanding of your victims experience, what they have gone through, and what they may experience into the future.
7.	**Identify what you have done to address your anger problem.** This requires not only identifying what type of educational services or treatment that you have participated in, but also outlining what additional services or support you may seek at a later time. This includes any ancillary issues addressed (e.g., chemical dependency, mental illness, depression, medication).
8.	**Identify what outcome you are hoping for and willing to do to make amends.** This means clearly identifying what changes you are hoping to make or have already made (e.g., sobriety, attending support groups, taking time-outs, exercising, engaging in hobbies). Also, identifying how you can make direct amends to all of your victims.
9.	**If you have not already done so, share your Cues, Offense Grid, and Offense Cycle with your victims, family, and support network**. This is an important step that demonstrates your willingness to openly admit and discuss your offense behavior with loved ones and support people. Sharing these assignments with others not only helps you to hold yourself accountable, but allows your support people to help as well. It also allows for others to understand that you chose to be abusive, and that the victims were not in any way deserving or asking to be abused. Humility is necessary for lasting change to occur.
10.	**Apologize for your abusive and violent behavior.** You have apologized before and it was basically meaningless because you were not sincere. Now you need to make genuine promises and do whatever it takes to keep them. Invest in change. You need to demonstrate genuine remorse for your abusive and violent behavior.
11.	**Share and demonstrate your commitment to be nonviolent and nonabusive.** This is not about your promises or words. It involves demonstrating appropriate coping skills and not choosing to be violent or abusive. You need to "practice what you preach."

Source: Scott A. Johnson, M.A., L.P., DABPS, DAACCE, ACFE , 2000©, Mending Path Relationship Center, Inc.

52 Dealing with Parents of the Offender

INTRODUCTION

There are times when the offender's parents are involved in the assessment or treatment process. Sometimes this is helpful, especially when the parents understand and accept that the offense occurred and when they hold the offender fully responsible for all of the abusive and violent behavior that has occurred. In approximately half of the cases, however, parents offer more resistance and challenges than helpful and supportive behavior. After releases of information are obtained, the parents offer some degree of insight into the offender's personality and traits. For cooperative parents, they can be as involved in the assessment or treatment process as the clinician deems appropriate. However, the parents who fail to hold the offender fully responsible for the offense behavior or who minimize the offender's offense history need to be addressed upfront and early in the stages of intervention.

DEALING WITH RESISTIVE PARENTS

1. First and foremost, I inform the parents that their child, the offender, is, in fact, guilty of the offense and of the offense details. I explain that the courts found him guilty or the offender pled guilty, and, therefore, the innocence of the offender will not be a topic of discussion.
2. I explain that if they interfere in the offender's assessment or treatment, the offender is the one who will pay the price in consequences. I review that, as a result of being supported in their delusion of innocence, the offender is more likely to reoffend and far less likely to benefit from treatment intervention. I reiterate that, in the long run, if the offender is not held accountable, more serious offenses are likely to occur, and the offender is more likely to spend time incarcerated as well as to spend far more money in legal and treatment fees.
3. I gently assess the probability that parents who condone their child's offense behavior or who fail to hold the offender 100% accountable for the offense behavior often have guilt and shame about parenting issues and may well have been abusive towards the offender when the offender was a child. The mother may well have been the victim of abuse from her spouse (offender's father) and, therefore, may find it more difficult to accept their child's subsequent violent behavior.
4. I set limits as necessary. I had one male sex offender who was in his early 20s whose mother was interfering with his treatment. She was eventually banned from the premise and from making any phone calls to the program and was directed to the offender's supervising agent for any questions or concerns. Interestingly, after about a year in sex offender treatment and making marginal progress, it was revealed that this mother was continuing to sexually abuse the offender on a daily basis. Her interference in the offender's treatment program came into full understanding. This will not likely be the case for most interfering parents, but the more common ongoing abuse would likely be verbal in nature and a history of physical or sexual abuse within the family directed at other family members or the offender.

5. I provide education to the parents to help them understand that, at the very least, their child will benefit from treatment if everyone supports the end goals of, at least, learning new coping skills, abstaining from substance abuse, and not engaging in abusive or violent behavior. The goal is not to change the family system, as that would likely take years, and the parents and other family members will not likely participate in lengthy treatment or insight orientated intervention.

53 Differentiating between Conflicting Goals

INTRODUCTION

It is important to differentiate the goals of an interrogation, an assessment for the court, and treatment. All of these have different goals and require a different focus. The approach to each interview differs accordingly and it is imperative to understand that possessing a doctorate degree does not instantly qualify an individual to conduct an interrogation, forensic assessment, or treatment assessment (intake). It takes specialized training, supervision, and years of experience to successfully conduct any of the interviews. Let's review some of the basics.

INTERROGATION

The primary role of the interrogator is to obtain a useable confession that will stand up in court. The interrogator is not concerned about possible treatment interventions and not concerned with mental health diagnosis. The goal is to get the offender to admit his involvement in a specific crime and to gather any other useful admissions about prior offense behavior.

FORENSIC ASSESSMENT

The forensic assessment has the responsibility to address for the court the issue at hand. The examiner may be asked to provide a professional opinion, with supporting facts, as to whether an individual is competent to stand trial, is mentally ill and in need of custodial case, or to provide custody recommendations. Our concern in this book is related to forensic assessments that address civil commitment for sexual predators (e.g., sexually dangerous persons, sexual psychopathic personalities) as well as risk assessments with treatment recommendations. I have addressed the assessment and interview process in detail Chapter 37. However, it is important to differentiate the goals of forensic assessments.

First, the primary question(s) must be clearly identified and shared with the court and offender. The examiner has no business conducting a forensic examination without a clear focus. The question may be what risk level the offender presents with in regards to engaging in future acts of violent behavior or violent sexual behavior. In this case, any sexual offense-based behavior (e.g., sexual assault, child molestation, rape) is considered "violent." In cases of assessing the risk for violent reoffense for batterers, *violent* includes psychological/verbal abuse, physical abuse, and sexual abuse. Another question might be what appropriate treatment options are available? Treatment options include residential, prison-based, state hospital-based, and out patient programs. In addition, it is important to assess all areas of intervention need; that is, to conduct a competent and thorough forensic assessment for the court, all related areas of need are to be clearly described and addressed. This may include the need to complete substance abuse treatment, which always precedes any other treatment program. The offender should not be involved in an anger management or other therapy programs before the substance abuse has been adequately addressed using standardized and empirically supported interventions (e.g., structured group treatment rather than individual, frequent use of urine drug screens, and support group attendance). The examiners must hold themselves to a high level of competence and identify clearly all aspects of the offender's life that either suggest lessened or increased risk for violent reoffense.

It is never the forensic examiner's place to advocate on behalf of the offender or victim. It is the examiner's job to assess, gather relevant information, and put the puzzle together to form a clinically sound picture of the offender related to the questions being asked. The examiner does not advocate for or against legal sanctions nor does the examiner associate with or take the cause of the offender or the offender's family. Only the facts are addressed and addressed as neutrally as possible.

A forensic examiner may become the provider of treatment only after (1) the forensic assessment has been officially completed and the report sent to the appropriate people, (2) the report has been accepted by the court and sentencing or recommendations by the court are completed, and (3) the offender has been provided a list of other competent treatment providers without prejudice. The role of the examiner is to assess and make recommendations. Only after formally terminating from that role can an examiner become the treatment provider, at which point the focus is not on information gathering, but rather on the clinically relevant treatment interventions. No favoritism can be shown to an offender who chooses to engage in treatment with the examiner who performed the forensic examination. After careful deliberation and consultation with numerous professionals, it does appear ethical for the forensic examiner to provide treatment after the above and other considerations are made.

Treatment Assessment (Intake Assessment)

The intake assessment involves gathering relevant information and facts to determine what type of treatment would be most appropriate for the offender and if the offender would fit into the criteria of the program for treatment. If the offender appears to require services that another program offers more competently, then the offender should be referred to that program. If the offender appears appropriate for the program conducting the intake assessment, the offender and his family are interviewed, the forensic report is obtained and reviewed, testing occurs as indicated, and the supervising agent is consulted throughout the treatment process. The offender then begins treatment and is monitored throughout the treatment duration for risk issues as well as compliance with any and all recommendations (e.g., completing a substance abuse program, maintaining abstinence and attending support groups, following all medication prescriptions, following all adjunctive therapy interventions). The treating clinician has the arduous and, at times, draining task of monitoring the offender's ongoing risk for any type of violent reoffense, whether it be verbal, physical, or sexual, or whether the offender is engaging in *seemingly unimportant decisions* that are a precursor for reoffense.

Several ethical issues are important to mention again here. Unfortunately, these are common ethical dilemmas that treating clinicians confront.

1. It is never the job of the treating clinician to advocate for reduced sentencing or lesser legal charges. That is the sole job of the offender's attorney.
2. It is not advisable to make special accommodations for offenders that are above and beyond what the clinician would do for all of his/her offenders. I am not referring to accommodations to address physical disabilities or language barriers. If an interpreter is needed, then the interpreter should attend the weekly group with the offender. It is not advisable to substitute individual therapy for structured group intervention. Giving into the offender's request for special treatment exceptions only services to allow the offender to be in control of the consequences of his offense behavior, and the offender then is in control of his treatment, which should never be allowed to occur. That is, the treatment staff is in control over program expectations and are fully responsible for holding each

and every offender accountable for complying with all program expectations without exception.

3. Some offenders request special considerations due to their financial or employment situations. However, offenders are expected to comply with recommended treatment intervention, and it is imperative that the treatment interventions follow standardized and empirically supported methodologies without exception. An offender may be at risk of losing his job by taking time off to attend a program; however that is simply one aspect of his own consequence for having committed the crime. Everyone is employable, maybe not at the same pay scale or degree of management, but they are employable. Employment demands never dictate the treatment regime, ever. If he misses work, then that is what is necessary.
4. Another common complaint that offenders present for consideration of special treatment is the cost. Put simply, the cost of treatment is another natural consequence of the offender's behavioral choice. Yes, the entire family feels the financial burden, but that it the way it is. If financial considerations become a reason to deter from following standardized treatment protocol, then the clinician is not providing the best and most effective treatment interventions to his/her patient and these clinicians have no business treating offenders.
5. One last common complaint offenders make is that they do not function well in a group setting. Again, however, individual therapy has not demonstrated any significant success in addressing abuse or sex offense behavior. Offenders that find it difficult to relate to others need to develop improved social and communication skills, and group therapy is the perfect safe place to do so. The offender may have some individual sessions to prepare for group, but the offender should be expected, like everyone else, to participate in the group setting. No exceptions to the rule. Consider referring such an offender for a psychiatric assessment to determine if an anti-anxiety medication would be appropriate.
6. The treating clinician should never, ever provide any forensic assessment or risk assessment on the offender they have treated. This is because the treating clinicians develop an obvious bias and are far more likely to underestimate risk for their clients and are more likely to make exceptions to rules. It is the treating clinicians' job, however, to hold the offenders totally accountable for complying with all program expectations, supervision requirements, adjunct therapies, substance abuse interventions and support group attendance, and to report any and all infractions or suspected infractions to the offender's supervising agent. In addition, the treating clinician is mandated to inform the victim of any possible impending risk of harm due to the offender's direct or indirect statements of intending to harm anyone. I have seen several clinicians allowing offenders to violate restraining orders, failing to report ongoing abuse, and failing to inform the supervising agent of risk for harm issues or of the ongoing abuse. Some clinicians even fail to file mandated reports on their offenders for fear of jeopardizing the therapeutic relationship. However, any attempt of the treating clinician to fail to hold the offender accountable for his behavior or failing to make mandated reports are colluding with the offender and, in my opinion, are equally responsible for any abuse or violence that occurs as a result of failing to report. There is also no empirical support that holding the offender accountable or filing mandated reports causes harm to the therapeutic relationship. Treating violent offenders, regardless of the type of abuse (e.g., psychological, verbal, physical, or sexual) requires specialized skills and training, and a competent clinician never fails to hold the offender completely responsible for his behavior and always keeps the supervising agent and victim (when allowed) updated about risk concerns, and follows ethical guidelines and laws pertaining to the report of threatening or actual harm.
7. The treating clinician should never provide any forensic assessment or risk assessment on the offenders they have treated. Even if the treating clinician performed the forensic

assessment, clinicians can never return to the role of forensic examiners for any offenders they have treated. It would be exceptionally difficult for a clinician in this situation to be neutral enough to conduct a forensic assessment, even if he/she last worked with the offender years ago. In my professional opinion, it is never competent or ethical for a clinician, who treated the offender or the offender's family, to ever become the forensic examiner for the offender or his family. It is very poor judgment and boundaries on the clinician's part to even contemplate doing so.

54 Why Interview the Victims or Partners of Violent Offenders?

INTRODUCTION

The victim's/partner's expectation of being abused again was found to be highly reliable.

Bowen et al. (2005) reported that the predictive accuracy of the partner's expectation of offender reoffending significantly improved prediction. Weisz, Tolman, and Saunders (2000) found that women's expectations regarding how likely it was that their partner would become violent during a dispute in the future was the single best predictor of severe violence. Heckert and Gondolf (2004) also found that the women's perception of risk significantly improved the prediction of repeat assult over prediction with men and women's characteristics alone. Therefore, it is imperative to include the partner/victim's report in assessments to provide more reliable assessments of reoffending and for risk of violent reoffense.

INTERVIEW WITH VICTIM

Whenever possible and when the victim is willing, conduct an interview. This provides an opportunity to obtain the victim's version of the offense and other pertinent information, such as the abuse history. However, if victims are not related to the offender, it is not acceptable to pressure their cooperation. However, if the victims are family or the partners/spouses of the offenders, then they should be strongly encouraged to participate.

CONSIDERATIONS FOR INTERVIEWING VICTIMS

Interviewing victims requires compassion and an understanding of the full scope of the offense. Victims were victimized at the time of the offense. Victims may or may not have been groomed, may or may not have witnessed others being abused, and may or may not have been abused over time on multiple occasions.

Interviewing the victim or any crime is in and of itself a revictimization. The victim has usually told and retold her story to several different professionals, including the 911 dispatcher, police officers arriving at the scene, subsequent questioning by officers and detectives several days following the initial offense, prosecuting attorneys, family and friends, and medical staff (if the victim was physically injured), and advocates and clinicians providing psychological support. And now the forensic psychologist or intake therapist is interviewing the victim once again, and often the interview is for the assessment of the abuser — thereby once again putting the victim in a situation where they are directly impacted by the offender's violent behavior.

However, the impact of being interviewed by the examiner is often far less significantly stressful and less likely to result in further injury to the victim. In fact, when questioned by a seasoned and competent clinician, victims are likely to have found the interview helpful; helpful because they were asked and allowed to share their version of the offense and offense history, and that they are believed.

Interviewing victims of sexual assault and child molestation has been written about, and I recommend *Assessing Allegations of Child Sexual Abuse* by K. Kuehnle.

As for victims of domestic abuse, however, there is less available. Therefore, I will focus on interviewing the victims of domestic assault. The victim may experience a sense of revictimization when being interviewed; however, there is no empirical evidence to suggest that this sense of victimization actually causes or exacerbates the degree of harm the victim experienced at the hands of the abuser. And again, if the interviewer is experienced and appreciates the subtle and overt abuse the victim has withstood, then the interview will not likely result in any harm to the victim.

When interviewing the victim, I find it most useful to begin with an informed consent and utilize an informed consent form, basically the same as the one used with the abuser or sex offender. It is important that the victim clearly understand how the information may be used and that information the victim reveals may result in adverse action against him/her (e.g., information disclosed resulting in a mandated report being created). Using an informed consent with the victim and/or spouse also demonstrates respect by allowing them to have the power to participate or not in the interview. Last, information they reveal needs to be incorporated into the offender's report and the information appears more credible when informed consent was obtained.

It is important to allow the victim to openly and freely disclose any information about the abuse that has occurred. The interviewer should validate the victim's experience as being factual because it is, in fact, the victim's perception of the facts. It is not the clinician's responsibility to prove or disprove any information the victim discloses.

It is common that victims will report that having the opportunity to disclose and discuss the abuse or sexual assault was helpful in keeping things in context, as well as helpful in that the clinician listened to what was being said differently than the others who may have questioned the victim (e.g., police officers, prosecutors).

When interviewing victims:

1. Ask open-ended questions.
2. Help the victim to identify when the abuse began. Generally the abuse began months before the actual first physically violent incident in the form of jealousy, control, belittling statements, and other forms of psychological abuse.
3. Explain any information that may need to be reported due to mandated reporting laws and explain the process of investigation as clearly as possible. This will lessen the victim's experience of being tricked and lessen the fear associated with mandated reports.
4. Explain that the information she shares will be included in the offender's assessment report and how this helps to hold the offender accountable for all of the actual abuse that has occurred, not simply the current offense situation.

55 When Offenders Complain

INTRODUCTION

Batterers and sex offenders tend to complain about the forensic assessment and treatment clinician for a variety of reasons. Perhaps due to the underlying psychopathology or perhaps because of their narcissism they will tend to complain. The complaints may pertain to the requirements necessary to complete a forensic assessment or treatment or the outcome of assessment results. The more thorough the forensic assessment, the more offenders tend to complain. As expected, the more offenders are expected to comply with treatment recommendations and are held accountable for their past, current, and future behavior, the more they may complain as well. Offenders have a right to complain. They are under scrutiny and are expected to invest a significant amount of time, money, and emotional to complete treatment and supervision. However, their complaints should not influence the clinician's decisions about treatment recommendations, expectations, or the need to interview and believe the credibility of collateral records or the interview data from others (e.g., victim, partner, and children).

INTERVIEWING OF THE VICTIM, PARTNER, AND FAMILY

It is not uncommon for batterers and sex offenders to complain that they do not want the victim interviewed because it is distressing (for the victim, offender, and family), forces the victim to relive the offense, and a time constraint for the victim. All of these are valid concerns. However, none of these excuses justify failing to interview the victim. The victim already experienced the worst of her experience and that occurred at the time of the offense. Further discussing the incident with an experienced clinician will rarely result in injuring the victim further. The same is true when the offender makes the identical argument for interviewing their current partner, family, and children. Rarely if ever will interviewing someone cause significant or permanent harm. Generally speaking, however, the offender does not want anyone interviewed that has information that might contradict his statements or that may report ongoing abuse. Experienced clinicians differentiate genuine from feigned concern and remorse, though this is a very difficult distinction even for experienced clinicians to make. But to err on the side of safety is to assume ulterior motives of the offender for most all of his requests or resistance.

The comments I have received from victims, partners, and children are that most did not want to be interviewed for the offender's forensic assessment or for ongoing risk assessment while the offender was involved in treatment. Interestingly, however, is that after the initial interview, I found that the vast majority of victims, partners, and children commented that they found it helpful to have been included in the offender's risk and forensic assessment, and that they felt they were heard and their experiences appreciated as a result. They also felt stronger and safer by having other professionals involved who they could report concerns to as needed. The isolation and secrets were no longer barriers for their safety.

DISCUSSING AND USING PRIOR CRIMINAL HISTORY

It is common for offenders to argue that their past criminal history, prior violent offenses, allegations made against them but never resulting in arrests or prosecution, or childhood records of any kind, should not be considered in their forensic assessment. But a forensic assessment must and should include all available information, and *all available information* does not mean simply what is currently in the clinician's possession. Whenever appropriate, the clinician should request any records necessary to obtain the most accurate and complete understanding of the offender.

When I conducted civil commitment assessments on sexual predators for the State of Minnesota, I routinely requested records from law enforcement from any area that an offender had resided in his life. I sent out teletypes and called any law enforcement agency where the offender ever lived. The result was that I received police records (most never resulting in an arrest or conviction) indicating robust histories of violence and criminal activity that was not previously known about the offender. What appalled me was that, as a forensic psychologist, I had obtained records of numerous sexual and nonsexual violent crimes that proved a pattern of violence for a specific offender and yet the prosecutors and law enforcement involved in the offender's more recent offenses failed to research and gather the same information. The result was that many predatory violent offenders were never profiled as the recidivistic violent predatory offenders that they actually were. So I say it again, invest the time to gather all available records. I found relevant prior violent criminal histories on approximately 7 out of 10 offenders I took the time to request records on.

Basically, any historical information that can be discovered should be used to profile the offender. All information is helpful to determine the underlying psychopathology and to accurately diagnose requires an exploration of historical information.

Assessment Results

Batterers and sex offenders often complain about the result of testing and actuarial risk prediction. Sometimes testing fails to identify psychopathology that exists, and at times may exaggerate the problems. It is the experienced clinician who must determine what the testing and assessment data should mean without reinterpreting the meaning. Psychopathic offenders may or may not score high for personality disorders on the Minnesota Multiphasic Personality Inventory (MMPI-2) or Millon Clinical Multiaxial Inventory (MCMI-III). The fact is that some will not have prior criminal or mental health records and even for those who do, the tests may or may not accurately identify pathology accurately. For example, it is not uncommon for a psychopathic offender to not care about the quality of their intimate relationship or to lack any significant sense of stress or anxiety. So the testing results may be misleading, that is, the results may not identify anxiety or relational problems yet the offender may well have very problematic relationships with others. They simply do not care. Another problem is that the offender may argue that the results of the assessment rely on historical factors that may or may not be changing or changeable. Perhaps the offender has not engaged in violent reoffense for a year. However, the risk for reoffense remains for the offender's lifetime, not a circumscribed shorter period of time.

Jealousy

Batterers and sex offenders alike tend to exhibit problems with jealousy and control. When the clinician interacts with the offender's victim or intimate partner, the offender may become jealous and misinterpret the clinician's behavior as having some romantic meaning rather than strictly professional. This becomes even more pronounced when the partner decides to separate from the offender, ends the intimate relationship or divorces, or when the partner obtains a restraining order/order for protection. As the offender begins to lose control over his victim or partner, the clinician may become the object of projection and anger. When the offender's partner or victim regains power and the isolation is decreased or ended, the offender has no one to blame but the clinician from his delusional point of view. So it is important to bear in mind that it is common for offenders to accuse clinicians of having romantic intentions involving the offender's partner when, in fact, this is rarely if ever the case.

How to Handle the Offender's Complaint

Certainly offenders have the right to complain. However, the veracity of their claims must be explored. In most circumstances, the offender's complaints are unfounded and based on delusional thinking or involving the defense of projection. Anyone who demonstrates support for the offender's partner or victim in protecting themselves becomes an adversary to the offender's controlled and violent life.

SUMMARY

Offenders will complain when they do not get their way, and will complain when the clinician obtains information that contradicts their statements or allegations. However, all information is relevant when conducting a forensic assessment and the clinician should make all attempts to gather appropriate information. This is even truer when attempting to civilly commit a sexual offender or violent offender to a state hospital or life in prison. The public has to rely on forensic professionals to adequately and competently do the job entrusted to them, which is to gather the most accurate and detailed facts about the offender and present the facts appropriately.

Section V

Investigation Strategies for Law Enforcement, Probation, and Parole Officers

56 Searching the Residence of a Batterer or Sex Offender

SEARCH CAN PROVIDE VALUABLE INFORMATION

Anytime an investigation occurs related to abuse or a sex offense, the offender's residence may be searched. When a search occurs, it is important to understand the typologies, which in summary provide valuable information about what fuels the offender's violent behavior and fantasies and, therefore, helps to better profile and understand the offender. As I have stated before, the majority of batterers will have pornographic material in some form (e.g., Internet, magazines, movies) as will sex offenders. In addition, many batterers and sex offenders will possess magazines related to detective and police work, as well as books on criminology and psychology. Below is what I find helpful for probation and parole officers, law enforcement, and forensic psychologist who have an opportunity to search the offender's environment.

When I worked as a forensic psychologist for the State of Minnesota, I had the opportunity to assist in the search of the cells of sexual predators. Interestingly, most of the searches resulted in finding pornographic material or material that could be used as pornography (e.g., for child molesters, finding store ads from the newspaper or department store catalogs with children in underwear or swimsuits). What was found often helped in not only profiling the sexual offender but understanding how engrained the deviant and violent fantasy life was despite being locked up in prison.

Examples of What to Look for in the Offender's Residence

Be aware of any of the following:

1. Any type of pornographic material. This includes magazines designated as pornography. Pornographic movies include those designated as pornography or adult only, and those that involve sexual content, including both softcore and hardcore pornography. Also included here are adult sex comics and sex lines of any kind.
2. Material similar to pornography. This includes magazines such as *Cosmopolitan*, *Teen*, *Sports Illustrated* swimsuit issue, *Vogue*, *Vanity Fair*, *True Detective*, and any other magazine that includes photos, drawings, or cartoons of people partially clothed, naked, bound, in swim wear, or in poses depicting intimate or sexual acts. Movies include R-rated movies that depict violence, especially those that involve violence against females, torture, mutilation, or sexual content in which nudity is shown or depicted.
3. Any type of teaching materials, including magazines, catalogs that depict or include photographs or drawings of anyone in lingerie, swimsuit, or partially naked. This includes newspaper ads, department store catalogs, etc. This is included because possession of these materials should be considered suspicious when in the possession of some child molesters and pedophiles. Teaching materials may be used for the offender's sexual fantasy life and/or as a means of interacting with potential victims. Many offenders that molest children gain compliance from the victim and the victim's family by establishing a relationship of trust as a coach, tutor, or leader of a social group (e.g., Boy or Girl Scouts).

4. When investigating a sex-based crime involving children, the presence of any type of children's entertainment materials, including children's movies, books, toys, stuffed animals, or children's clothing should be suspect.
5. Both batterers and sex offenders find new victims, sex partners, and a wide variety of legal and illegal pornographic material on the Internet. Make no mistake; even viewing pornography is a form of an affair. Batterers often hide this from their partners because they are aware of how they are harming their partners. Any type of computer software or accessing sites on the Internet, such as bulletin boards, multimedia materials, CD Roms, or any other material, programs, or interactive accesses that pertain to leaving or receiving messages, or anything that pertains to computerized pornography, sex, or dating services.
6. Any type of dating or "meet people" services. This includes responding to or placing singles ads, telephone dating services, etc. Check their telephone bill, cable bill, or Internet bill to see the sites subscribed to or used. In addition, viewing the history on the offender's computer offers evidence of at least some of the sites visited. Even when the offenders clean their computers by eliminating traces or cookies, it may still be possible to trace their Internet visits.
7. Access to or use of adult cable stations, including movies mentioned in the above paragraph pertaining to the use of pornographic material. Check their cable bill.
8. Any sexual device that is specifically designed to be used as part of sexual acts, or objects that depict genitalia or can be used for sexual acts with a child or adolescent, a vulnerable person, or the use of bondage or force. Included here would be inflatable dolls that resemble a child or adolescent, handcuffs, or whips. This is not an exclusive list.
9. Binoculars, telescope, or ability to view into the homes of others.

When examining magazines or pornography, try to identify which photos the offender used the most. What was the photo depicting (bondage, fear, sadomasochism, fear, rape, or consent). The most used photos will be on the pages that are creased or bent, or noticeably worn, or, in some cases, even blatantly marked.

The presence of any of the above items does not necessarily indicate that there is a problem, but certainly warrants further investigation.

57 Sex Offender Profile and Modus Operandi

The following is an example of a form I developed to be completed on each sex offender. The purpose is to provide the probation or parole officers, as well as family members, with a clear picture of the most important factors to watch for. Should the offender engage in any contact with the identified potential victim pool or risk situations, immediate intervention is warranted. This is by no means a complete profile or modus operandi, but rather a beginning. Only those factors or situations associated with the specific offender are identified; for example, by checking the items related to the offender or by crossing out items not pertaining to the offender. This is a quick way to identify risk factors specific to each offender.

NAME OF OFFENDER:

DATE OF BIRTH:

ADDRESS:

OTHER CONTACT ADDRESSES:

PHONE: (include all numbers to reach offender)

OTHER CONTACT PERSONS OFFENDER ASSOCIATES WITH (include contact phone numbers and addresses):

SPECIFIC TYPOLOGY OF OFFENDER:

GENDER OF VICTIM/VICTIMS

(This offender may offend against victims of any gender. However, the following genders have been identified as related to this offender's known offense patterns.)

Male	
Female	
Both male and female	

AGE OF VICTIM/VICTIMS:

(This offender may offend against a person who is about the same age of his victim, or may offend against someone who is significantly older or younger. The following victims' ages have been identified as related to this offender's known offense patterns.)

NUMBER OF KNOWN VICTIMS:

RELATIONSHIP WITH VICTIM:

(This offender may offend against others who are related in some way other than the way his current victim was related. The following offender/victim relationships have been identified as related to this offender's known offense patterns.)

Stranger/none (includes having met victim immediately prior to offense)	
Family	
Step-parent	
Grandparent	
Friend of family	
Employer	
Co-worker/classmate	
Acquaintance	
Dating or sexual relationship	
Engaged or married	

VIOLENT AND SEXUALLY DEVIANT BEHAVIOR ENGAGED IN:

(This offender may have engaged in other sexually abusive behavior. The following sexual behaviors have been identified as related to this offender's known offense patterns.)

Fondled victim's vagina/penis over clothing	
Fondled victim's vagina/penis under clothing	
Fondled victim's buttocks and/or anus over clothing	
Fondled victim's buttocks and/or anus under clothing	
Fondled victim's breasts over clothing	
Fondled victim's breasts under clothing	
Rubbed own penis against victim's vagina	
Rubbed own penis against victim's butt/anus	
Digitally penetrated victim's vagina	
Digitally penetrated victim's anus	
Penetrated victim's vagina with penis	
Penetrated victim's anus with penis	
Had victim perform fellatio/cunnilingus	
Performed fellatio on victim	
Inserted objects into victim's anus or vagina	
Physically assaulted victim before, during, or after assault	

Engaged in gratuitous violence (more than required to obtain victim compliance)	
Engaged in sadistic behavior (the offender becoming aroused to any behavior inflicted on the victims that was intended to or actually does cause the victim pain, suffering, orhumiliation).	
Tortured victim	
Mutilated victim	
Killed victim	
Exposed	
Voyeurism	
Sex with animals	
Group sex/with accomplice	

TIME OF OFFENSE:

Morning (5 A.M. to 12 P.M.)	
Afternoon (12:01 P.M. to 5 P.M.)	
Evening (5:01 P.M. to 10 P.M.)	
Late night (10:01 P.M. to midnight)	
Early morning (12:01 A.M. to 4:59 A.M.)	
Before work	
During work hours	
After work	
Weekends	

PRECURSORS TO OFFENDING:

(This offender may offend in response to other stressors. The following strategies have been identified as related to this offender's known offense patterns.) Moderate degrees of stress

Problems in relationship with significant other/wife/family	
Decreased attendance at support groups or treatment	
Decreased participation in support groups or treatment	
Loss (e.g., of relationship, job, residence)	
Depression	
Experiencing feelings of abandonment, isolation, rejection, or lack of intimacy	
Withholding sex from partner or engaging in assaultive behavior toward partner	
Avoiding support network	
Engaging in risk taking or high reoffense-based behavior (e.g., gambling, reckless driving, criminal activity, cruising, Internet use)	
Use of pornography or related material	
Use of alcohol or drugs	
Failing to comply with psychotropic medications	
Failing to comply with supervision, no contact orders, or treatment	

ENGAGEMENT STRATEGIES UTILIZED:

(This offender may utilize some, all, or none of these strategies to procure a victim. The following strategies have been identified as related to this offender's known offense patterns.)

Establishes relationship with vulnerable adults	
Establishes relationship with parents who have children or adolescents	
Establishes trusting relationship with victim	
Threatens victim	
Offers money, food, or gifts	
Provides drugs or alcohol	
Assertively/aggressively approaches victim	
Stalks victim	
Breaks into residence	
Engages in playful behavior that involves touches (hugs, wrestling, tickling)	

Offers ride to or escorts victim	
Has accomplice or accomplices	
Entices victim with false promises (e.g., claims to be a photographer for model agency)	
Kidnaps victim	
Cruises for victims	
Socializes with potential victims	
Hangs out where potential victim congregate	
Shows sexually explicit materials or pornography to potential victims or leaves these materials where potential victims may access them	
Approaches victims when they are sleeping or impaired in some way	
Utilizes physical force or a weapon against victim	
Forces way into victim's car or forces victim into his car	

DEGREE OF PLANNING SEXUAL ASSAULT:

(This offender may utilize the same or similar degrees of planning in another offense. The following have been identified as related to this offender's known offense patterns.)

Opportunistic (did not necessarily set out to offend, opportunity presented itself and offender acted on it; would not likely have offended at that time if situation did not occur)

Impulsive (may/may not have planned to offend in advance, acts on impulse, no specific victim necessary, acts upon experiencing desires)

Premeditated (planned offense in advance, selected specific victim, fantasized about offense prior to offending)

TYPE OF FORCE UTILIZED:

(This offender may utilize some, all, or none of these types of force against a victim. The following types of force have been identified as related to this offender's known offense patterns.) Physical Force:

Grabbing	
Pushing	
Grabbing/pulling hair	
Burning	
Restraining by using body weight	
Restraining by holding down wrists/arms	
Tying up	
Hitting	
Punching	
Slapping	
Pinching	
Throwing objects at victim	
Use of gun	
Use of knife	
Use of other object	
Psychological Force/Coercion:	
Threats	
Of further harm	
Of harming others	
Of killing	
Of using weapon or object	
Shows weapon or object that will be used to injure or kill victim	
Intimidation	
Isolating victim from others	
Driving to isolated location	
Taking to unknown area or home (perpetrator's house, for example)	
Raising voice	

Had accomplices, regardless of whether accomplices actively participated (includes perpetrator having friends near when assaulting victim)	
Game Playing/Manipulation/Emotional Blackmail	
Providing alcohol or drugs	
Taking advantage of emotional state of victim (depressed, sad)	
Mind games	
Victim sleeping	
Victim bathing or dressing	
Tricked victim	
Seduced/manipulated/enticed victim	
Lying/Deceit	
Made promises never intended to keep	
Led the victim on	
Dishonest about consequences or who would receive consequences (e.g., told victims they or their parents would get into trouble if they told)	
Boundary Violation	
Took advantage of position of power or authority	
Took advantage of relationship with victim (for example, friend)	
Forced Prostitution/ Dynamics of Pimping	
Coerced/forced victim to engage in sexual contact with others, regardless of whether money was exchanged for sex	
Coerced or forced victim to steal objects or sell drugs for money	

DRUG AND/OR ALCOHOL USE:

(This offender may or may not utilize alcohol or drugs as part of a new offense. The following chemical use pattern has been identified as related to this offender's known offense patterns.)

Consumed alcohol/drugs prior to offense	
Consumed alcohol/drugs during offense	
Clamed drugs/alcohol for offense	
Provided alcohol or drugs to victim	

HISTORY OF MENTAL ILLNESS OR MENTAL DISORDERS:

Personality Disorder: specify:	
Mood Disorder: Specify:	
Psychotic or Delusional Disorder: Specify:	
Intermittent Explosive Disorder	
Schizophrenia or related disorder: Specify:	
Substance Abuse or Dependency Disorder: Specify:	
Other:	

SITUATIONS PERPETRATOR SHOULD AVOID:

(This offender could offend in conditions other than those identified. The following situations have been identified as related to this offender's known offense patterns.)

Any contact with his victim	
Any contact with males who are under age 18	
Any contact with females who are under age 18	
Any contact with males and females who are under age 18	
Any position of power or authority over males who are under age 18	
Any position of power or authority over females who are under age 18	
Any position of power or authority over males or females who are under age 18	
Any unsupervised contact with males who are under age 18	
Any unsupervised contact with females who are under age 18	
Any unsupervised contact with males and females who are under age 18	

Work or activities that place him in areas where children or adolescents are	
Use of drugs or alcohol	
Any setting where drugs or alcohol is being consumed or available	
Malls, carnivals	
Bars/nightclubs/dance halls	
Work that puts perpetrator in contact with the above identified people (e.g., bus driver, teacher, coach, counselor, aide, babysitter)	
Residing in area or apartment where children or adolescents live or play	
Any relationship with women who are vulnerable	
Any relationship with women who have children or adolescents, regardless of whether the children or adolescents reside with him/her (unless he/she has no contact of any kind with the children or adolescents)	

The above profile represents a guide to the perpetrator's offending patterns that have been identified to date. When the perpetrator experiences any of the stressors or engages in any of the behaviors listed above, he is at greater risk for reoffense.

Name of Evaluator____________________ Date________________

Source: Scott A. Johnson, MA, LP, DABPS, DAACCE, ACFE, 1996/2005©

References

Abbey, A. (1991). Acquaintance rape and alcohol consumption on college campuses: How are they linked? *Journal of American College Health, 39*, 165–169.

Abbey, A.A., McAuslan, P., & Ross, L.T. (1998). Sexual assault perpetration by college men: The role of alcohol, misperception of sexual intent, and sexual beliefs and experiences. *Journal of Social and Clinical Psychology, 17*, 167–195.

Abel, E.M. (1999). Who are women in batterer intervention program: Implication for practice. Paper presented at the 6th International Family Violence Research Conference, Durham, NH.

Abel, G.G. (1995). Abel assessment for sexual interest: Therapist product information. Atlanta, GA: Abel Screening, Inc.

Abel, G.G., & Blanchard, E.B. (1974). The role of fantasy in the treatment of sexual deviation. *Archives of General Psychiatry, 30*, 467–475.

Abel, G.G. & Osborn, C. (1992). The paraphilias: The extent and nature of sexually deviant and criminal behavior. In J.M.W. Bradford (Ed.), *Psychiatric clinics of North America, 15*(3), (pp. 675–687). Philadelphia, PA: W.B. Saunders Company.

Abel, G.G., Barlow, D.H., Blanchard, E.B., & Guild, D. (1977). The components of rapists' sexual arousal. *Archives of General Psychology, 34*, 895–908.

Abel, G.G., Barlow, D.H., Blanchard, E.B., & Mavissakalin, M. (1975). Measurements of sexual arousal in male homosexual: The affects of instructions and stimulus modality. *Archives of Sexual Behavior, 4*, 623–629.

Abel, G.G., Becker, J.V., Mittleman, M., Cunningham-Rathner, J., Rouleau, J.L., & Murphy, W.D. (1987). Self-reported sex crimes of nonincarcerated paraphiliacs. *Journal of Interpersonal Violence, 2*, 1, 3–25.

Abel, G.G., Becker, J.V., Murphy, W.D., & Flanagan, B. (1981). Identifying dangerous child molesters. In R.B. Stuart (Ed.), *Violent behavior: Special learning approaches to prediction, management and treatment* (pp. 116–137). New York: Brunner/Mazel.

Abel, G.G., Hoffman, J., Warberg, B., & Holland, C.L. (1998). Visual reaction time and plethysmography as measures of sexual interest in child molesters. *Sexual Abuse: A Journal for Research and Treatment, 10:* 2, 81–95.

Abel, G.G., Lawry, S.S., Karlstrom, E., Osborn, C.A., & Gillespie, C.F. (1994). Screening tests for pedophilia. *Criminal Justice and Behavior, 21*, 1, 115–131.

Abel, G.G., Mittleman, M.S., & Becker, J.V. (1985). Sexual offender: Results of assessment and recommendations for treatment. In M.H. Ben-Aron, S.J. Hucker, & C.D. Webster (Eds.), *Clinical criminology: The assessment and treatment of criminal behavior.* Toronto: M & M Graphics.

Abidin, R.R. (1983, 1990). *Parenting stress index test* (3rd ed.). VA: Pediatric Psychology Press.

Adams, D. (1988). Treatment models for men who batter: A profeminist approach. In K. Yllo & M. Bograd (Eds.). *Feminist perspectives on wife abuse.* Newbury Park, CA: Sage Publications.

Adams, D. (1989). Feminist-based interventions for battering men. In P.L. Caesar & L.K. Hamberger, (Eds.), *Treating men who batter: Theory, practice, and programs* (pp. 3–23). New York: Springer.

Adams, P.J. (1999). It takes two to tango: A language-based approach to interventions with men who abuse women. In N. Robertson, (1999), Stopping violence programmes: Enhancing the safety of battered women or producing better educated batters? *New Zealand Journal of Psychology, 28*(2): 68–79.

Adams, P.J., Towns, A., & Gravey, N. (1995). Dominance and entitlement: The rhetoric men use to discuss their violence toward women. *Discourse and Society, 6*, 387–406.

Aldarando, E. (1996). Cessation and persistence of wife assault: A longitudinal analysis. *American Journal of Orthopsychiatry, 66*, 141–151.

Aldarondo, E., & Kaufman-Kantor, G. (1997). Social predictors of wife assault cessation. In G. Kaufman-Kanter & J.L. Jasinski (Eds.), *Out of the darkness: Contemporary perspectives on family violence.* (pp. 183–193). Thousand Oaks, CA: Sage Publications.

Aldarondo, E., & Sugarman, D.B. (1996). Risk marker analysis of the cessation and persistence of wife assault. *Journal of Consulting and Clinical Psychology, 64*(5), 1010–1019.

Alexander, M.A. (November 1994). Sex offender treatment: A response to the Furby et al., 1989 quasi meta-analysis II. Paper presented at the Association for the Treatment of Sexual Abusers 13th Annual Conference, San Francisco, CA.

Allen, C.M. (1990). Women as perpetrators of child sexual abuse: Recognition barriers. In A.L. Horton, B.L. Johnson, L.M. Roundy, & D. Williams (Eds.), *The incest perpetrator: A family member no one wants to treat.* Newbury Park, CA: Sage Publications.

Allen, C.M. (1991). *Women and men who sexually abuse children: A comparative analysis.* Orwell, VT: Safer Society Press.

Allen, H. (1987). *Justice unbalanced: Gender, Psychiatry and judicial decisions.* Milton Keynes, U.K.: Open Univeristy Press.

American Personnel and Guidance Association (1981). *Ethical standards.* Falls Church, VA: American Personnel and Guidance Association..

American Personnel and Guidance Association (1982). *Guidepost, 25*, 9.

American Psychiatric Association (APA) (1994). *Diagnostic and statistical manual of mental disorder,* 4th ed. (*DSM-IV*). Washington, D.C.: APA.

American Psychiatric Association (2000). *Diagnostic and statistical manual of mental disorder*, 4th ed., *Text Revised* (*DSM-IV-TR*). Washington, D.C.: APA.

American Psychological Association. (1992). Ethical principles of psychologists and code of conduct. *American Psychologist, 47*, 1597–1611.

American Psychological Association (1996). *Violence and the family.* Washington, D.C.: American Psychological Association.

American Psychological Association (2002). Ethical standards of psychologists. Washington, D.C.: American Psychological Association.

American Psychological Association Presidential Task Force on Violence and the Family (1996). Violence and the family: Report of the American Psychological Association Presidential Task Force on Violence and the Family. Washington, D.C.: American Psychological Association.

Anderson, P.B. (1996). Correlates of college women's self-reports of heterosexual aggression. *Sexual Abuse, 8*, 167–195.

Anderson, P.B. (1998). Women's motives for sexual initiation and aggression. In P.B. Anderson & C. Struckman-Johnson (Eds.), *Sexually aggressive women* (pp. 79–93). New York: Guilford Press.

Anderson, P.B., & Aymami, R. (1993). Reports of female initiation of sexual contact: Male and female differences. *Archives of Sexual Behavior, 22*, 335–342.

Anderson, P.B., & Struckman-Johnson, C. (Eds.). (1998). *Sexually aggressive women.* London: Guilford Press.

Andrews, D.A., & Bonta, J. (1994). *The psychology of criminal conduct.* Cincinnati, OH: Anderson Publishing.

Andrews, D.A., & Bonta, J. (2003). *The psychology of criminal conduct* (3rd ed.). Cincinnati, OH: Anderson Publishing.

Appel, A.E., & Holden, G.W. (1998). The co-occurrence of spouse and physical child abuse: A review and appraisal. *Journal of Family Psychology, 12*, 578–599.

Archer, J. (2000). Sex differences in aggression between heterosexual partners: A meta-analytic review. *Psychological Bulletin, 126*, 651–680.

Arkow, P. (1996). The relationships between animal abuse and other forms of family violence. *Family Violence and Sexual Assault Bulletin, 12*, 29–34.

Aronson, E., Wilson, T.D., & Akert, R.M. (2004). *Social psychology* (4th ed.). Upper Saddle River, NJ: Prentice-Hall.

Ascione, F.R. (1993). Children who are cruel to animals: A review of literature and implications for developmental psychopathology. *Anthrozoos, 6*, 226–247.

Ascione, F.R. (1998). Battered women's reports of their partner's and their children's cruelty to animals. *Journal of Emotional Abuse, 1*(1), 199–133.

Ascione, F.R. (2000). What veterinarians need to know about the link between animal abuse and interpersonal violence. Paper presented to the annual meeting of the American Veterinary Medical Association, Salt Lake City, UT.

Ascione, F.R. & Arkow, P. (Eds.). (1999). *Child abuse, domestic violence, and animal abuse.* West Lafayette, IN: Purdue University Press.

Association for the Treatment of Sexual Abusers (ATSA). (1993). *The ATSA practitioner's handbook.* Lake Oswego, OR: ATSA.

Astin, M.C., Lawrence, K.J., & Foy, D.W. (1993). Posttraumatic stress disorder among battered women: Risk and resiliency factors. *Violence and Victims, 8*, 17–28.

Avery-Clark, C.A. & Laws, D.R. (1984). Differential erection response patterns of child sexual abusers to stimuli describing activities with children. *Behavior Therapy, 15,* 71–83.

Babcock, J.C. & Steiner, R. (1999). The relationship between treatment, incarceration and recidivism of battering: A program evaluation of Seattle's coordinated community response to domestic violence. *Journal of Family Psychology, 13*(1), 46–59.

Babcock, J.C., Canady, B.E., Senior, A., & Eckhardt, C. (2005). Applying the transtheoretical model to female and male perpetrators of intimate partner violence: Gender differences in stages and processes of change. *Violence and Victims, 20*(1), 235–251.

Babcock, J., Green, C., & Robie, C. (2004). Does batterer treatment work? A meta-analytical review of domestic violence treatment. *Clinical Psychology Review, 23*, 1023.

Babcock, J.C., Miller, S.A., & Siard, C. (2003). Towards a typology of abusive women: Differences between partner-only and generally violent women in the use of violence. *Psychology of Women Quarterly, 27*, 153–161.

Bailey, D.S. & Taylor, S.P. (1991). Effects of alcohol and aggressive disposition on human physical aggression. *Journal of Research in Personality*, 25, 334–342.

Bandura, A. (1979). The social learning perspective: Mechanisms of aggression. In A. Toch (Ed.), *Psychology of crime and criminal justice* (pp. 298–336). New York: Holt, Rinehart and Winston.

Banning, A. (1989). Mother–son incest: Confronting prejudice. *Child Abuse and Neglect, 13,* 563–570.

Barbaree, H.E. (1991). Denial and minimization among sex offenders: Assessment and treatment outcome. *Forum on Corrections Research, 3*, 30–33.

Barbaree, H.E., & Marshall, W.L. (1988). Deviant sexual arousal, offense history and demographic variables as predictors of reoffense among child molesters. *Behavioral Sciences and the Law, 6*, 267–280.

Barbaree, H.E., Marshall, W.L., & Hudson, S.M. (1993). *The juvenile sex offender.* New York: The Guilford Press.

Barbaree, H.E., Seto, M.C., Langton, C.M., & Peacock, E.J. (2001). Evaluating the predictive accuracy of six risk assessment instruments for adult sex offenders. *Criminal Justice and Behavior, 28*, 4, 490–521.

Barnett, O.W., Lee, C.Y., & Thelen, R.E. (1997). Gender differences in attributions of self-defense and control in interpersonal aggression. *Violence Against Women, 3*, 162–181.

Barnett, O.W., Martinez, T.E., & Bluestein, B.W. (1995). Jealousy and romantic attachment in martially violent and nonviolent men. *Journal of Interpersonal Violence, 10*(4), 473–486.

Barnett, O.W., Miller-Perrin, C.L., & Perrin, R.D. (1997). *Family violence across the lifespan.* Thousand Oaks, CA: Sage Publications.

Baumgartner, M.P. (1993). Violent networks: The origins and management of domestic conflict. (pp. 209–231). In R. Felson & J. Tedeschi (Eds.). *Aggression and violence: Social interactionalist perspectives.* Washington, D.C.: American Psychological Association.

Bavolek, S.J. (1989). Assessing and treating high-risk parenting attributes. In J.T. Pardeck (Ed.), *Child abuse and neglect: Theory, research, and practice* (pp. 97–110). NewYork: Gordon and Breach Publishing.

Beach, S.R., Fincham, F.D., & Katz, J. (1998). Marital therapy in the treatment of depression: Toward a third generation of therapy and research. *Clinical Psychology Review, 18*, 635–661.

Beasley, R. & Stoltenberg, C.D. (1992). Personality characteristics of male spouse abusers. *Professional Psychology: Research and Practice, 23*, 310–327.

Beauregard, E., Lussier, P., & Proulx, J. (2004). An exploration of developmental factors related to deviant sexual preferences among adult rapists. *Sexual Abuse: A Journal of Research and Treatment, 16*, 151–162.

Beauregard, E., Lussier, P., & Proulx, J. (2005). The role of sexual interests and situational factors on rapists' modus operandi: Implications for offender profiling. *Legal and Criminological Psychology, Leicester, 10*(2), 265–279.

Becker, J.V. & Kaplan, M. (1990). Assessments of the adult sex offender. In P. Reynolds, J. Rosen, & C. Chelune (Eds.), *Advances in psychological assessment* (vol. 7, pp. 261–283). New York: Plenum Publishers.

Becker, J.V. & Quinsey, V.L. (1993). Assessing suspected child molesters. *Child Abuse and Neglect, 17*, 169–174.

Beech, A. & Ward, T. (in press). The integration of etiology and risk in sexual offenses: A theoretical framework. *Aggression and Violent Behavior.*

Beitchman, J.H., Zucker, K.J., Hood, J.E., DaCosta, G.A., & Akman, D. (1991). A review of the short-term effects of child sexual abuse. *Child Abuse and Neglect, 15*, 537–556.

Beitchman, J.H., Zucker, K.J., Hood, J.E., DaCosta, G.A., Akman, D., & Cassavia, E. (1992). A review of the long-term effects of child sexual abuse. *Child Abuse and Neglect, 16*, 101–118.

Bennett, L.W. (1995). Substance abuse and the domestic assault of women. *Social Work, 40*(6), 760–772.

Bennett, L. & Williams, O. (2004). Controversies and recent studies of batterer intervention program effectiveness. University of Minnesota, Applied Research Forum.

Berk, R.A., Berk, S.R., Loseke, D.R., & Rauma, D. (1983). Mutual combat and other family violence myths. In D. Finkelhor, R. Gelles, G. Hotaling, & M.A. Straus (Eds.), *The dark side of families: Current family violence research* (pp. 197–212). Newbury Park, CA: Sage Publications.

Berk, R., Campbell, A., Klap, R., & Western, B. (1992). The deterrent effect of arrest in incidents of domestic violence: A Bayesian analysis of four field experiments. *American Sociological Review, 57*, 698–708.

Berkowitz, L. (1973). The case for bottling up rage. *Psychology Today, 7*, 24–31.

Berkowitz, L. (1993). *Aggression: Its causes, consequences, and control.* Philadelphia: Temple University Press.

Bernard, J., Bernard, S., & Bernard, M. (1985). Courtship violence and sex-typing. *Family Relations, 34*, 573–576.

Biderman, A.D. (1975). Notes on the significance of measurements of events and of conditions by criminal victimization surveys. BSSR 0003-58. Washington, D.C.: Bureau of Social Science Research.

Biggio, G.F., Fadda, A., & Gessa, G.L. (1974). Rapid depletion of serum tryptophan, brain tryptophan, serotonin and 5-hydroxyindoleacetic acid by a tryptophan-free diet. *Life Science, 14*, 1321–1329.

Blackburn, R. (1975). An empirical classification of psychopathic personality. *British Journal of Psychiatry, 127*, 456–460.

Bland, R., & Orn, H. (1986). Family violence and psychological disorder. *Canadian Journal of Psychiatry, 31*, 129–137.

Blum, R.H. (1981). Violence, alcohol, and setting: An unexplored nexus. In J.J. Collins (Ed.). *Drinking and crime: Perspectives on the relationship between alcohol consumption and criminal behavior* (pp. 110–142). New York: Guilford Press.

Boat, B.W. (1995). The relationship between violence to children and violence to animals: An ignored link. *Journal of Interpersonal Violence, 10*, 229–235.

Boat, B.W., & Everson, M.D. (1988). Use of anatomical dolls among professionals in sexual abuse evaluations. *Child Abuse & Neglect, 12*, 171–179.

Boer, D.P., Wilson, R.J., Gauthier, C.M., & Hart, S.D. (1997). *Manual for the sexual violence risk-20.* Vancouver, British Columbia: The Mental Health, Law and Policy Institute and Simon Fraser University.

Bograd, M.J. (1988). Feminist perspectives on wife abuse: An introduction. In K. Yllo & M. Bograd (Eds.), *Feminist perspectives on wife abuse* (pp. 11–26). Beverly Hills, CA: Sage Publications.

Bonta, J., Law, M., & Hanson, R.K. (1996). *The prediction of criminal and violent recidivism among mentally disordered offenders: A meta-analysis.* Ottawa, Ontario: Solicitor General Canada.

Bonta, J., Law, M., & Hanson, R.K. (1998). The prediction of criminal and violent recidivism among mentally disordered offenders: A meta-analysis. *Psychological Bulletin, 123*, 123–142.

Bowen, E., Gilchrist, E.A., & Beech, A.R. (2005). An examination of the impact of community-based rehabilitation on the offending behavior of male domestic violence offenders and the characteristics associated with recidivism. *Legal and Criminological Psychology, 10*(2), 189–210.

Bower, G.H. (1981). Mood and memory. *American Psychologist, 36*, 129–148.

Bowes, M.P. (Ed.) (2005, June). The etiology of alcohol: Abuse and dependence: What happens in the brain? *Psychiatric Times, Reporter* (Suppl.), 3–6.

Bowes, M.P. (Ed.) (2005, June). Treating alcohol abuse and dependence. *Psychiatric Times, Reporter* (Suppl.), 7–11.

Bowker, L. (1983). *Beating wife beating*. Toronto: Lexington Books.

Bowlby, J. (1988). *A secure base: Parent–child attachment and healthy human development.* New York: Basic Books.

Boyatzis, R.E. (1974). The effects of alcohol consumption on the aggressive behavior of men. *Quarterly Journal of Studies on Alcohol*, 35, 959–972.

Boyatzis, R.E. (1975). The predisposition toward alcohol-related interpersonal aggression in men. *Journal of Studies on Alcohol, 36*, 1196–1207.

Boyle, D.J. & Vivian, D. (1996). Generalized versus spouse-specific anger/hostility and men's violence against intimates. *Violence and Victims, 11*, 293–317.

Briddel, D.W., Rimm, D.C., Caddy, G.R., Krawitz, G., Sholis, D., & Wunderin, R.J. (1978). Effects of alcohol and cognitive set on sexual arousal to deviant stimuli. *Journal of Abnormal Psychology, 87*, 418–430.

Briere, J.N. (1992). Child abuse trauma: *Theory and treatment of the lasting effects*. Newbury Park, CA: Sage Publications.

Briere, J.N., & Malamuth, N.M. (1993). Predicting self-reported likelihood of sexually abusive behavior: Attitudinal versus sexual explanations. *Journal of Research in Personality, 17,* 315–323.

Briere, J.N. & Runtz, M. (1990). Differential adult symptomology associated with three types of child abuse histories. *Child Abuse and Neglect, 14*, 357–364.

Briere, J., & Smiljanich, K. (1993). Childhood sexual abuse and subsequent aggression against adult women. Paper presented at the 101st Annual Convention of the American Psychological Association, Toronto, Canada.

Brittain, R.P. (1970). The sadistic murderer. *Medicine, Science and the Law, 10*, 198–207.

Broadhurst, R.G. & Mailer, R.A. (1992). The recidivism of sex offenders in the Western Australian prison population. *British Journal of Criminology, 32*(1), 54–77.

Brown, M. (2005). Brain impairments associated with addictive disorders. Presentation at the 3rd Annual Keys to Neuropsychiatric Care conference, Key West, Florida, December 2005.

Brown, M.E., Hull, L.A., & Panesis, S.K. (1984). *Women who rape.* Boston: Massachusetts Trial Court.

Brown, S., Goldman, M., Inn, A., & Anderson, L. (1980). Expectations of reinforcement from alcohol: Their domain and relation to drinking patterns. *Journal of Consulting and Clinical Psychology, 48*, 419–426.

Browne, A. (1987). *When battered women kill.* New York: Free Press.

Browne, K. (1994). Child sexual abuse. In J. Archer (Ed.), *Male violence* (pp. 210–230). London: Routledge Publishers.

Browne, K.D., Foreman, L., & Middleton, D. (1998). Predicting treatment dropout in sex offenders. *Child Abuse Review, 7*, 402–419.

Bruun, K. (1959). Significance of role and norms in the small group for individual behavioral changes while drinking. *Quarterly Journal of Studies on Alcohol, 35*, 959–972.

Buckley, P.F. (1998). Novel antipsychotic medications and the treatment of comorbid substance abuse in schizophrenia. *Journal of Substance Abuse Treatment, 15*, 113–116.

Burgess, A.W., Groth, A.N., Holmstrom, L.L., & Sgroi, S.S. (1978). *Sexual assault of children and adolescents*. Lexington, MA: Lexington Books.

Burgess, A.W., Hartman, C.R., Ressler, R.K., Douglas, J.E., & McCormack, A. (1986). Sexual homicide: A motivational model. *Journal of Interpersonal Violence, 1*, 251–272.

Burgess, A.W., Hazelwood, R.R., Rokous, F.E., Hartman, C.R., & Burgess, A.G. (1988). Serial rapists and their victims: Reenactment and repetition. In R.A. Prentky & V.L. Quinsey (Eds.), *Human sexual aggression: Current perspectives* (vol. 528, pp. 277–295). New York: Annals of the New York Academy of Sciences.

Burke, P., Stets, J., & Pirog-Good, M. (1989). Gender identity, self-esteem, and physical and sexual abuse in dating relationships. In M. Pirog-Good & J. Stets (Eds.), *Violence in dating relationships: Emerging social issues* (pp. 72–93). New York: Praeger.

Burns, T. F. (1980). Getting rowdy with the boys. *Journal of Drug Issues, 10*(2), 273–286.

Bushman, B., & Cooper, H. (1990). Effects of alcohol on human aggression: An integrative research review. *Psychological Bulletin, 107*, 341–354.

Butcher, J.N., Graham, J.R., Ben-Porath, Y.S., Tellegen, A., & Dahlstrom, B. (1989). *Multiphasic personality inventory-2.* Minneapolis: University of Minnesota Press.

Buydens-Branchey, L., Branchey, M.H., Noumair, D., & Lieber, C.S. (1989). Age of alcoholism onset II: Relation to susceptibility to serotonin precursor availability. *Archives of General Psychiatry, 46*, 231–236.

Byers, E.S. & Wilson, P. (1985). Accuracy of women's expectations regarding men's response to refusals of sexual advances in dating situations. *International Journal of Women's Studies, 8*, 376-387.

Caces, M.F., Stinson, F.S., & Harford, T.C. (1991). Alcohol use and physically risky behavior among adolescents. *Alcohol Health and Research, 15*, 228–233.

Cadsky, O. & Crawford, M. (1988). Establishing batterer typologies in a clinical sample of men who assault their female partners. Special issue: Wife battering: A Canadian perspective. *Canadian Journal of Community Mental Health, 7*, 119–127.

Caesar, P.L. (1988). Exposure to violence in the families-of-origin among wife abusers and martially nonviolent men. Special issue: Wife assaulters. *Violence and Victims, 3*, 49–63.

Campbell, A. (1981). *Girl delinquents*. New York: St. Martin's Press.

Campbell, J.C. (1989). Women's responses to sexual abuse in intimate relationships. *Women's Health Care International, 8*, 335–347.

Campbell, J.C. (1998). Making the healthcare system an empowerment zone for battered women. In J.C. Campbell, (Ed.), *Empowering survivors of abuse: Health care for battered women and their children* (pp. 3–22). Thousand Oaks, CA: Sage Publications.

Campbell, J.C. & Soeken, K.L. (1999). Forced sex and intimate partner violence: Effects of women's risk and women's health. *Violence Against Women, 5,* 1017–1035.

Campbell, J.C., Webster, D., Koziol-McLain, J., Block, C.R., Campbell, D., Curry, M.A. et al. (2002). Risk factors for femicide in abusive relationships: Results from a multi-site case study control study. *American Journal of Public Health, 93*, 1089–1097.

Capsi, A., Moffit, T.E., Silva, P.A., Stouthamer-Loeber, M., Kreuger, R.F., & Schmutte, P.S. (1994). Are some people crime-prone? Replications of the personality-crime relationship across countries, genders, races, and methods. *Criminology, 32*, 163–195.

Carden, A.D. (1994). Wife abuse and the wife abuser: Review and recommendations. *Counseling Psychologist, 22*, 539–582.

Carlin, K. (1988). Talking to the community about batterers. *Voice*: *The newsletter of the National Coalition Against Domestic Abuse.*

Carlson, B.E. (1991). Outcome of physical abuse and observation of marital violence among adolescents in placement. *Journal of Interpersonal Violence, 6*, 526–534.

Cascardi, M., & Vivian, D. (1995). Context for specific episodes of marital violence: Gender and severity of violence differences. *Journal of Family Violence, 10*, 265–293.

Cascardi, M., Langhinrichsen, J., & Vivian, D. (1992). Marital aggression: Impact, injury, and health correlates of husbands and wives. *Archives of Internal Medicine, 152*, 1178–1184.

Cesta, M. L. (2005, March 4). Personal communication.

Chamberlain, B., Ervin, F., Pihl, R., & Young, S. (1987). The effect of raising or lowering tryptophan levels on aggression in vervet monkeys. *Psychopharmacology Biochemistry and Behavior, 28*, 503–510.

Champion, D.J. (1994). *Measuring offender risk: A criminal justice sourcebook.* Westport, CT: Greenwood Press.

Chaplin, T.C., Rice, M.E., & Harris, G.T. (1995). Salient victim suffering and the sexual responses of child molesters. *Journal of Consulting and Clinical Psychology, 63*, 249–255.

Chase, K.A., O'Leary, K.D., & Heyman, R.E. (2001). Categorizing partner-violent men within the reactive-proactive typology model. *Journal of Consulting and Clinical Psychology, 69*(3), 567–572.

Chen, H., Bersani, C., Myers, S.C., & Denton, R. (1989). Evaluating the effectiveness of a court-sponsored abuser treatment program. *Journal of Family Violence, 4*(4), 309–322.

Chene, S. (2000). Le processus d'aggravation du viol au meurtre sexuel: Intensions de l'agresseur et donnees situationelles. Referenced in E. Beauregard, P. Lussier, & J. Proulx, The role of sexual interests and situational factors on rapists' modus operandi: Implications for offending. *Legal and Criminological Psychology, 10*(2), 265–279.

Chermack, S.T. & Giancola, P. (1997). The relation between alcohol and aggression: An integrated biopsychosocial approach. *Clinical Psychology Review, 17*, 621–649.

Chermack, S.T. & Taylor, S.P. (1995). Alcohol and human physical aggression: Pharmacological versus expectancy effects. *Journal of Studies on Alcohol, 56*, 449–456.

Chesney-Lind, M. & Shelden, R. (1992). *Girls, delinquency, and juvenile justice.* Pacific Grove, CA: Brooks/Cole.

Choquet, M., Menke, H., & Manfredi, R. (1991). Interpersonal aggressive behavior and alcohol consumption among young urban adolescents in France. *Alcohol and Alcoholism, 26*, 381–390.

Clark, C.R. (1997). Sociopathy, malingering, and defensiveness. In R. Rogers (Ed.), *Clinical assessment of malingering and deception*, (pp. 68–84). New York: Guilford Press.

Cleckley, H. (1976, 1941/1998). *The mask of insanity,* 5th ed. St. Louis, MO: Mosby Publishing.

Coker, A.L., Davis, K.E., Arias, I., Desai, S., Sanderson, M., & Brandt, H.M. (2002). Physical and mental health effects of intimate partner violence for men and women. *American Journal of Preventative Medicine, 23*, 260–268.

Coleman, E. & Dwyer, M. (1990). Proposed standards of care for the treatment of adult sexual offenders. *Journal of Offender Rehabilitation, 16*, 93–106.

Collins, J.J. (1981). Alcohol use and criminal behavior: An empirical, theoretical, and methodological overview. In J.J. Collins & M.E. Wolfgang (Eds.), *Drinking and crime: Perspectives on the relationships between alcohol consumption and criminal behavior* (pp. 288–338). New York: Guilford Press.

Collins, J.J. (1983). Alcohol use and expressive interpersonal violence: A proposed explanatory model. In E. Gottheil, K.A. Druley, T.E. Skoloda, & H.M. Waxman (Eds.), *Alcohol, drug abuse and aggression* (pp. 5–25). Springfield, IL: Charles C Thomas Publisher.

Collins, J.J. (1993). Drinking and violence: An individual offender focus. In S.E. Martin (Ed.), *Alcohol and interpersonal violence: Fostering multidisciplinary perspectives* (Research Monograph No. 24) (pp. 221–235). Rockville, MD: National Institutes of Health.

Collins, J.J. & Schlenger, W.E. (1988). Acute and chronic effects of alcohol use on violence. *Journal of Studies on Alcohol, 49*, 516–521.

Cooke, D. & Michie, C. (2001). Refining the construct of psychopathy: Towards a hierarchical model. *Psychological Assessment, 13,* 171–188.

Corne, S., Briere, J., & Esses, L.M. (1992, December). Women's attitudes and fantasies about rape as a function of early exposure to pornography. *Journal of Interpersonal Violence 7*(4), 454–461.

Cortoni, F. & Marshall, W.L. (1991). Sex as a coping strategy and its relationship to juvenile sexual history and intimacy in sexual offenders. *Sexual Abuse: A Journal of Research and Treatment, 13,* 27–43.

Counts, D.A., Brown, J.K., & Campbell, J. (1992). *Sanctions and sanctuary: Cultural perspectives on the beating of wives.* Boulder, CO: Westview Publishing.

Courtois, C.A. (1988). *Healing the incest wound: Adult survivors in therapy.* New York: W.W. Norton & Company.

Craig, L.A., Browne, K.D., Stringer, I., & Beech, A. (2004). Limitations in actuarial risk assessment of sexual offenders: A methodological note. *The British Journal of Forensic Practice, 6*(1), 16–33.

Craig, L.A., Browne, K.D., Stringer, I., & Beech, A. (in press). Sexual recidivism: A review of static, dynamic and actuarial predictors. *Journal of Sexual Aggression.*

Cramer, E. & McFarlane, J. (1994, August). Pornography and abuse of women. *Public Health Nursing 11*(4), 268–272.

Crawford, A. (1984a). Alcohol and expectancy I. Perceived differences in the effects of drinking. *Alcohol and Alcoholism, 19*, 63–69.

Crawford, A. (1984b). Alcohol and expectancy II. Perceived sex differences in the role of alcohol as a source of aggression. *Alcohol and Alcoholism, 19*, 71–75.

Croop, R.S., Faulkner, E.B., & Labriola, D.F. (1997). The safety of naltrexone in the treatment of alcoholism. Results from a multicenter usage study. The Naltrexone Usage Study Group. *Archive of General Psychiatry, 54*(12), 1130–1135.

Crowell, N. & Burgess, A.W. (Eds.). (1996). *Understanding violence against women.* Washington, D.C.: National Academy Press.

Cummings, J.G., Pepler, D.J., & Moore, T.E. (1999). Behavior problems in children exposed to wife abuse: Gender differences. *Journal of Family Violence, 14*, 133–156.

Currie, D.W. (1988). *The abusive husband: An approach to intervention.* Ottawa: National Clearinghouse for Family Violence, Health and Welfare.

Daly, K. (1994). *Gender, crime, and punishment.* New Haven, CT: Yale University Press.

Dansky, B.S., Byrne, C.A., & Brady, K.T. (1999). Intimate violence and post-traumatic stress disorder among individuals with cocaine dependence. *American Journal of Drug and Alcohol Abuse, 25,* 257–268.

Dasgupta, S.D. (1999). Just like men? A critical review of violence by women. In M.F. Shepard & E.L. Pense (Eds.), *Coordinating community responses to domestic violence: Lessons from Duluth and beyond* (pp. 195–222). Thousand Oaks, CA: Sage Publications.

Davidson, J.R.T., Rothbaum, B.O., van der Kolk, B.A., Sikes, C.R., & Farfel, G.M. (2001). Multicenter, double-blind comparison of Sertraline and placebo in the treatment of posttraumatic stress disorder. *Archives of General Psychiatry, 58*, 485–492.

Davis, R.C., Taylor, B.G., & Maxwell, C.M. (2000). (The Brooklyn Experiment). Does batterer treatment reduce violence: A randomized experiment in Brooklyn. Final report to the National Institute of Justice by Victim Services (now Safe Horizon), New York City.

Decker, D.J. (1999). *Stopping the violence. A group model to change men's abusive attitudes and behaviors.* New York: The Haworth Maltreatment and Trauma Press, Inc.

DeMaris, A. & Jackson, J.K. (1987). Batterer's reports of recidivism after counseling. *Social Casework: 68*, 458–465.

Dengerink, H. (1971). Anxiety, aggression, and physiological arousal. *Journal of Experimental Research in Personality, 5*, 223–232.

Denno, D. (1994). Gender, crime, and criminal law defenses. *Journal of Criminal Law, 85*(1), 80–180.

Denov, M. (2000). *Perspectives on female sex offending: A culture of denial.* Unpublished Ph.D. Thesis, Institute of Criminology, University of Cambridge, U.K..

Denov, M.S. (2001). A culture of denial: Exploring professional perspectives on female sex offending. *Canadian Journal of Criminology, 43*(3), 303–330.

Denov, M.S. (2003). The myth of innocence: Sexual scripts and the recognition of child sexual abuse by female perpetrators. *The Journal of Sex Research, 40*(3), 303–317.

Denov, M.S. (2004). The long-term effects of child sexual abuse by female perpetrators: A qualitative study of male and female victims. *Journal of Interpersonal Violence, 19*(10), 1137.

DePaulo, B.M., Lindsey, J.L., Malone, B.E., Muhlenbruck, L., Charlton, K., & Cooper, H. (2003). Cues to deception. *Psychological Bulletin, 129*(1), 74–103.

Dermen, K.H. & George, W.H. (1988). Alcohol expectancy and the relationship between drinking and physical aggression, *Journal of Psychology, 123*, 153–161.

Deschner, J.P. (1984). *How to end the hitting habit: Anger control for couples.* New York: The Free Press.

DeViney, E., Dickert, J., & Lockwood, R. (1983). The care of pets within child abusing families. *The International Journal for the Study of Animal Related Problems, 4*, 321–329.

Dickey, B., Normand, S-L.T., Weiss, R.D., Drake, R.E., & Azeni, H. (2000). Medical morbidity, mental illness, and substance use disorders. *Psychiatric Services, 53,* 861–867.

Dobash, R.P. & Dobash, R.E. (2001). Violence against women: A review of recent Anglo-American research. *Journal of Conflict and Violence Research, 3*, 5–22.

Dobash, R.P. & Dobash, R.E. (2004). Women's violence to men in intimate relationships. *The British Journal of Criminology, 44*(3), 324–349.

Dobash, R.E., Dobash, R.P., & Cavanagh, K. (1985). The contact between battered women and social and medical agencies. In J. Pahl (Ed.), *Private violence and public policy.* (pp. 142–165). London: Routledge and Kegan Paul.

Dobash, R.P., Dobash, R.E., Cavanagh, K., & Lewis, R. (2000). *Changing violent men.* Thousand Oaks, CA: Sage Publications.

Dobash, R.P., Dobash, R.E, Wilson, M., & Daly, M. (1992). The myth of sexual symmetry in marital violence. *Social Problems, 39*, 71–91.

Dobson, J. (1994). Interview in 1989 with Ted Bundy just before his execution. In K.L. Swisher, & C. Wekesser, (Eds.). *Violence against women* (pp. 122). San Diego, CA: Current Controversies.

Dodge, K.A. (1991). The structure and function of reactive and proactive aggression. In D. J. Pepler & K.H. Rubin (Eds.), *The development and treatment of childhood aggression* (pp. 201–218). Hillsdale, NJ: Erlbaum.

Donaldson, M.A. & Gardner, R. (1985). Diagnosis and treatment of traumatic stress among women after childhood incest. In C.R. Figley (Ed.), *Trauma and its wake: The study and treatment of post-traumatic stress disorder* (pp. 333–347). New York: Brunner/Mazel.

Donnerstein, E. (1980). Aggressive erotica and violence against women. *Journal of Personality and Social Psychology, 39*, 269–277.

Doren, D.M. (1987). *Understanding and treating the psychopath.* New York: John Wiley & Sons.

Doren, D.M. (1998). Recidivism base rates: Predictions of sex offender recidivism and the sexual predator commitment laws. *Behavioral Sciences and the Law, 16*, 197–114.

Dowd, L.S. (2001). Female perpetrators of partner aggression: Relevant issues and treatment. *Journal of Aggression, Maltreatment, and Trauma,* Special Issue on Batterer Treatment Standards, *5*(2), 73–104.

Dowd, L.S., Leisring, P.A., & Rosenbaum, A. (2005). Partner aggressive women: Characteristics and treatment attrition. *Violence and Victims, 20*(1), 219–234.

Duke, R.L. & Kean, R.B. (1983). An experimental study of gender and situation in the perception and reportage of child abuse. *Child Abuse and Neglect, 13,* 351–360.

Dunford, F.W. (2000). The San Diego experiment: An assessment of interventions for men who assault their wives. *Journal of Consulting and Clinical Psychology, 68*, 468.

Dutton, A.D., Kaltman, S., Goodman, L.A., Weinfurt, K., & Vankos, N. (2005). Patterns of intimate partner violence: Correlates and outcomes. *Violence and Victims, 20*(5), 483–498.

Dutton, D.G., (1986). The outcome of court-mandated treatment for wife assault: A quasi-experimental evaluation. *Violence and Victims, 1,* 163–175.

Dutton, D.G. (1987). The criminal justice system response to wife assault. *Law and Human Behavior, 11,* 189–206.

Dutton, D.G. (1988). Profiling of wife assaulters: Preliminary evidence for a trimodal analysis. *Violence and Victims, 3*, 5–29.

Dutton, D.G. (1992). *Empowering and healing the battered main.* New York: Springer.

Dutton, D.G. (1995a). *The domestic assault of women: Psychological and criminal justice perspectives* (2nd ed.). Vancouver, Canada: University of British Columbia Press.

Dutton, D.G. (1995b). Trauma symptoms and PTSD-like profiles in perpetrators of intimate abuse. *Journal of Traumatic Stress, 8*, 299–316.

Dutton, D.G. (1998). *The abusive personality: Violent and control in intimate relationships.* New York: Guilford Press.

Dutton, D.G. & Hart, S.D. (1992a). Evidence for long-term, specific effects of childhood abuse on criminal behavior in men. *International Journal of Offender Therapy and Comparative Criminology, 36,* 129–137.

Dutton, D.G. & Hart, S.D. (1992b). Risk markers for family violence in a federally incarcerated population. *International Journal of Law and Psychiatry, 15,* 101–112.

Dutton, D.G., Bodnarchuk, M., Kropp, R., Hart, S.D., & Ogloff, J.P. (1997). Wife assault treatment and criminal recidivism: An 11-year follow-up. *International Journal of Offender Therapy and Comparative Criminology, 41*(1), 9–23.

Dutton, D.G., Saunders, D.G., Starzomski, A.J., & Bartholomew, K. (1994). Intimacy-anger and insecure attachment as precursors of abuse in intimate relationships. *Journal of Applied Social Psychology, 24*(15), 1367–1386.

Dutton, D.G. & Starzomski, A.J. (1993). Borderline personality in perpetrators of psychological and physical abuse. *Violence and Victims, 8*, 327–337.

Dyck, N. (1980). Booze, barrooms and scrapping: Masculinity and violence in a western Canadian town. *Canadian Journal of Anthropology, 1*, 191–198.

Eckhardt, C.I. & Jamison, T.R. (2002). Articulated thoughts of male dating violence perpetrators during anger arousal. *Cognitive Therapy and Research, 26*(3), 289–308.

Eckhardt, C.I., Barbour, K.A., & Davidson, G.C. (1998). Articulated thoughts of martially violent and non-violent men during anger arousal. *Journal of Consulting and Clinical Psychology, 66*(2), 259–269.

Edguer, N. & Janisse, M. (1994). Type A behaviour and aggression: Provocation, conflict and cardiovascular responsivity in the Buss teacher-learner paradigm. *Personality and Individual Differences, 17,* 377–393.

Edleson, J.L. (1995, March). Do batterer's programs work? Paper presented at the International Study Group on the Future of Intervention with Battered Women and Their Families. Haifa, Israel.

Ehlers, A., Mayou, R.A., & Bryant, B. (1998). Psychological predictors of chronic posttraumatic stress disorder after motor vehicle accidents. *Journal of Abnormal Psychology, 107*(3), 508–519.

Eisikovits, Z.C. & Edleson, J.L. (1989, September). Intervening with men who batter: A critical review of the literature. *Social Services Review*, 384–414.

Ekman, P. (1992). *Telling lies: Clues to deceit in the marketplace, politics, and marriage.* New York: W.W. Norton & Co.

Ekman, P. & Friesen, W.V. (1969). Nonverbal leakage and clues to deception. *Psychiatry, 32*, 88–106.

Ekman, P., Friesen, W.V., & Simons, R.C. (1985). Is the startle reaction an emotion? *Journal of Personality and Social Psychology, 49*, 1416–1426.

Epperson, D.L., Kaul, J.D., & Hesselton, D. (1998). Final report of the development of the Minnesota sex offender screening tool-revised (MnSOST-R). Presentation at the 17th Annual Research and Treatment Conference of the Association for the Treatment of Sexual Abusers, Vancouver, B.C., Canada.

Fagan, J. (1990). Intoxication and aggression. In M. Tonry & J.Q. Wilson (Eds.), *Drugs and Crime* (pp. 241–320). Chicago: University of Chicago Press.

Fagan, J. (1993). Set and setting revisited: Influences of alcohol and illicit drugs on the social context of violent events. In S.E. Martin (Ed.), *Alcohol and interpersonal violence: Fostering multidisciplinary perspectives* (vol. 24, pp. 160–192). Rockville, MD: National Insitutes of Health.

Fagan, J. (1996). The criminalisation of domestic violence: Promises and limits. National Institute of Justice Research Report.

Fagan, J. & Browne, A. (1994). Violence between spouses and intimates. In A. Reiss & J. Roth (Eds.), *Understanding and preventing violence: Vol. 3. Social Influences* (pp. 115–292), Washington, D.C.: National Academy Press.

Fagan, J., Stewart, D., & Hansen, K. (1983). Violent men or violent husbands? Background factors and situational correlates. In D. Finkelhor, R. Gelles, G. Hotaling, and M. Straus, (Eds.), *The dark side of families: Current family violence research* (pp. 49–67). Beverly Hills, CA: Sage Publications.

Fagan, R.W., Barnett, O.W., & Patton, J.B. (1988). Reasons for alcohol use in martially violent men. *American Journal of Drug and Alcohol Abuse, 14*, 371–392.

Faller, K.C. (1987). Women who sexually abuse children. *Violence and Victims, 2*(4), 263–276.

Faller, K.C. (1988). The spectrum of sexual abuse in daycare. *Journal of Family Violence, 3*(4), 283–298.

Faller, K. (1989). Characteristics of a sample of sexually abused children: How boy and girl victims differ. *Child Abuse and Neglect, 13*, 281–291.

Faller, K.C. (1990). *Understanding child sexual abuse maltreatment*. Newbury Park, CA: Sage Publications.

Faller, K.C. (1995). A clinical sample of women who have sexually abused children. *Journal of Child Sexual Abuse, 4*(3), 13–31.

Fantuzzo, J.W. & Lindquist, C.U. (1989). The effects of observing conjugal violence on children: A review and analysis of research methodology. *Journal of Family Violence, 4,* 77–94.

Fantuzzo, J.W., Boruch, R., Beriama, A., Atkins, M., & Marcus, S. (1997). Domestic violence and children: Prevalence and risk in five major U.S. cities. *Journal of the American Academy of Child and Adolescent Psychiatry, 36*, 116–122.

Faulk, M. (1974). Men who assault their wives. *Medicine, Science and the Law,* 14, 180–183.

Faver, C.A. & Strand, E.B. (2003). Domestic violence and animal cruelty: Untangling the web of abuse. Special section: Domestic violence and social work education, *Journal of Social Work Education, 39*(2), 237–248.

Feazelle, C.S., Mayers, R.S., & Deschner, J.P. (1984). Services for men who batter: Implications for programs and policies. *Family Relations, 33*, 217–223.

Feder, L.R. & Dugan, L. (2002). A test of the efficacy of court-mandated counseling for domestic violence offenders: The Broward County experiment. *Justice Quarterly, 19*(2), 343.

Feder. L.R. & Wilson, D.B. (2005). A meta-analytical of court-mandated batterer intervention programs: Can courts affect abusers' behavior? *Journal of Experimental Criminology, 1*, 239–262.

Federoff, J.P., Wisner-Carslon, R., Dean, S., & Berlin, F.S. (1992). Medroxy-progesterone acetate in the treatment of paraphilic sexual disorder: Rate of relapse in paraphilic men treated in long-term group psychotherapy with or without medroxy-progesterone acetate. *Journal of Offender Rehabilitation, 18,* 109–123.

Feeny, N.C., Zoellner, L.A., & Foa, E.B. (2000). Anger, dissociation, and posttraumatic stress disorder among female assault victims. *Journal of Traumatic Stress, 13*, 89–100.

Feld, S.L., & Straus, M.A. (1989). Escalation and resistance of wife assault in marriage. *Criminology, 27*, 141–161.

Feldman, C.M. & Ridley, C.A. (1995). The etiology and treatment of domestic violence between adult partners. *Clinical Psychology: Science and Practice, 2*, 317–348.

Felthous, A.R. & Kellert, S.R. (1987). Childhood cruelty to animals and later aggression against people: A review. *American Journal of Psychiatry, 144,* 710–717.

Finkelhor, D. (1979). *Sexually victimized children*. New York: Free Press.

Finkelhor, D. & Yllo, K. (1985). *License to rape: Sexual abuse of wives*. New York: Holt, Rinehart and Winston.

Finkelhor, D., Williams, L.M., & Burns, N. (1988). *Nursery crimes: Sexual abuse in day care*. Newbury Park, CA: Sage Publications.

Fisher, D., Beech, A., & Browne, K. (2000). The effectiveness of relapse prevention training in a group of incarcerated child molesters. *Psychology, Crime & Law, 00*, 1–5.

Flynn, C.P. (2000a). Why family professionals can no longer ignore violence towards animals. *Family Relations, 49*(1), 87–95.

Flynn, C.P. (2000b). Woman's best friend: Pet abuse and the role of companion animals in the lives of battered women. *Violence Against Women, 6*(2), 162–177.

Foa, E.B., Cascardi, M., Zoellner, L.A., & Feeny, N.C. (2000). Psychological and environmental factors associated with partner violence. *Trauma, Violence, and Abuse, 1*, 67–91.

Follingstad, D.R., Rutledge, L.L., Berg, B.J., Hause, E.S., & Polek, D.S. (1990). The role of emotional abuse in physically abusive relationships. *Journal of Family Violence, 5*, 107–120.

Follingstad, D.R., Wright, S., Lloyd, S., & Sebastian, J.A. (1991). Sex differences in motivations and effects in dating violence. *Family Relations, 40*, 51–57.

Foo, L. & Margolin, G. (1995). A multivariate investigation of dating aggression. *Journal of Family Violence, 10*, 351–377.

Ford, D. & Regoli, M.J. (1993). The criminal prosecution of wife assaulters. In Z. Hilton (Ed.), *Legal response to wife assault: Current trends and evaluation* (pp. 127–164). Newbury Park, CA: Sage Publications.

Forth, A. & Kroner, D. (1994). *The factor structure of the revised psychopathy checklist with incarcerated rapists and incest offenders*. Unpublished manuscript.

Foss, R.D. & Perrine, M.W.B. (1993). Predictors of impaired driving. In H.D. Utzelmann, G. Berghaus, & G. Krol (Eds.), *Alcohol, drugs and traffic safety* (vol. 2, pp. 1139–1144). Cologne: Rheinland GmbH, Koln, Verlag TUV.

Freeman-Longo, R.E., & Knopp, F.H. (1992). State of the art sex offender treatment: Outcome and issues. *Annals of Sex Research, 5*, 141–160.

Freund, K. (1965). Diagnosing heterosexual pedophilia by means of a test for sexual interest. *Behaviour Research and Therapy, 3*, 229–234.

Frieze, I.H., Parsons, J.E., Johnson, P.B., Ruble, D.N., and Zellman, G.L. (1978). *Women and sex roles: A social psychological perspective*. New York: W.W. Norton and Company.

Frisbie, L.V. & Dondis, E.H. (1965). Recidivism among treated sex offenders. California Mental Health Research Monogram (No. 5). California: State of California Department of Mental Hygiene. In R.K. Hanson & M.T. Bussiere. (1996). *Predictors of sexual offender recidivism: A meta analysis*. Ottawa: Ministry of the Solicitor General of Canada.

Fritz, G.S., Stoll, K., & Wagner, N.N. (1981). A comparison of males and females who were sexually molested as children. *Journal of Sex and Marital Therapy, 7*, 54–59

Fromm, E. (1956). *The art of loving*. New York: Harper & Row.

Fromuth, M. & Burkehart, B. (1989). Long-term psychological correlated of childhood sexual abuse in two samples of college men. *Child Abuse and Neglect, 13*, 533–542.

Fruend, K. & Blanchard, R. (1989). Phallometric diagnosis of pedophilia. *Journal of Consulting and Clinical Psychology, 57*, 100–105.

Furby, L., Weinrott, M.R., & Blackshaw, L. (1989). Sex offender recidivism: A review. *Psychological Bulletin, 105*, 3–30.

Gantner, A.B. & Taylor, S.P. (1988). Human physical aggression as a function of diazepam. *Personality and Social Psychology Bulletin, 14*, 479–484.

Gantner, A.B. & Taylor, S.P. (1992). Human physical aggression as a function of alcohol and threat of harm. *Aggressive Behavior, 18*, 29–36.

Gebhard, P.H., Gagnon, J.H., Pomeroy, W.B., & Christenson, C.V. (1965). *Sex offenders*. New York: Harper & Row Publishers.

Geller, D.A., Biederman, J., Reed, E.D., Spencer, T., & Wilens, T.E. (1995). Similarities in response to fluoxetine in the treatment of children and adolescents with obsessive-compulsive disorder. *Journal of the American Academy of Child and Adolescent Psychiatry, 34*(1), 36–44.

Gelles, R.J. (1979). *Family violence*. Beverly Hills, CA: Sage Publications.

Gelles, R.J. (1987). The family and its role in the abuse of children. *Psychiatric Annals, 17*, 229–232.

Gelles, R.J. (1995). *Violence toward men: Fact or fiction?* (Report prepared for the American Medical Association, Council on Scientific Affairs). Kingston, RI: Family Violence Research Program, University of Rhode Island.

Gelles, R.J. (1997). *Intimate violence in families*, 3rd ed. Thousand Oaks, CA: Sage Publications.

Gelles, R.J. & Straus, M.A. (1979). Determinants of violence in the family: Toward a theoretical integration. In W.R. Burr, R. Hill, F.I. Nye, & I.L. Reiss (Eds.), *Contemporary theories about the family* (pp. 549–581). New York: Free Press.

Gendreau, P., Little, T., & Goggin, C. (1996). A meta-analysis of the predictors of adult offender recidivism: What works! *Criminology, 34*, 575–607.

George, W.H. & Marlatt, G.A. (1986). The effects of alcohol and anger on interest in violence, erotica, and deviance. *Journal of Abnormal Psychology, 95*, 150–158.

George, W.H., Dermen, K.H., & Nochajski, T.H. (1989). Expectancy set, self-reported expec-tancies and predispositional traits: Predicting interest in violence and erotica. *Journal of Studies on Alcohol, 50*, 541–551.

George, W.H., Stoner, S. A., Norris, J., Lopez, P.A., & Lehman, G.L. (2000, January). Alcohol expectancies and sexuality: A self-fulfilling prophecy analysis of dyadic perceptions and behavior. *Journal of Studies on Alcohol, 61*(1), 168–176.

Gergen, M. (1990). Beyond the evil empire: Horseplay and aggression, *Aggressive Behavior, 16*, 381–398.

Gershuny, B.S., Baer, L., Jenike, M.A., Minichiello, W.E., & Wilhelm, S. (2002). Comorbid posttraumatic stress disorder: Impact on treatment outcome for obsessive compulsive disorder. *American Journal of Psychiatry, 159*, 852–854.

Giancola, P.R. & Zeichner, A. (1995a). Alcohol-related aggression in males and females: Effects of blood alcohol concentration, subjective intoxication, personality, and provocation, *Alcoholism: Clinical and Experimental Research, 19*, 130–134.

Giancola, P. & Zeichner, A. (1995b). An investigation of gender differences in alcohol related aggression. *Journal of Studies on Alcohol, 56*, 573–579.

Giancola, P.R. & Zeichner, A. (1997). The biphasic effects of alcohol on human physical aggression, *Journal of Abnormal Psychology, 106*, 598–607.

Giancola, P.R., Reagan, C.M., van Weenan, R.V., & Zeichner, A. (1998, October). Alcohol-induced stimulation and sedation: Relation to physical aggression. *The Journal of General Psychology, 125*, 297–305.

Gibbens, T.C.N., Soothill, K.L., & Way, C.K. (1981). Sex offenses against young girls: A long-term record study. *Psychological Medicine, 11*, 351–357.

Giles-Sims, J. (1983). *Wife battering: A systems theory approach*. New York: Guilford Press.

Gilfus, M. (1992). From victims to survivors to offenders: Women's routes of entry and immersion into street crime. *Women's Criminal Justice, 4*, 63–89.

Giordano, P.C., Millhollin, T.J., Cernkovich, S.A., Pugh, M.D., & Rudolph, J.L. (1999). Delinquency, identity, and women's involvement in relationship violence. *Criminology, 37*, 17–40.

Golding, J. (1999). Intimate partner violence as a risk factor for mental disorders: A meta-analysis. *Journal of Family Violence, 14*, 99–132.

Goldsmith, H.R. (1990). Men who abuse their spouses: An approach to assessing future risk. *Journal of Offender Counseling, Services and Rehabilitation, 15*, 45–56.

Gondolf, E.W. (1985). Fighting for control: A clinical assessment of men who batter. Social casework: *The Journal of Contemporary Social Work, 66*(1), 48–54.

Gondolf, E.W. (1988). Who are those guys? Toward a behavioral typology of batterers. *Violence and Victims, 3*, 187–203.

Gondolf, E.W. (1997a). Batterer program: What we know and need to know. *Journal of Interpersonal Violence, 12*, 83–98.

Gondolf, E.W. (1997b). Patterns of reassault in batterer programs. *Violence and Victims, 12*, 373–387.

Gondolf, E.W. (2000). A 30-month follow-up of court-referred batterers in four cities. *International Journal of Offender Therapy and Comparative Criminology, 44*, 111–128.

Gondolf, E.W. (2002). Batterer intervention systems safe. Beverly Hills, CA: Sage.

Gondolf, E.W. & Russell, D. (1986). The case against anger control treatment programs for batterers. *Response to the Victimisation of Women and Children, 9*(3), 2–5.

Goodchilds, J.D. and Zellman, G.L. (1984). Sexual signalling and sexual aggression in adolescent relationships. In N. Malamuth and E. Donnerstein (Eds.), *Pornography and sexual aggression*. Orlando, FL: Academic Press.

Goodman, M.S. & Fallon, B.C. (1995). *Pattern changing for abused women: An educational program*. Thousand Oaks, CA: Sage Publications.

Gottfredson, D.M. & Hirschi, T. (1990). *A general theory of crime*. Stanford, CA: Stanford University Press.

Gottfredson, D.M. & Tonry, M. (Eds.) (1987). *Prediction and classification: Criminal justice decision-making*. Chicago: University of Chicago Press.

Gottman, J.M. (1999). *The marriage clinic*. New York: W.W. Norton and Company.

Gottman, J.M., Jacobson, N.S., Rushe, R.H., Shortt, J.W., Babcock, J., La Taillade, J.J., & Waltz, J. (1995). The relationship between heart rate reactivity, emotionally aggressive behavior, and general violence in batterers. *Journal of Family Psychology, 9*, 227–248.

Graham, D.L.R. & Rawlings, E.I. (1987). Bonding with abusive dating partners: Dynamics of Stockholm Syndrome. In B. Levy (Ed.), *Dating violence: Young women in danger* (1991), (pp. 110–135). Seattle, WA: Seal Press.

Graham, D.L.R., Rawlings, E.I., & Rimini, N. (1988). Survivors of terror: Batterered women, hostages and the Stockholm Syndrome. In K. Yllo and B. Bograd (Eds.), *Feminist perspectives on wife abuse* (pp. 217–233). Newbury Park, CA: Sage Publications.

Graham, K. & Wells, S. (2001, Fall). I'm OK. You're drunk!: Self-other differences in the perceived effects of alcohol in real-life incidents of aggression. *Contemporary Drug Problems*, 28.

Graham, K., Leonard, K.E., Room, R., Cameron Wild, T., Pihl, R.O., Bois, C, and Single, E. (1998). Current directions in research in understanding and preventing intoxicated aggression. *Addiction, 93(*5), 659–676.

Graham, K., Schmidt, G., & Gillis, K. (1996). Circumstances when drinking leads to aggression: An overview of research findings. *Contemporary Drug Problems*, *23*(3), 49, 65.

Graham, K., Wells, S., and West P. (1997, Winter). A framework for applying explanations of alcohol-related aggression to naturally occurring aggressive behavior. *Contemporary Drug Problems, 24,* 625–667.

Gray, A. (1994). Intervention programmes for domestic violence abusers: A literature review. Family Violence Prevention Coordinating Committee: Wellington, New Zealand.

Grayston, A.D. & De Luca, R.V. (1999). Female perpetrators of child sexual abuse: A review of the clinical and empirical literature. *Aggression and Violent Behavior, 4,* 93–106.

Green, A.H. (1999). Female sex offenders. In J.A. Shaw (Ed.), *Sexual aggression* (pp. 195–210). Washington, D.C.: American Psychiatric Press.

Groth, A.N. (1979/1981). Sexual trauma histories of rapists and child molesters, *Victimology*, *4*, 10–16.

Groth, A.N., Burgess, A.W., & Holmstrom, L.L. (1978). Rape: Power, anger, and sexuality. *American Journal of Psychiatry, 134(*11), 1239–1243.

Groth, A.N. & Birnbaum, H.J. (1978). Adult sexual orientation and attraction to underage persons. *Archives of Sexual Behavior*, *7*, 175–181.

Groth, A.N. & Birnbaum, H.J. (1979). *Men who rape: The psychology of the offender*. New York: Plenum Press.

Groth, A.N. (1981). Sexual offenders against children. Handout.

Grubin, D. (1998). *Sex offending against children: Understanding the risk*. Police Research Series Paper 99. London: Home Office.

Gruszynski, R.J. & Carillo, T.P. (1988). Who completes batterer's treatment groups? An empirical investigation. *Journal of Family Violence, 3,* 141–150.

Gudjonsson, G. (1992). *The psychology of interrogations, confessions and testimony*. West Sussex, England: John Wiley & Sons.

Gustafson, R. (1987). Lack of correspondence between alcohol-related aggressive expectancies for self and others. *Psychological Reports*, *60*, 707–710.

Gustafson, R. (1990, Winter). Wine and physical aggression. *Journal of Drug Issues*, *20*(1), 75.

Gustafson, R. (1991). Aggressive and nonaggressive behavior as a function of alcohol intoxication and frustration in women. *Alcoholism: Clinical and Experimental Research*, *15*, 886–892.

Gustafson, R. (1992). Alcohol and aggression: A replication study controlling for potential confounding variables. *Aggressive Behavior*, *18*, 21–28.

Gustafson, R. (1993). What do experimental paradigms tell us about alcohol-related aggressive responding? *Journal of Studies on Alcohol* (Suppl. 11), 20–29.

Hagen, M.A. (1997). *Whores of the court: The fraud of psychiatric testimony and the rape of American justice.* New York: Regan/Harper Collins.

Hagen, M.P. & Gust-Brey, K.L. (1999). A ten-year longitudinal study of adolescent rapists upon return to the community. *International Journal of Offender Therapy and Comparative Criminology, 43*(4), 448–458.

Hall, G.C.N. (1988). Criminal behavior as a function of clinical and actuarial variables in a sexual offender population. *Journal of Consulting and Clinical Psychology, 56*, 773–775.

Hall, G.C.N. (1995). Sexual offender recidivism revisited: A meta-analysis of recent treatment studies. *Journal of Consulting and Clinical Psychology, 63*, 802–909.

Hall, G.C.N. & Proctor, W.C. (1987). Criminological predictors of recidivism in a sexual offender population. *Journal of Consulting and Clinical Psychology, 55*, 111–112.

Hamberger, L.K. (1997). Female offenders in domestic abuse: A look at actions in their context. *Journal of Aggression, Maltreatment, and Trauma, 1*, 117–130.

Hamberger, L.K. (2005). Men's and women's use of intimate partner violence in clinical samples: Toward a gender-sensitive analysis. *Violence and Victims, 20*(1), 131–152.

Hamberger, L.K. & Ambuel, B. (1997). Training psychology students and professionals to recognize and intervene into partner violence: Borrowing a page from medicine. *Psychotherapy*, 34, 375–385.

Hamberger, L.K. & Guse, C. (2000). What domestically violent men say about their partner's violence. Paper presented at the meeting of the Association for the Advancement of Behavior Therapy. New Orleans, LA.

Hamberger, L.K. & Guse, C. (2002). Men's and women's use of intimate partner violence in clinical samples. *Violence Against Women, 8*, 1301–1331.

Hamberger, L.K. & Hastings, J.E. (1986). Personality correlates of men who abuse their partners: A cross-validation study. *Journal of Family Violence, 1*, 323–341.

Hamberger, L.K. & Hastings, J.E. (1988). Characteristics of male spouse abusers consistent with personality disorders. *Hospital and Community Psychiatry, 39*, 763–770.

Hamberger, L.K., & Hastings, J.E. (1989). Counseling male spouse abuser: Characteristics of treatment completers and dropouts. *Violence and Victims, 4*(4), 275–286.

Hamberger, L.K. & Hastings, J.E. (1990). Recidivism following spouse abatement counseling: Treatment program implications. *Violence and Victims, 5*, 157–170.

Hamberger, L.K. & Hastings, J.E. (1993). Court-mandated treatment of men who assault their partner: Issues, controversies, and outcomes. In N.Z. Hilton, (Ed.), *Legal responses to wife assault: Current trends and evaluation* (pp. 188–229). Newbury Park, CA: Sage Publications.

Hamberger, L.K. & Potente, T. (1994). Counseling heterosexual women arrested for domestic violence: Implications for theory and practice. *Violence and Victims, 9*, 125–137.

Hamberger, L.K., Lohr, J.M., & Bonge, D. (1994). The intended function of domestic violence is different for arrested male and female perpetrators. *Family Violence and Sexual Assault Bulletin, 10*, 40–44.

Hamberger, L.K., Lohr, J.M., Bonge, D., & Tolin, D.F. (1996). A large sample empirical typology of male spouse abusers and its relationship to dimensions of abuse. *Violence and Victims, 11*, 277–292.

Hamberger, L.K., Lohr, J.M., Bonge, D., & Tolin, D.F. (1997). An empirical classification of motivations for domestic violence. *Violence Against Women, 3*, 401–423.

Hamilton, C.J. & Collins, J.J. (1981). The role of alcohol in wife beating and child abuse: A review of the literature. In J.J. Collins (Ed.), *Drinking and crime: Perspectives on the relationship between alcohol consumption and criminal behavior* (pp. 253–287). New York: Guilford Press.

Hanson, R.K. (1997). The development of a brief actuarial risk assessment scale for sexual offense recidivism. (User Report 97-04). Ottawa, Ontario: Department of the Solicitor General of Canada.

Hanson, R.K.. (1999). STATIC 99. Ontario: Public Works and Government Services Canada.

Hanson, R.K. & Bussiere, M.T. (1996). Predictors of sexual offender recidivism: A meta-analysis (User Report 96-04). Ottowa, Onterio: Ministry of the Solicitor General of Canada.

Hanson, R.K. & Bussiere, M.T. (1998). Predicting relapse: A meta-analysis of sexual offender recidivism studies. *Journal of Consulting and Clinical Psychology, 66*(2), 348–362.

Hanson, R.K. & Harris, A.J.R. (1998). Dynamic predictors of sexual recidivism. (User Report 1998-01). Ottawa, Ontario: Department of the Solicitor General of Canada.

Hanson, R.K. & Harris, A.J.R. (2000). Where should we intervene? Dynamic predictors of sexual recidivism. *Criminal Justice and Behavior, 27*, 6–35.

Hanson, R.K. & Morton-Bourgon, K. (2004). Predictors of sexual recidivism: An updated meta-analysis. Public Safety and Emergency Preparedness Canada. Ottawa, Ontario: Public Works and Government Services.

Hanson, R.K. & Thornton, D. (2000). Improving risk assessment for sex offenders: A comparison of three actuarial scales. *Law and Human Behavior, 24*(1), 119–136.

Hanson, R.K. & Wallace-Capretta, S. (2000). Predicting recidivism among male batterers. Ottawa, Ontario: Solicitor General, Canada.

Hanson, R.K., Broom, I., & Stephenson, M. (2004). Evaluating community sex offender treatment programs: A 12-year follow-up of 724 offenders. *Canadian Journal of Behavioural Science, 36*(2), 87–97.

Hanson, R.K., Cadsky, O., Harris, A., & Lalonde, C. (1997). Correlates of battering among 997 men: Family history, adjustment, and attitudinal differences. *Violence and Victims, 12*, 191–208.

Hansen, R.K.., Haraway, M., & Cervantes N. (1991). Therapist's perceptions of severity in cases of family violence. *Violence and Victims*, 6, 225–234.

Hanson, R.K., Scott, H., & Steffy, R.A. (1995). A comparison of child molesters and non-sexual criminals: Risk predictors and long-term recidivism. *Journal of Research in Crime and Delinquency, 32(*3), 325–337.

Hanson, R.K., Steffy, R.A., & Gauthier, R. (1993). Long-term recidivism of child molesters. *Journal of Consulting and Clinical Psychology, 61*, 646–652.

Happel, R.M. & Auffrey, J.J. (1995). Sex offender assessment: Interrupting the dance of denial. *American Journal of Forensic Psychology, 13*(2), 5–22.

Hare, R.D. (1982). Psychopathy and physiological activity during anticipation of an aversive stimulus in a distraction paradigm. *Psychophysiology, 15*, 165–172.

Hare, R.D. (1991). *Hare psychopathy checklist-Revised (PCL-R).* Toronto, ON: Multi-Health Systems, Inc.

Hare, R.D. (1995a). Psychopathy: A clinical construct whose time has come. *Criminal Justice and Behavior, 23*, 25–54.

Hare, R.D. (1995b). Antisocial personality disorder: Psychopaths: New trends in research. *Harvard Mental Health Letter, 12*, 4–5.

Hare, R.D., Frazelle, J., and Cox, D.N. (1978). Psychopathy and physiological responses to threat of an aversive stimulus. *Psychophysiology, 19*, 266–271.

Hare, R.D., Harpur, T., Hakstian, A., Forth, A., Hart, S., & Newman, J. (1990). The revised psychopathy checklist: Descriptive statistics, reliability, and factor structure. *Psychological Assessment, 2*, 338–341.

Hare, R.D., Hart, S.D., & Harpur, T.J. (1991). Psychopathy and the *DSM-IV* criteria for antisocial personality disorder. *Journal of Abnormal Psychology, 100*, 391–398.

Harned, MS. (2001). Abused women or abused men? An examination of the context and outcomes of dating violence. *Violence and Victims, 16*(3), 269–285.

Harper, S. (1979). Violent men and fathers. Paper presented at the conference of Battered Women: Learning how to help. University of Texas at Arlington, May 1979. In J.P. Deschner (1984), *How to end the hitting habit: Anger control for couples* (pp. 39–40). New York: The Free Press.

Harpur, T.J. & Hare, R.D. (1994). Assessment of psychopathy as a function of age. *Journal of Abnormal Psychology, 103*, 604–609.

Harpur, T.J., Hare, R.D., & Hakstian, R.A. (1989). Two-factor conceptualization of psychopathy: Construct validity and assessment implications. *Psychological Assessment: A Journal of Consulting and Clinical Psychology, 1*, 6–7.

Harrell, A. (1991). Evaluation of court-ordered treatment for domestic violence offenders. Washington D.C.: The Urban Institute.

Harris, G.T. & Rice, M.E. (1996). The science of phallometric measurement of male sexual interest. *Current Directions in Psychological Science, 5*, 156–160.

Harris, G.T., Rice, M.E., & Quinsey, V.L. (1993). Violent recidivism of mentally disordered offenders: The development of a statistical prediction instrument. *Criminal Justice and Behavior, 20*, 315–335.

Harris, G.T., Rice, M.E., & Quinsey, V.L. (1998). Appraisal and management of risk in sexual aggressors: Implications for criminal justice policy. *Psychology, Public Policy, and Law, 4*($^1/_2$), 73–117.

Harris, G.T., Rice, M.E., Quinsey, V.L., & Chaplin, T.C. (1996). Viewing time as a measure of sexual interest among child molesters and normal heterosexual men. *Behaviour Research and Therapy, 34*, 389–394.

Harris, G.T., Rice, M.E., Quinsey, V.L., Chaplin, T.C., & Earls, C. (1992). Maximizing the discriminate validity of phallometric assessment data. *Psychological Assessment, 4*(4), 502–511.

Hart, S.D., Dutton, D.G., & Newlove, T. (1993). The prevalence of personality disorder amongst wife assaulters. *Journal of Personality Disorders, 7*, 329–341.

Hazelwood, R.R. (1995). Analyzing the rape and profiling the offender. In R.R. Hazelwood & A.W. Burgess (Eds.), *Practical aspects of rape investigation: A multidisciplinary approach* (pp. 155–181). Boca Raton, FL: CRC Press.

Hazelwood, R.R. (1998, January 29 and 30). *The sexually violent offender*. San Diego: Training for Specialized Training Services.

Hazelwood, R.R. & Burgess, A.W. (1995). *Practical aspects of rape investigation: A multidisciplinary approach*. Boca Raton, FL: CRC Press.

Hazelwood, R.R. & Warren, J.L. (1995). The relevance of fantasy in serial sexual crime investigations. In R.R. Hazelwood & A.W. Burgess (Eds.), *Practical aspects of rape investigation: A multidisciplinary approach*, 2nd ed. (pp. 127–137). Boca Raton, FL: CRC Press.

Hazelwood, R. & Warren, J. (2000). The sexually violent offender: Impulsive or ritualistic? *Aggression and Violent Behavior, 5*, 267–279.

Hazelwood, R.R., Reboussin, R., & Warren, J.I. (1989). Serial rape: Correlates of increased aggression and the relationship of offender pleasure to victim resistance. *Journal of Interpersonal Violence, 4*, 65–78.

Hazzard, A. & Rupp, G. (1986). A note on the knowledge and attitudes of professional groups toward child abuse. *Journal of Community Psychology, 14*, 219–223.

Healy, K., Smith, C., & O'Sullivan, C. (1998). *Batterer intervention: Program approaches and criminal justice strategies*. (NCJ 168638). Washington, D.C.: National Institute of Justice.

Heckert, D.A. & Gondolf, E.W. (2004). Battered women's perceptions of risk versus risk factors and instruments in predicting reassault. *Journal of Interpersonal Violence, 19*(7), 778–800.

Heckert, D.A. & Gondolf, E.W. (2005). Do multiple outcomes and conditional factors improve prediction of batterer reassault? *Violence and Victims, 20*(1), 3–25.

Hemphill, J.E., Hare, R.D., & Wong, S. (1998). Psychopathy and recidivism: A review. *Legal and Criminological Psychology, 3*(1), 139–170.

Hendricks-Mathews, M. (1982). The battered woman: Is she ready for help? *Social Casework, 63*, 131–137.

Henning, K., & Feder, L. (2004). A comparison of men and women arrested for domestic abuse: Who presents the greater threat? *Journal of Family Violence, 19*(4), 60–80.

Herzberger, S.D. & Tennen, H. (1985). "Snips and snails and puppy dog tails": Gender of agent, recipient, and observer as determinants of perceptions of discipline. *Sex Roles, 12*, 853–865.

Hilton, Z., Harris, G.T., & Rice, M.E. (2001). Predicting violence by serious wife assaulters. *Journal of Interpersonal Violence, 76*(5), 408–423.

Hogben, M., Byrne, D., & Hamburger, M.E. (1996). Coercive heterosexual sexuality in dating relationships of college students: Implications of differential male–female experiences. In E.S. Byers & L.F. O'Sullivan (Eds.), *Sexual coercion* (pp. 68–78). New York: Haworth Press.

Holden, G.W. & Ritchie, K.L. (1991). Linking extreme marital discord, child rearing, and child behavior problems: Evidence from battered women. *Child Development*, 62, 311–327.

Holtzworth-Munroe, A. & Anglin, K. (1991). The competency of responses given my martially violent versus nonviolent men to problematic marital situations. *Violence and Victims*, *6*, 257–269.

Holtzworth-Munroe, A. & Stuart, G.L. (1994). Typologies of male batterers: Three subtypes and the differences among them. *Psychological Bulletin, 116*, 476–497.

Holtzworth-Munroe, A., Bates, L., Smutzler, N., & Sandin, E. (1997). A brief review of the research on husband violence. Part 1. Maritally violent versus nonviolent men. *Aggression and Violent Behavior, 2*(1), 65–69.

Holtzworth-Munroe, A., Beatty, S.B., & Anglin, K. (1995). The assessment and treatment of marital violence: An introduction for the marital therapist. In N.S. Jacobson & A.S. Gurman (Eds.), *Clinical handbook of couple therapy* (pp. 317–339). New York: Guilford Press.

Holtzworth-Munroe, A., Meehan, J.C., Herron, K., Rehman, U., and Stuart, G.L. (2003). Do subtypes of martially violent men continue to differ over time? *Journal of Consulting and Clinical Psychology, 71*(4), 728–740.

Holtzworth-Munroe, A., Waltz, J., Jacobson, N.S., Monaco, V., Fehrenbach, P.A., & Gottman, J.M. (1992). Recruiting nonviolent men as control subjects for research on marital violence: How easily can it be done? *Violence and Victims*, *7*, 79–88.

Hood, R., Shute, S., Feilzer, M., & Wilcox, A. (2002). Sex offenders emerging from long-term imprisonment: A study of their long-term reconviction rates and of parole board members' judgments of their risk. *British Journal of Criminology, 41(2)*, 371–394.

Horton, D. (1943). The function of alcohol in primitive societies: A cross cultural study. *Quarterly Journal of Studies on Alcohol, 4*, 199–320.

Hotaling, G.T. & Sugarman, D.B. (1986). An analysis of risk markers in husband-to-wife violence: The current state of knowledge. *Violence and Victims, 1*, 101–124.

Hughes, H.M. (1988). Psychological and behavioral correlates of family violence in child witnesses and victims. *American Journal of Orthopsychiatry, 58*, 77–90.

Hull, J.G. (1981). A self-awareness model of the causes and effects of alcohol consumption. *Journal of Abnormal Psychology, 90*, 586–600.

Hungund, B.L. & Basavarajappa, B.S. (2004). Role of endocannabinoids and cannabinoid CB1 receptors in alcohol-related behaviors, *Annals of the New York Academy of Sciences, 1025*, 515–527.

Hyden, M. (1995). Verbal aggression as prehistory of woman battering. *Journal of Family Violence, 10*, 55–71.

Inbau, F.E., Reid, J.E., & Buckley, J.P. (1962). *Criminal interrogation and confessions*. Baltimore: William and Wilkins.

Inbau, F.E., Reid, J.E., & Buckley, J.P. (1986). *Criminal investigation and confessions,* 3rd ed. Baltimore: Williams and Wilkins.

Irving, B. & Hilgendorf, L. (1980). Police interrogation: The psychological approach. *Research Studies, 1*, HMSO: London.

Isaacs, M. (1977). Stereotyping by children of the effects of drinking on adults. *Journal of Studies on Alcohol*, 38, 913–921.

Isaacs, M. (1979). College student's expectations of the results of drinking. *Journal of Studies on Alcohol, 40*, 476–479.

Island, D. & Letelleir, P. (1991). *Men who beat the men who love them: Batterered gay men and domestic violence.* New York: Haworth Press.

Itzin, C. (1993). *Pornography: Women, violence and civil liberties, a radical view.* New York: Oxford University Press.

Jacobson, N.A. & Gottman, J.M. (1998). *When men batter women: New insights into ending abusive relationships.* New York: Simon and Schuster.

Jacobson, N.A., Gottman, J.M., & Shortt, J.W. (1995). The distinction between type 1 and type 2 batterers: Further consideration: Reply to Ornduff et al. (1995), Margolin et al. (1995), and Walker (1995). *Journal of Family Psychology, 9*(3), 272–279.

Jacobson, N.A., Gottman, J.M., Gartner, E., Berns, S., & Shortt, J.W. (1996). Psychological factors in the longitudinal course of battering: When do couples split up? When does the abuse decrease? *Violence and Victims, 11*, 371–392.

Jacobson, N.A., Gottman, J.M., Waltz, J., Rushe, R., Babcock, J., & Holtzworth-Munroe, A. (1994). Affect, verbal content, and psychophysiology in arguments of couples with a violent husband. *Journal of Consulting and Clinical Psychology, 62,* 982–988.

Jacoby, J.H. & Galanter, M. (1986). Alcohol idiosyncratic intoxication and other alcohol-related states of acute behavioral disinhibition. In R.E. Meyer (ed.), *Psychopathology and addictive disorders* (pp. 238–259). New York: Guilford Press.

Jaffe, P., Hurley, D., & Wolfe, D.A. (1990). Children's observations of violence: Critical issues in child development and intervention planning. *Canadian Journal of Psychiatry, 35,* 466–470.

Jaffe, P., Wolfe, D., Wilson, S.K., & Zac, L. (1986). Similarities in behavioral and social maladjustment among child victims and witnesses to family violence. *American Journal of Orthopsychiatry, 56*, 142–146.

Janus, E.A. & Meehl, P.E. (1997). Assessing the legal standard for predictions of dangerousness in sex offender commitment proceedings. *Psychology, Public Policy and Law, 3,* 33–64.

Jennings, J.L. (1987). History and issues in the treatment of battering men: A case for unstructured group therapy. *Journal of Family Violence, 2*, 193–214.

Johnson, B.A. (2004). An overview of the development of medications including novel anticonvulsants for the treatment of alcohol dependence. *Expert Opinion Pharmacotherapy*, *5*(9), 1943–1955.

Johnson, H. (2003). The cessation of assaults on wives. *Journal of Comparative Family Studies, 34(*1), 75–87.

Johnson, M.P. (1995). Patriarchal terrorism and common couple violence: Two forms of violence against women. *Journal of Marriage and the Family, 57*, 283–294.

Johnson, M.P. (2000a). Conflict and control: Images of symmetry and asymmetry in domestic violence. In A. Booth, A.C. Crouter & M. Clements (Eds.), *Couples in conflict*. Mahwah, NJ: Lawrence Erlbaum.

Johnson, M.P. (2000b). Domestic violence is not a unitary phenomenon. A major flaw in domestic violence literature. Unpublished manuscript.

Johnson, M.P. & Ferraro, K.J. (2000). Research on domestic violence in the 1990's: Making distinctions. *Journal of Marriage and the Family, 62*(4), 948–964.

Johnson, S.A. (1992). *Man-to-man: When your partner says no*. Vermont: Safer Society Press.

Johnson, S.A. (1993). *When "I love you" turns violent: Emotional and physical abuse in dating relationships*. Far Hills, New Jersey: New Horizon Press.

Johnson, S.A. (1995). The overlooked force in sexual assault: Implications for recovery. *Acta Sexologica, 1*(3), 268–302.

Johnson, S.A. (1997). The overlooked force in sexual assault: Implications for recovery. In B.K. Schwartz & H. R. Cellini, (Eds.), *The sex offender: New insights, treatment innovations and legal developments* (pp. 17-1–17-11). Kingston, NJ: Civic Research Institute.

Johnson, S.A., (1998). An overlooked factor in sexual abuse: Psychological and physical force examined. *Journal of Offender Rehabilitation, 28*(1), 141–151.

Johnson, S.A. (2001a). Relationship violence and domestic abuse program outline and curriculum. Unpublished.

Johnson, S.A. (2001b). The Johnson Force Assessment. Unpublished.

Johnson, S.A. (2005). Forensic considerations for assessing violence. *The Forensic Examiner, 14*(3), 6–12.

Johnson, S.A. & Listiak, A. (1999). The measurement of sexual preference: A preliminary comparison of phallometry and the Abel Assessment. In B.K. Schwartz (Ed.), *The sexual offender: Theoretical advances, treating special populations, and legal developments*. (pp. 26-1–26-20). Kingston, NJ: Civic Research Institute, Inc.

Johnston, C. (1975). Parental characteristics and parent-child interactions in families of nonproblem children and ADHD children with higher and lower levels of oppositional-defiant disorder. *Journal of Abnormal Psychology, 24*, 85–104.

Jones, B. & Jones, M. (1976, Fall). States of consciousness and alcohol. *Alcohol Health and Research World*, 1015.

Jouriles, E.N., Murphy, C.M., & O'Leary, K.D. (1989). Interspousal aggression, marital discord, and child problems. *Journal of Consulting and Clinical Psychology, 57*, 453–455.

Kafka, M.P. (1997). Hypersexual desire in males: An operational definition and clinical implications for men with paraphilias and paraphilia-related disorders. *Archives of Sexual Behavior, 26*, 505–526.

Kamali, M., Kelly, L., Gervin, M., Browne, S., Larkin, C., & O'Callaghan, E. (2001). Insight and comorbid substance misuse and medication compliance among patients with schizophrenia. *Psychiatric Services, 52*, 161–163, 166.

Kane, T.A., Stanger, P.K., & Riciardelli, L.A. (2001). Male domestic violence: Attitudes, aggression, and interpersonal dependency. *Journal of Interpersonal Violence, 15*(1), 16–29.

Kantor-Kaufman, G.K. & Straus, M.A. (1987). The "drunken bum" theory of wife beating. *Social Problems, 34*, 213–230.

Kantor, G.K. & Straus, M.A. (1989). Substance use as a precipitant of wife abuse victimizations. *American Journal of Drug and Alcohol Abuse, 15*, 173–189.

Karpman, B. (1941). On the need for separating psychopathy into two distinct clinical types: Symptomatic and idiopathic. *Journal of Criminology and Psychopathology, 3*, 112–137.

Kaufman, G. (1992). The mysterious disappearance of battered women in family therapist's offices: Male privilege colluding with male violence. *Journal of Marital and Family Therapy, 18*, 233–243.

Kaufman, G. & Zigler, E. (1987). Do abused children become abusive parents? *American Journal of Orthopsychiatry, 57*, 186–192.

Kaufman, K.L., Hilliker, D.R., Lanthrop, P., & Daleiden, E.L. (1993). Assessing sexual offenders' modus operandi: Accuracy in self-reported use of threats and coercion. *Annals of Sex Research, 6*, 213–229.

Kellner, R., Hernandez, J., & Pathak, D. (1992). Self-rated inhibited anger, somatization and depression. *Psychotherapy and Psychosomatics, 57*, 102–107.

Kelly, T.H., Cherek, D.R., Steinberg, J.L., & Robinson, D. (1988). Effects of provocation and alcohol on human aggressive behavior, *Drug and Alcohol Dependence, 21*, 105–112.

Keltner, D. & Buswell, B.N. (1996). Evidence for the distinctness of embarrassment, shame, and guilt: A study of recalled antecedents and facial expressions of emotions. *Cognition and Emotion, 10*, 155–171.

Keltner, D. & Harker, L.A. (1998). Forms and functions of the nonverbal signal of shame. In P. Gilbert & B. Andrews (Eds.), *Interpersonal approaches to shame* (pp. 78–98). Oxford, U.K.: Oxford University Press.

Kendall-Tackett, K.A. & Watson, M.W. (1991). Factors that influence professionals' perceptions of behavioral indication of child sexual abuse. *Journal of Interpersonal Violence, 6,* 385–395.

Kenna, G.A., McGeary, J.E., & Swift, R.M. (2004). Pharmacotherapy, pharmacogenomics, and the future of alcohol dependence treatment, part 1. *American Journal of Health Systems Pharmacology, 61*(21), 2272–2279.

Kennedy, H.G. & Grubin, D.H. (1992). Patterns of denial in sex offenders. *Psychological Medicine, 22*, 191–196.

Kessler, R.C., McGonagle, K.A., Zhao, S., Nelson, C.B., Hughes, M., & Eschelman, S. (1994). Lifetime and 12-month prevalence of DMS-III-R psychiatric disorders in the United States: Results from the National Comorbidity Survey. *Archives of General Psychiatry, 51*, 8–19.

Kienlen, K.K. (1998). Developmental and social antecedents of stalking. In J.R. Meloy (Ed.), *The psychology of stalking: Clinical and forensic perspectives* (p. 51). San Diego, CA: Academic Press.

Kitzmann, K.M., Gaylord, N.K., Holt, A.R., & Kenny, E.D. (2003). Child witnesses to domestic violence: A meta-analytic review. *Journal of Consulting and Clinical Psychology, 71*(2), 339–352.

Klein, A., Wilson, D., Crowe, A., & DeMichele, M. (2005). *Evaluation of the Rhode Island probation specialized domestic violence supervision unit*. National Institute of Justice, Office of Justice Programs, U.S. Department of Justice, Washington, D.C.

Klein, H. & Pittman, D.J. (1990). Perceived consequences associated with the use of beer, wine, distilled spirits, and wine coolers. *International Journal of the Addictions, 25*, 471–493.

Klinteberg, B., Andersson, T., Magnusson, D., & Stattin, H. (1993). Hyperactive behavior in children as related to subsequent alcohol problems and violent offending: A longitudinal study of male subjects. *Personality and Individual Differences, 15*, 381–388.

Knapp, M., McCrone, P., Fombonne, E., Beecham, J., & Wostear, G. (2002). The Maudsley long-term follow-up of child and adolescent depression. *British Journal of Psychiatry, 180,* 19–23.

Knight, R.A., & Prentky, R. A. (1990). Classifying sexual offenders: The development and corroboration of taxonomic models. In W.L. Marshall, D.R. Laws, & H.E. Barbaree (Eds.). *Handbook of Sexual Assault* (pp. 23–52). New York: Plenum Press.

Knight, R.A. & Sims-Knight, J.E. (2003). The developmental antecedents of sexual coercion against women: Testing alternative hypothesis with structural equation modeling. In R.A. Prentky, E.S. Janus, & M. Seto (Eds.), *Sexually coercive behavior: Understanding and management* (pp. 72–85). Annals of the New York Academy of Sciences, Vol. 989.

Knopp, F.H., Freeman-Longo, R., & Stevenson, W.F. (1992). *Nationwide survey of juvenile and adult sex offender treatment programs and models.* Brandon, VT: Safer Society.

Kocot, T. & Goodman, L. (2003). The roles of coping and social support in battered women's mental health. *Violence Against Women, 9*, 322–346.

Kolko, J.R., Blakely, E.H., & Engleman, D. (1996). Children who witness domestic violence: A review of empirical literature. *Journal of Interpersonal Violence, 11*, 281–293.

Kopelman, M.D. (1987). Amnesia: Organic and psychogenic. *British Journal of Psychiatry, 150*, 428–442.

Koss, M.P. (1987). Outrageous acts and everyday seduction: Sexual aggression and victimization among college students. In B. Levy (Ed.), *Dating violence: Young women in danger* (pp. 87–96). Seattle, WA: Seal Press.

Koss, M. (1988). The reality of acquaintance rape. In R. Warshaw, *I never called it rape: The Ms. Report on recognizing, fighting, and surviving date and acquaintance rape*. New York: Harper & Row.

Kossen, D.S., Steuerwald, B.L., Forth, A.E., & Kirkhart, K.J. (1997). A new method for assessing the interpersonal behavior of psychopathic individuals: Preliminary validation studies. *Psychological Assessment, 9*, 1–13.

Krahe, B., Waizenhofer, E., & Moller, I. (2003). Women's sexual aggression against men: Prevalence and predictors. *Sex roles, 49*(5), 219–233.

Kreutzer, J.S., Schneider, H.G., & Myatt, C.R. (1984). Alcohol, aggression and assertiveness in men: Dosage and expectancy effects. *Journal of Studies on Alcohol, 45*, 275–278.

Kropp, P.R. (1994). The relationship between psychopathy and malingering of mental illness. *Dissertation Abstracts International, 54-B*, 5945–5946.

Kropp, P.R., Hart, S. D., Webster, C.D., and Eaves, D. (1999). *Spousal assault risk assessment guide.* New York: Multi-Health Systems, Inc. (MHS) & B.C. Institute Against Family Violence

Kuehnle, K. (1996). *Assessing allegations of child sexual abuse.* Sarasota, FL: Professional Resource Press.

Kuleshnyk, I. (1984). The Stockholm Syndrome: Toward an understanding. *Social Action and the Law, 10*(2), 37–42.

Labriola, M., Rempel, M., & Davis, R.C. (2005, November). Testing the effectiveness of batterer programs and judicial monitoring. Final report submitted to the National Institute of Justice, New York, NY.

Laffaye, C., Kennedy, C., & Stein, M.B. (2003). Post-traumatic stress disorder and health-related quality of life in female victims of intimate partner violence. *Violence and Victims, 18,* 227–238.

Lalumiere, M.L. & Quinsey, V.L. (1994). The discriminability of rapists from non-sex offenders using phallometric measurements: A meta-analysis. *Criminal Justice and Behavior, 21*(1), 150–175.

Land, W.B. (1995). Psychopharmacological options for sex offenders. In B.K. Schwartz, & H.R. Cellini (Eds.), *The sex offender: corrections, treatment and legal practice* (chap. 18). Kingston, NJ: Civic Research Institute, Inc.

Lang, A.R., Goeckner, D.J., Adesso, V.J., & Marlatt, G.A. (1975). Effects of alcohol on aggression in male social drinkers. *Journal of Abnormal Psychology, 5*, 508–518.

Lang, A.R. (1993). Alcohol-related violence: Psychological perspectives. In S. E. Martin (Ed.), *Alcohol and interpersonal violence: Fostering multidisciplinary perspectives* Research Monograph No. 24 (pp. 121–148). Rockville, MD: National Institutes of Health.

Lang, A.R. & Frenzel, R.R. (1988). How sex offenders lure children. *Annals of Sex Research, 1*, 303–317.

Langevin, R. (Ed.) (1985). *Erotic preference, gender identity, and aggression in men: New research studies.* Mahwah, NJ: Lawrence Erlbaum Associates.

Langevin, R. (1988). Defensiveness in sex offenders. In R. Rogers (Ed.), *Clinical assessment of malingering and deception,* 1st ed. (pp. 260–290). New York: Guilford Press.

Langhinrichsen-Rohling, J., Neidig, P., & Thorn, G. (1995). Violent marriages: Gender differences in levels of current violence and past abuse. *Journal of Family Violence, 10*, 159–176.

Lanning, K.V. (1995). Child molestation: A law enforcement typology. In R.R. Hazelwood & A.W. Burgess (1995). *Practical aspects of rape investigation: A multidisciplinary approach.* Boca Raton, FL: CRC Press.

Laplace, A.C., Chermack, S.T., & Taylor, S.P. (1994). Effects of alcohol and drinking experience on human physical aggression. *Personality and Social Psychology Bulletin, 20*, 439–444.

Larimer, M.E., Lydum, A.R., Anderson, B.K., & Turner, A.P. (1999). Male and female recipients of unwanted sexual contact in a college student sample: Prevalence rates, alcohol use, and depressive symptoms. *Sex Roles, 40*, 295–308.

Lau, M.A., Pihl, R.O., & Peterson, J.B. (1995). Provocation, acute alcohol intoxication, cognitive performance, and aggression. *Journal of Abnormal Psychology, 1*, 150–155.

Laws, D.R. & Holmes, M.L. (1978). Sexual response faking by pedophiles. *Criminal Justice and Behavior, 5,* 343–356.

Lawson, C. (1993). Mother-son sexual abuse: Rare or underreported? A critique of the research. *Child Abuse and Neglect, 17*(2), 261–269.

Lawson, D. (2003, Winter). Incidence, explanations, and treatment of partner violence. *Journal of Counseling and Development: JCD, 81*, 1.

Lawson, D.M., Dawson, T.E., Kieffer, K.M., Perez, L.M., Burke, J., & Kier, F.J. (2001). An integrated feminist/cognitive-behavioral and psychodynamic group treatment model for men who abuse their partners. *Psychology of Men and Masculinity, 2*, 86–99.

Leavitt, S.B. (2002, March). Naltrxone clinical update: Evidence for the efficacy of naltrexone in the treatment of alcohol dependence (alcoholism). *Addiction Treatment Forum,* 1–8.

Lee, J., Jackson, H., Pattison, P., & Ward, T. (2002). Developmental risk factors for sexual offending. *Child Abuse and Neglect, 26,* 254–274.

Leibsohn, M.T., Oetting, E.R., & Deffenbacher, J.L. (1994). Effects of trait anger on alcohol consumption and consequences. *Journal of Child and Adolescent Substance Abuse, 3*, 17–32.

Leigh, B.C. (1987). Beliefs about the effects of alcohol on self and others. *Journal of Studies on Alcohol, 48*, 467–475.

Leisring, P.A., Dowd, L., & Rosenbaum, A. (1999). Characteristics of women mandated to anger management treatment. Paper presented at the Sixth International Family Violence Research Conference, Durham, NH.

Leisring, P.A., Dowd, L., & Rosenbaum, A. (2000, November). Trauma histories, PTSD symptoms, and aggression in partner-abusive women. Paper presented at the 34th Annual Meeting of the Association for the Advancement of Behavior Therapy, New Orleans, LA. November.

Leonard, K. & Senchak, M. (1993). Alcohol and premarital aggression among newlywed couples. *Journal of Studies on Alcohol, 11*(Suppl.), 96–108.

Leonard, K.E. (1984). Alcohol consumption and escalatory aggression in intoxicated and sober dyads. *Journal of Studies on Alcohol, 45*, 75–80.

Leonard, K.E. & Blane, H.T. (1992). Alcohol and marital aggression in a national sample of young men, *Journal of Interpersonal Violence*, *7*, 19–30.

Lerman, L.G. (1992). The decontextualisation of domestic violence. *Criminal Law and Criminology, 83*, 217–240.

Leshner, A.I. & Koob, G.F. (1999). Drugs of abuse and the brain: *Proc. Association American Physicians, 111*(2), 99–108.

Letellier, P. (1994). Gay and bisexual domestic violence victimization: Challenges to feminist theory and response to violence. *Violence and Victims*, *9*, 95–106.

Levenson, R., Sher, K., Grossman, L., Newman, J., & Newlin, D. (1980). Alcohol and stress response dampening: Pharmacological effects, expectancy, and tension reduction. *Journal of Abnormal Psychology*, *89*, 528–538.

Levine, M. (1975). Interparental violence and its effects on the children: A study of 50 families in general practice. *Medical Science Law, 15*, 172–176.

Levinson, D. (1983). Social setting, cultural factors and alcohol-related aggression. In E. Gottheil, K.A. Druley, T.E. Skoloda, & H.M. Waxman (Eds.), *Alcohol, Drug Abuse, and Aggression* (pp. 41–58). Springfield, IL: Charles C Thomas.

Levinson, D. (1989). *Family violence in cross-cultural perspective*. Newbury Park, CA: Sage Publications.

Levy, B. (Ed.) (1991). *Dating violence: Young women in danger*. Seattle, WA: The Seal Press.

Lindman, R.E. & Lang, A.R. (1994). The alcohol-aggression stereotype: A cross-cultural comparison of beliefs. *International Journal of the Addictions*, *29*, 1–13.

Lindqvist, P. (1991). Homicides committed by abusers of alcohol and illicit drugs. *British Journal of Addiction*, *86*, 321–326.

Linehan, M.M. (1993). *Cognitive-behavioral treatment of borderline personality disorder.* New York: Guilford Press.

Lisak, D. & Roth, S. (1988). Motivational factors in nonincarcerated sexually aggressive men. *Journal of Personality and Social Psychology, 55*, 795–802.

Listiak, A. (2005, December 15). Personal communication.

Livingston, L.R. (1986). Measuring domestic violence in an alcoholic population. *Journal of Sociology and Social Welfare, 13*, 934–954.

Lockhart, L.L., White, B.W., Causby, V., & Isaac, A. (1994). Letting out the secret: Violence in lesbian relationships. *Journal of Interpersonal Violence*, *9*, 469–492.

Lockwood, R. & Ascione, F.R. (Eds.) (1998). *Cruelty to animals and interpersonal violence: Readings in research and application.* West Lafayette, IN: Purdue University Press.

Loftus, E.F. (1980, March). "Did I really say that last night?" Alcohol, marijuana and memory. *Psychology Today, 92*, 42–56.

Lykken, D.T. (1995). *The antisocial personalities*. Mahwah, NJ: Lawrence Erlbaum Associates.

MacCulloch, M.J., Snowden, P.R., Wood, P.J.W., & Mills, H.E. (1983). Sadistic fantasy, sadistic behavior and offending. *British Journal of Psychiatry, 143*, 20–29.

MacQueen, G.M. & Young, L.T. (2001). Bipolar II disorder: Symptoms, course, and response to treatment. *Psychiatric Services, 52*, 358–361.

Magdol, L., Moffitt, T.E., Caspi, A., Newman, D.L., Fagan, J., & Silva, P.A. (1997). Gender differences in partner violence in a birth order cohort of 21-year olds: Bridging the gap between clinical and epidemiological approaches. *Journal of Consulting and Clinical Psychology, 65*, 68–78.

Magdol, L., Moffitt, T.E., Caspi, A., & Silva, P.A. (1998). Developmental antecedents of partner abuse: A prospective-longitudinal study. *Journal of Abnormal Psychology, 107*(3), 375–389.

Maiuro, R.D., Cahn, T.S., Vitaliano, P.P., Wagner, B.C., & Zegree, J.B. (1988). Anger, hostility, and depression in domestically violent versus generally assaultive men and nonviolent control subjects. *Journal of Consulting and Clinical Psychology, 56*, 17–23.

Makepeace, J.M. (1981) Courtship violence among college students. *Family Relations, 30*, 97–102.

Makepeace, J.M. (1986). Gender differences in courtship violence victimization. *Family Relations, 35*, 383–388.

Makepeace, J.M. (1987). Social factors and victim-offender differences in courtship violence. *Family Relations, 36*, 87–91.

Malamuth, N.M. (1981). Rape proclivity among males. *Journal of Social Issues, 37*(4), 138–157.

Malamuth, N.M. (1986). Predictors of naturalistic sexual aggression. *Journal of Personality and Social Psychology, 5*, 953–962.

Malamuth, N.M. (1989). The attraction to sexual aggression scale: Part 2. *Journal of Sex Research, 26*, 324–354.

Malamuth, N.M. (1996). The confluence model of sexual aggression: Feminist and evolutionary perspectives. In D. M. Buss & N.M. Malamuth (Eds.), *Sex, power, conflict: Evolutionary and feminist perspective* (pp. 269–295). New York: Oxford University Press.

Malamuth, N.M. (2003). Criminal and noncriminal sexual aggressors: Integrating psychpathy in hierarchical-mediational confluence model. In R.A. Prentky, E.S. Janus, & M. Seto (Eds.), *Sexually coercive behavior: Understanding and management* (pp. 33–58). Annals of the New York Academy of Sciences, vol. 989.

Malamuth, N.M., Check, J, & Briere, J. (1986). Sexual arousal in response to aggression: Ideological, aggressive, and sexual correlates. *Journal of Personality and Social Psychology, 50*, 330–340.

Malamuth, N.M., Heavy, C.L., & Linz, D. (1993). Predicting men's antisocial behavior against women: The interaction model of sexual aggression. In G.C.N. Hall, R. Hirschman, J.R. Graham, & M.S. Zaragoza (Eds.), *Sexual aggression: Issues in etiology, assessment, and treatment* (pp. 63–97). Washington, D.C.: Taylor & Francis.

Malamuth, N.M., Heim, M., & Feshbach, S. (1980, March). Sexual responsiveness of college students to rape depiction: Inhibitory and disinhibitory effects. *Journal of Personality & Social Psychology, 38*(3), 399–408.

Malamuth, N.M., Sockloskie, R.J., Koss, M.P., & Tanaka, J.S. (1991). Characteristics and aggressors against women: Testing a model using a national sample of college students. *Journal of Consulting and Clinical Psychology, 59*, 670–681.

Maletzky, B.M. (1976). The diagnosis of pathological intoxication. *Journal of Studies on Alcohol, 37*, 1215–1228.

Maletsky, B.M. (1991). *Treating the sexual offender.* Newbury Park, CA: Sage Publications.

Malinosky-Rummell, R. & Hanson, D.J. (1993). Long-term consequences of childhood physical abuse. *Psychological Bulletin, 114*, 68–79.

Mann, J., Arango, V., & Underwood, M. (1990). Serotonin and suicidal behavior. *Annals of the New York Academy of Sciences, 600*, 476–485.

Mann, S., Vrij, A., & Bull, R. (2004). Detecting true lies: Police officer's ability to detect suspect's lies. *Journal of Applied Psychology, 89*(1), 137–149.

Margolin, G. (1988). Interpersonal and intrapersonal factors associated with marital violence. In G.T. Hotaling, D. Finkelhor, J.T. Kirkpatrick, & M.A. Straus (Eds.), *Family abuse and its consequences: New directions in research* (pp. 203–217). Newbury Park, CA: Sage Publications.

Margolin, G., Gordis, E.B., Oliver, P.H., & Raine, A. (1995). A physiologically based typology of batters — Promising but preliminary: Comment on Gottman et al. (1995). *Journal of Family Psychology, 9*(3), 253–263.

Marques, J.K., Day, D.M., Nelson, C., & West, M.A. (1994). Effects of cognitive-behavioral treatment on sex offender recidivism: Preliminary results of a longitudinal study. *Criminal Justice and Behavior, 21*, 28–54.

Marshall, W.L. (1994). Treatment effects on denial and minimization in incarcerated sex offenders. *Behavior Research and Therapy, 32*(5), 559–564.

Marshall, W.L. & Barbaree, H.E. (1990). An integrated theory of the etiology of sexual offending. In W.L. Marshall, D.R. Laws, & H.E. Barbaree (Eds.), *Handbook of sexual assault: Issues, theories, and treatment of the offender.* New York: Plenum Press.

Marshall, W.L., Abel, G.G., & Quinsey, V.L. (1983). The assessment and treatment of sexual offenders. In S. Simon Jones (Ed.), *Sexual aggression and the law* (pp. 43–52). Burnaby, B.C., Canada: Criminology Research Centre, Simon Fraser University.
Marshall, W.L., Barbaree, H.E., & Christopher, D. (1986). Sexual offenders against female children: Sexual preferences for age of victims and type of behavior. *Canadian Journal of Behavioral Science, 18,* 424–439.
Marshall, W.L., Earls, C.M., Segal, Z., & Darke, J.L. (1983). A behavioral program for the assessment and treatment of sexual aggressors. In K.D. Graig & R.J. McMahon (Eds.), *Advances in clinical behavior therapy* (pp. 147–174). New York: Brunner/Mazel.
Marshall, W.L., Jones, R., Ward, T., Johnson, P., & Barbaree, H.E. (1991). Treatment outcome with sex offenders. *Clinical Psychology Review, 11*, 465–485.
Marshall, W.L., Laws, D. R., & Barbaree, H.E. (1990). *Handbook of sexual assault: Issues, theories, and treatment of the offender.* New York: Plenum Press.
Martin, C., Earleywine, M., Musty, R., Perrine, M., & Swift, R. (1993). Development and validation of the biphasic alcohol effects scale. *Alcoholism: Clinical and Experimental Research, 17*, 140–146.
Mason, B.J. (2001). Treatment of alcohol-dependent outpatients with acamprosate: A clinical review. *Journal of Clinical Psychiatry, 62*(Suppl. 20), 42–48.
Matthews, R., Matthews, J.K., and Speltz, K. (1989). *Female sex offenders: An exploratory study.* Orwell, VT: Safer Society Press.
Matthews, R., Matthews, J.K., and Speltz, K. (1990). Female sex offenders. In M. Hunter (Ed.), *The sexually abused male: Prevalence, impact, and treatment* (pp. 275–293). New York: Lexington Books.
Mayfield, D. (1976). Alcoholism, alcohol intoxication and assaultive behavior. *Diseases of the Nervous System, 3*, 288–291.
Mayou, R.A., Ehlers, A., & Bryant, B. (2002). Posttraumatic stress disorder after motor vehicle accidents: Three-year follow-up of a prospective longitudinal study. *Behavior Research and Therapy, 40,* 665–675.
McCarty, L.M. (1986). Mother-child incest: Characteristics of the offender. *Child Welfare, 65*, 447–458.
McCloskey, L.A., Figueredo, A.J., & Koss, M.P. (1995). The effects of systematic family violence on children's mental health. *Child Development, 66*, 1239–1261.
McDonald, R. & Jouriles, E.N. (1991, September). Marital aggression and child behavior problems: Research findings, mechanisms, and intervention. *The Behavior Therapist, 14*, 189–192.
McGee, R.A., Wolfe, D.A., & Wilson, S.K. (1997). Multiple maltreatment experiences and adolescent behavior problems: Adolescents' perspectives. *Development and Psychopathology, 9*, 131–149.
McGrath, R.J. & Purdy, L.A. (1999). Referring sex offenders for psychosexual evaluation: A review. *Journal of Addictions and Offender Counseling, 19*(2), 62–76.
McKibben, A., Proulx, J., & Lusignan, R. (1994). Relationships between conflict, affect, and deviant sexual behaviors in rapists and pedophiles. *Behaviour Research and Therapy, 13*, 571–575.
McLelland, D.C., Davis, W.N., Kalin, R., & Wanner, E. (1972). *The drinking man — alcohol and human motivation.* Toronto, Canada: Collier-Macmillan Canada.
McNeil, D.E., Sandberg, D.A., & Binder, R.L. (1998). The relationship between confidence and accuracy in clinical assessments of psychiatric patients' potential for violence. *Law and Human Behavior, 22*, 655–669.
Meehan, J.C. & Holtzworth-Munroe, A. (2001). Heart rate reactivity in male batterers: Reply to Gottman (2001) and a second look at the evidence. *Journal of Family Psychology, 15*(3), 415–424.
Meloy, J.R. (1988). *The psychopathic mind: Origins, dynamics, and treatment.* Northvale, NJ: Jason Aronson Publishers.
Meloy, J.R. (1996). Stalking (obsessional following): A review of some preliminary studies. *Aggression and Violent Behavior, 1*, 147–162.
Meloy, J.R. (1998). *The psychology of stalking: Clinical and forensic perspectives.* San Diego, CA: Academic Press.
Meloy, J.R. (2000). The nature and dynamics of sexual homicide: An integrative review. *Aggression and Violent Behavior, 5*, 1–22.
Meloy, J.R. (2002). The "polymorphously perverse" psychopath: Understanding a strong empirical relationship. *Bulletin of the Menninger Clinic, 66*(3), 273–290.

Meloy, J.R. & Gacano, C.B. (1995). Assessing the psychopathic personality. In J.N. Butcher (Ed.), *Clinical personality assessment* (pp. 410–422). New York: Oxford University Press.

Meloy, J.R. & Gothard, S. (1995). A demographic and clinical comparison of obsessional followers and offenders with mental disorders. *American Journal of Psychiatry, 152*, 258–263.

Melton, G.V., Petrila, J., Poythress, N.G., & Slobogin, C. (1987). *Psychological evaluation for the courts: A handbook for mental health professionals*. New York: Guilford.

Mendel, M.P. (1995). *The male survivor: The impact of sexual abuse*. London: Sage Publications.

Meyer, R.G. & Deitsch, S.M. (1995). The assessment of malingering in psychodiagnostic evaluations: Research-based concepts and methods for consultants. *Consulting Psychology Journal, 47*(4), 234–235.

Meyer, R.G. & Deitsch, S.E. (1996). *The clinician's handbook: Integrated diagnostics, assessment, and intervention in adult and adolescent psychopathology*. Boston: Allyn & Bacon.

Miccio-Fonseca, L.C. (2000). Adult and adolescent female sex offenders: Experiences compared to other female and male sex offenders. *Journal of Psychology and Human Sexuality, 11*, 75–88.

Miller, B. (1988). Date rape: Time for a new look at prevention. *Journal of College Student Development, 29*, 553–555.

Miller, B. & Marshall, J. (1987). Coercive sex on the university campus. *Journal of College Student Personnel, 28*(1), 38–47.

Miller, E. & Porter, C. (1983). Self-blame in victims of violence. *Journal of Social Issues, 39*, 139-152.

Miller, B., Downs, W., & Gondoli, D. (1989). Spousal violence among alcoholic women as compared to a random household sample of women. *Journal of Studies on Alcohol*, 50, 533–540.

Millon, T. (1983). Manual for the *Millon clinical multiaxial inventory (MCMI)*, 3rd ed. Minneapolis, MN: National Computer Systems.

Milner, J.S. (1986). *The child abuse potential inventory manual*, 2nd ed. Webster, NC: Psytec.

Milner, J.S. (1994). Assessing physical child abuse risk: The Child Abuse Potential Inventory. *Clinical Psychology Review, 14*, 547–583.

Milner, J.S. & Murphy, W.D. (1995). Assessment of child physical and sexual abuse offenders. *Family Relations, 44*(4), 478–489.

Monahan, J. (1981). *Predicting violent behavior: An assessment of clinical techniques*. Beverly Hills, CA: Sage Publications.

Monahan, J. (1996). Violence prediction: The past twenty and the next twenty years. *Criminal Justice and Behaviour, 23*(1), 107–120.

Monrand, C., Young, S.N., & Ervin, F.R. (1983). Clinical responses of aggressive schizophrenics to oral tryptophan. *Biological Psychiatry, 18*, 575–578.

Moreno, J.K., Fuhriman, A., & Selby, M.J. (1993). Measurement of hostility, anger, and depression in depressed and nondepressed subjects. *Journal of Personality Assessment, 61*(3), 511–523.

Morris, J.L., Johnson, C.F., & Clasen, M. (1985). To report or not report: Physicians' attitudes toward discipline and child abuse. *American Journal of Diseases of Children, 139*, 194–197.

Mosher, D.L. & Sirkin, M. (1984). Measuring a macho personality constellation. *Journal of Research in Personality, 18*, 150–163.

Muehlenhard, C.L. (1988). Misinterpreted dating behaviors and the risk of date rape. *Journal of Social and Clinical Psychology, 6*(1), 20–37.

Muehlenhard, C.L. & Cook, S.W. (1988). Men's self-reports of unwanted sexual activity. *Journal of Sex Research, 24*, 58–72.

Muehlenhard, C.L. & Linton, M. (1987). Date rape and sexual aggression in dating situations: Incidence and risk factors. *Journal of Counseling Psychology, 34*, 186–196.

Muller, R.T., Caldwell, R.A., & Hunter, J.E. (1993). Child provocativeness and gender as factors contributing to the blaming of victims of physical child abuse. *Child Abuse and Neglect, 7*, 249–260.

Murdoch, D. & Pihl, R.O. (1988a). The influence of dose, beverage type and sex of interactor on female bar patrons' verbal aggression. *International Journal of Addictions, 23*, 953–966.

Murdoch, D. & Pihl, R.O. (1988b). The influence of beverage type, BAL and sex of confederate on aggression in males in the natural setting. *Aggressive Behavior, 14*, 325–336.

Murdoch, D., Pihl, R.O., & Ross, D. (1990). Alcohol and crimes of violence: Present issues. *The International Journal of the Addictions, 25*, 1065–1081.

Murphey, C., Meyer, S.L., & O'Leary, K. (1994). Dependency characteristics of partner assaultive men. *Journal of Abnormal Psychology, 103*, 729–735.
Murphy, W.D., & Barbaree, H.E. (1994). *Assessments of sex offenders by measures of erectile response: Psychometric properties and decision-making* (rev. and update). Brandon, VT: Safer Society Press.
Murphy, W.D. & Peters, J.M. (1992). Profiling child sexual abusers: Psychological considerations. *Criminal Justice and Behavior, 19*, 24–37.
Myers, D.L. (1995). Eliminating the battering of women by men: Some considerations for behavior analysis. *Journal of Applied Behavior Analysis, 28*, 493–507.
Myers, T. (1986). An analysis of context and alcohol consumption in a group of criminal events. *Alcohol and Alcoholism, 21*, 389–395.
Neidig, P.H. & Friedman, D.H. (1984). *Spouse abuse: A treatment program for couples*. Champaign, IL: Research Press.
Neidig, P.H. & Friedman, D.H. (1985). Domestic conflict containment: A spouse abuse treatment programme. *The Journal of Contemporary Social Casework, 66*, 195–204.
Nemeroff, C.B. (2002). Comorbidity of mood and anxiety disorders: The rule, not the exception? Editorial: *American Journal of Psychiatry, 159*, 3–4.
Nichols, H.R. & Molinder, I. (1984). *Multiphasic Sex Inventory*. Fircrest, WA: Nichols & Molinder Assessments.
Norris, J. & Kerr, K.L. (1993). Alcohol and violent pornography: Response to permissive and nonpermissive cues. *Journal of Studies on Alcohol* (Suppl. 11), 118–127.
Nuttall, R. & Jackson, H. (1994). Personal history of childhood abuse among clinicians. *Child Abuse and Neglect, 8*, 455–472.
Oates, M.R. (1979) A classification of child abuse and its relation to treatment and prognosis. *Child Abuse and Neglect, 3*, 907–915.
O'Connor, A. (1987). Female sex offenders. *British Journal of Psychiatry, 150*, 615–620.
Ogloff, J.R., Wong, S., & Greenwood, A. (1990). Treating criminal psychopaths in a therapeutic community program. *Behavioral Sciences and the Law, 8*, 27–43.
O'Hagan, K. (1989). *Working with child sexual abuse*. Milton Keynes, U.K.: Open University Press.
O'Keefe, M. (1994). Linking marital violence, mother-child/father-child aggression, and child behavior problems. *Journal of Family Violence, 9*, 63–78.
O'Leary, K.D. (1993). Through a psychological lens: Personality traits, personality disorders, and levels of violence. In R.J. Gelles & D.R. Loeske (Eds.), *Current controversies on family violence* (pp. 7–30). Newbury Park, CA: Sage Publications.
O'Leary, K.D. (1995, July). Assessment and treatment of partner abuse. *Clinician's Research Digest, 12*(Suppl.), 1–3.
O'Leary, K.D. & Vivian, D. (1992). Assessment of physical aggression against women in marriage: The need for a multimodal assessment. *Behavioral Assessment, 14*, 5–14.
O'Leary, K.D., Barling, J., Arias, I., Rosenbaum, A., Malone, J., & Tyree, A. (1989). Prevalence and stability of physical aggression between spouses: A longitudinal analysis. *Journal of Consulting and Clinical Psychology, 57*, 263–268.
Olson, P.E. (1990). The sexual abuse of boys: A study of the long-term psychological effects. In M. Hunter (Ed.), *The sexually abused male: Prevalence, impact and treatment*. Lexington, MA: Lexington Books.
Osanka, F.M. & Johann, S.L. (1989). In F.M. Osanka & S.L. Johann, (Eds.). *Sourcebook on pornography*. Lexington Books. Referenced in K.L., Swisher, & C., Wekesser (Eds.). (1994). *Violence Against Women*. San Diego, California: Current Controversies.
O'Sullivan, L.R., Byers, E.S., & Finkelman, L. (1998). A comparison of male and female college students' experiences of sexual coercion. *Psychology of Women Quarterly, 22*, 177–195.
Ouimet, M., Guay, J.R., & Proulx, J. (2000). Analyse de la gravite des aggressions sexuelles de femmes adultes et de ses determinants. *Revue Internationale de Criminologie et de Police Technique et Scientifique, 2*, 157–172.
Pagelow, M.D. (1981). *Woman-battering: Victims and their experiences*. Newbury Park, CA: Sage Publications.
Paglia, A. & Room, R. (1999). Expectancies about the effects of alcohol on self and others as determinants of alcohol policy attitudes. *Journal of Applied Social Psychology, 29*, 2632–2651.
Painter, S. & Dutton, D. (1985). Patterns of emotional bonding in battered women: Traumatic bonding. *International Journal of Women's Studies, 8*, 363–375.

Palmer, S., Brown, R., & Barrera, M. (1992). Group treatment program for abusive husbands: Long-term evaluation. *American Journal of Orthopsychiatry, 62,* 276–283.

Parker, R. (1993). The effects of context on alcohol and violence. *Alcohol Health and Research World, 17,* 117–122.

Patterson, G.R. (1982). *Coercive family process.* Eugene, OR: Castalia Publications.

Patterson, G.R., Dishion, T., & Reid, J. (1992). *Antisocial boys.* Eugene, OR: Castilia Publications.

Paulhaus, D.L. (1984). Two-component models of socially desirable responding. *Journal of Personality and Social Psychology, 46,* 598–609.

Pence, E.L. (1989). *The justice system's response to domestic assault cases: A guide for policy development.* Revised ed. Duluth, MN: Minnesota Program Development, Domestic Abuse Intervention Project.

Pence, E.L. & McDonnell, C. (1999). Developing policies and protocols. In E.L. Pence & M. Shepard (Eds.), *Coordinated community responses to domestic violence: Lessons from Duluth and beyond.* Thousand Oaks, CA. Sage Publications.

Pence, E.L. & Paymar, M. (1993). *Education groups for men who batter. The Duluth model.* New York: Springer Publishing.

Pernanen, K. (1976) Alcohol and crimes of violence. In B. Kissin & H. Begleiter (Eds.), *The biology of alcoholism* (pp. 351–444). New York: Plenum Publications.

Pernanen, K. (1991). *Alcohol in human violence.* New York: Guilford Press.

Pernanen, K. (1993). Alcohol-related violence: Conceptual models and methodological issues. In S. E. Martin (Ed.), *Alcohol and interpersonal violence: Fostering multidisciplinary perspectives,* 24th ed. (vol. 93-3496, pp. 37–70). Rockville, MD: National Institutes of Health.

Peters, J.J. (1976). Children who are victims of sexual assault and the psychology of offenders. *American Journal of Psychotherapy, 30,* 398–421.

Petrovich, M. & Templar, D. (1984). Heterosexual molestation of children who later became rapists. *Psychological Reports, 54,* 8–10.

Phelan, M.B., Hamberger, L.K., Hare, S., & Edwards, S. (2002). Impact of partner violence on male and female emergency room department patients. Paper presented at the meeting of the American Psychological Association, Chicago, IL.

Phil, R., Peterson, J., & Lau, M. (in press). A biosocial model of the alcohol-aggression relationship. *Journal of Studies on Alcohol.*

Philipchalk, R.P. (1995). *Invitation to social psychology.* Fort Worth, TX: Harcourt Brace.

Pierce, R. & Pierce, L. (1985). The sexually abused child: A comparison of male and female victims. *Child Abuse and Neglect, 9,* 191–199.

Pihl, R.O. & Peterson, J.B. (1992). Etiology. *Annual Review of Addictions Research and Treatment,* 153–175.

Pihl, R.O. & Peterson, J.B. (1993a). Alcohol/drug use and aggressive behavior. In S. Hodgins (Ed.), *Mental disorder and crime* (pp. 263–283). Newbury Park, CA: Sage Publications.

Pihl, R.O. & Peterson, J.B. (1993b). Alcohol and aggression: Three potential mechanisms of the drug effect. In S. E. Martin (Ed.), *Alcohol and interpersonal violence: Fostering multidisciplinary perspectives* (Research Monograph No. 24) (pp. 149–159). Rockville, MD: National Institutes of Health.

Pihl, R.O. & Peterson, J.B. (1993c). Alcohol, serotonin, and aggression. *Alcohol Health and Research World, 17*(2), 113–116.

Pihl, R.O. & Ross, D. (1987). Research on alcohol-related aggression: A review and implications for understanding aggression. *Drugs and Society, 1*(4), 105–126.

Pihl, R.O., Smith, M., & Farrell, B. (1984) Alcohol and aggression in men: A comparison of brewed and distilled beverages, *Journal of Studies on Alcohol, 45,* 278–282.

Pinccinelli, M., Pini, S., Bellantuono, C., & Wilkinson, G. (1995). Efficacy of drug treatment in obsessive-compulsive disorder: A meta-analysis review. *British Journal of Psychiatry, 166,* 424–443.

Pini, S., Cassano, G.B., Simonini, E., Savino, M., Russo, A., & Montgomery, S.A. (1997). Pervalence of anxiety disorders comorbidity in bipolar depression, unipolar depression and dysthymia, *Journal of Affective Disorders, 42,* 145–153.

Pithers, W.D. & Laws, D.R. (1995). Phallometric assessment. In B.K. Schwartz & H.R. Cellini (Eds.), *The sex offender: Corrections, treatment, and legal practice* (vol. I, pp. 12-1–12-18). Kingston, NJ: Civic Research Institute.

Pithers, W.D., Beal, L.S., Armstrong, J., & Petty, J. (1989). Identification of risk factors through clinical interviews and analysis of records. In D.R. Laws, (Ed.), *Relapse prevention with sex offenders* (pp. 77–87). New York: Guildford Press.

Pithers, W.D., Kashima, K., Cummings, G.F., Beal, L.S., & Buell, M. (1988). Relapse prevention of sexual aggression. In R.A. Prentky & V.L. Quinsey, (Eds.), *Human sexual aggression: Current perspectives,* (pp. 244–260). New York: New York Academy of Sciences.

Pliner, P. & Cappell, H. (1974). Modification of affective consequences of alcohol: A comparison of social and solitary drinking. *Journal of Abnormal Psychology, 83*, 418–425.

Poppen, P.J. & Segal, N.J. (1988). The influence of sex and sex role orientation on sexual coercion. *Sex Roles, 19*, 689–701.

Posternak, M.A. & Zimmerman, M. (2002). Partial validation of the atypical features subtype of major depressive disorder. *Archives of General Psychiatry, 59*, 70–76.

Prentky, R.A. & Knight, R.A. (1986). Impulsivity in the lifestlye and criminal behavior of sexual offenders. *Criminal Justice and Behavior, 13*, 141–164.

Prentky, R.A. & Knight, R.A. (1990). Classifying sexual offenders: The developent and corroboration of taxonomic models. In W.L. Marshall, D.R. Laws, & H.E. Barbaree (1990). *Handbook of sexual assauilt: Issues, theories, and treatment of the offender* (pp. 23–52). New York: Plenum Publications.

Prentky, R.A. & Knight, R.A. (1991). Identifying critical dimensions for discriminating among rapists. *Journal of Consulting and Clinical Psychology, 59*, 643–661.

Prentky, R.A., Burgess, A.W., Rokous, F., Lee, A., Hartman, C., Ressler, R., & Douglas, J. (1989). Serial vs. solo sexual homicide: The role of fantasy. *American Journal of Psychiatry, 146,* 887–891.

Prentky, R.A., Knight, R.A., & Lee A.F.S. (1991). Impulsivity as a typological discriminator for recidivism among rapists. In R.A. Prentky & R.A. Knight (1991). Identifying critical dimensions for discriminating among rapists. *Journal of Consulting and Clinical Psychology, 59*, 643–661.

Prentky, R.A., Knight, R.A., & Lee, A.F.S. (1997). Risk factors associated with recidivism among extrafamilial child molesters. *Journal of Consulting and Clinical Psychology, 65,* 141–149.

Prentky, R.A., Knight, R.A., Lee A.F.S., & Cerce, D.D. (1995). Predictive validity of lifestyle impulsivity for rapists. *Criminal Justice and Behavior, 22*, 106–128.

Prentky, R.A., Lee, A.F.S., Knight, R.A., & Cerce, D. (1998). Long-term comparison of rate and force of recidivism between child molesters and rapists: A methodological analysis. Referenced in G.T. Harris, M.E. Rice, & V.L. Quinsey. Appraisal and management of risk in sexual aggressors: Implications for criminal justice policy, *Law and Human Behavior, 4*, 73–115.

Prescott, C.A., Aggen, S.H., & Kendler, K.S. (2000). Sex-specific genetic influences on the comorbidity of alcoholism and major depression in a population-based sample of U.S. twins. *Archives of General Psychiatry, 57*, 803–811.

Pressman, B. & Sheps, A. (1994). Treating wife abuse: An integrated model. *International Journal of Group Psychotherapy, 44*, 477–498.

Proulx, J., McKibben, A., & Lusignan, R. (1996). Identification of risk factors through clinical interviews and analysis of records. In D.R. Laws (Ed.), *Relapse prevention with sex offenders* (pp. 77–87). New York: Guildford Press.

Proulx, J,. Pellerin, B., McKibben, A., Aubut, J., & Ouimet, M. (1997). Static and dynamic risk predictors of recidivism in sexual offenders. *Sexual Abuse: A Journal of Research and Treatment, 1*, 7–27.

Proulx, J., St-Yves, M., Guay, J.R., & Ouimet, M. (1999). Les agresseurs sexuels de femmes: scenarios delictuels et troubles de la personnalite. In J. Proulx, M. Cusson, & M. Ouimet (Eds.), *Les volences criminelles* (pp. 157–185). Saint Nicholas, QC: Les Presses de Universite Laval.

Pulkkinen, L. & Pitkanen, T. (1993). Continuities in aggressive behavior from childhood to adult. *Aggressive Behavior, 19*, 249–263.

Quinlisk, J.A. (1999). Animal abuse and family violence. In F.R. Ascione & P. Arkow (Eds.), *Child abuse, domestic violence, and animal abuse* (pp. 169–175). West Lafayette, IN: Purdue University Press.

Quinsey, V.L. (1984). Sexual aggression: Studies of offenders against women. In D. Weisstub, (Ed.), *Law and mental health: International perspectives* (pp. 84–121). New York: Pergamon Press.

Quinsey, V.L. (1986). Men who have sex with children. In D.N. Weisstub (Ed.), *Law and mental health: International perspective* (vol. 2, pp. 140–172). New York: Pergamon Press.

Quinsey, V.L. & Bergersen, S.G. (1976). Instructional control of penile circumference in assessments of sexual preference. *Behavior Therapy, 7*, 489–493.

Quinsey, V.L. & Earls, C.M. (1990). The modification of sexual preferences. In W.L. Marshall, D.R. Laws, & H.E. Barbaree (Eds.), *Handbook of sexual assault: Issues, theories, and treatment of the offender* (pp. 279–295). New York: Plenum.

Quinsey, V.L. & Lalumiere, M.L. (1996). *Assessment of sexual offenders against children.* Thousand Oaks, CA: Sage Publications.

Quinsey, V.L., Chaplin, T.C., & Upfold, D. (1984). Sexual arousal to nonsexual violence and sadomasochistic themes among rapists and non sex-offenders. *Journal of Consulting and Clinical Psychology, 52*, 651–657.

Quinsey, V.L., Coleman, G., Jones, B., & Altrows, I. (1997). Proximal antecedents of eloping and reoffending among supervised mentally disordered offenders. *Journal of Interpersonal Violence, 12*, 794–813.

Quinsey, V.L., Harris, G.T., Rice, M.E., & Cormmier, C.A. (1998a). Static and dynamic predictors of recidivism in sexual offenders. *Sexual Abuse, 9*, 7–28.

Quinsey, V.L., Harris, G.T., Rice, M.E., & Cormmier, C.A. (1998b). *Violent offenders: Appraising and managing risk.* Washington, D.C.: American Psychological Association.

Quinsey, V.L., Lalumiere, M.L., Rice, M.E., & Harris, G.T. (1995). Predicting sexual offenses. In J.C. Campbell (Ed.), *Assessing dangerousness: Violence by sexual offenders, batterers, and child abusers* (pp.114–137). Thousand Oaks, CA: Sage Publications.

Quinsey, V.L., Rice, M.E., & Harris, G.T. (1995). Actuarial prediction of sexual recidivism. *Journal of Interpersonal Violence, 10*, 85–105.

Raine, A., Venables, P.H., & Williams, M. (1990). Relationships between central and autonomic measures of arousal at age 15 years and criminality at age 24 years. *Archives of General Psychiatry, 47*, 1003–1007.

Rand, M.R. & Saltzman, L.E. (2003). The nature and extent of recurring intimate partner violence against women in the United States. *Journal of Comparative Family Studies, 34*(1), 137–146.

Ray, A. & Gold, S. (1996). Gender roles, Aggression, and Alcohol Abuse. *The Journal of Sex Research, 33*(1), 47–55.

Raymond, N.C., Coleman, E., Ohlerking, F., Christenson, G.A., & Miner, M. (1999). Psychiatric comorbidity in pedophilic sex offenders. *The American Journal of Psychiatry, 156*(5), 786–789.

Register, E. (1993). Feminism and recovering from battering: Working with the individual woman. In M. Hansen & M. Harway (Eds.), *Battering and family therapy: A feminist perspective* (pp. 93–104). Thousand Oaks, CA: Sage Publications.

Reinhardt, J.M. (1957). *Sex perversions and sex crimes: A psychocultural examination of the causes, nature and criminal manifestations of sex perversions.* Police Science Series. Springfield, IL: Charles C Thomas.

Reinhart, M. (1987). Sexually abused boys. *Child Abuse and Neglect, 11*, 229–235.

Renzetti, C.M. (1992). *Violent betrayal: Partner abuse in lesbian relationships.* Newbury Park, CA: Sage Publications.

Ressler, R.K., Burgess, A.W., & Douglas, J.E. (1988). *Sexual homicide: Patterns and motives.* New York: The Free Press.

Revitch, E. (1965). Sex murder and the potential sex murderer. *Diseases of the Nervous System, 26*, 640–648.

Revitch, E. (1980). Gynocide and unprovoked attacks on women. *Correctional and Social Psychiatry, 26*, 6–11.

Rice, M.E. & Harris, G.T. (1995). Violent recidivism: Assessing predictive validity. *Journal of Consulting and Clinical Psychology, 63*, 737–748.

Rice, M.E. & Harris, G.T. (1997). Cross validation and extension of the violence risk appraisal guide for child molesters and rapists. *Law and Human Behavior, 21*, 213–241.

Rice, M.E., Harris, G.T., & Quinsey, V.A. (1990). A follow-up of rapists assessed in a maximum security psychiatric facility, *Journal of Interpersonal Violence, 5*, 435–448.

Rice, M.E., Quinsey, V.A., & Harris, G.T. (1991). Sexual recidivism among child molesters released from a maximum security psychiatric institution. *Journal of Consulting and Clinical Psychology, 59*, 381–386.

Rigdon, L.D. & Tapia, F. (1977). Children who are cruel to animals — A follow-up study. *Journal of Operational Psychiatry, 8*, 27–36.

Riggs, D., Dancu, C.V., Gershuny, B.S., Greenberg, D., & Foa, E.B. (1992). Anger and posttraumatic stress disorder in female crime victims. *Journal of Traumatic Stress, 5*(4), 613–625.

Riggs, D., Murphy, C., & O'Leary, K. (1989). Intentional falsification in reports of interpartner aggression. *Journal of Interpersonal Violence, 4*, 220–232.

Riley, W.T., Treiber, F.A., & Woods, G. (1989). Anger and hostility in depression. *Journal of Nervous and Mental Disease, 177*, 668–674.

Ritmeester, T. (1993). Batterer's programs, battered women's movement, and issues of accountability. In E. Pence & M. Paymar (Eds.), *Education groups for men who batter: The Duluth model.* New York: Springer.

Roberts, C.F., Doren, D.M., & Thornton, D. (2002). Dimensions associated with assessments of sex offender recidivism risk. *Criminal Justice and Behavior, 29*, 569–589.

Robertson, N. (1999), Stopping violence programmes: Enhancing the safety of battered women or producing better educated batters? *New Zealand Journal of Psychology, 28*(2), 68–79.

Robinson, B. (2001, October 15). The unusual suspects: Female sex offenders are rare, driven by more than sex. Retrieved June 3, 2002 from http://www.ABCnews.com.

Rogan, A. (1985–1986). Domestic violence and alcohol: Barriers to cooperation. *Alcohol Health and Research World, 9*, 22–27.

Rogers, R. (1990a). Development of a new classificatory model of malingering. *Bulletin of the American Academy of Psychiatry and Law, 18*, 323–333.

Rogers, R. (1990b). Models of feigned mental illness. *Professional Psychology, 21*, 181–188.

Rogers. R. (1995). *Diagnostic and structured interviewing: A handbook for psychologists.* Odessa, FL: Psychological Assessment Resources.

Rogers, R. (Ed.). (1997). *Clinical assessment of malingering and deception.* NewYork: Guilford Press.

Rogers, R. & Dickey, R. (1991). Denial and minimization among sex offenders: A review of competing models of deception. *Annals of Sex Research, 4*, 49–63.

Rogers, R. & Mitchell, C.N. (1991). *Mental health experts and the criminal courts: A handbook for lawyers and clinicians*. Toronto, Ontario: Carswell Publishing.

Rogers, G. & Renshaw, K., (1993). Covert communication between sex offenders and their victims. *Annals of Sex Research, 6*, 185–196.

Rohsenow, D.J. (1983). Drinking habits and expectancies about alcohol's effects for self versus others. *Journal of Consulting and Clinical Psychology, 51*, 752–756.

Rohsenow, D.J. & Bachorowski, J. (1984). Effects of alcohol and expectancies on verbal aggression in men and women. *Journal of Abnormal Psychology, 93*, 418–432.

Rosen, R. & Beck, J.G. (1988). *Patterns of sexual arousal: Psychophysiological processes and clinical applications.* New York: Guilford Press.

Rosenbaum, A. & Leisring, P.A. (2003). Beyond power and control: Towards an understanding of partner abusive men. *Journal of Comparative Family Studies, 34*(1), 7–19.

Rosenbaum, A. & Maiuro, R.D. (1990). Perpetrators of spousal abuse. In R.T. Ammerman & M. Hersen (Eds.), *Treatment of family violence: A sourcebook* (pp. 280–309). New York: Plenum Press.

Rosenbaum, A. & O'Leary, K.D. (1981). Children: The unintended victims of marital violence. *American Journal of Orthopsychiatry, 51*, 692–699.

Rosenbaum, A., Gearan, P.J., & Ondovic, C. (2001). Completion and recidivism among court- and self-referred batterers in a psychoeducational group treatment program: Implications for intervention and public policy. *Journal of Aggression, Maltreatment, and Trauma, 5*(2), 199–220.

Rosenberg, M. (1987). Children of battered women: The effects of witnessing violence on their social problem-solving abilities. *Behavior Therapist, 10*, 85–89.

Rosenberg, R., Knight, R.A., Prentky, R.A., & Lee, A. (1988). Validating the components of a taxonomic system for rapists: A path analytic approach. *Bulletin of American Academy of Psychiatry and the Law, 16*, 169–185.

Rosenfeld, B. (1992). Court-ordered treatment of spouse abuse. *Clinical Psychology Review, 12*, 205–226.

Rouse, L.P. (1984). Models, self-esteem and locus of control as factors contributing to spouse abuse. *Victimology: An International Journal, 9*(1), 130–141.

Rowen, E., Rowan, J., & Langelier, P. (1990). Women who molest children. *Bulletin of the American Academy of Psychiatry and Law, 18*(1), 79–83.

Rushe, R.H. (1995). *Tactics of power and influence in violent marriages*. Unpublished doctoral dissertation, University of Washington, Seattle.

Russell, D.E.H. (1980). Pornography and violence: What does the new research say? In L. Lenderer (Ed.) *Take back the Night: Women on pornography.* New York: Morrow, 1980.

Russell, D.E.H. (Ed.) (1993). *Making pornography sexy: Feminist views on pornography.* New York: Teachers College Press, and Buckingham, U.K.: Open University Press.

Russell, D.E.H. (1994a). *Against pornography: The evidence of harm.* Berkeley, CA: Russell Publications.

Russell, D.E.H. (1994b). *Dangerous relationships: Pornography, misogyny, and rape.* Newbury Park, CA: Sage Publications, 1998. *Against pornography: The evidence of harm.* (Revised/expanded ed.), Berkeley, CA: Russell Publications.

Russell, D.E.H. (1999, September 22–25). *Against pornography: The evidence of harm.* Presentation at The Association for the Treatment of Sexual Abusers, 18th Annual Research and Treatment Conference. Orlando, FL.

Russell, M. (1988). Wife assault theory, research, and treatment: A literature review. *Journal of Family Violence, 3,* 193–208.

Salloum, I.M., Cornelius, J.R., Thase, M.E., Daley, D.C., Kirisci, L., & Spotts, C. (1998). Naltrexone utility in depressed alcoholics. *Psychopharmacology Bulletin, 34(*1), 111–115.

Saunders, D.G. (1986). When battered women use violence: Husband-abuse or self-defense? *Violence and Victims, 1*, 47–60.

Saunders, D.G. (1992a). A typology of men who batter women: Three types derived from cluster analysis. *American Osteopsychiatry, 62*, 264–275.

Saunders, D.G. (1992b). Woman battering. In R.T. Ammerman & M. Hersen (Eds.), *Assessment of family violence: A clinical and legal sourcebook* (pp. 208–235). New York: John Wiley & Sons.

Saunders, D.G. (1993). Husbands who assault: Multiple profiles requiring multiple responses. In N.Z. Hilton (Ed.), *Legal responses to wife assault: Current trends and evaluation* (pp. 9–34). Newbury Park, CA: Sage Publications.

Saunders, D.G. (1994). Posttraumatic stress symptom profiles of battered women: A comparison of survivors in two settings. *Violence and Victims, 9*, 125–138.

Saunders, D.G. (1995). Prediction of wife assault. In J. Campbell (Ed.), *Assessing dangerousness: Violence by sexual offenders, batterers, and child abusers* (pp. 68–95). Thousand Oaks, CA: Sage Publications.

Saunders, D.G., Lynch, A.B., Grayson, M., & Linz, D. (1987). The inventory of beliefs about wife beating: The construction and initial validation of a measure of beliefs and attitudes. *Violence and Victims, 2(*1), 39–57.

Sayette, M.A., Wilson, T., & Elias, M.J. (1993). Alcohol and aggression: A social information processing analysis. *Journal of Studies on Alcohol, 54*(4), 399–407.

Scarpa, A. & Raine, A. (1997). Psychophysiology of anger and violent behavior. *Psychiatric Clinics of North America, 20*, 375–394.

Schater, D.L., Wang, P.L., Tulving, E., & Freedman, M. (1982). Functional retrograde amnesia: A quantitative case study. *Neuropsychologia, 20*, 523–532.

Schlank, A.M. & Shaw, T. (1997). Treating sexual offenders who deny: A review. In B.K. Schwartz & H.R. Cellini (pp. 6-1–6-7), *The sexual offender: New insights, treatment innovations and legal developments.* Kingston, NJ: Civic Research Institute, Inc.

Schlesinger, L.B. (2000). Serial homicide: Sadism, fantasy, and a compulsion to kill. In L.B. Schlesinger (Ed.), *Serial offenders: Current thought, recent findings* (pp. 3–22). Boca Raton, FL: CRC Press.

Schmutte, G. & Taylor, S. (1980). Physical aggression as a function of alcohol and pain feedback. *Journal of Social Psychology, 110*, 235–244.

Schroffel, A. (2004). Characteristics of female perpetrators of domestic violence in group therapy. *Smith College Studies in Social Work, 74*(3), 505–525.

Schuckit, M.A. (2005, June). Neurobiology and genetics of alcohol dependence: Etiology and current treatment. *In The etiology of alcohol: Abuse and dependence: what happens in the brain? Psychiatric Times, Reporter*, 3–6.

Schwartz, B.K. (1995). Group therapy. In B.K. Schwartz, & H.R. Cellini, (Eds.), *The sexual offender: Corrections, treatment and legal practice* (Chap. 14-1–14-15. Kingston, NJ: Civic Research Institute, Inc.

Schwartz, B.K., & Cellini, H.R. (Eds.) (1995). Female sex offenders. In B.K. Schwartz & H.R. Cellini, (Eds.), *The sexual offender: Corrections, treatment and legal practice* (5-1–5-22). NJ: Civic Research Institute, Inc.

Schwartz, M.D. (1987). Gender and injury in marital assault. *Sociological Focus, 20*, 61–75.

Seidman, B.T., Marshall, W.L., Hudson, S.M., & Robertson, P.J. (1994). An examination of intimacy and loneliness in sex offenders. *Journal of Interpersonal Violence, 9*, 518–534.

Seligman, M. (1975). *Helplessness: On depression, development, and death.* San Francisco, CA: W.H. Freeman and Company.

Senn, C.Y., Desmarais, S., Vernberg, N., & Wood, E. (2000). Predicting coercive sexual behavior across the lifespan in a random sample of Canadian men. *Journal of Social and Personal Relationships, 17*, 95–113.

Serin, R., Mallioux, D., & Malcolm, P. (2001). Psychopathy, deviant sexual arousal and recidivism among sexual offenders. *Journal of Interpersonal Violence, 16,* 234–246.

Seto, M. & Barbaree, H. (1999). Psychopathy, treatment behavior, and sex offender recidivism. *Journal of Interpersonal Violence, 14*, 1235–1248.

Sewell, K.W. & Salekin, R.T. (1997). Understanding and detecting dissimulation in sex offenders. In R. Rogers (Ed.), *Clinical assessment of malingering and deception* (pp. 328–350). New York: Guilford Press.

Shea, M.C. (1998). When the tables are turned: Verbal sexual coercion among college women. In R.B. Anderson & C. Struckman-Johnson (Eds.), *Sexually aggressive women* (pp. 96–104). New York: Guilford Press.

Sherman, L. (1992). *Policing domestic violence*. New York: Free Press.

Shuntich, R. & Taylor, S. (1972). The effects of alcohol on human physical aggression. *Journal of Experimental Research in Personality, 6*, 34–38.

Siegel, J.A. (2000). Aggressive behavior among women sexually abused as children. *Violence and Victims, 15*, 235–255.

Smallbone, S.W. & Dadds, M.R. (1998). Childhood attachment and adult attachment in incarcerated adult male sex offenders. *Journal of Interpersonal Violence, 13*, 555–573.

Smith, J., Berthelsen, D., & O'Connor, I. (1997). Child adjustment in high conflict families. *Child, Care, Health, and Development, 23*, 113–133.

Smith, S.S. & Newman, J.P. (1990). Alcohol and drug abuse-dependence in psychopathic and nonpsychopathic criminal offenders. *Journal of Abnormal Psychology, 99*, 430–439.

Sonkin, D.J. (1987). The assessment of court-mandated male batterers. In D.J. Sonkin (Ed.), *Domestic violence on trial: Psychological and legal dimensions of family violence* (pp. 174–196). New York: Springer.

Sonkin, D.J., Martin, D., & Walker, L. (1985). *The male batterer: A treatment approach*. New York: Springer.

Sorrel, P.M. & Masters, W.H. (1982). Sexual assault on men by women. *Archives of Sexual Behavior, 11*, 117–131.

Spaccarelli, S., Sandler, I.N., & Roosa, M. (1994). History of spousal violence against mother: Correlated risks and unique effects in child mental health. *Journal of Family Violence, 9*, 79–98.

Spencer, T., Biederman, J., Coffey, B., Geller, D., Crawford, M., Bearman, S.K., Tarazi, R., & Faraone, S.V. (2002). A double-blind comparison of desipramine and placebo in children and adolescents with chronic tic disorder and comorbid attention deficit hyperactivity disorder. *Archives of General Psychiatry, 59*, 649–656.

Sperberg, E.D. & Stabb, S.D. (1998). Depression in women as related to anger and mutuality in relationships. *Psychology of Women Quarterly, 22*, 223–238.

Spielberger, C.D. (1999). *State-trait anger expression inventory-2 (STAXI-2)*. Odessa, FL: Psychological Assessment Resources, Inc.

Spielberger, C.D., Gorsuch, R., & Lushere, R. (1970). *Manual for the state-trait anxiety scale*. Palo Alto, CA: Consulting Psychologists Press.

Spoont, M.R. (1992). Modulatory role of serotonin in neutral information processing: Implications for human psychopathology. *Psychological Bulletin, 112*, 330–350.

Steele, C.M. & Josephs, R.A. (1990). Alcohol myopia: Its prized and dangerous effects, *American Psychologist, 45*, 921–933.

Steffensmeier, D. & Allan, E. (1996). Gender and crime: Toward a gendered theory of female offending. *Annual Review of Sociology, 22*, 459–488.

Sternberg, K.L., Lamb, M.E., Greenbaum, C., Cicchetti, D., Dawud, S., & Cortes, R. M. (1993). Effects of domestic violence on children's behavior problems and depression. *Developmental Psychology, 29*, 44–52.

Stets, J.E. & Straus, M.A. (1990). Gender differences in reporting marital violence and its medical and psychological consequences. In M.A. Straus & R.J. Gelles, (Eds.), *Physical violence in American families: Risk factors and adaptations to violence in 8,145 families* (pp. 151–165). New Brunswick, NJ: Transaction Publishers.

Stith, S.M. & Farley, S.C. (1993). A predictive model of male spousal violence. *Journal of Family Violence, 8*, 183–201.

Stith, S.M., Crossman, R.K., & Bischof, G.P. (1991). A comparative study of men in alcohol treatment programs and batterer treatment programs. *Alcoholism Treatment Quarterly, 8*, 3–20.

Stith, S.M., Rosen, K.H., Middleton, K.M., Busch, A., Lundeburg, K., & Carlton, R. (2000). The intergenerational transmission of spouse abuse: A meta-analysis. *Journal of Marriage and the Family, 62*, 640–654.

Stith, S.M., Rosen, K.H., & McCollum, E.E. (2003). Effectiveness of couples treatment for spouse abuse. *Journal of Marital and Family Therapy, 29*(3), 407–425.

Stith, S.M., Rosen, K.H., McCollum, E.E., & Thomsen, C.J. (2004a). Treating intimate partner violence within intact couple relationships: Outcomes of multi-couple versus individual couple therapy. *Journal of Marital and Family Therapy, 30*(3), 305–318.

Stith, S.M., Smith, D.B., Penn, C.E., Ward, D.B., & Tritt, D. (2004b). Intimate partner physical abuse perpetration and victimization risk factors: A meta-analytic review. *Aggression and Violent Behavior, 10*, 65–98.

Straus, M. (1979). Measuring intrafamilial conflict and violence: The conflicts tactics (CT) scale. *Journal of Marriage and the Family, 41*, 75–88.

Straus, M. & Steinmetz, S. (1974). *Violence in the family*. New York: Harper & Row.

Straus, M.A. (1990a). The conflicts tactics scale and its critics: An evaluation and new data on validity and reliability. In M.A. Straus & R.J. Gelles (Eds.), *Physical violence in American families: Risk factors and adaptations to violence in 8,145 families* (pp. 95–112). New Brunswick, NJ: Transaction Publishers.

Straus, M.A. (1990b). Social stress and marital violence in a national sample of American families. In M.A. Straus & R.J. Gelles (Eds.), *Physical violence in American families: Risk factors and adaptations to violence in 8,145 families* (pp. 403–424). New Brunswick, NJ: Transaction Publishers.

Straus, M.A. (1993). Physical assaults by wives: A major social problem. In R.J. Gelles & D.R. Loseke (Eds.), *Current controversies on family violence* (pp. 67–87), Newbury Park, CA: Sage Publications.

Straus, M.A. (1999). The controversy over domestic violence by women: A methodological, theoretical, and sociology of science analysis. In X.B. Arriaga & S. Oskamp (Eds.), *Violence in intimate relationships* (pp. 17–44). Thousand Oaks, CA: Sage Publications.

Straus, M.A. & Gelles, R.J. (1990a). Societal change and change in the family from 1975 to1985 as revealed by two national surveys. In M.A. Straus & R.J. Gelles (Eds.), *Physical violence in American families: Risk factors and adaptations in 8,145 families* (pp. 113–131). New Brunswick, NJ: Transaction Publishers.

Straus, M.A. & Gelles, R.J. (1990b). How violent are American families? In M.A. Straus & R.J. Gelles (Eds.), *Physical violence in American families: Risk factors and adaptations in 8,145 families* (pp. 113–131). New Brunswick, NJ: Transaction Publishers.

Straus, M.A., Gelles, R.J., & Steinmetz, S.K. (1980). *Behind closed doors: Violence in the American family*. New York: Anchor Books.

Street, A.E., King, L.A., King, D.W., & Riggs, D.S. (2003). The associations among male-perpetrated partner violence, wives' psychological distress and children's behavior problems: A structural equation model analysis. *Journal of Comparative Family Studies, 34*(1), 23–37.

Struckman-Johnson, C. (1988). Forced sex on dates: It happens to men, too. *Journal of Sex Research, 24*, 234–241.

Struckman-Johnson, C. & Struckman-Johnson, D. (1994). Men pressured and forced into sexual experience. *Archives of Sexual Behavior, 23*, 93–114.

Struckman-Johnson, C. & Struckman-Johnson, D. (1998). The dynamics and impact of sexual coercion of men by women. In P.B. Anderson & C. Struckman-Johnson (Eds.), *Sexually aggressive women* (pp. 121–143). New York: Guilford Press.

Stuart, E.P. & Campbell, J.C. (1989). Assessment of patterns of dangerousness with battered women. *Issues in Mental Health Nursing, 10*, 245–260.

Sugarman, D.B. & Hotaling, G.T. (1980). Dating violence: A review of contextual and risk factors. In B. Levy (Ed.), *Dating violence: Young women in danger* (pp. 100–118). Seattle, WA: Seal Press.

Swan, S.C., Gambone, L.J., Fields, A.M., Sulivan, T.P., & Snow, D.L. (2005). Women who use violence in intimate relationships: The role of anger, victimization, and symptoms of posttraumatic stress and depression. *Violence and Victims, 20*(3), 267–286.

Swisher, K.L. & Wekesser, C. (Eds.) (1994). *Violence against women*. San Diego, CA: Current Controversies.

Takala, M., Pihkanen, T., & Maren, T. (1957). *The effects of distilled and brewed beverages: a physiological, neurological and psychological study*, Publication 4. Helsinki, Finland: Finnish Foundation for Alcohol Studies.

Tapia, E.H. (1971). Children who are cruel to animals. *Child Psychiatry and Human Development, 2*, 70–77.

Tavris, C. (1992). *The mismeasure of women*. New York: Simon & Schuster.

Taxman, F. (2002). Supervision: Exploring the dimensions of effectiveness. *Federal Probation, 66,* 14–27.

Taylor, P.J & Kopelman, M.D. (1984). Amnesia for criminal offences. *Psychological Medicine, 14*, 581–588.

Taylor, S. & Gammon, C. (1975). Effects of type and dose of alcohol on human physical aggression. *Journal of Personality and Social Psychology, 32*, 169–175.

Taylor, S. & Gammon, C. (1976). Aggressive behavior of intoxicated subjects: The effect of third-party intervention. *Journal of Studies on Alcohol, 37*, 917–930.

Taylor, S.P. & Leonard, K.E. (1983). Alcohol and human physical aggression. In R.G. Geen & R.G. Donnerstein (Eds.), *Aggression: Theoretical and Empirical Reviews* (vol. 2, pp. 77–101). New York: Academic Press.

Taylor, S., Gammon, C., & Capasso, D. (1976). Aggression as a function of the interaction of alcohol and threat. *Journal of Personality and Social Psychology, 34*, 938–941.

Taylor, S.P. (1993). Experimental investigation of alcohol-induced aggression in humans. *Alcohol Health and Research World, 17*, 2.

Theodore, R.M. (1992). The relationship between locus of control and level of violence in married couples. In E.C. Viano (Ed.), *Intimate violence: Interdisciplinary perspectives.* Washington, D.C.: Taylor & Francis.

Thomas, S.P. (1991). Toward a new conceptualization of women's anger. *Issues in Mental Health Nursing, 12,* 31–49.

Thornton, D., (1997, September). Structured anchor clinical judgment risk assessment (SACJ). *Proceedings of the NOTA Conference*, Brighton, U.K.

Tifft, L.L. (1993). *Battering of women: The failure of intervention and the case for prevention.* Boulder, CO: Westview Press.

Tingle, D., Barnard, G.W., Robbins, G., Newman, G., & Hutchinson, D. (1986). Childhood and adolescent characteristics of pedophiles and rapists. *International Journal of Law and Psychiatry, 9,* 103–116.

Tinklenberg, J. (1973). Alcohol and violence. In P. Bourne (Ed.), *Alcoholism: Progress in research and treatment* (pp. 195–210). New York: Academic Press.

Tjaden, P. & Thoennes, N. (1997). Stalking in America: Findings from the National Violence Against Women Survey, Denver, CO: Center for Policy Research. In J.R. Meloy (Ed.), *The psychology of stalking: Clinical and forensic perspectives* (p. 6). San Diego, CA: Academic Press.

Tolman, R.M. & Bennett, L.W. (1990). A review of research on men who batter. *Journal of Interpersonal Violence, 5*, 87–118.

Tolmen, R.M. & Edleson, J.L. (1995). Intervention with men who batter: A review of research. In S.R. Stith & M.A. Straus (Eds.). *Partner violence: Prevalence, causes and solutions*. Minneapolis, MN: National Council on Family Relations.

Tomsen, S. (1997). A top night out: Social protest, masculinity and the culture of drinking violence. *British Journal of Criminology, 37*, 990–1102.

Toone, S. (1992, March 28). Women's experience of violence: Men's accountability to battered women. Speech to Men for Non-Violence National Conference, Palmerston North.

Travin, S. & Protter, B. (1984). Malingering and malingering-like behavior: Some clinical and conceptual issues. *Psychiatric Quarterly, 56*, 189–197.

Turkkan, J., Stitzer, M., & McCaul, M. (1988). Psychophysiological effects of oral ethanol in alcoholics and social drinkers. *Alcoholism: Clinical and Experimental Research, 12*, 3–38.

Tweed, R.G. & Dutton, D.G. (1998). A comparison of impulsive and instrumental subgroups of batterers. *Violence and Victims, 13*, 217–230.

United States Department of Justice. (2002). Crime in the United States 2001. Washington, D.C.: Federal Bureau of Investigation.

Valliere, V. (1997). Relationships between alcohol use, alcohol expectancies, and sexual offenses in convicted offenders. In B.K. Schwartz and H.R. Cellini (Eds.), *The Sexual Offender*, vol II. Kingston, NJ: Civic Research Institute.

van der Kolk, B.A. (1988). Trauma in men: Effects on family life. In M.B. Straus (Ed.), *Abuse and victimization across the life span* (pp. 170–185). Baltimore, MD: Johns Hopkins University Press.

Vassileva, J., Kosson, D.S., Abramowitz, C., & Conrad, P. (2005). Psychopathy versus psychopathies in classifying criminal offenders. *Legal and Criminological Psychology, 10*(1), 27–44.

Vieta, E., Colom, F., Martinez-Aran, A., Benabarre, A., Reinares, M., & Gasto, C. (2000). Bipolar II disorder and comorbidity. *Comprehensive Psychiatry, 41*, 339–343.

Virkkunen, M. (1974). Alcohol as a factor precipitating aggression and conflict behaviour leading to homicide. *British Journal of Addiction, 69*, 149–154.

Virkkunen, M. & Linnoila, M. (1993a). Brain serotonin, type II alcoholism and impulsive violence. *Journal of Studies on Alcohol, 11* (Suppl.), 163–169.

Virkkunen, M. & Linnoila, M. (1993b). Serotonin in personality disorders with habitual violence and impulsivity. In S. Hodgkins (Ed.), *Mental Disorders and Crime*. Newbury Park, CA: Sage Publications.

Vivian, D. (1998). The context and characteristics of partner violence in couples seeking marital therapy. Paper presented at the meeting of the Association for the Advancement of Behavior Therapy.

Volavka, J., Crowner, M., Brizer, D., Conuit, A., Van Praag, H., & Suckow, R. (1990). Tryptophan treatment of aggressive psychiatric inpatients. *Biological Psychiatry, 25*, 728–732.

Wagner, W.G., Aucoin, R., & Johnson, J.T. (1993). Psychologists' attitudes concerning child sexual abuse: The impact of sex of perpetrator, sex of victim, age of victim, and victim response. *Journal of Child Sexual Abuse, 2*(2), 61–74.

Waldner-Haugrud, L.K. & Magruder, B. (1995). Male and female sexual victimization in dating relationships: Gender differences in coercion techniques and outcomes. *Violence and Victims, 10*, 203–215.

Walker, L.E. (1979). *The battered woman*. New York: Harper Perennial.

Walker, L.E. (1984). *The battered woman syndrome*. New York: Springer-Verlag.

Walker, L.E. (1989). Psychology and violence against women. *American Psychologist, 44,* 695–702.

Walker, L.E. (1995, September 9). Current perspectives on men who batter women: Implications for intervention and treatment to stop violence against women: Comment on Gottman et al. *Journal of Family Psychology, 3*, 264.

Walker, L.E. & Meloy, J.R. (1998). Stalking and domestic abuse. In J.R. Meloy (Ed.), *The psychology of stalking: Clinical and forsenic perspectives* (pp. 139–161). San Diego, CA: Academic Press.

Walker, L.E., Rowe, R.C., & Quinsey, V.L. (1993). Authoritarianism and sexual aggression. *Journal of Personality and Social Psychology, 65*, 1036–1045.

Waltz, J., Babcock, J.C., Jacobson, N.S., & Gottman, J.M. (1995). Reports of behavior problems in batterers and battered women. In J.M. Gottman, N.S. Jacobsen, R.H. Rushe, J.W. Shortt, J. Babcock, J.J. LaTaillade, & J. Waltz. The relationship between heart rate reactivity, emotionally aggressive behavior, and gerneral violance in batterers. *Journal of Family Psychology, 9*, 227–248.

Waltz, J., Babcock, J.C., Jacobson, N.S., & Gottman, J.M. (2000). Testing a typology of batterers. *Journal of Consulting and Clinical Psychology, 68*(4), 658–669.

Ward, T., Hudson, S.M., & Marshall, W.L. (1995). Cognitive distortions and affective deficits in sex offenders. *Sexual Abuse: A Journal of Research and Treatment, 7*, 67–83.

Ward, T., Hudson, S.M., Marshall, W.L., & Siegert, R. (1995). Attachment style and intimacy deficits in sex offenders: A theoretical framework. *Sexual Abuse: A Journal of Research and Treatment, 7*, 317–335.

Warren, J.I., Reboussin, R., Hazelwood, R.R., & Wright, J.A. (1991). Prediction of rapist type and violence from verbal, physical, and sexual scales. *Journal of Interpersonal Violence, 6,* 55–67.

Waterman, C.K. & Foss-Goodman, D. (1994). Child molesting: Variables relating to attribution of fault to victims, offenders and nonparticipating parents. *Journal of Sex Research, 20,* 329–349.

Waterman, C.K., Dawson, M.A., & Bologna, M.J. (1989). Sexual coercion in gay and lesbian relationships: Predicators of gay rape. *Journal of Sex Research, 8,* 245–308.

Webster, C.D., Dickens, B.M., & Addario, S. (1985). *Constructing dangerousness: Scientific, legal and policy implications.* Toronto, Ontario: Center of Criminology, University of Toronto.

Webster, C.D., Douglas, K.S., Eaves, D., and Hart, S.D. (1997). *HCR-20: Assessing Risk for Violence.* Burnaby, British Columbia: Mental Health, Law, and Policy Institute, Simon Fraser University.

Weisman, A.M. & Taylor, S.P. (1993). Effect of alcohol and risk of physical harm on human physical aggression. *The Journal of General Psychology, 121*, 67–76.

Weisz, A., Tolman, R., & Saunders, D. (2000). Assessing the risk of severe domestic violence: The importance of survivor's predictions. *Journal of Interpersonal Violence, 15*, 75–88.

Wellman, M.M. (1993). Child sexual abuse and gender differences: Attitudes and prevalence. *Child Abuse and Neglect, 15*, 539–547.

Welte, J.W. & Miller, B.A. (1987). Alcohol use by violent and property offenders. *Drug and Alcohol Dependence, 19*, 313–324.

West, J.C., Zarin, D.A., & Pincus, H.A. (1997). Treatment issues in clinical psychopharmacology: Clinical and psychopharmacologic practice patterns of psychiatrists in routine practice. *Psychopharmacology Bulletin, 33*(1), 79–85.

Whitaker-Azmitia, P. & Peroutka, S.J. (1990, October 15). (Eds.) The neuropharmacology of serotonin. *Annals of the New York Academy of Sciences, 600.*

White, H.R., Brick, J., & Hansell, S. (1993). A longitudinal investigation of alcohol use and aggression in adolescence. *Journal of Studies on Alcohol, 11(Suppl.)*, 62–76.

White, H.R., Hansell, S., & Brick, J. (1993). Alcohol use and aggression among youth. *Alcohol Health and Research World, 2*, 144–150.

White, J. (1983). Sex and gender issues in aggression research. In R. Green & E. Donnerstein (Eds.), *Aggression: Theoretical and empirical reviews* (vol. 2, pp. 1–26). New York: Academic Press.

White, J.W. & Humphrey, J.A. (1994). Women's aggression in heterosexual conflicts. *Aggressive Behavior, 20*, 195–202.

White, J.W., Smith, P.H., Koss, M.P., & Figueredo, A.J. (2000). Intimate partner aggression: What have we learned? Comment on Archer. *Psychological Bulletin, 126*(5), 690–696.

Wiehe, V.R. (1990). *Sibling abuse.* New York: Lexington Books.

Wiley, J.A. & Weisner, C. (1995) Drinking in violent and nonviolent events leading to arrest: evidence from a survey of arrestees, *Journal of Criminal Justice, 23*, 461–476.

Wilson, M. & Daly, M. (1993). Spousal homicide risk and estrangement. *Violence and Victims, 8*, 3–16.

Wilson, R.J. (1999). Emotional congruence in sexual offenders against children. *Sexual Abuse: A Journal of Research and Treatment, 11*, 33–47.

Windom, C. (1989). Child abuse, neglect and violent criminal behavior. *Criminology, 27*, 251–271.

Windom, C.S. (1978). Towards an understanding of female criminality. *Progress in Experimental Psychology Research, 8*, 245–308.

Windom, C.S. (1999). Posttraumatic stress disorder in abused and neglected children grown up. *American Journal of Psychiatry, 156*, 1223–1229.

Woffardt, S., Mihalic, D.E., & Menard, S. (1994). Continuities in marital violence. *Journal of Family Violence, 9*, 195–225.

Wolf, F.A. (1985, March). *Twelve female sexual offenders.* Presentation to the conference on Next Step in Research on the Assessment and Treatment of Sexually Aggressive Persons (Paraphilias), St. Louis, MO.

Yates, H.E., Barbaree, H.E., & Marshall, W.L. (1984). Anger and deviant sexual arousal. *Behavior Therapy, 15*, 287–294.

Yochelson, S. & Samenow, S. (1976). *The criminal personality A profile for change* (vol. 1). New York: Jason Aronson Publishers.

Yuille, J.C. & Tollestrup, P.A. (1990). Some effects of alcohol on eyewitness memory. *Journal of Applied Psychology, 75*, 268–273.

Zanarini, M.C., Williams, A.A., Lewis, R.E., Reich, R.B., Vera, S.C., & Marino, M.E. (1997). Reported pathological childhood experiences associated with the development of borderline personality disorder. *American Journal of Psychiatry, 154*(8), 1101–1106.

Zeanah, C.H. & Zeanah, P.D. (1989). Intergenerational transmission of maltreatment: Insights from attachment theory and research. *Psychiatry, 52*, 177–195.

Zeichner, A. & Pihl, R.O. (1979). Effects of alcohol and behavior contingencies on human aggression, *Journal of Abnormal Psychology, 88*, 153–160.

Zeichner, A., Pihl, R.O., Niaura, R., & Zacchia, C. (1982). Attentional processes in alcohol-mediated aggression. *Journal of Studies on Alcohol, 43*, 714–724.

Zelas, K. (1995). Child protection or another "ultimate weapon?" *Butterworths Family Law Journal, 1,* 210.

Zellman, G.L. (1992). The impact of case characteristics on child abuse reporting decisions. *Child Abuse and Neglect, 14*, 57–74.

Zillmann, D. (1988). Cognitive-excitation interdependencies in aggressive behavior. *Aggressive Behavior 14,* 51–64.

Zillmann, D. & Bryant, J. (1986). Pornography, sexual callousness, and the trivialization of rape. *Journal of Communication, 32*, 4.

Zlontnick, C., Kohn, R., Peterson, J., & Pearlstein, T. (1998). Partner physical victimization in a national sample of American families: Relationship to psychological functioning, psychosocial factors, and gender. *Journal of Interpersonal Violence*, *13*, 156–166.

Zona, M., Lane, J., & Palarea, R. (1997). The psychodynamics of stalking. Paper presented at the Seventh Annual Threat Management Conference, Los Angeles, CA. In J.R. Meloy (Ed.), *The psychology of stalking: Clinical and forensic perspectives* (pp. 51). San Diego, CA: Academic Press.

Zucker, R.A., Fitzgerald, H.E., & Moses, H.D. (1975). Emergence of alcohol problems and the several alcoholisms: A developmental perspective on etiological theory and life course trajectory. In D. Cicchetti & D.J. Cohen (Eds.), *Developmental psychopathological: Risk, disorder, and adaptation* (vol. 2, pp. 677–711). New York: John Wiley & Sons.

Zuckerman, M., Depaulo, B.M., & Rosenthal, R. (1981). Verbal and nonverbal communication of deception. In L. Berkowitz (Ed.), *Advances in experimental social psychology* (vol. 14, pp. 1–59). New York: Academic Press.

Index

A

D

E

J

L

M

N

Q

R

S

T